Methods in Enzymology

Volume 410
DNA MICROARRAYS PART A: ARRAY PLATFORMS AND WET-BENCH PROTOCOLS

METHODS IN ENZYMOLOGY

EDITORS-IN-CHIEF

John N. Abelson   Melvin I. Simon

DIVISION OF BIOLOGY
CALIFORNIA INSTITUTE OF TECHNOLOGY
PASADENA, CALIFORNIA

FOUNDING EDITORS

Sidney P. Colowick and Nathan O. Kaplan

Methods in Enzymology

Volume 410

DNA Microarrays Part A: Array Platforms and Wet-Bench Protocols

EDITED BY

Alan Kimmel & Brian Oliver

NIDDK
NATIONAL INSTITUTES OF HEALTH
BETHESDA, MARYLAND

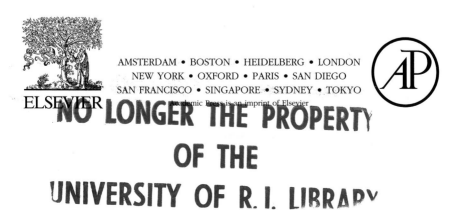

AMSTERDAM • BOSTON • HEIDELBERG • LONDON
NEW YORK • OXFORD • PARIS • SAN DIEGO
SAN FRANCISCO • SINGAPORE • SYDNEY • TOKYO

ELSEVIER

Academic Press is an imprint of Elsevier

Academic Press is an imprint of Elsevier
525 B Street, Suite 1900, San Diego, California 92101-4495, USA
84 Theobald's Road, London WC1X 8RR, UK

This book is printed on acid-free paper. ∞

Permissions may be sought directly from Elsevier's Science & Technology Rights
Department in Oxford, UK: phone: (+44) 1865 843830, fax: (+44) 1865 853333,
E-mail: permissions@elsevier.com. You may also complete your request on-line
via the Elsevier homepage (http://elsevier.com), by selecting "Support & Contact"
then "Copyright and Permission" and then "Obtaining Permissions."

For information on all Elsevier Academic Press publications
visit our Web site at www.books.elsevier.com

ISBN-13: 978-0-12-182815-8
ISBN-10: 0-12-182815-8

PRINTED IN THE UNITED STATES OF AMERICA
06 07 08 09 9 8 7 6 5 4 3 2 1

Table of Contents

Section II. Wet-Bench Protocols

Contributors to Volume 410

Article numbers are in parentheses and following the name of contributors. Affiliations listed are current.

JUSTEN ANDREWS (5), *Department of Biology, Drosophila Genomics Resource Center, Center for Genomics and Bioinformatics, Indiana University, Bloomington, Indiana*

DANIEL H. APPELLA (9), *Laboratory of Bioorganic Chemistry, NIDDK, NIH, DHHS Bethesda, Maryland and Department of Chemistry, Northwestern University, Evanston, Illinois*

TOMAS BABAK (14), *Banting and Best Department of Medical Research, Department of Medical and Molecular Genetics, University of Toronto, Toronto, Ontario, Canada*

DAVID BARKER (3, 17), *Illumina, Inc., San Diego, California*

MICHEL BELLIS (15), *Centre de Recherche en Biochimie Macromoleculaire, Montpellier, Cedex 05, France*

MARINA BIBIKOVA (3), *Illumina, Inc., San Diego, California*

BENJAMIN J. BLENCOWE (14), *Banting and Best Department of Medical Research, Department of Medical and Molecular Genetics, Microbiology, University of Toronto, Toronto, Ontario, Canada*

KEVIN BOGART (5), *Drosophila Genomics Resource Center, Center for Genomics and Bioinformatics, Indiana University, Bloomington, Indiana*

MICHAEL BROWNSTEIN (11), *The J. Craig Venter Institute, Rockville, Maryland*

DAVE BULLIS (17), *Illumina, Inc., San Diego, California*

MARTHA L. BULYK (13), *Division of Genetics, Departments of Medicine and Pathology, Harvard-MIT Division of Health Sciences & Technology (HST), Brigham & Women's Hospital, Harvard Medical School, Boston, Massachusetts*

ANGELA BURR (5), *Drosophila Genomics Resource Center, Center for Genomics and Bioinformatics, Indiana University, Bloomington, Indiana*

GIACOMO CAVALLI (15), *Chromatin and Cell Biology Lab, Institute of Human Genetics – CNRS, Montpellier, Cedex 05, France*

WEIHUA CHANG (17), *Illumina, Inc., San Diego, California*

JING CHEN (3), *Illumina, Inc., San Diego, California*

DAVID J. CLARK (21), *Laboratory of Molecular Growth Regulation, NICHD, National Institutes of Health, Bethesda, Maryland*

PATRICK J. COLLINS (2), *Microarray Quality Agilent Technologies, Inc., Santa Clara, California*

JASON CONATY (5), *Drosophila Genomics Resource Center, Center for Genomics and Bioinformatics, Indiana University, Bloomington, Indiana*

DENNISE D. DALMA-WEISZHAUSZ (1), *Expression Product Development, AFFYMETRIX, INC., Santa Clara, California*

ANNIEK DE WITTE (2), *Microarray Quality Agilent Technologies, Inc., Santa Clara, California*

JEREMY N. ERICKSON (18), *Model System Genomics, Department of Biology, Duke University, Durham, North Carolina*

JIAN-BING FAN (3), *Illumina, Inc., San Diego, California*

ELIZA WICKHAM GARCIA (3), *Illumina, Inc., San Diego, California*

REED A. GEORGE (6), *Howard Hughes Medical Institute, Janelia Farm Research Campus, Ashburn, Virginia*

FRAUKE GREIL (16), *Division of Molecular Biology, Netherlands Cancer Institute, Amsterdam, The Netherlands*

KEVIN L. GUNDERSON (3, 17), *Illumina, Inc., San Diego, California*

RICHELE M. GWIRTZ (12), *Biology Division 156-29, Caltech, Pasadena, California*

JANET HAGER (7), *W.M. Keck Biotechnology Resource Laboratory, Yale University, New Haven, Connecticut*

JÉRÔME HENNETIN (15), *Centre de Recherche en Biochimie Macromoleculaire, Montpellier, Cedex 05, France*

STEPHEN M. HEWITT (20), *Tissue Array Research Program, Laboratory of Pathology, CCR, NCI, NIH TARP Lab, Advanced Technology Center, Bethesda, Maryland*

SHAWNA L. HILEY (14), *Banting and Best Department of Medical Research, University of Toronto, Toronto, Ontario, Canada*

LEROY E. HOOD (8), *Institute for Systems Biology, Seattle, Washington*

TIMOTHY R. HUGHES (14), *Banting and Best Department of Medical Research, Department of Medical and Molecular Genetics, University of Toronto, Toronto, Ontario, Canada*

CHRISTINE KING (17), *Illumina, Inc., San Diego, California*

DAVID P. KREIL (4), *Department of Biotechnology, University of Natural Resources and Applied Life Sciences, Vienna, Austria*

KENNETH M. KUHN (17), *Illumina, Inc., San Diego, California*

STEPHEN R. LASKY (8), *Institute for Systems Biology, Seattle, Washington*

MARC LAURENT (3), *Illumina, Inc., San Diego, California*

CHRISTOPHER G. LAUSTED (8), *Institute for Systems Biology, Seattle, Washington*

SERGEY LAVROV (15), *Chromatin and Cell Biology Lab, Institute of Human Genetics – CNRS, Montpellier, Cedex 05, France*

LORI L. LEBRUSKA (3, 17), *Illumina, Inc., San Diego, California*

ANNE B. LUCAS (2), *Microarray Quality Agilent Technologies, Inc., Santa Clara, California*

GEORGES LUTFALLA (19), *UMR5124, cc86 CNRS/Université Montpellier II, Montpellier, Cedex 05, France*

C. GARRETT MIYADA (1), *Expression Product Development, AFFYMETRIX, INC., Santa Clara, California*

CELINE MOORMAN (16), *Division of Molecular Biology, Netherlands Cancer Institute, Amsterdam, The Netherlands*

JOE MUSMACKER (17), *Illumina, Inc., San Diego, California*

SCOTT J. NEAL (10), *Department of Zoology, University of Toronto, Canadian Drosophila Microarray Centre, Mississauga, Ontario, Canada*

NICOLAS NÈGRE (15), *Chromatin and Cell Biology Lab, Institute of Human Genetics – CNRS, Montpellier, Cedex 05, France*

PAULINE NG (17), *Illumina, Inc., San Diego, California*

ARNOLD OLIPHANT (17), *Illumina, Inc., San Diego, California*

JONATHAN K. POKORSKI (9), *Laboratory of Bioorganic Chemistry, NIDDK, NIH, DHHS Bethesda, Maryland and Department of Chemistry, Northwestern University, Evanston, Illinois*

HONGI REN (17), *Illumina, Inc., San Diego, California*

ROSLIN R. RUSSELL (4), *Department of Genetics, University of Cambridge, Cambridge, United Kingdom*

STEVEN RUSSELL (4), *Department of Genetics, University of Cambridge, Cambridge, United Kingdom*

ARNEET L. SALTZMAN (14), *Banting and Best Department of Medical Research, Department of Medical and Molecular Genetics, Microbiology, University of Toronto, Toronto, Ontario, Canada*

KAREN W. SHANNON (2), *Microarray Quality Agilent Technologies, Inc., Santa Clara, California*

CHANG-HUI SHEN (21), *Institute for Macromolecular Assemblies, CUNY, Department of Biology, College of Staten Island, Staten Island, New York*

RICHARD SHEN (3, 17), *Illumina, Inc., San Diego, California*

ERIC P. SPANA (18), *Model System Genomics, Department of Biology, Duke University, Durham, North Carolina*

FRANK J. STEEMERS (17), *Illumina, Inc., San Diego, California*

EUGENE Y. TANIMOTO (1), *Expression Product Development, AFFYMETRIX, INC., Santa Clara, California*

CHAN TSAN (17), *Illumina, Inc., San Diego, California*

GILLES UZE (19), *UMR5124, cc86 CNRS/ Université Montpellier II, Cedex 05, France*

BAS VAN STEENSEL (16), *Division of Molecular Biology, Netherlands Cancer Institute, Amsterdam, The Netherlands*

CHARLES B. WARREN (8), *Institute for Systems Biology, Seattle, Washington*

JANET WARRINGTON (1), *Expression Product Development, AFFYMETRIX, INC., Santa Clara, California*

J. TIMOTHY WESTWOOD (10), *Department of Zoology, University of Toronto, Canadian Drosophila Microarray Centre, Mississauga, Ontario, Canada*

BRIAN A. WILLIAMS (12), *Biology Division 156-29, Caltech, Pasadena, California*

MARK A. WITSCHI (9), *Laboratory of Bioorganic Chemistry, NIDDK, NIH, DHHS Bethesda, Maryland and Department of Chemistry, Northwestern University, Evanston, Illinois*

PAUL K. WOLBER (2), *Microarray Quality Agilent Technologies, Inc., Santa Clara, California*

BARBARA J. WOLD (12), *Biology Division 156-29, Caltech, Pasadena, California*

JOANNE M. YEAKLEY (3), *Illumina, Inc., San Diego, California*

LIXIN ZHOU (17), *Illumina, Inc., San Diego, California*

METHODS IN ENZYMOLOGY

VOLUME XVI. Fast Reactions
Edited by KENNETH KUSTIN

VOLUME XVII. Metabolism of Amino Acids and Amines
(Parts A and B)
Edited by HERBERT TABOR AND CELIA WHITE TABOR

VOLUME XVIII. Vitamins and Coenzymes (Parts A, B, and C)
Edited by DONALD B. MCCORMICK AND LEMUEL D. WRIGHT

VOLUME XIX. Proteolytic Enzymes
Edited by GERTRUDE E. PERLMANN AND LASZLO LORAND

VOLUME XX. Nucleic Acids and Protein Synthesis (Part C)
Edited by KIVIE MOLDAVE AND LAWRENCE GROSSMAN

VOLUME XXI. Nucleic Acids (Part D)
Edited by LAWRENCE GROSSMAN AND KIVIE MOLDAVE

VOLUME XXII. Enzyme Purification and Related Techniques
Edited by WILLIAM B. JAKOBY

VOLUME XXIII. Photosynthesis (Part A)
Edited by ANTHONY SAN PIETRO

VOLUME XXIV. Photosynthesis and Nitrogen Fixation (Part B)
Edited by ANTHONY SAN PIETRO

VOLUME XXV. Enzyme Structure (Part B)
Edited by C. H. W. HIRS AND SERGE N. TIMASHEFF

VOLUME XXVI. Enzyme Structure (Part C)
Edited by C. H. W. HIRS AND SERGE N. TIMASHEFF

VOLUME XXVII. Enzyme Structure (Part D)
Edited by C. H. W. HIRS AND SERGE N. TIMASHEFF

VOLUME XXVIII. Complex Carbohydrates (Part B)
Edited by VICTOR GINSBURG

VOLUME XXIX. Nucleic Acids and Protein Synthesis (Part E)
Edited by LAWRENCE GROSSMAN AND KIVIE MOLDAVE

VOLUME XXX. Nucleic Acids and Protein Synthesis (Part F)
Edited by KIVIE MOLDAVE AND LAWRENCE GROSSMAN

VOLUME XXXI. Biomembranes (Part A)
Edited by SIDNEY FLEISCHER AND LESTER PACKER

VOLUME XXXII. Biomembranes (Part B)
Edited by SIDNEY FLEISCHER AND LESTER PACKER

VOLUME XXXIII. Cumulative Subject Index Volumes I-XXX
Edited by MARTHA G. DENNIS AND EDWARD A. DENNIS

VOLUME XXXIV. Affinity Techniques (Enzyme Purification: Part B)
Edited by WILLIAM B. JAKOBY AND MEIR WILCHEK

VOLUME XXXV. Lipids (Part B)
Edited by JOHN M. LOWENSTEIN

VOLUME XXXVI. Hormone Action (Part A: Steroid Hormones)
Edited by BERT W. O'MALLEY AND JOEL G. HARDMAN

VOLUME XXXVII. Hormone Action (Part B: Peptide Hormones)
Edited by BERT W. O'MALLEY AND JOEL G. HARDMAN

VOLUME XXXVIII. Hormone Action (Part C: Cyclic Nucleotides)
Edited by JOEL G. HARDMAN AND BERT W. O'MALLEY

VOLUME XXXIX. Hormone Action (Part D: Isolated Cells, Tissues, and Organ Systems)
Edited by JOEL G. HARDMAN AND BERT W. O'MALLEY

VOLUME XL. Hormone Action (Part E: Nuclear Structure and Function)
Edited by BERT W. O'MALLEY AND JOEL G. HARDMAN

VOLUME XLI. Carbohydrate Metabolism (Part B)
Edited by W. A. WOOD

VOLUME XLII. Carbohydrate Metabolism (Part C)
Edited by W. A. WOOD

VOLUME XLIII. Antibiotics
Edited by JOHN H. HASH

VOLUME XLIV. Immobilized Enzymes
Edited by KLAUS MOSBACH

VOLUME XLV. Proteolytic Enzymes (Part B)
Edited by LASZLO LORAND

VOLUME XLVI. Affinity Labeling
Edited by WILLIAM B. JAKOBY AND MEIR WILCHEK

VOLUME XLVII. Enzyme Structure (Part E)
Edited by C. H. W. HIRS AND SERGE N. TIMASHEFF

VOLUME XLVIII. Enzyme Structure (Part F)
Edited by C. H. W. HIRS AND SERGE N. TIMASHEFF

VOLUME XLIX. Enzyme Structure (Part G)
Edited by C. H. W. HIRS AND SERGE N. TIMASHEFF

VOLUME L. Complex Carbohydrates (Part C)
Edited by VICTOR GINSBURG

VOLUME LI. Purine and Pyrimidine Nucleotide Metabolism
Edited by PATRICIA A. HOFFEE AND MARY ELLEN JONES

VOLUME LII. Biomembranes (Part C: Biological Oxidations)
Edited by SIDNEY FLEISCHER AND LESTER PACKER

VOLUME LIII. Biomembranes (Part D: Biological Oxidations)
Edited by SIDNEY FLEISCHER AND LESTER PACKER

VOLUME LIV. Biomembranes (Part E: Biological Oxidations)
Edited by SIDNEY FLEISCHER AND LESTER PACKER

VOLUME LV. Biomembranes (Part F: Bioenergetics)
Edited by SIDNEY FLEISCHER AND LESTER PACKER

VOLUME LVI. Biomembranes (Part G: Bioenergetics)
Edited by SIDNEY FLEISCHER AND LESTER PACKER

VOLUME LVII. Bioluminescence and Chemiluminescence
Edited by MARLENE A. DELUCA

VOLUME LVIII. Cell Culture
Edited by WILLIAM B. JAKOBY AND IRA PASTAN

VOLUME LIX. Nucleic Acids and Protein Synthesis (Part G)
Edited by KIVIE MOLDAVE AND LAWRENCE GROSSMAN

VOLUME LX. Nucleic Acids and Protein Synthesis (Part H)
Edited by KIVIE MOLDAVE AND LAWRENCE GROSSMAN

VOLUME 61. Enzyme Structure (Part H)
Edited by C. H. W. HIRS AND SERGE N. TIMASHEFF

VOLUME 62. Vitamins and Coenzymes (Part D)
Edited by DONALD B. MCCORMICK AND LEMUEL D. WRIGHT

VOLUME 63. Enzyme Kinetics and Mechanism (Part A: Initial Rate and Inhibitor Methods)
Edited by DANIEL L. PURICH

VOLUME 64. Enzyme Kinetics and Mechanism
(Part B: Isotopic Probes and Complex Enzyme Systems)
Edited by DANIEL L. PURICH

VOLUME 65. Nucleic Acids (Part I)
Edited by LAWRENCE GROSSMAN AND KIVIE MOLDAVE

VOLUME 66. Vitamins and Coenzymes (Part E)
Edited by DONALD B. MCCORMICK AND LEMUEL D. WRIGHT

VOLUME 67. Vitamins and Coenzymes (Part F)
Edited by DONALD B. MCCORMICK AND LEMUEL D. WRIGHT

VOLUME 68. Recombinant DNA
Edited by RAY WU

VOLUME 69. Photosynthesis and Nitrogen Fixation (Part C)
Edited by ANTHONY SAN PIETRO

VOLUME 70. Immunochemical Techniques (Part A)
Edited by HELEN VAN VUNAKIS AND JOHN J. LANGONE

VOLUME 71. Lipids (Part C)
Edited by JOHN M. LOWENSTEIN

VOLUME 72. Lipids (Part D)
Edited by JOHN M. LOWENSTEIN

VOLUME 73. Immunochemical Techniques (Part B)
Edited by JOHN J. LANGONE AND HELEN VAN VUNAKIS

VOLUME 74. Immunochemical Techniques (Part C)
Edited by JOHN J. LANGONE AND HELEN VAN VUNAKIS

VOLUME 75. Cumulative Subject Index Volumes XXXI, XXXII, XXXIV–LX
Edited by EDWARD A. DENNIS AND MARTHA G. DENNIS

VOLUME 76. Hemoglobins
Edited by ERALDO ANTONINI, LUIGI ROSSI-BERNARDI, AND EMILIA CHIANCONE

VOLUME 77. Detoxication and Drug Metabolism
Edited by WILLIAM B. JAKOBY

VOLUME 78. Interferons (Part A)
Edited by SIDNEY PESTKA

VOLUME 79. Interferons (Part B)
Edited by SIDNEY PESTKA

VOLUME 80. Proteolytic Enzymes (Part C)
Edited by LASZLO LORAND

VOLUME 81. Biomembranes (Part H: Visual Pigments and Purple Membranes, I)
Edited by LESTER PACKER

VOLUME 82. Structural and Contractile Proteins (Part A: Extracellular Matrix)
Edited by LEON W. CUNNINGHAM AND DIXIE W. FREDERIKSEN

VOLUME 83. Complex Carbohydrates (Part D)
Edited by VICTOR GINSBURG

VOLUME 84. Immunochemical Techniques (Part D: Selected Immunoassays)
Edited by JOHN J. LANGONE AND HELEN VAN VUNAKIS

VOLUME 85. Structural and Contractile Proteins (Part B: The Contractile Apparatus and the Cytoskeleton)
Edited by DIXIE W. FREDERIKSEN AND LEON W. CUNNINGHAM

VOLUME 86. Prostaglandins and Arachidonate Metabolites
Edited by WILLIAM E. M. LANDS AND WILLIAM L. SMITH

VOLUME 87. Enzyme Kinetics and Mechanism (Part C: Intermediates, Stereo-chemistry, and Rate Studies)
Edited by DANIEL L. PURICH

VOLUME 88. Biomembranes (Part I: Visual Pigments and Purple Membranes, II)
Edited by LESTER PACKER

VOLUME 89. Carbohydrate Metabolism (Part D)
Edited by WILLIS A. WOOD

VOLUME 125. Biomembranes (Part M: Transport in Bacteria, Mitochondria, and Chloroplasts: General Approaches and Transport Systems)
Edited by SIDNEY FLEISCHER AND BECCA FLEISCHER

VOLUME 126. Biomembranes (Part N: Transport in Bacteria, Mitochondria, and Chloroplasts: Protonmotive Force)
Edited by SIDNEY FLEISCHER AND BECCA FLEISCHER

VOLUME 127. Biomembranes (Part O: Protons and Water: Structure and Translocation)
Edited by LESTER PACKER

VOLUME 128. Plasma Lipoproteins (Part A: Preparation, Structure, and Molecular Biology)
Edited by JERE P. SEGREST AND JOHN J. ALBERS

VOLUME 129. Plasma Lipoproteins (Part B: Characterization, Cell Biology, and Metabolism)
Edited by JOHN J. ALBERS AND JERE P. SEGREST

VOLUME 130. Enzyme Structure (Part K)
Edited by C. H. W. HIRS AND SERGE N. TIMASHEFF

VOLUME 131. Enzyme Structure (Part L)
Edited by C. H. W. HIRS AND SERGE N. TIMASHEFF

VOLUME 132. Immunochemical Techniques (Part J: Phagocytosis and Cell-Mediated Cytotoxicity)
Edited by GIOVANNI DI SABATO AND JOHANNES EVERSE

VOLUME 133. Bioluminescence and Chemiluminescence (Part B)
Edited by MARLENE DELUCA AND WILLIAM D. MCELROY

VOLUME 134. Structural and Contractile Proteins (Part C: The Contractile Apparatus and the Cytoskeleton)
Edited by RICHARD B. VALLEE

VOLUME 135. Immobilized Enzymes and Cells (Part B)
Edited by KLAUS MOSBACH

VOLUME 136. Immobilized Enzymes and Cells (Part C)
Edited by KLAUS MOSBACH

VOLUME 137. Immobilized Enzymes and Cells (Part D)
Edited by KLAUS MOSBACH

VOLUME 138. Complex Carbohydrates (Part E)
Edited by VICTOR GINSBURG

VOLUME 139. Cellular Regulators (Part A: Calcium- and Calmodulin-Binding Proteins)
Edited by ANTHONY R. MEANS AND P. MICHAEL CONN

VOLUME 140. Cumulative Subject Index Volumes 102–119, 121–134

VOLUME 226. Metallobiochemistry (Part C: Spectroscopic and
Physical Methods for Probing Metal Ion Environments in Metalloenzymes
and Metalloproteins)
Edited by JAMES F. RIORDAN AND BERT L. VALLEE

VOLUME 227. Metallobiochemistry (Part D: Physical and Spectroscopic
Methods for Probing Metal Ion Environments in Metalloproteins)
Edited by JAMES F. RIORDAN AND BERT L. VALLEE

VOLUME 228. Aqueous Two-Phase Systems
Edited by HARRY WALTER AND GÖTE JOHANSSON

VOLUME 229. Cumulative Subject Index Volumes 195–198, 200–227

VOLUME 230. Guide to Techniques in Glycobiology
Edited by WILLIAM J. LENNARZ AND GERALD W. HART

VOLUME 231. Hemoglobins (Part B: Biochemical and Analytical Methods)
Edited by JOHANNES EVERSE, KIM D. VANDEGRIFF, AND ROBERT M. WINSLOW

VOLUME 232. Hemoglobins (Part C: Biophysical Methods)
Edited by JOHANNES EVERSE, KIM D. VANDEGRIFF, AND ROBERT M. WINSLOW

VOLUME 233. Oxygen Radicals in Biological Systems (Part C)
Edited by LESTER PACKER

VOLUME 234. Oxygen Radicals in Biological Systems (Part D)
Edited by LESTER PACKER

VOLUME 235. Bacterial Pathogenesis (Part A: Identification and Regulation of
Virulence Factors)
Edited by VIRGINIA L. CLARK AND PATRIK M. BAVOIL

VOLUME 236. Bacterial Pathogenesis (Part B: Integration of Pathogenic
Bacteria with Host Cells)
Edited by VIRGINIA L. CLARK AND PATRIK M. BAVOIL

VOLUME 237. Heterotrimeric G Proteins
Edited by RAVI IYENGAR

VOLUME 238. Heterotrimeric G-Protein Effectors
Edited by RAVI IYENGAR

VOLUME 239. Nuclear Magnetic Resonance (Part C)
Edited by THOMAS L. JAMES AND NORMAN J. OPPENHEIMER

VOLUME 240. Numerical Computer Methods (Part B)
Edited by MICHAEL L. JOHNSON AND LUDWIG BRAND

VOLUME 241. Retroviral Proteases
Edited by LAWRENCE C. KUO AND JULES A. SHAFER

VOLUME 242. Neoglycoconjugates (Part A)
Edited by Y. C. LEE AND REIKO T. LEE

VOLUME 243. Inorganic Microbial Sulfur Metabolism
Edited by HARRY D. PECK, JR., AND JEAN LEGALL

VOLUME 262. DNA Replication
Edited by JUDITH L. CAMPBELL

VOLUME 263. Plasma Lipoproteins (Part C: Quantitation)
Edited by WILLIAM A. BRADLEY, SANDRA H. GIANTURCO, AND JERE P. SEGREST

VOLUME 264. Mitochondrial Biogenesis and Genetics (Part B)
Edited by GIUSEPPE M. ATTARDI AND ANNE CHOMYN

VOLUME 265. Cumulative Subject Index Volumes 228, 230–262

VOLUME 266. Computer Methods for Macromolecular Sequence Analysis
Edited by RUSSELL F. DOOLITTLE

VOLUME 267. Combinatorial Chemistry
Edited by JOHN N. ABELSON

VOLUME 268. Nitric Oxide (Part A: Sources and Detection of NO; NO Synthase)
Edited by LESTER PACKER

VOLUME 269. Nitric Oxide (Part B: Physiological and Pathological Processes)
Edited by LESTER PACKER

VOLUME 270. High Resolution Separation and Analysis of Biological Macromolecules (Part A: Fundamentals)
Edited by BARRY L. KARGER AND WILLIAM S. HANCOCK

VOLUME 271. High Resolution Separation and Analysis of Biological Macromolecules (Part B: Applications)
Edited by BARRY L. KARGER AND WILLIAM S. HANCOCK

VOLUME 272. Cytochrome P450 (Part B)
Edited by ERIC F. JOHNSON AND MICHAEL R. WATERMAN

VOLUME 273. RNA Polymerase and Associated Factors (Part A)
Edited by SANKAR ADHYA

VOLUME 274. RNA Polymerase and Associated Factors (Part B)
Edited by SANKAR ADHYA

VOLUME 275. Viral Polymerases and Related Proteins
Edited by LAWRENCE C. KUO, DAVID B. OLSEN, AND STEVEN S. CARROLL

VOLUME 276. Macromolecular Crystallography (Part A)
Edited by CHARLES W. CARTER, JR., AND ROBERT M. SWEET

VOLUME 277. Macromolecular Crystallography (Part B)
Edited by CHARLES W. CARTER, JR., AND ROBERT M. SWEET

VOLUME 278. Fluorescence Spectroscopy
Edited by LUDWIG BRAND AND MICHAEL L. JOHNSON

VOLUME 279. Vitamins and Coenzymes (Part I)
Edited by DONALD B. MCCORMICK, JOHN W. SUTTIE, AND CONRAD WAGNER

VOLUME 280. Vitamins and Coenzymes (Part J)
Edited by DONALD B. MCCORMICK, JOHN W. SUTTIE, AND CONRAD WAGNER

VOLUME 281. Vitamins and Coenzymes (Part K)
Edited by DONALD B. MCCORMICK, JOHN W. SUTTIE, AND CONRAD WAGNER

VOLUME 282. Vitamins and Coenzymes (Part L)
Edited by DONALD B. MCCORMICK, JOHN W. SUTTIE, AND CONRAD WAGNER

VOLUME 283. Cell Cycle Control
Edited by WILLIAM G. DUNPHY

VOLUME 284. Lipases (Part A: Biotechnology)
Edited by BYRON RUBIN AND EDWARD A. DENNIS

VOLUME 285. Cumulative Subject Index Volumes 263, 264, 266–284, 286–289

VOLUME 286. Lipases (Part B: Enzyme Characterization and Utilization)
Edited by BYRON RUBIN AND EDWARD A. DENNIS

VOLUME 287. Chemokines
Edited by RICHARD HORUK

VOLUME 288. Chemokine Receptors
Edited by RICHARD HORUK

VOLUME 289. Solid Phase Peptide Synthesis
Edited by GREGG B. FIELDS

VOLUME 290. Molecular Chaperones
Edited by GEORGE H. LORIMER AND THOMAS BALDWIN

VOLUME 291. Caged Compounds
Edited by GERARD MARRIOTT

VOLUME 292. ABC Transporters: Biochemical, Cellular, and
Molecular Aspects
Edited by SURESH V. AMBUDKAR AND MICHAEL M. GOTTESMAN

VOLUME 293. Ion Channels (Part B)
Edited by P. MICHAEL CONN

VOLUME 294. Ion Channels (Part C)
Edited by P. MICHAEL CONN

VOLUME 295. Energetics of Biological Macromolecules (Part B)
Edited by GARY K. ACKERS AND MICHAEL L. JOHNSON

VOLUME 296. Neurotransmitter Transporters
Edited by SUSAN G. AMARA

VOLUME 297. Photosynthesis: Molecular Biology of Energy Capture
Edited by LEE MCINTOSH

VOLUME 298. Molecular Motors and the Cytoskeleton (Part B)
Edited by RICHARD B. VALLEE

VOLUME 299. Oxidants and Antioxidants (Part A)
Edited by LESTER PACKER

VOLUME 300. Oxidants and Antioxidants (Part B)
Edited by LESTER PACKER

VOLUME 301. Nitric Oxide: Biological and Antioxidant Activities (Part C)
Edited by LESTER PACKER

VOLUME 302. Green Fluorescent Protein
Edited by P. MICHAEL CONN

VOLUME 303. cDNA Preparation and Display
Edited by SHERMAN M. WEISSMAN

VOLUME 304. Chromatin
Edited by PAUL M. WASSARMAN AND ALAN P. WOLFFE

VOLUME 305. Bioluminescence and Chemiluminescence (Part C)
Edited by THOMAS O. BALDWIN AND MIRIAM M. ZIEGLER

VOLUME 306. Expression of Recombinant Genes in
Eukaryotic Systems
Edited by JOSEPH C. GLORIOSO AND MARTIN C. SCHMIDT

VOLUME 307. Confocal Microscopy
Edited by P. MICHAEL CONN

VOLUME 308. Enzyme Kinetics and Mechanism (Part E: Energetics of
Enzyme Catalysis)
Edited by DANIEL L. PURICH AND VERN L. SCHRAMM

VOLUME 309. Amyloid, Prions, and Other Protein Aggregates
Edited by RONALD WETZEL

VOLUME 310. Biofilms
Edited by RON J. DOYLE

VOLUME 311. Sphingolipid Metabolism and Cell Signaling (Part A)
Edited by ALFRED H. MERRILL, JR., AND YUSUF A. HANNUN

VOLUME 312. Sphingolipid Metabolism and Cell Signaling (Part B)
Edited by ALFRED H. MERRILL, JR., AND YUSUF A. HANNUN

VOLUME 313. Antisense Technology (Part A: General Methods, Methods of
Delivery, and RNA Studies)
Edited by M. IAN PHILLIPS

VOLUME 314. Antisense Technology (Part B: Applications)
Edited by M. IAN PHILLIPS

VOLUME 315. Vertebrate Phototransduction and the Visual Cycle (Part A)
Edited by KRZYSZTOF PALCZEWSKI

VOLUME 316. Vertebrate Phototransduction and the Visual Cycle (Part B)
Edited by KRZYSZTOF PALCZEWSKI

VOLUME 371. RNA Polymerases and Associated Factors (Part D)
Edited by SANKAR L. ADHYA AND SUSAN GARGES

VOLUME 372. Liposomes (Part B)
Edited by NEJAT DÜZGÜNEŞ

VOLUME 373. Liposomes (Part C)
Edited by NEJAT DÜZGÜNEŞ

VOLUME 374. Macromolecular Crystallography (Part D)
Edited by CHARLES W. CARTER, JR., AND ROBERT W. SWEET

VOLUME 375. Chromatin and Chromatin Remodeling Enzymes (Part A)
Edited by C. DAVID ALLIS AND CARL WU

VOLUME 376. Chromatin and Chromatin Remodeling Enzymes (Part B)
Edited by C. DAVID ALLIS AND CARL WU

VOLUME 377. Chromatin and Chromatin Remodeling Enzymes (Part C)
Edited by C. DAVID ALLIS AND CARL WU

VOLUME 378. Quinones and Quinone Enzymes (Part A)
Edited by HELMUT SIES AND LESTER PACKER

VOLUME 379. Energetics of Biological Macromolecules (Part D)
Edited by JO M. HOLT, MICHAEL L. JOHNSON, AND GARY K. ACKERS

VOLUME 380. Energetics of Biological Macromolecules (Part E)
Edited by JO M. HOLT, MICHAEL L. JOHNSON, AND GARY K. ACKERS

VOLUME 381. Oxygen Sensing
Edited by CHANDAN K. SEN AND GREGG L. SEMENZA

VOLUME 382. Quinones and Quinone Enzymes (Part B)
Edited by HELMUT SIES AND LESTER PACKER

VOLUME 383. Numerical Computer Methods (Part D)
Edited by LUDWIG BRAND AND MICHAEL L. JOHNSON

VOLUME 384. Numerical Computer Methods (Part E)
Edited by LUDWIG BRAND AND MICHAEL L. JOHNSON

VOLUME 385. Imaging in Biological Research (Part A)
Edited by P. MICHAEL CONN

VOLUME 386. Imaging in Biological Research (Part B)
Edited by P. MICHAEL CONN

VOLUME 387. Liposomes (Part D)
Edited by NEJAT DÜZGÜNEŞ

VOLUME 388. Protein Engineering
Edited by DAN E. ROBERTSON AND JOSEPH P. NOEL

VOLUME 389. Regulators of G-Protein Signaling (Part A)
Edited by DAVID P. SIDEROVSKI

VOLUME 390. Regulators of G-Protein Signaling (Part B)
Edited by DAVID P. SIDEROVSKI

VOLUME 391. Liposomes (Part E)
Edited by NEJAT DÜZGÜNEŞ

VOLUME 392. RNA Interference
Edited by ENGELKE ROSSI

VOLUME 393. Circadian Rhythms
Edited by MICHAEL W. YOUNG

VOLUME 394. Nuclear Magnetic Resonance of Biological Macromolecules
(Part C)
Edited by THOMAS L. JAMES

VOLUME 395. Producing the Biochemical Data (Part B)
Edited by ELIZABETH A. ZIMMER AND ERIC H. ROALSON

VOLUME 396. Nitric Oxide (Part E)
Edited by LESTER PACKER AND ENRIQUE CADENAS

VOLUME 397. Environmental Microbiology
Edited by JARED R. LEADBETTER

VOLUME 398. Ubiquitin and Protein Degradation (Part A)
Edited by RAYMOND J. DESHAIES

VOLUME 399. Ubiquitin and Protein Degradation (Part B)
Edited by RAYMOND J. DESHAIES

VOLUME 400. Phase II Conjugation Enzymes and Transport Systems
Edited by HELMUT SIES AND LESTER PACKER

VOLUME 401. Glutathione Transferases and Gamma Glutamyl Transpeptidases
Edited by HELMUT SIES AND LESTER PACKER

VOLUME 402. Biological Mass Spectrometry
Edited by A. L. BURLINGAME

VOLUME 403. GTPases Regulating Membrane Targeting and Fusion
Edited by WILLIAM E. BALCH, CHANNING J. DER, AND ALAN HALL

VOLUME 404. GTPases Regulating Membrane Dynamics
Edited by WILLIAM E. BALCH, CHANNING J. DER, AND ALAN HALL

VOLUME 405. Mass Spectrometry: Modified Proteins and Glycoconjugates
Edited by A. L. BURLINGAME

VOLUME 406. Regulators and Effectors of Small GTPases: Rho Family
Edited by WILLIAM E. BALCH, CHANNING J. DER, AND ALAN HALL

VOLUME 407. Regulators and Effectors of Small GTPases: Ras Family
Edited by WILLIAM E. BALCH, CHANNING J. DER, AND ALAN HALL

VOLUME 408. DNA Repair (Part A)
Edited by JUDITH L. CAMPBELL AND PAUL MODRICH

VOLUME 409. DNA Repair (Part B)
Edited by JUDITH L. CAMPBELL AND PAUL MODRICH

Section I

Array Platforms

[1]   The Affymetrix GeneChip® Platform: An Overview

By DENNISE D. DALMA-WEISZHAUSZ, JANET WARRINGTON,
EUGENE Y. TANIMOTO, and C. GARRETT MIYADA

Abstract

The intent of this chapter is to provide the reader with a review of GeneChip technology and the complete system it represents, including its versatility, components, and the exciting applications that are enabled by this platform. The following aspects of the technology are reviewed: array design and manufacturing, target preparation, instrumentation, data analysis, and both current and future applications. There are key differentiators between Affymetrix' GeneChip technology and other microarray-based methods. The most distinguishing feature of GeneChip microarrays is that their manufacture is directed by photochemical synthesis. Because of this manufacturing technology, more than a million different probes can be synthesized on an array roughly the size of a thumbnail. These numbers allow the inclusion of multiple probes to interrogate the same target sequence, providing statistical rigor to data interpretation. Over the years the GeneChip platform has proven to be a reliable and robust system, enabling many new discoveries and breakthroughs to be made by the scientific community.

Introduction

Starting in the 1990s, a genomic revolution, propelled by major technological advances, has enabled scientists to complete the sequences of a variety of organisms, including viruses, bacteria, invertebrates, and culminating in the full draft sequence of the human genome (Lander *et al.*, 2001). In the wake of this flood of sequence information, scientists are currently faced with the daunting task of translating genomic sequence information into functional biological mechanisms that will allow a better understanding of life and its disease states and hopefully offer better diagnostics and novel therapeutic interventions. High-density microarrays are uniquely qualified to tackle this daunting task and have therefore become an essential tool in life sciences research. They provide a reliable, fast, and cost-effective method that effectively scales with the ever-increasing amounts of genomic information.

METHODS IN ENZYMOLOGY, VOL. 410   0076-6879/06 $35.00
                  DOI: 10.1016/S0076-6879(06)10001-4

In the last decade, there has been an immense growth in the use of high-throughput microarray technology for three major genetic explorations: the genome-wide analysis of gene expression, SNP genotyping, and resequencing. While many of these studies have focused on human subjects and diseases, microarrays are also being used to study the gene expression and sequence variation of a variety of model organisms, such as yeast, *Drosophila*, mice, and rats. New applications are rapidly emerging, such as the discovery of novel transcripts (from coding and noncoding regions), the identification of novel regulatory sequences, and the characterization of functional domains in the RNA transcript. Integrating all of the information emanating from whole-genome studies will undoubtedly allow a more global understanding of the genome and the regulatory circuits that govern its activity.

The comparison of genome-wide expression patterns provides researchers with an objective and hypothesis-free method to better understand the dynamic relationship between mRNA content and biological function. This method has enabled scientists to discover, for example, the genetic pathways that are changed and disrupted in a wide range of diseases, from cancer (Armstrong *et al.*, 2002; Huang *et al.*, 2004; Yeoh *et al.*, 2002) to multiple sclerosis (Steinman and Zamvil, 2003). Across multiple disciplines, whole-genome expression analysis is helping scientists to stratify disease states, predict patient outcome, and make better therapeutic choices. Some of the recent examples of scientific and medical findings utilizing this technology include the identification of murine longevity genes and the discovery of novel transcripts that question our basic understanding of gene expression (Kapranov *et al.*, 2002).

The most recent generation of GeneChip microarrays for DNA sequence analysis allows scientists to genotype single nucleotide polymorphisms on a genome-wide scale (Kennedy *et al.*, 2003; Matsuzaki *et al.*, 2004a,b). The ability to quickly genotype over 100,000 single nucleotide polymorphisms (SNPs) distributed across the human genome has allowed researchers to conduct linkage analysis and genetic association studies. These new tools for disease mapping studies have already helped scientists pinpoint genes linked to diseases such as sudden infant death syndrome (Puffenberger *et al.*, 2004), neonatal diabetes (Sellick *et al.*, 2003), and bipolar disorder (Middleton *et al.*, 2004). The technology has proven to be scalable, and assays that cover 500,000 SNPs are now available.

Microarrays have revolutionized basic scientific research and are constantly challenging our view of the genome and its complexity. They are finding their way from the research laboratory to the clinic, where they promise the same kind of revolution in patient care. Microarrays used in clinical research and clinical applications promise to help scientists develop more accurate diagnostics and create novel therapeutics. By standardizing microarray data and integrating it with a patient's existing medical records, physicians can offer more tailored and more successful therapies. The combination of a patient's genetic and clinical data will allow for personalized medicine, which is where GeneChip technology holds the greatest promise to improve health.

GeneChip Microarrays, a Flexible Platform

GeneChip arrays are the result of the combination of a number of technologies, design criteria, and quality control processes. In addition to the arrays, the technology relies on standardized assays and reagents, instrumentation (fluidics system, hybridization oven and scanner), and data analysis tools that have been developed as a single platform. The key assay steps are outlined in Fig. 1 and are discussed in greater detail in later sections along with array design and manufacturing. The considerable flexibility of the GeneChip system and the manufacturing technology allows the design of the arrays to be dictated by their intended use, such as whole-genome transcriptome mapping, gene expression profiling, or custom genotyping. In addition to GeneChip catalogue microarrays (over 50 arrays and array sets are currently available), a custom program exists, where researchers can design their own arrays for organisms not covered by existing products and for specialized or directed studies. These designs may be based on many of the same design features and manufacturing techniques available in catalogue arrays (probe selection algorithms, manufacturing control tests, etc.) and are expected to provide customers equivalent performance to their commercial counterparts.

Array Manufacturing

Adapting technologies used in the semiconductor industry, GeneChip array manufacturing begins with a 5-in.[2] quartz wafer (Fodor et al., 1991; McGall and Christians, 2002). This substrate is first modified covalently with a silane reagent to produce a stable surface layer of hydroxyalkyl groups. Linker molecules with photolabile-protecting groups are then attached covalently to this layer to create a surface that may be spatially activated by light (Fig. 2). A photolithographic mask set that represents the sequence information content on the array is carefully designed. Each mask is manufactured with windows that either block or permit the transmission of ultraviolet light. These windows are distributed over the mask based on the desired sequence of each probe. The mask is carefully aligned

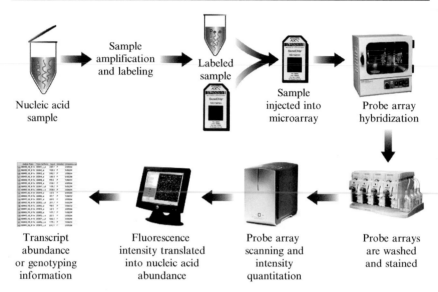

FIG. 1. Flowchart of a GeneChip System microarray experiment. Once the nucleic acid sample has been obtained, target amplification and labeling result in a labeled sample. The labeled sample is then injected into the probe array and allowed to hybridize overnight in the hybridization oven. Probe array washing and staining occur on the fluidics station, which can handle four probe arrays simultaneously. The probe array is then ready to be scanned in the Affymetrix GeneChip scanner, where the fluorescence intensity of each feature is read. Data output includes an intensity measurement for each transcript or the detailed sequence or genotyping (SNP) information.

with the quartz wafer, which ensures that oligonucleotide synthesis is only activated at precise locations on the wafer. When near-ultraviolet light shines through the mask, terminal hydroxyl groups on the linker molecules in exposed areas of the wafer are deprotected, thereby activating them for nucleotide coupling, while linkers in unexposed regions remain protected and inactive. A solution containing a deoxynucleoside phosphoramidite monomer with a light-sensitive protecting group is flushed over the surface of the wafer, and the nucleoside attaches to the activated linkers (coupling step), initiating the synthesis process.

Oligonucleotide synthesis proceeds by repeating the two basic steps: deprotection and coupling. For each round of synthesis, deprotection generally uses a unique mask from the designed set. The coupling steps alternate through the addition of A-, C-, G-, or T-modified nucleotides. The deprotection and coupling cycle is repeated until all of the full-length probe sequences, usually 25-mers, are completed. Algorithms that optimize

A Photolithography

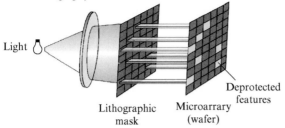

Light

Lithographic
mask

Microarrary
(wafer)

Deprotected
features

B Chemical synthesis cycle

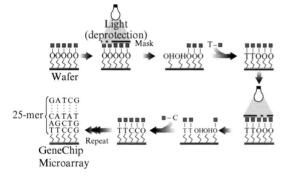

Light
(deprotection)
Mask

T-■

Wafer

25-mer {
GATCG
CATAT
AGCTG
TTCCG

Repeat

TTCCO

■-C

TT OHOHO

TTOOO

GeneChip
Microarray

C Dicing and cartridge assembly

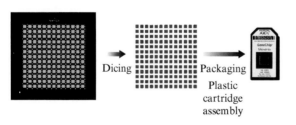

Dicing

Packaging

Plastic
cartridge
assembly

GeneChip
Microarray

Fig. 2. Manufacture of a GeneChip probe array. (A) Photolithography. (Left) Near-ultraviolet light is passed through a mask containing open windows. The size and the location of each open window delineate the surface on the quartz wafer that will be activated for chemical synthesis. The use of sequential masks in conjunction with the chemical synthesis creates a cycle that directs the precise sequence synthesis of oligonucleotides that compose the array. (Right) The photolithographic process. (B) (Left) Schematic representation of the nucleic acid synthesis cycle. Light removes protecting groups (squares) at defined areas on the array. A single nucleotide is washed over the array and couples to the deprotected areas. Through successive steps, any oligonucleotide sequence can be built on each feature of the array. The number of steps required to build a 25 nucleotide sequence on the array is 100, although the optimization of mask usage has lowered that number to ~75 steps. (Right) The chemical synthesis station, where nucleotide binding occurs. (C) (Left) Complete synthesis on

mask usage allow the creation of the arrays in significantly fewer than the 100 cycles that would normally be required to synthesize all possible 25-mer sequences (Lipshutz *et al.*, 1995). The information density of the array depends on the spatial resolution of the photolithographic process.

Once oligonucleotide synthesis is complete, wafers can be diced in a variety of array sizes and packaged individually into cartridges. Generally, each 5-in. square wafer can yield between 49 and 400 identical GeneChip microarrays, depending on the amount of genetic information required. A typical 1.28-cm^2 array (49-format), for example, will contain more than 1.4 million different probe locations, or features, assuming the features are spaced 11 μm apart. Each of these features contains millions of identical DNA molecules. A reduction of the feature spacing to 5 μm (as available on the Mapping 500K Array Set released in September of 2005) produces over 6.5 million different features on the same 1.28-cm^2 array—an exponential increase in the available data from a single experiment. This demonstrates the power of "feature shrink" on the Affymetrix microarray platform. The manufacturing process ends with a comprehensive series of quality control tests to ensure that GeneChip arrays deliver accurate and reproducible data.

Array Design

Array design is closely coupled to sample preparation and the biological question to be addressed. Specific examples are described in greater detail for expression and genotyping applications. Almost all of the designs utilize two types of probes: (1) probes that have complete complementarity to their target sequence [perfect match probe (PM)] and (2) probes with a single mismatch to the target, centered in the middle of the oligonucleotide [mismatch probe (MM), Fig. 3]. The number of probes used to interrogate a specific SNP or transcript is selected to meet specific performance criteria for each assay.

In addition to the probes specific for a particular assay, arrays contain a number of different control probes. There are probes specific for quality control assays. Another set of probes is arranged in checkerboard patterns on the array. These probes bind to a specific biotinylated oligonucleotide included in the hybridization cocktail. Following scanning, these

the wafer results in many (49–400) identical high-density oligonucleotide microarrays in one wafer. Dicing of the wafer into individual microarrays occurs, and each microarray is inserted into a plastic cartridge. (Right) Machinery used to incorporate the diced microarray into the plastic cartridge.

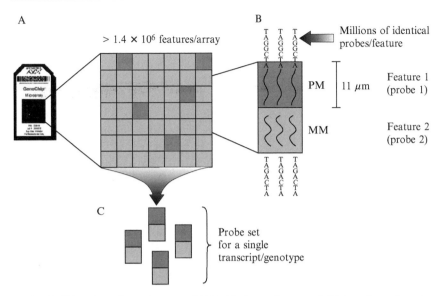

FIG. 3. Dissection of a probe array. (A) Inside the probe array (left) is a piece of quartz, generally containing a synthesis area of 1.28 cm² and carrying more than a million different features, assuming 11-μm feature spacing. Each feature, in turn, is composed of millions of oligonucleotide sequences. (B) For every perfect match (PM) feature, a mismatch (MM) feature is included, which is identical to the PM sequence, except for a nucleotide transversion on the 13th nucleotide, the central nucleotide. (C) A probe set refers to all features (PM and MM) that interrogate the same target sequence.

checkerboard patterns provide a means to ensure that signal intensities are properly assigned to the correct feature on the array. Other probes can detect specific controls that are added during sample preparation, providing evidence that the upstream assay was performed properly.

Array Designs for Gene Expression

The probe selection strategy used for gene expression arrays is dictated by the intended use of the array. For example, probes can be selected that identify unique transcripts, common transcript sequence segments, multiple splice sites, or polyadenylation variants. Bioinformatics techniques are used to assemble sequences from various public sources such as GenBank, dbEST, and RefSeq. Genome sequence alignments allow the selection of high-quality sequence data, as well as the consolidation of redundant transcripts and the identification of splice variants. The use of cDNA assemblies over exemplar sequences results in a higher quality design based on all of the empirical sequence data. cDNA sequence orientation is determined using a probabilistic model applied to genetic annotations, genomic splice-site usage, polyadenlyation sites, and sequence observations. This combination of metrics ensures that probes are selected against the correct strand. Annotations are generated for each target and are then prioritized for inclusion in the final array design.

In the *in vitro* transcription (IVT) assay the probe selection region is typically defined as the first 600 bases proximal to the polyadenylation site (3′ end) (Fig. 3). Probe selection requires applying a multiple linear regression model to identify those probes whose hybridization intensities respond in a linear fashion to the relative abundance of the target (Mei *et al.*, 2003). The algorithm is based on a thermodynamic model of nucleic acid duplex formation modified with empirically derived parameters. Probes are also selected to minimize the effects of cross hybridization and to maximize spacing between the probes. Typically 11 probe pairs are selected per 600-bp probe selection region. A probe pair consists of a PM probe and its corresponding MM probe (Fig. 3).

Expression assays based on random priming methods can be applied to both prokaryotes and eukaryotes, and probe selection regions need not be restricted to the 600 bases proximal to the 3′ end of the gene. In the case of prokaryotes, arrays are usually designed using open reading frames as the probe selection region. For eukaryotes the probe selection region is defined by potential exons. This type of design permits expression analysis over the

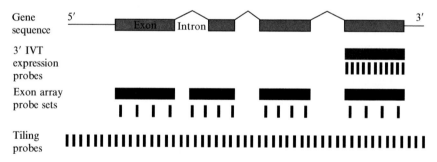

FIG. 4. Gene expression array design strategies. The different expression strategies for probe selection are represented. The gene sequence shown at the top represents an example of a target transcript. Rectangles represent exons, while the connecting lines represent introns. The 3′ IVT expression probes target sequences are at the extreme 3′ end and are adjacent to the poly(A) tail of the mRNA. This strategy is the most commonly used for commercial whole genome transcriptome designs. Exon array probe sets include probes that are within exon sequences. For tiling arrays, probes are placed sequentially throughout the genome at the same approximate distance from each other.

entire transcript and allows the identification of alternatively spliced transcripts. Such a design is illustrated in Fig. 4.

A third type of design is used in expression analysis. Tiling arrays interrogate the genome at regular intervals without regard to gene annotations (Fig. 4). Originally this type of design was applied to a pair of human chromosomes. Currently the entire human genome can be interrogated at 35-bp intervals (measured center to center from adjacent probes) using 14 arrays that contain features spaced $5\mu m$ apart. This type of design has proven useful in the identification of novel transcripts but its utility stretches beyond RNA mapping. For example, the chromosomal location of binding sites for DNA-binding proteins have been identified by applying chromatin-immunoprecipitated material to these arrays. It should be noted that with the exception of tiling, the array designs can be improved with better gene annotations.

Array Design for DNA Analysis

High-density oligonucleotide arrays enable rapid analysis of sequence variation (resequencing) and analysis of single nucleotide polymorphisms (genotyping). A different set of strategies is used to select probes for DNA analysis. The design of the array relies on multiple probes to interrogate individual nucleotides in a sequence. For sequence variation analysis (or resequencing), the identity of a target base can be deduced using four identical probes that vary only in the target position, each containing one of the four possible bases. For SNP genotyping, arrays with many probes for each allele can be created to provide redundant information. The probe tiling strategy for SNP genotyping is provided in greater detail later.

For any given SNP with alleles A and B, probes are synthesized on the array to represent both potential variants (Fig. 5). Each SNP is represented on the array by a probe set that consists of multiple probe pairs. The probe pairs differ in the location of the SNP within the oligonucleotide sequence. In addition to the PM and MM pair that contain the SNP on the central position of the probe (position 0), there are probes for each SNP that are shifted either upstream ($+1$, $+3$, $+4$ nucleotides) or downstream (-1, -2, -4 nucleotides) relative to the probe containing the SNP at the central position. Each of the 7 probes is empirically tested on a pilot microarray, and a total of 5 probe pairs are ultimately selected for inclusion on the final array product. Additionally, for each position, probes are included from the sense and the antisense strand. Therefore, there are a total of 20 probes interrogating each allele for a total of 40 probes per SNP. Following hybridization to the arrays, one can determine the identity of the particular SNP location as homozygous (AA or BB) or heterozygous (AB).

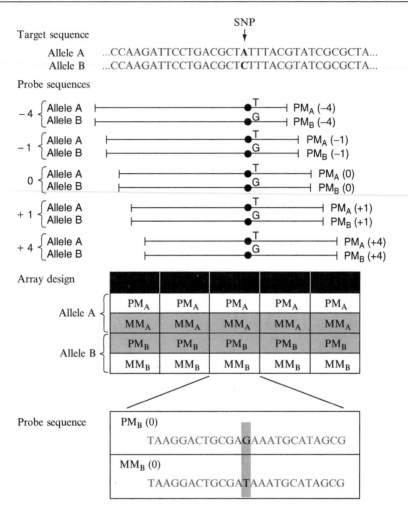

FIG. 5. Array design for DNA analysis. The top of the figure shows two possible alleles of the nucleic acid sequence to be analyzed (target). The probe sequence selection strategy for SNP genotyping includes probes that are centered on the SNP location (0), as well as probes that are shifted to the left (-4, -1) and to the right ($+1$, $+4$) of the central SNP location. The array design contains interrogation probes for both alleles and, similar to expression designs, includes a PM and MM probe pair. Depicted at the bottom of this schematic are two features representing the B allele, one harboring the centered PM probe and the one below representing its partner MM probe.

Another type of array for DNA analysis is used for resequencing. Some experimental approaches, such as sequencing large genomic regions, analyzing the sequence variants of a candidate gene, analyzing the genetic variability within a clinical trial population, and even assessing the sequence alterations among the genome of a pathogen are well served by this array design. This design provides a highly efficient analysis of up to 30 kb of double-stranded sequence, for a total of 60 kb. The array design includes tiling four different probes for each base interrogated per strand, for a total of eight probes per nucleotide position, which provides the redundancy for analysis of sequence variation and genotype determination.

Other Array Designs

As the foregoing demonstrates, Affymetrix core technology may be used to interrogate genetic material in numerous different assays to answer a broad range of different biological questions. In addition to the current gene expression and genotyping assays, two additional assays that demonstrate the flexibility of the technology are worth describing. Both of these are generic arrays to which a number of different assay or targets can be applied.

The GenFlex probe array contains over 2000 generic capture probes, which were selected for their lack of homology to existing genomic and cDNA sequences and for their similar hybridization behavior. This idea has been expanded up to 20,000 capture probes in the universal tag arrays, which are designed to work with the molecular inversion probe assay (Hardenbol *et al.*, 2003), which is designed to genotype flexible panels of SNPs that can be selected by the researcher.

Another generic array is the all *n*-mer design (Lipshutz *et al.*, 1995). For example, all possible 10-mer sequences can be synthesized in 40 steps on a single (1.28-cm^2) array with 12-μm feature spacing. These arrays may be used for differentiating variants of a known sequence.

Target Preparation

Most target preparation protocols start with a purified nucleic acid sample that is usually amplified and then labeled and fragmented. RNA targets are prepared by *in vitro* transcription, which provides amplification of the target. Biotinylated nucleotides or analogues are incorporated into the target during the IVT process. The labeled RNA is then purified and fragmented by hydrolysis. In the case of DNA targets the purified material is first purified and then fragmented by DN'ase I. At this point the DNA fragments are labeled with terminal deoxynucleotidyl transferase (TdT) and a biotinlyated nucleotide analogue. The material is now used directly in hybridizations.

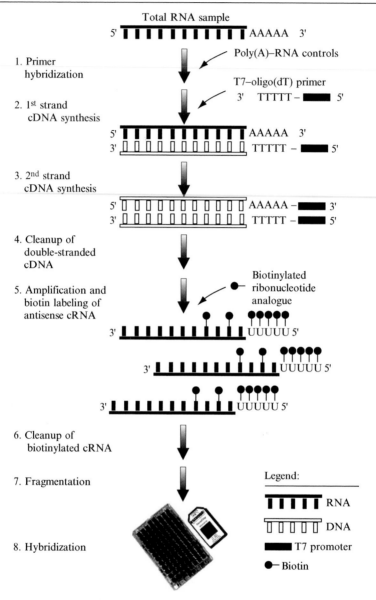

1. Primer hybridization

2. 1st strand cDNA synthesis

3. 2nd strand cDNA synthesis

4. Cleanup of double-stranded cDNA

5. Amplification and biotin labeling of antisense cRNA

6. Cleanup of biotinylated cRNA

7. Fragmentation

8. Hybridization

Fig. 6. One-cycle sample preparation for gene expression profiling. The flowchart depicts the steps by which eukaryotic samples are prepared for gene expression profiling. Briefly, total RNA or poly(A)–RNA is isolated. A primer that includes a poly(T) tail and a T7 polymerase-binding site [T7–oligo(dT) primer] is used for reverse transcription, resulting in synthesis of the first strand complementary DNA (cDNA). The second cDNA strand is completed, resulting in a double-stranded cDNA. In the one-cycle method, the double-stranded cDNA is

Gene expression assays have made use of both RNA and DNA targets. The most widely used sample preparation for gene expression utilizes the IVT reaction as originally described by Eberwine and colleagues (Van Gelder *et al.*, 1990). In this assay cDNA synthesis is initiated from an oligo(dT) primer that is also coupled to a T7 RNA polymerase primer. In this case cDNA synthesis starts adjacent to the poly(A) tail of the mRNA. After second strand synthesis, a double-stranded cDNA copy of each mRNA is created attached to the T7 RNA polymerase primer. An IVT reaction is then carried out to create a biotinylated RNA target. A schematic of the assay is shown in Fig. 6. A variation of this technique utilizes two rounds of IVT amplification and is used to create a target from very small amounts (100 ng or less) of starting material.

Gene expression assays have also been described that utilize random priming of cDNA synthesis for target preparation. This style of target preparation is used in the case of prokaryotic expression, where mRNAs lack poly(A) tails and in instances where the entire transcript is interrogated. Examples of the latter include targets for either tiling or exon designs. The final target after random priming is either single- or double-stranded cDNA. In either case the target is fragmented by DN'ase I digestion and labeled using TdT and a biotinylated nucleotide analogue.

Chromatin immunoprecipitation (ChIP) represents another sample preparation technique where the final product may be applied to tiling arrays. Proteins are first cross-linked to chromosomal DNA by formaldehyde. The cross-linked chromatin is then fragmented and immunoprecipitated with antibodies specific for the protein of interest. The associated DNA fragments are released from the immunoprecipitated material, purified, and amplified by a polymerase chain reaction (PCR). The PCR products are labeled using techniques described previously and the final target is hybridized to the array.

The whole genome sampling analysis assay for SNP analysis does not require site-specific primers, is highly scalable, and enables the creation of hybridization target starting with as little as 250 ng of chromosomal DNA (Fig. 7). The assay starts with the digestion of the DNA sample with a single restriction enzyme, followed by ligation of a common primer and amplification by PCR. The PCR conditions are optimized for the selective amplification of fragments that are 250–2000 nucleotides in length. The

---

used as a template for *in vitro* transcription with biotinylated ribonucleotides, resulting in a biotin-labeled RNA sample. After cRNA fragmentation, the sample is ready to be hybridized to the array.

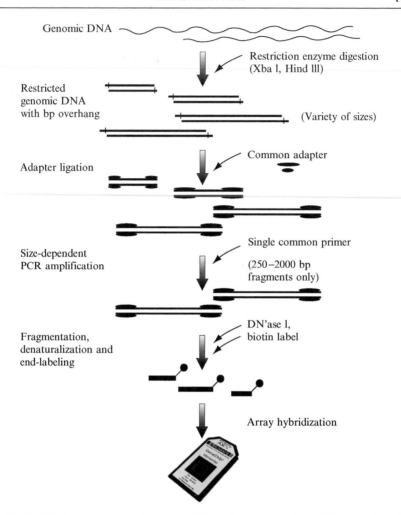

Genomic DNA

Restriction enzyme digestion
(Xba l, Hind lll)

Restricted
genomic DNA
with bp overhang

(Variety of sizes)

Common adapter

Adapter ligation

Size-dependent
PCR amplification

Single common primer

(250–2000 bp
fragments only)

Fragmentation,
denaturalization and
end-labeling

DN'ase l,
biotin label

Array hybridization

FIG. 7. Whole genome sampling assay. Schematic representation of the experimental procedure used to create a sample amenable to SNP analysis. Genomic DNA is subjected to restriction enzyme digestion, which results in varied size fragments. A common adapter is linked to the restriction overhangs and is used as a primer for PCR. PCR is conducted under controlled conditions were only fragments of 250–2000 bp are amplified, which results in a dramatic reduction of genome complexity. PCR product fragmentation and labeling result in a sample that is ready for microarray hybridization.

combination of restriction digestion and size-selective PCR amplification creates a sample of reduced complexity relative to the entire genome, which results in more accurate genotyping. The amplicons are fragmented and labeled as described previously for other DNA targets.

Resequencing assays start with a PCR amplicon or amplicons specific for the region of interest. When combining amplicons, the relative molar amount of each amplicon can be normalized to ensure a relatively uniform signal over the array. The PCR products are fragmented and labeled as described previously prior to hybridization.

GeneChip Instrument Components and Associated Assay Steps

The GeneChip instrument components include a hybridization oven, the fluidics station, an optional autoloader, and a scanner. All of these instruments are designed to work together and, with the exception of the hybridization oven, are directed by the GeneChip operating software (GCOS). The hybridization oven can hold up to 64 probe arrays and provides continuous rotation and consistent temperatures over the 16 h that are typically required for hybridization. The temperature is tunable to cover the different array applications and is usually selected between 40 and 50°.

After hybridization the arrays are transferred to the fluidics station. The fluidics station performs washing and staining operations for GeneChip microarrays, a crucial step in the assay that impacts data consistency and reproducibility. It washes and stains up to four probe arrays simultaneously. Unbound nucleic acid is washed away through a combination of low and high stringency washes. The stringency of the wash is determined by the salt concentration of the buffer and the temperature and duration of the wash, with the temperature and duration controlled by the fluidics station. The fluidics station contains inlets for two different buffers and heats buffers up to 50°, permitting temperature-controlled washes.

In the next step, bound target molecules are "stained" with a fluorescent streptavidin–phycoerythrin conjugate (SAPE), which binds to the biotins incorporated during target amplification. Most protocols also include an additional signal amplification process where biotinylated anti-streptavidin antibodies are bound to the initial SAPE molecules and then stained with a second SAPE addition. In the latest fluidics station, the 450 Model, wash and stain steps proceed in an automated fashion, ending with an array that is ready for scanning. The fluidics station is controlled by a computer workstation running GCOS. Different array applications require predetermined fluidics scripts, which can also be modified for custom protocols.

The AutoLoader is a front-loading sample carousel that can be added to the latest generation scanners as an option. The AutoLoader increases throughput by permitting unattended scanning for up to 48 arrays. Arrays are maintained at 15° prior to and after scanning. The instrument also includes a bar code reader that identifies the arrays, permits sample tracking, and aids in high-throughput analysis.

The current scanner is a wide-field, epifluorescent, confocal microscope that uses a solid-state laser to excite fluorophores bound to hybridized nucleic acids. The scanning mechanism incorporates a "flying objective," which employs a large numerical aperture objective that eliminates the need for multiple array scans. The most recent version of the scanner has a pixel resolution of 0.7 μm and is able to scan features with 5-μm spacing. The scanner can resolve more than 65,000 different fluorescence intensities. During the scan process a photomultiplier tube collects and converts fluorescence values into an electronic signal, which is then converted into the corresponding numerical values. These numerical values represent the fluorescence intensities, which are stored as pixel values that comprise the image data file (.dat file).

Image and Data Analysis

The next step in analysis is the assignment of pixels that make up the image (.dat) file to the appropriate feature. Previous methods have used a global gridding method in which the four corners of the array, defined by checkerboard patterns, serve as anchors for the grid. Features are then created by evenly dividing the area defined by the anchored corners into the known number of features for a given array. As the number of pixels per feature continued to decrease, an additional step called Feature Extraction was implemented to assign pixels to features in a more robust manner. In Feature Extraction the original pixels assigned to a feature are shifted as a block, a pixel at a time, and the coefficient of variation (CV) of pixel intensities for the shifted feature is computed. After allowing the feature pixels to shift up to a predetermined distance, the feature is defined where the pixel intensity CV is a minimum. Following Feature Extraction the intensity of each feature is calculated and stored in a .CEL file.

Regardless of application, the feature intensities found in .CEL files are used by analysis software to detect sequence variation or to differentiate gene expression levels of transcripts. During analysis, the use of multiple probes per genotype or gene is combined with standard statistical methods to provide a transparent and robust conversion of probe intensities to biological information.

For gene expression, a variety of algorithms exist to summarize multiple probe intensities (including PM or MM probes in a probe set) into an aggregate signal estimate that is correlated to the relative abundance of the transcript in the experimental sample. Detection calls are made by

Affymetrix software through an arithmetic vote of probe pairs within a set designed to detect a specific transcript (GeneChip MAS5 and GCOS software). More widely used is an estimation of relative transcript abundance by a probe set signal and the trend has shifted away from median probe intensity-based algorithms such as MAS5 to probe modeling algorithms such as dCHIP (Schadt *et al.*, 2004), RMA (Irizarry *et al.*, 2003), and PLIER Estimation (Affymetrix Technical Note, 2005). The probe modeling analysis software considers intrinsic probe behavior to account for systematic nonsystematic biases, error, and allows for true replicate analysis.

It is still common for algorithms to use both PM and MM probes; however, PM-only algorithms are popular. Subtraction of MM probe intensity from the PM intensity or subtraction of modeled background estimates in PM-only analyses serves the same purpose, which is to estimate the true probe intensity by subtraction of background from the raw PM probe intensity. Raw probe intensity (PM or MM) is the sum of a true hybridization signal, specific cross-hybridization signal, nonspecific binding signal, and small amounts of signal generated by system noise. Background consists of everything but the true signal, and most would agree that, for an unbiased or true measurement of probe intensity, background must be subtracted from the raw perfect match probe intensity. For most Affymetrix expression measurements, subtraction of MM probe intensity is an accurate method to remove background. However, background can also be estimated in the absence of MM probes, for example, RMA. Continued discussion around this topic is indicative of the maturing thought in this area. Despite differences in precision, accuracy, and bias, most signal estimate-algorithms (PM, MM, or PM only) result in similar biological interpretations from the same data sets.

For genotypic sequence variation detection, a dynamic model-mapping algorithm has been developed by Affymetrix. In recent applications as few as six probe quartets (24 probes) are used to generate a genotype call and confidence score for all genotypes called. The dynamic model-based approach provides a highly accurate genotype calling method, is effective for SNP screening, is robust against changes in experimental conditions, is flexible to experiment designs, and is scalable to more SNPs (Di *et al.*, 2005). For resequencing, a unique base-calling algorithm derived from the work of Cutler *et al.* (2001) is employed.

Current Applications

To date, there are more than 12,000 peer-reviewed publications based on microarray technology. Given that the use of microarrays became feasible in the late 1990s, it is easy to imagine how researchers from a myriad of fields have quickly leveraged this technology for a variety of scientific endeavors. This section touches briefly on some examples of the applications of this technology, initially those based on gene expression profiling, and later on those based on whole-genome DNA analysis, or genotyping.

Expression

Gene expression profiling studies are performed with the goal of comparing tissues, tissue types, and cellular responses to a variety of stimuli such as altered growth conditions, cancer, and infectious processes to gain biological insight into basic biochemical pathways or molecular mechanisms of disease and its regulatory circuits. To date, whole-genome expression analysis has already helped scientists stratify disease, predict patient outcome, compare strains with varying virulence, study the relationship between host and parasite, and understand the affected molecular pathways of certain diseases.

Cancer research is one of the clinical fields in which microarrays have had an unquestionable impact. Whole genome expression profiles of cancerous cells have already allowed scientists to classify cancer subtypes, predict a patient's prognosis, select between alternate therapies, and even identify new classes of tumors. In a now classic example, Armstrong and colleagues (2002) studied the gene expression profile of cells isolated from patients with acute lymphoblastic leukemia (ALL) and acute myelogenous leukemia (AML). The current diagnostic methodology for these diseases includes a microscopic assessment of the morphology of the cells. Given that the morphology of these two cell types can sometimes be very similar, it is difficult to differentiate ALL from AML. Gene expression profiling of these two cell populations resulted in a unique molecular signature for each one. Even more surprising was this group's finding of a unique molecular signature that was distinct from that of ALL and of AML among the diagnosed patients. This unique signature corresponded to a new leukemia subtype, namely mixed lineage leukemia (MLL). Upon review of the MLL patients' clinical histories it was noted that not only had all of them failed standard ALL therapy, but they also had a poor prognosis compared to ALL patients. The latter was the first whole-genome transcriptome study that showed the effects that a translocation, such as that of the MLL gene, can specify a unique expression signature. This information has allowed this research group to expand on their studies and, for example, study the effect of this translocation on the hematopoietic properties of granulocyte/macrophage progenitors (Wang et al., 2005).

Another novel application of high-density gene expression microarrays includes the unbiased study of the transcription that occurs throughout the genome, independent of considerations such as open reading frames and annotations. Most genetic studies have focused on regions that code for proteins, which compose around 2% of the human genome. However, given the 3.1 billion base pairs in our genome, it now seems striking that the rest of the 98% of the genome would be nonfunctional. There are new and collaborative efforts, such as the ENCODE (Encyclopedia of DNA Elements) project that are now attempting to study those neglected regions of DNA. For example, a group led by Thomas Gingeras has used tiling arrays on chromosomes 21 and 22 and discovered that there is widespread transcription, that is, they found far more transcriptional activity than could be accounted for by known genes that express proteins. This work has raised the possibility that genome function and regulation is far more complex than previously thought (Kapranov *et al.*, 2002).

Additionally, this group was able to identify transcription-binding sites on all the nonrepetitive sequences of these two chromosomes. They were able to identify a large and unexpected number of binding sites for three common transcription factors, Sp1, cMyc, and p53, distributed across chromosomes 21 and 22, suggesting a far more complex network of transcriptional regulation (Cawley *et al.*, 2004). Most of the transcriptional binding sites were not located at random, but rather at the start of novel, noncoding transcripts, embedded within or between known coding genes. These novel transcripts are expressed simultaneously with the coding transcripts and are regulated similarly, suggesting that coding and noncoding counterparts function in concert. The group of transcripts may actually be the genetic functional unit. As additional transcription factor-binding sites are studied across the whole genome, a better understanding of the complex regulatory networks that govern genome function will undoubtedly be discovered.

In addition to gene expression profiling in cancer and in the basic study of the genome function, there is an extensive collection of exciting examples covering fields such as infectious disease (Apidianakis *et al.*, 2005; Comer *et al.*, 2005; Fan *et al.*, 2005), cardiovascular disease (Boerma *et al.*, 2005; Kong *et al.*, 2005), and psychiatric disorders (Hekmat-Scafe et al., 2005; Iwamoto *et al.*, 2005). Additionally, there are numerous examples of gene expression profiling applications based on a variety of different species, such as *Drosophila* (Girardot *et al.*, 2004; Hekmat-Scafe *et al.*, 2005), *Caenorhabditis elegans* (Dinkova *et al.*, 2005; Reinke *et al.*, 2004), and *Arabidopsis* (Davletova *et al.*, 2005; Gomez-Mena *et al.*, 2005). [The reader is invited to visit the Affymetrix Web site, where a database of all applications based on Affymetrix GeneChip technology are listed and classified.]

Genotyping

The following selection of applications is based on DNA analysis (genotyping) rather than on expression profiling. There are an estimated 3×10^6 nucleotide differences between any two humans, which only accounts for 1 out of every 1000 bases in the human genome. The ability to analyze 100,000–500,000 SNPs at once across the whole genome enables scientists to create detailed genetic maps and, among other things, discover the gene(s) responsible for disease. Additionally, pharmacogenomics—the use of genomic information to study a patient's response to drugs—has also been enabled by high-density DNA analysis microarrays. An understanding of the enzymatic mechanisms underlying the pharmacology and pharmacokinetics associated with every drug for each individual patient allows a more personalized approach to the administration of pharmacologic treatments and could potentially avoid the trial-and-error process currently employed for drug selection.

The GeneChip Mapping 100K Set is already bearing its scientific fruit. Klein and colleagues (2005) used this array set to study age-related macular degeneration (AMD), a major cause of blindness in the elderly. Even though family-based and candidate gene studies had been undertaken, causative genes or gene mutations were hard to find. Because performing an association study requires typing hundreds of thousands of SNPs, Klein and colleagues (2005) used the high-density microarrays to study the whole genome SNP variations between AMD patients and healthy subjects. This study led to the identification of an intronic and common variant in the complement factor gene (CFH) that puts patients at higher risk for AMD. This gene is located on chromosome 1, consistent with chromosomal regions previously identified as being linked to AMD. The identification of this risk factor may be used in the future, for example, for diagnostics and for preventive therapies in patients at high risk of AMD.

Many diseases include an alteration in the normal number of chromosomes or chromosome segments, as well as mutations, deletions, or amplifications of more succinct sequence fragments. For example, Down syndrome results from a trisomy (triplication) of chromosome 21 (Korenberg, 1993), while a loss of a fragment of chromosome 17 (17q25.1) is characteristic of ovarian cancers (Presneau *et al.*, 2005). Information stemming from the characterization of this altered DNA copy number is crucial to the understanding of the mechanisms underlying the disease. There are two experimental approaches for DNA analysis that have been used to study the chromosomal stability of cancer biopsies: chromosomal copy number and loss of heterozygosity (LOH). Commonly used methods for addressing these issues include fluorescence *in situ* hybridization, Northern blotting, microsatellites, and comparative genomic hybridization (CGH), among others. However, high-density SNP microarrays enable scientists to interrogate the genome at far higher resolution than these techniques allow.

For example, it is known that chromosomal amplifications and deletions frequently contribute to cancer. Loss of heterozygosity refers to the loss of one allele caused by either a mutation or a deletion resulting in homozygosity. When this occurs at a tumor suppressor gene locus, for example, it may result in a neoplastic transformation.

One of the first studies to use SNP arrays to study genomic alteration such as LOH was conducted by the Meyerson group at the Dana Farber Cancer Institute. Their initial studies validated the large-scale genotyping of SNPs on small cell carcinoma cells on the first-generation high-density SNP array and showed that the loss of LOH data was consistent with previous CGH results (Lindblad-Toh et al., 2000).

Lung cancer is one of the leading causes of cancer deaths in the United States (Jemal et al., 2003). Many studies have focused on the LOH patterns of lung cancer; however, this is such a complex disease that a correlation between LOH analyses and clinical outcome has been challenging. Given that SNPs occur at a frequency of once every thousand base pairs, the study of their identity allows a higher LOH mapping resolution. A group of researchers at Harvard studied the LOH patterns in human cancer cell lines. By using the 10K SNP array, in conjunction with the dChipSNP informatics software package, these investigators were able to compare and confirm LOH patterns to those obtained previously with microsatellites. Moreover, this effort also resulted in the identification of previously undetected LOH regions that were smaller and unattainable by other methods (Janne et al., 2004).

More recent studies stemming from this research group detected genomic regions with an altered DNA copy number and LOH. By hybridizing breast and lung carcinoma cell DNAs and measuring the fluorescence intensity of the allele-specific hybridization to certain segments, genomic amplifications and deletions were identified, as well as some LOH events. Some of these alterations were consistent with previous data, although some were novel, and could serve as new diagnostic markers (Zhao et al., 2004). Studies such as the examples given earlier demonstrate that the combination of SNP analysis and copy number analysis provides insight into the genetic alterations and molecular mechanisms responsible for cancer. The advent of technologies such as the GeneChip Mapping 100K Set enables researchers to study whole genome chromosomal copy number changes, as well as LOH markers simultaneously, at an unprecedented efficiency.

Advancing the Future of Genomics

Since their inception, high-density microarrays have followed the same trend as computer microprocessors. In 1965, Moore predicted that the power of microprocessors would double every 18 months. This trend has held true for the computer industry, where much faster and smaller central processing units are constantly being produced. High-density oligonucleotide arrays have evolved similarly. More and more genetic information is being included into a smaller and smaller surface area. This allows scientists to analyze vast amounts of genetic information at an unprecedented efficiency. This trend, in combination with the wealth of information generated by the sequencing of the human and other genomes, has generated a unique opportunity in the advancement of clinical and life sciences research. Global views of the genome will undoubtedly accelerate the understanding of complex diseases such as psychiatric and cardiovascular ailments and drug response.

In addition to feature-size reduction, the overall microarray platforms are changing in other ways. High-throughput automation systems are currently being developed that provide the convenience of hybridizing 96 arrays at a time. Microarray systems have also been developed for diagnostic purposes, an example being the Roche system for the detection of polymorphisms in cytochrome P450 genes. This gene family controls how individuals respond to different drugs, and knowledge of an individual's genotype should aid in prescribing proper doses, reducing side effects. Controls and standard practices are also developing with microarrays in mind. The establishment of controls and standard practices will allow greater acceptance of microarray assays into the clinical and diagnostic fields.

In a little over a decade the microarray has evolved from a research publication to a mainstream tool of life science research. At the time of completion for this manuscript, several new applications are being introduced commercially: a 500K SNP genotyping assay and an exon-based expression assay. What remains certain is that microarray assays and technology will continue to evolve in the future and further expand from life sciences research into the clinical and diagnostic communities.

Acknowledgments

The authors thank Sean Walsh and Glenn McGall for critically reviewing this manuscript and Andy Lau and Dan Bartell for providing the graphics. In addition, the authors thank their many colleagues at Affymetrix for contributing the ideas, methods, and products that made this review possible.

References

Affymetrix Technical Note (2005). Guide to Probe Logarithmic Intensity Error (PLIER) Estimation, http://www.affymetrix.com.

Apidianakis, Y., Mindrinos, M. N., Xiao, W., Lau, G. W., Baldini, R. L., Davis, R. W., and Rahme, L. G. (2005). Profiling early infection responses: *Pseudomonas aeruginosa* eludes host defenses by suppressing antimicrobial peptide gene expression. *Proc. Natl. Acad. Sci. USA* **102**, 2573–2578.

Armstrong, S. A., Staunton, J. E., Silverman, L. B., Pieters, R., den Boer, M. L., Minden, M. D., Sallan, S. E., Lander, E. S., Golub, T. R., and Korsmeyer, S. J. (2002). MLL translocations specify a distinct gene expression profile that distinguishes a unique leukemia. *Nat. Genet.* **30**, 41–47.

Boerma, M., van der Wees, C. G., Vrieling, H., Svensson, J. P., Wondergem, J., van der Laarse, A., Mullenders, L. H., and van Zeeland, A. A. (2005). Microarray analysis of gene expression profiles of cardiac myocytes and fibroblasts after mechanical stress, ionising or ultraviolet radiation. *BMC Genom.* **6**, 6.

Cawley, S., Bekiranov, S., Ng, H. H., Kapranov, P., Sekinger, E. A., Kampa, D., Piccolboni, A., Sementchenko, V., Cheng, J., Williams, A. J., Wheeler, R., Wong, B., Drenkow, J., Yamanaka, M., Patel, S., Brubaker, S., Tammana, H., Helt, G., Struhl, K., and Gingeras, T. R. (2004). Unbiased mapping of transcription factor binding sites along human chromosomes 21 and 22 points to widespread regulation of noncoding RNAs. *Cell* **116**, 499–509.

Comer, J. E., Galindo, C. L., Chopra, A. K., and Peterson, J. W. (2005). GeneChip analyses of global transcriptional responses of murine macrophages to the lethal toxin of *Bacillus anthracis*. *Infect. Immun.* **73**, 1879–1885.

Cutler, D. J., Zwick, M. E., Carrasquillo, M. M., Yohn, C. T., Tobin, K. P., Kashuk, C., Mathews, D. J., Shah, N. A., Eichler, E. E., Warrington, J. A., and Chakravarti, A. (2001). High-throughput variation detection and genotyping using microarrays. *Genome Res.* **11**, 1913–1925.

Davletova, S., Rizhsky, L., Liang, H., Shengqiang, Z., Oliver, D. J., Coutu, J., Shulaev, V., Schlauch, K., and Mittler, R. (2005). Cytosolic ascorbate peroxidase 1 is a central component of the reactive oxygen gene network of Arabidopsis. *Plant Cell* **17**, 268–281.

Di, X., Matsuzaki, H., Webster, T. A., Hubbell, E., Liu, G., Dong, S., Bartell, D., Huang, J., Chiles, R., Yang, G., Shen, M. M., Kulp, D., Kennedy, G. C., Mei, R., Jones, K. W., and Cawley, S. (2005). Dynamic model based algorithms for screening and genotyping over 100 K SNPs on oligonucleotide microarrays. *Bioinformatics* **21**, 1958–1963.

Dinkova, T. D., Keiper, B. D., Korneeva, N. L., Aamodt, E. J., and Rhoads, R. E. (2005). Translation of a small subset of *Caenorhabditis elegans* mRNAs is dependent on a specific eukaryotic translation initiation factor 4E isoform. *Mol. Cell. Biol.* **25**, 100–113.

Fan, W., Bubman, D., Chadburn, A., Harrington, W. J., Jr., Cesarman, E., and Knowles, D. M. (2005). Distinct subsets of primary effusion lymphoma can be identified based on their cellular gene expression profile and viral association. *J. Virol.* **79**, 1244–1251.

Fodor, S. P., Read, J. L., Pirrung, M. C., Stryer, L., Lu, A. T., and Solas, D. (1991). Light-directed, spatially addressable parallel chemical synthesis. *Science* **251**, 767–773.

Girardot, F., Monnier, V., and Tricoire, H. (2004). Genome wide analysis of common and specific stress responses in adult *Drosophila melanogaster*. *BMC Genom.* **5**, 74.

Gomez-Mena, C., de Folter, S., Costa, M. M., Angenent, G. C., and Sablowski, R. (2005). Transcriptional program controlled by the floral homeotic gene AGAMOUS during early organogenesis. *Development* **132**, 429–438.

Hardenbol, P., Baner, J., Jain, M., Nilsson, M., Namsaraev, E. A., Karlin-Neumann, G. A., Fakhrai-Rad, H., Ronaghi, M., Willis, T. D., Landegren, U., and Davis, R. W. (2003).

Multiplexed genotyping with sequence-tagged molecular inversion probes. *Nat. Biotechnol.* **21**, 673–678.

Hekmat-Scafe, D. S., Dang, K. N., and Tanouye, M. A. (2005). Seizure suppression by gain-of-function escargot mutations. *Genetics* **169**, 1477–1493.

Huang, J., Wei, W., Zhang, J., Liu, G., Bignell, G. R., Stratton, M. R., Futreal, P. A., Wooster, R., Jones, K. W., and Shapero, M. H. (2004). Whole genome DNA copy number changes identified by high density oligonucleotide arrays. *Hum. Genom.* **1**, 287–299.

Irizarry, R. A., Bolstad, B. M., Collin, F., Cope, L. M., Hobbs, B., and Speed, T. P. (2003). Summaries of Affymetrix GeneChip probe level data. *Nucleic Acids Res.* **31**, e15.

Iwamoto, K., Bundo, M., and Kato, T. (2005). Altered expression of mitochondria-related genes in postmortem brains of patients with bipolar disorder or schizophrenia, as revealed by large-scale DNA microarray analysis. *Hum. Mol. Genet.* **14**, 241–253.

Janne, P. A., Li, C., Zhao, X., Girard, L., Chen, T. H., Minna, J., Christiani, D. C., Johnson, B. E., and Meyerson, M. (2004). High-resolution single-nucleotide polymorphism array and clustering analysis of loss of heterozygosity in human lung cancer cell lines. *Oncogene* **23**, 2716–2726.

Jemal, A., Murray, T., Samuels, A., Ghafoor, A., Ward, E., and Thun, M. J. (2003). Cancer statistics, 2003. *CA Cancer J. Clin.* **53**, 5–26.

Kapranov, P., Cawley, S. E., Drenkow, J., Bekiranov, S., Strausberg, R. L., Fodor, S. P., and Gingeras, T. R. (2002). Large-scale transcriptional activity in chromosomes 21 and 22. *Science* **296**, 916–919.

Kennedy, G. C., Matsuzaki, H., Dong, S., Liu, W. M., Huang, J., Liu, G., Su, X., Cao, M., Chen, W., Zhang, J., Liu, W., Yang, G., Di, X., Ryder, T., He, Z., Surti, U., Phillips, M. S., Boyce-Jacino, M. T., Fodor, S. P., and Jones, K. W. (2003). Large-scale genotyping of complex DNA. *Nat. Biotechnol.* **21**, 1233–1237.

Klein, R. J., Zeiss, C., Chew, E. Y., Tsai, J. Y., Sackler, R. S., Haynes, C., Henning, A. K., SanGiovanni, J. P., Mane, S. M., Mayne, S. T., Bracken, M. B., Ferris, F. L., Ott, J., Barnstable, C., and Hoh, J. (2005). Complement factor H polymorphism in age-related macular degeneration. *Science* **308**, 385–389.

Kong, S. W., Bodyak, N., Yue, P., Liu, Z., Brown, J., Izumo, S., and Kang, P. M. (2005). Genetic expression profiles during physiological and pathological cardiac hypertrophy and heart failure in rats. *Physiol. Genom.* **21**, 34–42.

Korenberg, J. R. (1993). Toward a molecular understanding of Down syndrome. *Prog. Clin. Biol. Res.* **384**, 87–115.

Lander, E. S., Linton, L. M., Birren, B., Nusbaum, C., Zody, M. C., Baldwin, J., Devon, K., Dewar, K., Doyle, M., FitzHugh, W., Funke, R., Gage, D., Harris, K., Heaford, A., Howland, J., Kann, L., Lehoczky, J., LeVine, R., McEwan, P., McKernan, K., Meldrim, J., Mesirov, J. P., Miranda, C., Morris, W., Naylor, J., Raymond, C., Rosetti, M., Santos, R., Sheridan, A., Sougnez, C., Stange-Thomann, N., Stojanovic, N., Subramanian, A., Wyman, D., Rogers, J., Sulston, J., Ainscough, R., Beck, S., Bentley, D., Burton, J., Clee, C., Carter, N., Coulson, A., Deadman, R., Deloukas, P., Dunham, A., Dunham, I., Durbin, R., French, L., Grafham, D., Gregory, S., Hubbard, T., Humphray, S., Hunt, A., Jones, M., Lloyd, C., McMurray, A., Matthews, L., Mercer, S., Milne, S., Mullikin, J. C., Mungall, A., Plumb, R., Ross, M., Shownkeen, R., Sims, S., Waterston, R. H., Wilson, R. K., Hillier, L. W., McPherson, J. D., Marra, M. A., Mardis, E. R., Fulton, L. A., Chinwalla, A. T., Pepin, K. H., Gish, W. R., Chissoe, S. L., Wendl, M. C., Delehaunty, K. D., Miner, T. L., Delehaunty, A., Kramer, J. B., Cook, L. L., Fulton, R. S., Johnson, D. L., Minx, P. J., Clifton, S. W., Hawkins, T., Branscomb, E., Predki, P., Richardson, P., Wenning, S., Slezak, T., Doggett, N., Cheng, J. F., Olsen, A., Lucas, S., Elkin, C., Uberbacher, E., Frazier, M., *et al.* (2001). Initial sequencing and analysis of the human genome. *Nature* **409**, 860–921.

Lindblad-Toh, K., Tanenbaum, D. M., Daly, M. J., Winchester, E., Lui, W. O., Villapakkam, A., Stanton, S. E., Larsson, C., Hudson, T. J., Johnson, B. E., Lander, E. S., and Meyerson, M. (2000). Loss-of-heterozygosity analysis of small-cell lung carcinomas using single-nucleotide polymorphism arrays. *Nat. Biotechnol.* **18,** 1001–1005.

Lipshutz, R. J., Morris, D., Chee, M., Hubbell, E., Kozal, M. J., Shah, N., Shen, N., Yang, R., and Fodor, S. P. (1995). Using oligonucleotide probe arrays to access genetic diversity. *Biotechniques* **19,** 442–447.

Matsuzaki, H., Dong, S., Loi, H., Di, X., Liu, G., Hubbell, E., Law, J., Berntsen, T., Chadha, M., Hui, H., Yang, G., Kennedy, G. C., Webster, T. A., Cawley, S., Walsh, P. S., Jones, K. W., Fodor, S. P., and Mei, R. (2004a). Genotyping over 100,000 SNPs on a pair of oligonucleotide arrays. *Nat. Methods* **1,** 109–111.

Matsuzaki, H., Loi, H., Dong, S., Tsai, Y. Y., Fang, J., Law, J., Di, X., Liu, W. M., Yang, G., Liu, G., Huang, J., Kennedy, G. C., Ryder, T. B., Marcus, G. A., Walsh, P. S., Shriver, M. D., Puck, J. M., Jones, K. W., and Mei, R. (2004b). Parallel genotyping of over 10,000 SNPs using a one-primer assay on a high-density oligonucleotide array. *Genome Res.* **14,** 414–425.

McGall, G. H., and Christians, F. C. (2002). High-density genechip oligonucleotide probe arrays. *Adv. Biochem. Eng. Biotechnol.* **77,** 21–42.

Mei, R., Hubbell, E., Bekiranov, S., Mittmann, M., Christians, F. C., Shen, M.-M., Lu, G., Fang, J., Liu, W.-M., Ryder, T., Kaplan, P., Kulp, D., and Webster, T. A. (2003). Probe selection for high density oligonucleotide arrays. *Proc. Natl. Acad. Sci. USA* **100,** 11237–11242.

Middleton, F. A., Pato, M. T., Gentile, K. L., Morley, C. P., Zhao, X., Eisener, A. F., Brown, A., Petryshen, T. L., Kirby, A. N., Medeiros, H., Carvalho, C., Macedo, A., Dourado, A., Coelho, I., Valente, J., Soares, M. J., Ferreira, C. P., Lei, M., Azevedo, M. H., Kennedy, J. L., Daly, M. J., Sklar, P., and Pato, C. N. (2004). Genomewide linkage analysis of bipolar disorder by use of a high-density single-nucleotide-polymorphism (SNP) genotyping assay: A comparison with microsatellite marker assays and finding of significant linkage to chromosome 6q22. *Am. J. Hum. Genet.* **74,** 886–897.

Moore, G. (1965). Cramming more components onto integrated circuits. *Electronics* **38,** 114–117.

Presneau, N., Dewar, K., Forgetta, V., Provencher, D., Mes-Masson, A. M., and Tonin, P. N. (2005). Loss of heterozygosity and transcriptome analyses of a 1.2 Mb candidate ovarian cancer tumor suppressor locus region at 17q25.1–q25.2. *Mol. Carcinog.* **43,** 141–154.

Puffenberger, E. G., Hu-Lince, D., Parod, J. M., Craig, D. W., Dobrin, S. E., Conway, A. R., Donarum, E. A., Strauss, K. A., Dunckley, T., Cardenas, J. F., Melmed, K. R., Wright, C. A., Liang, W., Stafford, P., Flynn, C. R., Morton, D. H., and Stephan, D. A. (2004). Mapping of sudden infant death with dysgenesis of the testes syndrome (SIDDT) by a SNP genome scan and identification of TSPYL loss of function. *Proc. Natl. Acad. Sci. USA* **101,** 11689–11694.

Reinke, V., Gil, I. S., Ward, S., and Kazmer, K. (2004). Genome-wide germline-enriched and sex-biased expression profiles in *Caenorhabditis elegans*. *Development* **131,** 311–323.

Schadt, E. E., Edwards, S. W., GuhaThakurta, D., Holder, D., Ying, L., Svetnik, V., Leonardson, A., Hart, K. W., Russell, A., Li, G., Cavet, G., Castle, J., McDonagh, P., Kan, Z., Chen, R., Kasarskis, A., Margarint, M., Caceres, R. M., Johnson, J. M., Armour, C. D., Garrett-Engele, P. W., Tsinoremas, N. F., and Shoemaker, D. D. (2004). A comprehensive transcript index of the human genome generated using microarrays and computational approaches. *Genome Biol.* **5,** R73.

Sellick, G. S., Garrett, C., and Houlston, R. S. (2003). A novel gene for neonatal diabetes maps to chromosome 10p12.1-p13. *Diabetes* **52,** 2636–2638.

Steinman, L., and Zamvil, S. (2003). Transcriptional analysis of targets in multiple sclerosis. *Nat. Rev. Immunol.* **3,** 483–492.

Van Gelder, R. N., von Zastrow, M. E., Yool, A., Dement, W. C., Barchas, J. D., and Eberwine, J. H. (1990). Amplified RNA synthesized from limited quantities of heterogeneous cDNA. *Proc. Natl. Acad. Sci. USA* **87,** 1663–1667.

Wang, J., Iwasaki, H., Krivtsov, A., Febbo, P. G., Thorner, A. R., Ernst, P., Anastasiadou, E., Kutok, J. L., Kogan, S. C., Zinkel, S. S., Fisher, J. K., Hess, J. L., Golub, T. R., Armstrong, S. A., Akashi, K., and Korsmeyer, S. J. (2005). Conditional MLL-CBP targets GMP and models therapy-related myeloproliferative disease. *EMBO J.* **24,** 368–381.

Yeoh, E. J., Ross, M. E., Shurtleff, S. A., Williams, W. K., Patel, D., Mahfouz, R., Behm, F. G., Raimondi, S. C., Relling, M. V., Patel, A., Cheng, C., Campana, D., Wilkins, D., Zhou, X., Li, J., Liu, H., Pui, C. H., Evans, W. E., Naeve, C., Wong, L., and Downing, J. R. (2002). Classification, subtype discovery, and prediction of outcome in pediatric acute lymphoblastic leukemia by gene expression profiling. *Cancer Cell* **1,** 133–143.

Zhao, X., Li, C., Paez, J. G., Chin, K., Janne, P. A., Chen, T. H., Girard, L., Minna, J., Christiani, D., Leo, C., Gray, J. W., Sellers, W. R., and Meyerson, M. (2004). An integrated view of copy number and allelic alterations in the cancer genome using single nucleotide polymorphism arrays. *Cancer Res.* **64,** 3060–3071.

[2] The Agilent *In Situ*-Synthesized Microarray Platform

By PAUL K. WOLBER, PATRICK J. COLLINS, ANNE B. LUCAS, ANNIEK DE WITTE, and KAREN W. SHANNON

Abstract

Microarray technology has become a standard tool in many laboratories. Agilent Technologies manufactures a variety of catalog and custom long-oligonucleotide (60-mer) microarrays that can be used in multiple two-color microarray applications. Optimized methods and techniques have been developed for two such applications: gene expression profiling and comparative genomic hybridization. Methods for a third technique, location analysis, are evolving rapidly. This chapter outlines current best methods for using Agilent microarrays, provides detailed instructions for the most recently developed techniques, and discusses solutions to common problems encountered with two-color microarrays.

Introduction

During the last decade, microarrays have evolved from a promising technology for exploring a variety of genomic problems (Hughes *et al.*, 2001; Kuhn *et al.*, 2001; Miki *et al.*, 2001; Nacht *et al.*, 1999) into a workhorse technology for investigating important questions in cancer research (Chang

METHODS IN ENZYMOLOGY, VOL. 410
0076-6879/06 $35.00
DOI: 10.1016/S0076-6879(06)10002-6

et al., 2005; Dai *et al.*, 2005; Golub *et al.*, 1999; Ramaswamy *et al.*, 2001; Segal *et al.*, 2004; Singh *et al.*, 2002; van de Vijver *et al.*, 2002; van't Veer *et al.*, 2003, 2005), toxicology (Waring *et al.*, 2001a,b), transcript annotation (Johnson *et al.*, 2005; Schadt *et al.*, 2004; Shoemaker *et al.*, 2001), gene silencing (Jackson *et al.*, 2003), alternative gene splicing (Castle *et al.*, 2003; Johnson *et al.*, 2003; Shoemaker *et al.*, 2001), polymorphism mapping (Barrett *et al.*, 2004; Greshock *et al.*, 2004; Hodgson *et al.*, 2001; Pinkel *et al.*, 1998; Pollack *et al.*, 1999; Snijders *et al.*, 2001), promoter mapping (Harbison *et al.*, 2004), and pathogen characterization (Wang *et al.*, 2002; Wilson *et al.*, 1999). Several manufacturers produce microarrays as commercial products; some investigators also manufacture their own microarrays using either commercially produced or home-built robots. Agilent Technologies manufactures both custom and catalog oligonucleotide microarrays on the same 1 × 3-in. microscope slide format preferred by users who manufacture their own microarrays.

Methods for designing, using, and interpreting microarrays have evolved along with the microarrays themselves. The methods are now beginning to stabilize, and there is a growing acceptance of best practices that are supported by the manufacturers and validated by their proven ability to yield useful results. This chapter captures the current state of practices used with Agilent microarrays and details the best practices currently known.

Platform Description

Technology

Agilent manufactures microarrays via an *in situ*-synthetic scheme based on inkjet printing of nucleotide precursors and common chemical processing of each added nucleotide layer (Blanchard *et al.*, 1996; Hughes *et al.*, 2001; Kronick, 2004). The standard probe length is 60 nucleotides. The method is effectively 5-ink (4 bases plus catalyst), 60-layer printing with reregistration at each layer. Because the technology synthesizes oligonucleotides on demand, based on a digital representation of the desired sequences and layout, it is well suited to manufacturing custom microarray designs or microarrays that mix standard and custom probe content. The current maximum feature count is 184,672 per 1 × 3-in. slide. The labeling, scanning, and data interpretation systems are built upon hybridization to two samples labeled with different fluorophores (two-color hybridization).

Probe design is initially performed *in silico* using proprietary techniques similar to the literature methods for designing long oligomer probes (Chou *et al.*, 2004; Gordon and Sensen, 2004; Nordberg, 2005; Rimour *et al.*, 2005; Shannon *et al.*, 2001). The detailed probe design methods are customized to the intended application (see later). For some catalog microarrays, probe sets are refined further by empirical observation of performance with real samples. Final microarray designs generally use one probe per gene (for expression microarrays) or region (for mapping microarrays).

Applications

Agilent microarrays are used in a number of different applications, and the use methods are application dependent. The current application areas (in order of decreasing maturity) are gene expression profiling, microarray comparative genomic hybridization (aCGH), and chromatin immunoprecipitation mapping of transcription factor-binding sites [also known as location analysis (LA) or "ChIP on Chip"].

The gene expression profiling application measures the relative levels of gene expression in two different samples on a gene-by-gene basis. In a two-color system (such as Agilent's), the two samples are labeled with distinct fluorophores (usually, Cy3 and Cy5) and hybridized to a single microarray. The final labeled sample can be either single-stranded cDNA or linearly amplified, single-stranded cRNA produced from the mRNA component of total cellular RNA. Amplified cRNA is the preferred target for the Agilent system. After hybridization, scanning, and data extraction, the microarrays yield a list of hybridization intensities in each color channel as an intermediate work product and a list of normalized expression ratios (or their logarithms) as a final work product. Aligent has recently introduced one-color gene expression profiling microarray applications, using the same arrays employed by two-color applications and samples labeled with Cy3 only (Aligent, 2005h).

Array comparative genomic hybridization is a two-color microarray method that maps local differences in the copy number of a particular genome from that of a reference genome. The aCGH method is particularly useful for detecting genetic amplifications and deletions in cancer cells (Barrett et al., 2004; Greshock et al., 2004; Hodgson et al., 2001; Pinkel et al., 1998; Pollack et al., 1999; Snijders et al., 2001). It also shows promise in mapping copy number polymorphisms in humans and other organisms (Pinkel et al., 1998; Sharp et al., 2005; Snijders et al., 2001). The labeled sample is usually amplified, double-stranded DNA produced from fragmented genomic DNA. The final work product is a map of relative copy number (or its logarithm) of the target genome, relative to a reference genome, ordered by a genome physical map location.

Location analysis is a method for mapping the binding sites of transcription factors to particular chromosomal locations (Harbison et al., 2004; Pokholok et al., 2005). It is a two-color microarray method in which a sample produced from chemically cross-linked, fragmented, immunoprecipitated chromatin from an actively transcribing cell type is compared to a control sample from the same chromatin preparation, with the immunoprecipitation step omitted. The antibodies used to perform the immunoprecipitation are directed against various components of transcriptional complexes. The final work product of LA is a map of immuno-enrichment (relative to the control sample), ordered by chromosomal location; the degree of enrichment indicates the degree to which a particular component of the transcriptional machinery was associated with that location.

Methods Descriptions

This chapter distinguishes between several categories of methods for performing various steps for each application area. The categories are as follow.

1. Preferred Methods: the "best" methods known to Agilent. The detailed methods described in this chapter are all preferred methods.
2. Supported Methods: methods that are well characterized by Agilent and that receive warranty support. This chapter summarizes the differences between preferred methods and other supported methods.
3. Other Published Methods: methods that do not automatically receive warranty support, but that have been described in the peer-reviewed microarray literature. This chapter briefly reviews important alternative methods that have been published.

Array Gene Expression Profiling

Array Design

Preferred/Supported Methods

Agilent recommends one of three methods for designing microarrays.

1. Catalog microarrays are designs that have undergone several rounds of *in silico* and experimental optimization (Kronick, 2004). The content of catalog microarrays is reannotated to the latest version of the target genome on a quarterly basis. Currently, Agilent offers catalog designs for both expression profiling and aCGH. The list of available catalog microarrays is published on the Agilent web site (Agilent, 2005c).
2. An eArray Portal is a web-based design service (Agilent, 2005g) that allows users to mix and match probes from catalog microarray

designs and new, user-driven *in silico* probe designs to specified sequences. At this time, the eArray web service is available for both expression profiling and aCGH applications.
3. The custom design service (Agilent, 2005e) performs *in silico* probe design to the user's specifications and is available for all microarray applications. Sharing of custom designs among specified user lists is supported. The full list of available custom microarray services is documented on the Agilent web site (Agilent, 2005d).

In all cases, Agilent adds a grid of quality control (QC) probes to the microarray design and randomizes probe placement across the microarray in order to avoid confounding data analysis with hybridization intensity gradients and other spatial artifacts.

Other Published Methods

Agilent will act as an "oligo foundry" and print microarrays from a customer-specified list of sequences. Support for this method of microarray production is offered on a case-by-case basis.

Sample Isolation, Labeling, and Quality Control

Preferred Methods

Agilent offers labeling kits that can be used to produce amplified, labeled cRNA targets. The kits are optimized to utilize either 50 ng to 5 μg of total RNA or 10–100 ng of poly(A)$^+$ RNA. Most methods of purifying RNA are supported. RNA samples should have an OD ratio value (A_{260}/A_{280}) between 1.8 and 2.0 and an A_{260}/A_{230} ratio greater than 2.0. In addition, Agilent recommends use of the Agilent 2100 bioanalyzer for evaluation of RNA integrity and purity (Agilent, 2005a).

The detailed protocol for use of the Agilent labeling kit is available on the World Wide Web (Agilent, 2005h) and is not duplicated here. The method is a variant of the linear amplification procedure of Eberwine *et al.* (1992) and Van Gelder *et al.* (1990). Briefly, cDNA is synthesized by an RNase H$^+$ reverse transcriptase, utilizing a poly(dT) primer with a VN 3' terminus [to anchor priming to the 5' end of the poly(A) tail] and a 5' extension that encodes a T7 promoter. Second-strand synthesis is believed to be self-priming (due to the RNase H activity of the reverse transcriptase), but can be optionally augmented by the addition of random hexamers. Finally, labeled cRNA is synthesized by thermally inactivating the reverse transcriptase, adding the T7-RNA polymerase and a master mix containing NTPs, CTP labeled with Cy3 or Cy5, and the other components required by the T7-RNA polymerase, and then incubating at 40° for 2 h.

Supported Methods

Agilent's labeling kits can also be used to produce labeled cDNA targets; the methods for doing so are also documented in detail in the kit protocols published on the web (Agilent, 2005h). The use of cDNA targets is not preferred because these targets are not fragmented before use (resulting in increased cross-hybridization between regions of the target distal to the microarray-bound probe and other unintended targets) and because the hybridization and washing conditions have been optimized for cRNA targets.

Other Published Methods

Multiple published studies have utilized either amplified cRNA targets that incorporate aminoallyl-UTP (Hughes *et al.*, 2001) or cDNA targets that incorporate aminoallyl-dUTP (Castle *et al.*, 2003; Shoemaker *et al.*, 2001). In both cases, the targets are subsequently labeled with Cy3 or Cy5 by coupling the aminoallyl nucleic acid to the fluorophore succinimide ester. After cleanup, the targets are used in the same manner as the corresponding directly labeled cRNA or cDNA targets.

Array Hybridization and Scanning

Preferred Methods

Agilent offers apparatus, reagents, and protocols for hybridizing labeled cRNA to microarrays. Detailed lists of required materials and preferred protocol are available on the World Wide Web (Agilent, 2005h). Basically, the hybridization and washing protocol consists of preparation of a mixture of two cRNA samples (one labeled with Cy3 and one with Cy5), cRNA fragmentation, an "overnight" (nominally 17 h) hybridization, a low-stringency wash to remove unhybridized sample, a high-stringency wash to diminish cross-hybridization of unintended targets to homologous probes, a drying step, and a scanning step. The preferred scanner is the Agilent DNA microarray scanner (G2565BA). The protocol includes the option of inserting an organic stabilization and drying wash between the high-stringency wash and slide drying; this wash includes an agent that protects the Cy5 fluorophore from damage by ozone (Fare *et al.*, 2003) and other airborne oxidants. The discussion of slide washing for aCGH includes a detailed description of the handling and use of this reagent. Agilent recommends the use of dye-swap pairs to maximize the accuracy of differential expression measurements.

Supported Methods

Use of the stabilization and drying solution is optional, provided that the wash, dry, and scan environment is ozone free. An ozone-free laboratory environment can be achieved either by filtration of the laboratory air supply through activated charcoal or by use of a contained, controlled atmosphere environment (such as a tent supplied with ozone-free air or N_2) during washing, drying, and scanning. See the Agilent web site (Agilent, 2004) for details. Agilent also supports the use of Axon microarray scanners (GenePix scanner series, Molecular Devices) to scan Agilent microarrays.

Other Published Methods

Multiple published studies have made use of a formamide-based hybridization buffer at a 40° hybridization temperature (Hughes *et al.*, 2001) and hybridization times as long as 48 h (Dai *et al.*, 2002). The formamide-based hybridization yields a stringency equivalent to the most recent non-formamide hybridization protocols released by Agilent (65°). Longer hybridization times have been proven to enhance performance, particularly for low-abundance messages. The choice of hybridization time must balance laboratory throughput and data quality; each laboratory needs to determine this balance based on the details of the research problems addressed.

Data Extraction

Preferred Methods

Agilent recommends use of its Feature Extraction Software (Agilent, 2005j) to extract data from scanned microarray images.

Supported Methods

Agilent supports the use of GenePix software with the Axon scanner system (Molecular Devices).

Other Published Methods

Several published studies have utilized data extraction software written and tested by the researchers themselves (Dudoit *et al.*, 2003; Hughes *et al.*, 2001; Marton *et al.*, 1998). User-written software has served as a rich source of new microarray data extraction and data reduction techniques; many of these innovations eventually find their way into commercially available applications.

Array Comparative Genomic Hybridization

Array Design

Preferred/Supported Methods

Agilent has developed software and experimental methods for the efficient design and testing of aCGH probes. This system is used to design the probes placed on Agilent's current catalog comparative genomic hybridization microarrays. The probe design rules balance the conflicting needs of probe hybridization properties, local increases in probe density in interesting genomic regions (e.g., coding regions, control elements), and even probe placement across the genome. The probes are then laid out on microarrays at random locations. In addition, a certain number of microarray locations are devoted to control probes that aid in background determination and QC of the sample labeling, hybridization, washing, and drying operations. The list of available catalog CGH microarrays is published on the Agilent web site (Agilent, 2005c).

Agilent also supports the *in silico* design of custom CGH microarrays via its custom microarray design service. The list of available custom CGH microarray services is documented on the Agilent web site (Agilent, 2005d). Web-based design of CGH microarrays via the eArray Portal is also supported.

Process Description

Agilent's aCGH application uses a "two-color" process to measure genomic alterations in an experimental sample relative to a reference sample. The most current methods for performing aCGH are documented in detail in the protocols published on the web (Agilent, 2005b). The following section documents the best method known at the time of this writing.

The type of sample used as a reference is a matter of experimental choice; however, many experimenters use normal genomic DNA as a commercially available reference sample. The preferred workflow for sample preparation and microarray hybridization includes (1) genomic DNA amplification of experimental and reference samples (optional), (2) restriction enzyme digestion of amplified DNAs, (3) cleanup of amplified/digested DNAs, (4) quantification of amplified/digested DNAs, (5) genomic DNA labeling of amplified DNAs using random priming and exo-Klenow to incorporate Cy3-dUTP or Cy5-dUTP, (6) combining of appropriate Cy3- and Cy5-labeled samples and cleanup of labeled DNA, (7) preparation of sample for

TABLE I
REQUIRED EQUIPMENT

Apparatus	Preference
UV-VIS spectrophotometer	Nanodrop ND-1000 or equivalent
Scanner	Agilent G2565BA
Powder-free gloves	No preferred source
Micropipettors	Pipetman P-10, -20, -200, -1000
Sterile, nuclease-free aerosol barrier pipette tips	No preferred source
Sterile, nuclease-free, boilable 1.5-ml microfuge tubes	Ambion P/N 12400
Vortex mixer	No preferred source
Ice bucket	No preferred source
Microcentrifuge (30 tube rotor)	Eppendorf 5417R or equivalent
Timer (h, min, s)	No preferred source
Clean forceps	No preferred source
QIAprep spin miniprep kit	Qiagen P/N 27106
Microcon YM-30 filter units	Millipore P/N 42410
60-mer *in situ* aCGH slides, each containing one 44K microarray	Agilent HGA44B or equivalent
Agilent hybridization chamber, stainless	Agilent P/N G2534A
Surehyb backings	Agilent P/N G2534–60003
Circulating water bath set to 30°	No preferred source
Circulating water bath set to 37°	No preferred source
Circulating water bath set to 65°	No preferred source
Circulating water bath set to 95°	No preferred source
Vacuum concentrator	Speed-Vac or equivalent
1.5-liter capacity dish	Pyrex 213-R or equivalent
250-ml capacity slide staining dish, with slide rack (×5)	Wheaton P/N 900200, or equivalent
Hybridization oven; temperature set at 65°	Agilent P/N G2505–80081
Oven hybridization rotator for Agilent microarray hybridization chambers	Agilent P/N G2530–60020
Oven; temperature set at 37°	Agilent P/N G2505–80081 or equivalent
Magnetic stir plate (×3)	No preferred source
Magnetic stir plate with heating element (×1)	No preferred source
Magnetic stir bar (×4)	No preferred source
Vacuum desiccator or N_2 purge box for slide storage	No preferred source

hybridization and assembly of hybridization sample, (8) 40 h of hybridization ($65°$, 750 mM total monovalent cations), (9) wash (including optional ozone protection), (10) scan, and (11) feature extraction. Agilent recommends the use of dye-swap pairs to maximize the accuracy of detection of genomic amplifications and deletions. The equipment, reagents, and computation resources employed during these steps are listed in Tables I, II, and III, respectively.

Sample Isolation, Labeling, and Quality Control

Preferred/Supported Methods

Genomic Amplification (Optional). The Repli-G amplification kit (Table II) provides a highly uniform amplification across the entire genome with minimal sequence bias. The method utilizes phi29 DNA polymerase and exonuclease-resistant primers in an isothermal amplification reaction

TABLE II
REQUIRED REAGENTS

Reagent	Preference
Human genomic DNA: Female (reference sample)	Promega P/N G1521
1 × TE (pH 8.0) molecular grade	Promega P/N V6231
100% ethanol	Amresco P/N E193
Repli-G amplification kit	Qiagen 59043
*Alu*I 10 U/ul	Promega P/N R6281
*Rsa*I 10 U/ul	Promega P/N R6371
Buffer C	Promega P/N R003A (supplied with *Rsa*I)
DNase/RNase-free distilled water	Invitrogen P/N 10977015
BioPrime array CGH genomic labeling module	Invitrogen P/N 18095–012
Cyanine-3-dUTP 1.0 mM	Perkin Elmer P/N NEL578
Cyanine-5-dUTP 1.0 mM	Perkin Elmer P/N NEL579
Human Cot-1 DNA 1.0 mg/ml	Invitrogen P/N 15279–011
aCGH hybridization kit	Agilent P/N 5188–5220
aCGH 10× blocking reagent	In Agilent aCGH hybridization kit
aCGH 2× hybridization buffer	In Agilent aCGH hybridization kit
aCGH wash buffer 1	Agilent P/N 5188–5221
aCGH wash buffer 2	Agilent P/N 5188–5222
Acetonitrile	Sigma P/N 271004–1L
Stabilization and drying solution	Agilent P/N 5185–5979

(Dean *et al.*, 2002; Hosono *et al.*, 2003). A typical Repli-G amplification reaction yields approximately 25–30 μg DNA, after digestion and clean-up, from 100 ng of input genomic DNA.

1. Add 100 ng genomic DNA to a 1.5-ml nuclease-free microfuge tube. Add nuclease-free water to bring to a final volume of 20 μl.
2. Thaw all Repli-G kit components immediately before use and maintain on ice. Caution: Repli-G kit components should be combined on ice immediately prior to addition to sample. The 4× mix may form a precipitate when it is thawed. The precipitate will fully dissolve upon vortexing for 10 s.
3. Mix the following components on ice in the order indicated (see Table IV).
4. Dispense 30-μl aliquots of Repli-G master mix into each reaction tube (50 μl volume total). Quick freeze the remainder of Repli-G kit components on dry ice and then return to –80°.
5. Transfer sample tubes to a circulating water bath at 30°. Incubate at 30° for 16 h.

TABLE III
REQUIRED COMPUTATIONAL RESOURCES

Item	Preference
PC, Pentium III 800 MHz or higher (Pentium 4 1.5 GHz or higher recommended)	No preferred source
512 MB RAM (1 GB recommended)	No preferred source
20 GB available disk space	No preferred source
Windows 2000 with SP2 (or SP3) or Windows XP with SP6	Microsoft
Feature Extraction software	Agilent, version 8.5

TABLE IV
REPLI-G REACTION MIXTURE

Component	Volume (μl)/reaction	Volume (μl)/12 reactions
Nuclease-free water	17.0	212.5
4× mix	12.5	156
DNA polymerase	0.5	6.25
Volume of Repli-G master mix	30.0	375

6. Transfer sample tubes to a circulating water bath at 65°. Incubate at 65° for 10 min and then move to ice.

Restriction Digestion of Amplified DNA or Genomic DNA. The amplification described in the previous step is an optional procedure that enables customers with limited amounts of DNA to generate sufficient template-specific DNA for the subsequent labeling reaction. Customers with larger amounts of genomic DNA (3.0 μg genomic DNA or more) may choose to proceed directly to the restriction digestion/labeling step.

1. For "direct labeling," add 3.0 μg genomic DNA to a 1.5-ml nuclease-free microfuge tube. Add nuclease-free water to bring to a final volume of 50 μl. For "amplification," use the entire amplification reaction (50 μl) for the restriction digestion.
2. Mix the components shown in Table V in the order indicated on ice.
3. Dispense 50-μl aliquots of *Alu*I/*Rsa*I mix into each reaction tube (100 μl volume total).
4. Transfer sample tubes to a circulating water bath at 37°. Incubate at 37° for 2 h and then move to ice.

Cleanup of Digested DNA. The QIAprep spin miniprep kit is recommended for the purification of digested genomic DNA.

1. Prepare buffer PE by adding ethanol (100%) to a buffer PE bottle (supplied with QIAprep miniprep columns) (see bottle label for volume). Mark appropriate checkbox to indicate that ethanol was added to bottle.
2. Add 500 μl of buffer PB (supplied with QIAprep miniprep columns) to each 100-μl sample.
3. Apply sample to QIAprep miniprep column. Centrifuge for 60 s at 17,900g (13,000 rpm in Eppendorf 5417R) in a microcentrifuge. Discard flow through.
4. Add 0.75 ml buffer PE to each spin column. Centrifuge for 60 s at 17,900g in a microcentrifuge. Discard flow through.

TABLE V
RESTRICTION MIXTURE

Component	Volume (μl)/reaction	Volume (μl)/12 reactions
Nuclease-free water	30.0	375
10× reaction buffer C	10.0	125
*Alu*I (10 U/μl)	5.0	62.5
*Rsa*I (10 U/μl)	5.0	62.5
Volume of *Alu*I/*Rsa*I mix	50.0	625

5. Centrifuge an additional 60 s at 17,900g in a microfuge to remove residual wash buffer. Caution: Residual wash buffer will not be completely removed unless the flow through is discarded before this additional centrifugation. Residual ethanol from buffer PE may inhibit subsequent enzymatic reactions.

6. Place the QIAprep spin column in a clean 1.5-ml microfuge tube. To elute DNA, add 50 μl of buffer EB (10 mM Tris–Cl, pH 8.5) to the center of each spin column. Let stand at room temperature for 60 s. Centrifuge for 60 s at 17,900g in a microfuge to collect purified DNA.

7. Samples may be stored at –20° prior to labeling.

Quantitation and Concentration of Digested DNA Product

1. Quantitate digested DNA using a Nanodrop ND-1000 UV-VIS spectrophotometer.
   a. Select "sample type" to be "DNA-50" on the NanoDrop instrument.
   b. Use 1.5 μl of buffer EB to blank instrument.
   c. Use 1.5 μl of each DNA to measure DNA concentration. Record the DNA concentration (ng/μl).
   d. Calculate amplification yield (μg) by multiplying DNA concentration (ng/μl) by the sample volume (50 μl) and dividing by 1000.

2. Concentrate the digested DNA product. Note: The concentration of amplified samples should not be necessary unless the DNA concentration is below 334 ng/μl. The nonamplified samples should be concentrated in a Speed-Vac unless the concentration is over 333 ng/μl. Note: Turn on the condenser unit 15 min before using the Speed-Vac.
   a. Place open tubes in the Speed-Vac rotor and check that the temperature setting is at 35°. Close the Speed-Vac and start the rotor on a manual run.
   b. Turn on the vacuum and concentrate the samples for 15 min.
   c. To stop the run, turn off the vacuum unit and then stop the manual run.
   d. Check the volumes in each tube to confirm that each is at or below 21 μl. If the volumes exceed 21 μl, continue to Speed-Vac for 5 min and recheck the sample volumes. If sample volumes are <21 μl, add nuclease-free water to a final volume of 21 μl.

Genomic DNA Labeling. The BioPrime Array CGH Genomic Labeling System (Table II) uses random octamers and exo-Klenow to label DNA samples with fluorescently labeled nucleotides (Feinberg and Vogelstein, 1983, 1984). For Agilent's aCGH application, the experimental sample is labeled using one dye, and the reference sample is labeled using the other dye. The "polarity" of the sample labeling is a matter of experimental choice. Agilent recommends the use of dye-swap pairs to maximize the accuracy of detection of genomic amplification and deletions. The mass of input template required by the labeling reaction is dependent on the type of template (non-amplified or amplified): 7.0 μg of digested and purified amplified genomic DNA is added to the labeling reaction *or* the entire 21 μl of digested, purified and concentrated nonamplified genomic DNA is required.

1. Prepare digested DNA sample for labeling by transferring 7 μg of digested and purified amplified genomic DNA to a nuclease-free 1.5-ml microfuge tube. Add nuclease-free water to bring to a final volume of 21 μl. Alternatively, proceed with the 21 μl of digested, purified, and concentrated nonamplified genomic DNA.
2. Dispense 20-μl aliquots of 2.5× random primers solution (supplied with the BioPrime kit) into each reaction tube.
3. Transfer sample tubes to a circulating water bath at 95°. Incubate at 95° for 5 min, move to ice, and incubate on ice for 5 min.
4. Mix the components in Table VI in the order shown, on ice.
5. Dispense 9.0-μl aliquots of exo-Klenow mix into each reaction tube. Note: Each 50-μl labeling DNA reaction contains 50 mM Tris–HCl (pH 6.8), 5 mM MgCl$_2$, 10 mM 2-mercaptoethanol, 300 μg/ml random octamers, 120 μM dATP, 120 μM dGTP, 120 μM dCTP, 60 μM dTTP, 60 μM Cy3-dUTP or Cy5-dUTP, and 40 U exo-Klenow.
6. Transfer sample tubes to a circulating water bath at 37°. Incubate at 37° for 2 h.
7. Remove sample tubes from the water bath. Add 5.0 μl of stop buffer (supplied with the BioPrime kit) to each reaction tube.
8. Reactions can be stored at –20° overnight in the dark.

TABLE VI
LABELING MIXTURE

Component	Volume (μl)/reaction	Volume (μl)/12 reactions
10× dUTP mix	5.0	62.5
Cy3-dUTP 1.0 mM or Cy5-dUTP 1.0 mM	3.0	37.5
Exo-Klenow	1.0	12.5
Volume of exo-Klenow mix	9.0	112.5

Cleanup of Labeled DNA Using Microcon YM-30 Filters

1. Combine appropriate Cy5-labeled sample and Cy3-labeled sample for a mixture volume of approximately 100 μl.
2. Add 400 μl of 1× TE (pH 8.0) to each reaction tube.
3. Place Microcon YM-30 filters in a 1.5-ml collection tube and load each sample into the filter. Centrifuge for 10 min at 7000g (8118 rpm in Eppendorf 5417R) in a microcentrifuge at room temperature. Discard flow through.
4. Add 480 μl 1× TE (pH 8.0) to each filter. Centrifuge for 10 min at 7000g in a microcentrifuge at room temperature. Discard flow through.
5. Invert the filter into a fresh 1.5-ml collection tube. Centrifuge for 1 min at 7000g in a microcentrifuge at room temperature to collect the purified sample.
6. Measure and record the volume (μl) of each eluate. Transfer the entire eluate to a fresh boilable 1.5-ml microfuge tube. Bring the total sample volume to 150 μl with nuclease-free water. Note: If the sample volume exceeds 150 μl, return the sample to its filter and centrifuge for 5 min at 7000g in a microcentrifuge at room temperature. Discard flow through. Repeat steps 5 and 6 until each sample volume <150 μl. Bring the total sample volume to 150 μl with nuclease-free water.
7. Reactions can be stored at –20° overnight in the dark.

Array Hybridization and Scanning

Preferred Methods

Prewarming of Wash Solutions. The wash 2 solution and the stabilization and drying solution should be prewarmed a day in advance of performing the wash procedure. Use of the stabilization and drying solution is optional, provided that the wash, dry, and scan environment is ozone free. An ozone-free laboratory environment can be achieved either by filtration of the laboratory air supply through activated charcoal or by use of a contained, controlled atmosphere environment (such as a tent supplied with ozone-free air or N_2) during washing, drying, and scanning. See the Agilent web site (Agilent, 2004) for details.

1. Warming of aCGH wash buffer 2. Note: The temperature of the aCGH wash buffer 2 solution must be 37° for optimal performance. Store the solution overnight in an incubator or circulating water bath set to 37°. Put a slide staining dish in a 1.5-liter dish filled with water and prewarm to 37° by storing overnight in an incubator set to 37°.

2. Warming of stabilization and drying solution. The Agilent stabilization and drying solution contains an ozone-scavenging compound dissolved in acetonitrile. The compound in solution is present in saturating amounts and may precipitate from the solution. If the solution shows visible precipitation, warming of the solution will be necessary to redissolve the compound. Washing slides using stabilization and drying solution showing visible precipitation will affect microarray performance adversely.

   a. Use extreme caution: The Agilent stabilization and drying solution is a flammable liquid, and warming the solution will increase the generation of ignitable vapors.

   b. Warm the solution slowly in a water bath in a vented fume hood at 40° in a closed container with sufficient head space to allow for expansion. Warning: Failure to follow the outlined process will increase the potential for fire, explosion, and possible personal injury. Agilent assumes no liability or responsibility for damage or injury caused by individuals performing this process. Note: The original container can be used to warm the solution. The container volume is 700 ml and contains 500 ml of liquid. If a different container is used, maintain or exceed this head space/liquid ratio. The time needed to completely redissolve the precipitate is dependent on the amount of precipitate present and may require overnight warming if precipitation is heavy. *Do not filter* the stabilization and drying solution.

   c. Gentle mixing may be necessary to obtain a homogeneous solution. If necessary, perform the mixing procedure under a vented fume hood away from open flames or other sources of ignition. Warm the solution only in a controlled and contained area that meets local fire code requirements.

   d. After the precipitate has been dissolved completely, let the covered solution stand at room temperature, allowing it to equilibrate to room temperature prior to use. Open and use the solution in a vented fume hood, as described later.

Hybridization

1. Preparation of 10× blocking reagent.

   a. Add 1.25 ml of nuclease-free water to a vial containing the lyophilized 10× blocking reagent. Store at room temperature for 60 min to reconstitute the sample.

   b. The 10× blocking reagent can be prepared in advance and stored at −20°.

2. Preparation of hybridization samples.
   a. Add the components shown in Table VII in the order indicated to reaction tubes containing 150 µl of the purified Cy5- and Cy3-labeled sample mixture.
   b. Pipette up and down to mix the sample and then quick spin in a microcentrifuge to drive the contents to the bottom of the reaction tube.
   c. Transfer sample tubes to a circulating water bath at 95°. Incubate at 95° for 3 min.
   d. Transfer sample tubes to a circulating water bath at 37°. Incubate at 37° for 30 min.
   e. Remove samples from water bath. Centrifuge for 1 min at 17,900g (13,000 rpm in Eppendorf 5417R) in a microfuge to collect the sample at the bottom of the tube.
3. Hybridization assembly.
   a. Load a clean SureHyb gasket slide into the chamber base with the label facing up and aligned with the rectangular section of the chamber base. Ensure that the gasket slide is flush with the chamber base and is not ajar.
   b. Use a pipettor to slowly dispense 490 µl of hybridization mixture onto the gasket well in a "drag-and-dispense" manner.
   c. Place a microarray active side down onto the SureHyb gasket so that the numeric bar code side is facing up. Assess that the sandwich pair is aligned properly.
   d. Place the chamber cover onto the sandwiched slides and slide the clamp assembly onto both pieces.
   e. Hand tighten the clamp onto the chamber.
   f. Vertically rotate the assembled chamber to wet the gasket and assess the mobility of the bubbles. Tap the assembly on a hard surface if necessary to move stationary bubbles.
   g. Set your hybridization rotator to rotate at 10 rpm. Note: A higher rotation speed (up to 20 rpm) can further enhance the overall assay signal intensity.
   h. Hybridize at 65° for 40 h.

TABLE VII
HYBRIDIZATION MIXTURE

Component	Volume (µl)
Cot-1 DNA (1.0 mg/ml)	50
10× blocking reagent	50
Agilent 2× hybridization buffer	250
Final hybridization sample volume	500

Wash

Note: Prepare aCGH wash buffer 2 and the stabilization and drying solution a day in advance of performing the wash procedure according to the instructions given earlier. The temperature of wash 2 solution must be 37° for optimal performance.

Note: Fresh oligonucleotide aCGH wash buffers 1 and 2 should be used for each wash group (up to 5 slides). The acetonitrile and stabilization and drying solution may be reused for washing of up to four groups of slide (20 slides total). Caution: The stabilization and drying solution must be set up in a fume hood. For practical reasons, wash 1 and wash 2 setup areas should be placed close to, or preferably in, the same fume hood.

1. Completely fill a first slide staining dish with aCGH wash buffer 1 at room temperature.

2. Place a slide rack into a second slide staining dish. Add a magnetic stir bar. Fill the slide staining dish with enough aCGH wash buffer 1 at room temperature to cover the slide rack. Place this dish on a magnetic stir plate.

3. Put the prewarmed 1.5-liter glass dish filled with water and containing the third slide staining dish on a magnetic stir plate with a heating element. Fill the slide staining dish approximately 75% full with aCGH wash buffer 2 (prewarmed to 37°). Add a magnetic stir bar. Turn on the heating element to maintain the temperature of aCGH wash buffer 2 at 37°, monitoring using a thermometer.

4. Fill a fourth staining dish approximately 75% full with acetonitrile. Add a magnetic stir bar and place this dish on a magnetic stir plate.

5. Fill a fifth staining dish approximately 75% full with stabilization and drying solution. Add a magnetic stir bar and place this dish on a magnetic stir plate.

6. Remove one hybridization chamber from the incubator and record time. Record whether bubbles formed during hybridization and if all bubbles are rotating freely.

7. Disassemble the hybridization chamber and place the microarray-gasket sandwich in the first slide staining dish filled with aCGH wash buffer 1 at room temperature. With the sandwich completely submerged in aCGH wash buffer 1, pry the sandwich open with blunt-ended tweezers. Remove the slide and place into the slide rack in the second slide staining dish containing aCGH wash buffer 1 at room temperature. Minimize exposure of the slide to air.

8. Repeat steps 6 and 7 for up to five hybridization chambers.

9. When all slides in the group are placed into the slide rack in the staining dish, stir for 5 min.

10. Transfer the slide rack to the slide staining dish containing aCGH wash buffer 2 at 37° and stir for 1 min.
11. Remove the slide rack from wash 2 and tilt the rack slightly to minimize wash buffer carryover. Immediately transfer the slide rack to the slide staining dish containing acetonitrile and stir for 1 min.
12. Transfer the slide rack to the slide staining dish filled with stabilization and drying solution and stir for 30 s.
13. Remove the slide rack slowly, trying to minimize droplets on the slides. It should take 5 to 10 s to remove the slide rack.
14. Discard used aCGH wash buffer 1 and aCGH wash buffer 2. The acetonitrile and the stabilization and drying solution may be reused for washing of up to four groups of slides. After use, rinse the slide rack and the slide staining dish that were in contact with the stabilization and drying solution with acetonitrile followed by a rinse in MilliQ water.
15. Scan slides immediately to minimize the impact of environmental oxidants on signal intensities. If necessary, store slides in original slide boxes in a N_2 purge box, in the dark, or in a vacuum desiccator prior to scanning.
16. Dispose of acetonitrile and stabilization and drying solution as a flammable solvent (organic) waste.

Scanning Using Agilent Scanner

1. Assemble slides into the appropriate slide holder. For version B slide holders, slides should be placed into the slide holder with the Agilent bar code facing up and the numeric bar code facing down.
2. Place assembled slide holders into the scanner carousel.
3. Verify default scan settings.
4. Choose settings and then modify default settings.
5. Verify that the scan region is set to scan area (61 × 21.6 mm), the dye channel is set to red and green, the scan resolution (μm) is set to 10, and the red photomultiplier tube (PMT) and green PMT are each set to 100%.
6. Select the following settings for the automatic file naming: instrument serial number for Prefix 1 and array bar code for Prefix 2.
7. Verify that the scanner status in the main window says scanner ready.
8. Click scan slot m-n on the scan control main window where the letter m represents the start slot where the first slide is located and the letter n represents the end slot where the last slide is located.

Supported Methods

Use of the stabilization and drying solution is optional, provided that the wash, dry, and scan environment is ozone free. An ozone-free laboratory environment can be achieved either by filtration of the laboratory air supply through activated charcoal or by use of a contained, controlled atmosphere environment (such as a tent supplied with ozone-free air or N_2) during washing, drying, and scanning. See the Agilent web site (Agilent, 2004) for details.

Data Extraction

Preferred/Supported Methods

Use the following method to extract Agilent HGA44B CGH microarrays using Agilent Feature Extraction (version 8.1.1).

1. Open the Feature Extraction 8.5 software.
2. Import .tif images to be extracted.
   a. Select edit and then add extraction set(s).
   b. Browse to the location of the .tif files.
   c. Use shift + Ctrl keys to select the multiple files.
   d. Select open to add image files to the FE project.
3. Set project properties.
   a. Select project properties tab.
   b. Select general. Enter your name in the operator field.
   c. Select input. Verify that the number of extraction sets included is correct.
   d. Select output folder. Verify that the same as the image file is set to true.
   e. Select output. Verify that the following selections are checked (with a checkmark): TEXT, Visual Results, Grid, QC Report. Uncheck the output options specifying GEML, MAGE, and JPEG files unless one or more of these output types is required by your downstream processing.
   f. Select automatic protocol assignment. Verify that highest priority default protocol is set as grid template default.
   g. Select automatic grid template assignment. Verify that use grid file, if available, is set to false.
   h. Select other. Verify that the external dyenorm list file is empty. Verify that overwrite previous results and send FTP results are each set to false.

TABLE VIII
FEATURE EXTRACTION SETTINGS FOR aCGH

Topic	Subtopic	Setting
Protocol properties	Place grid	Checked
	Find and measure spots	Checked
	Flag outliers	Checked
	Correct background and signal biases	Checked
	Correct dye biases	Checked
	Compute ratios and errors	Checked
	Calculate metrics	Checked
	Generate results	Checked
Place grid	Placement method	Allow some distortion
Find and measure spots	Use the nominal diameter from grid template	True
	Spot deviation limit	1.50
	Pixel outlier rejection method	Interquartile region
	Reject IQR feat	1.42
	Reject IQR BG	1.42
	Calculation of spot statistics method	Use cookie
	Cookie percentage	0.561
	Exclusion zone percentage	1.200
	Auto estimate the local radius	True
Flag outliers	Compute population outliers	True
	Minimum population	10
	IQ ratio	1.42
	Compute nonuniform outliers	True
	Feature: (%CV)^2 term	0.01440
	Poissonian noise term	320
	Background term	600
	Background: (%CV)^2 term	0.02250
	Poissonian noise term	320
	Background term	600
Correct background and signal biases	Background subtraction method	No background subtraction
	Feature significance: Two-sided t test	0.01
	Well AboveMulti	2.6
	Signal correction: Calculate surface fit	True
	Feature set for surface fit	Only negative controls
	Perform low-pass filtering	False
	Perform spatial detrending	True
	Adjust background globally	False
Correct dye biases	Dye normalization probe selection method	Use rank-consistent probes
	Rank tolerance	0.050
	Omit background population outliers?	False
	Allow positive and negative controls?	False
	Signal characteristics	Only positive and significant signals
	Normalization correction method	Linear

(*continued*)

TABLE VIII (*continued*)

Topic	Subtopic	Setting
Compute	Peg log ratio value	4.00
ratios and	Choose propagated error, univ error, or both	Most conservative
errors	MultiErrorGreen	0.1000
	MultiErrorRed	0.1000
	Auto estimate add error red	True
	Auto estimate add error green	True
	Negative control multiplier	1
	Surface fit RMS multiplier	0
Use	True	
surrogates	Spikein target used false	
Calculate	Min population for replicate stats?	3
metrics	PValue for differential expression?	0.01000
Generate	Generate single text file	True
results,	JPEG down sample factor	4
settings		

4. Extraction set configuration.
   a. Open grid template browser by selecting view and then grid template browser. Right click in the grid template browser window and then select import. Browse to location of .xml files for the AMADID's for the CGH microarrays to be extracted (e.g., 013282). Select open and template file will load into the database for future use.
   b. Select extraction set configuration tab. Verify that the .xml file for the correct AMADID is entered into the grid field for each microarray.
   c. Open FE protocol browser by selecting view and then FE protocol browser. Right click in the FE protocol browser window and then select add. Browse to location of the .xml file for oligoCGH FE8.5 protocol (44k_CGH_0605.xml, a copy of this protocol file can be found at the Agilent web site). Select open and the template file will load into the database for future use.
   d. Open FE protocol browser by selecting view and then the FE protocol browser. Double click on 44k_CGH_0605. Select protocol steps, verify that the parameters settings specified for place grid, find and measure spots, flag outliers, correct Bkgd and signal biases, correct dye biases, compute ratios and errors, calculate metrics, and generate results match the parameters specified in Table VIII.

e. Select extraction set configuration tab. Select the protocol field corresponding to the first microarray and use the pull-down menu to select 44k_CGH_0605 from the drop-down list. Repeat for each microarray.
5. Select file and then Save As. Browse to location for storage of extraction data. Specify a name for the FE project file (.fep) and then select save.
6. Verify that the icons for the image files in the left-hand window no longer have the red X through the icon.
7. Select project and then start extracting.
8. Once all extractions are complete, open the QC report for each extraction and determine whether the grid has been placed properly by inspection of spot finding at the four corners of the array.

Location Analysis

Location analysis is a new microarray application, and methods are still evolving rapidly. Current methods all come from the published location analysis literature (Harbison et al., 2004; Mukherjee et al., 2004; Odom et al., 2004; Pokholok et al., 2005). As Agilent develops optimized protocols for location analysis, they will be posted on the Agilent microarray web site (Agilent, 2005f).

Troubleshooting Guide

Accurate measurements of nucleic acid quality and quantity are crucial to the success of hybridizing labeled nucleic acids to Agilent microarrays. High-quality total RNA or genomic DNA should be free of contaminants such as carbohydrates, proteins, and traces of organic solvents and should also be minimally degraded.

Agilent recommends quantitating nucleic acid samples using the Nano-Drop ND-1000 UV-VIS spectrophotometer (or equivalent) to assess both concentration and purity. High-quality samples should have an A_{260}/A_{280} ratio of 1.8–2.0, indicating the absence of contaminating proteins. Many organic compounds such as guanidium thiocyanate, alcohol, and phenol, as well as cellular contaminants such as carbohydrates, absorb strongly at 230 nm. An A_{260}/A_{230} ratio >2.0 is indicative of pure genomic DNA or high-purity RNA. For DNA, a more accurate quantitation alternative to UV-spectrophotometry is the use of a fluorescent nucleic acid assay, such as the Quant-iT PicoGreen dsDNA assay kit (Invitrogen). This kit is highly sensitive and selective for double-stranded DNA.

Total RNA integrity should be measured using the Agilent bioana-lyzer, as described previously. To help standardize the interpretation of RNA integrity, Agilent has introduced the RNA integrity number (RIN) (Agilent, 2005i) for the purpose of standardizing the interpretation of RNA integrity. This tool provides a quantitative measure of integrity graded on a scale of 1 (poor) to 10 (high). A general "rule of thumb" is that the use of RNA preparations with RIN values of 5 or less will impact microarray data quality negatively.

Assessment of the intactness of genomic DNA (for aCGH) using agarose gel electrophoresis (or equivalent) is equally important. Intact genomic DNA should appear as a compact, high molecular weight band in an agarose gel, with no lower molecular weight smear. A small amount of degradation is not likely to interfere with the direct DNA-labeling procedure. However, it will have a negative impact on phi29-based ampli-fication results.

Problems commonly encountered in the generation of labeled cRNA for hybridization to gene expression microarrays include low cRNA yields and the generation of cRNA populations with dye incorporation rates outside the optimum range. The use of too little cRNA material in the microarray target or the use of target for which the specific activity is too low will result in diminished microarray signals. This will reduce the sensitivity of the system in detecting transcripts that are present at low abundance, as well as in detecting differential expression.

Low cRNA yields from labeling reactions can generally be attributed to poor input RNA quality. The presence of contaminating organic com-pounds or proteins, as described earlier, can result in the overestimation of RNA concentration and consequently in the input of too little template RNA to labeling reactions. Furthermore, these contaminants can be inhib-itory to the processing activity of reverse transcriptase and T7-RNA poly-merase. Yields of cRNA will also be decreased when RNA preparations with poor integrity are used.

Agilent recommends that labeled cRNA microarray targets should have dye incorporation rates in the range of 10–15 pmol of cyanine dye per microgram and should not be used if the value is less than 8. Low specific activities generally result when the ratio of T7 promoter primer to RNA template is too high, causing T7-RNA polymerase to generate primer-derived material with a low level of dye incorporation. The use of less RNA (or degraded RNA) or more primer than recommended in Agilent's optimized protocol may consequently reduce dye incorporation rates. Specific activities greater than 20 pmol of dye per microgram may also be problematic as they can be indicative of inadequate removal of unincorporated cyanine-CTPs during purification of the labeled material. The presence of high levels of unincorporated cyanine-CTPs in microarray targets can result in elevated microarray background signal intensities.

A variety of microarray hybridization and wash process-related problems can be encountered when generating gene expression data. Mixing of the hybridization solution in Agilent's hybridization chambers is dependent on the presence of a large mobile bubble. Small bubbles introduced during microarray loading may become trapped in the chamber corners and along edges, preventing exposure of localized regions of the microarray to the hybridization solution. These regions will appear as "bubble scars" with low signal intensities on microarray images. This problem can be avoided by ensuring that all bubbles are freely mobile before loading hybridization chambers into the oven. Incorrect loading of the chamber, or chamber leakage, can lead to inadequate hybridization solution being present to allow exposure of the entire microarray surface during the incubation. This may result in an area of dim signal in the center of the microarray image. Occasionally microarray features are observed to have highly nonuniform morphologies, i.e., nonuniform distribution of signal intensities across pixels within features. This is generally due to the microarrays being transferred between the various wash steps too slowly, allowing signal-quenching salt crystals to form within the features. Care must also be taken to ensure that excessive wash solution is not allowed to dry onto the slides after the final wash. Dried salts and detergents from the wash solution may fluoresce, adding contaminating signal to both microarray features and backgrounds.

Ozone degradation of the Cy5-derived signal in two-color microarray experiments is a common problem. Damage can occur either while the microarray is drying or after it has dried. Ozone damage during drying leads to diagnostic "measles" features: green or yellow spots with a small red disk at or near the center. Damage to dried microarrays manifests either as an overall reduction in the Cy5 signal (if ozone exposure is homogeneous) or a red/green color gradient (if exposure is inhomogeneous). Any of these symptoms indicates the presence of damaging levels of an airborne oxidant (usually ozone, although bleach fumes and other oxidants can cause similar phenomena). Alternatives for providing protection from such oxidants can be found on the Agilent web site (Agilent, 2004).

The most recent version of Agilent's Feature Extraction software produces a microarray quality report (Agilent, 2005k) as a part of its work product. The microarray quality report can be used to diagnose many common difficulties with microarray experiments.

Finally, microarrays sometimes display green streaks, swirls, finger-prints, speckles, and other blemishes. Most of these imperfections can be traced to one of three sources: a poor or hasty technique, poor equipment cleaning, or use of disposable equipment that releases fluorescent impurities. The exercise of a good technique is a matter of planning, practice, and patience. Allow sufficient time to complete washing, drying, and scanning of microarrays. Avoid distractions. Practice with some simple experiments before committing to processing a large number of microarrays or a set of irreplaceable samples. If possible, learn techniques from someone with a track record of successful microarray experiments.

Equipment should be cleaned regularly and rinsed with generous amounts of sterile (filtered) MilliQ or distilled water. Slide staining dishes that were used in the wash step for wash buffers 1 and 2 should be rinsed with MilliQ water. Slide racks and slide staining dishes that were in contact with the stabilization and drying solution should be rinsed with acetonitrile followed by a rinse in MilliQ water. Do not use detergents to clean slide staining dishes.

If possible, dry and store equipment in a clean, dust-free environment, such as a laminar flow hood. Clean work areas frequently. If using paper laboratory bench covers, make sure that they are lint free. Avoid powdered gloves or gloves containing dyes. Finally, qualify a source of rubber gloves that yield good results and then avoid changing sources.

References

Agilent (2004). Improving microarray results by preventing ozone-mediated fluorescent signal degradation. http://www.chem.agilent.com/temp/rad5B686/00047606.pdf.
Agilent (2005a). Agilent 2100 Bioanalyzer 2100 Expert User's Guide. http://www.chem. agilent.com/scripts/literaturepdf.asp?iWHID=40690.
Agilent (2005b). Array based CGH Procedures for Genomic DNA Analysis. http://www. chem.agilent.com/scripts/literaturePDF.asp?iWHID=39707.
Agilent (2005c). Catalog DNA Microarray Kits. http://www.chem.agilent.com/Scripts/PCol. asp?lPage=494.
Agilent (2005d). Custom DNA Microarrays. http://www.chem.agilent.com/Scripts/PCol.asp? lPage=494.
Agilent (2005e). DNA Microarray Design Services. http://www.chem.agilent.com/Scripts/ PDS.asp?lPage=2989.
Agilent (2005f). DNA Microarrays. http://www.chem.agilent.com/Scripts/PCol.asp?lPage=494.
Agilent (2005g). eArray 3.0. http://www.chem.agilent.com/scripts/pds.asp?lpage=29443.
Agilent (2005h). Microarray Protocols and Manuals. http://www.chem.agilent.com/scripts/ generic.asp?lpage=11617&indcol=N&prodcol=Y.
Agilent (2005i). The RNA Integrity Number (RIN). http://www.chem.agilent.com/Scripts/ Generic.ASP?lPage=14975&indcol=Y&prodcol=Y.
Agilent (2005j). A Showcase of Agilent's Feature Extraction Software 8.1. http://www.chem. agilent.com/scripts/generic.asp?lpage=10289&indcol=Y&prodcol=Y.

Agilent (2005k). Use of Agilent Feature Extraction Software (v8.1) QC Report to Evaluate Microarray Performance. http://www.chem.agilent.com/scripts/literaturepdf.asp?iWHID=40915.

Barrett, M. T., Scheffer, A., Ben-Dor, A., Sampas, N., Lipson, D., Kincaid, R., Tsang, P., Curry, B., Baird, K., Meltzer, P. S., Yakhini, Z., Bruhn, L., and Laderman, S. (2004). Comparative genomic hybridization using oligonucleotide microarrays and total genomic DNA. *Proc. Natl. Acad. Sci. USA* **101**, 17765–17770.

Blanchard, A. P., Kaiser, R. J., and Hood, L. E. (1996). High-density oligonucleotide arrays. *Biosens. Bioelectron.* **11**, 687–690.

Castle, J., Garrett-Engele, P., Armour, C. D., Duenwald, S. J., Loerch, P. M., Meyer, M. R., Schadt, E. E., Stoughton, R., Parrish, M. L., Shoemaker, D. D., and Johnson, J. M. (2003). Optimization of oligonucleotide arrays and RNA amplification protocols for analysis of transcript structure and alternative splicing. *Genome Biol.* **4**, R66.

Chang, H. Y., Nuyten, D. S., Sneddon, J. B., Hastie, T., Tibshirani, R., Sorlie, T., Dai, H., He, Y. D., van't Veer, L. J., Bartelink, H., van de Rijn, M., Brown, P. O., and van de Vijver, M. J. (2005). Robustness, scalability, and integration of a wound-response gene expression signature in predicting breast cancer survival. *Proc. Natl. Acad. Sci. USA* **102**, 3738–3743.

Chou, H. H., Hsia, A. P., Mooney, D. L., and Schnable, P. S. (2004). Picky: Oligo microarray design for large genomes. *Bioinformatics* **20**, 2893–2902.

Dai, H., Meyer, M., Stepaniants, S., Ziman, M., and Stoughton, R. (2002). Use of hybridization kinetics for differentiating specific from non-specific binding to oligonucleotide microarrays. *Nucleic Acids Res.* **30**, e86.

Dai, H., van't Veer, L., Lamb, J., He, Y. D., Mao, M., Fine, B. M., Bernards, R., van de Vijver, M., Deutsch, P., Sachs, A., Stoughton, R., and Friend, S. (2005). A cell proliferation signature is a marker of extremely poor outcome in a subpopulation of breast cancer patients. *Cancer Res.* **65**, 4059–4066.

Dean, F. B., Hosono, S., Fang, L., Wu, X., Faruqi, A. F., Bray-Ward, P., Sun, Z., Zong, Q., Du, Y., Du, J., Driscoll, M., Song, W., Kingsmore, S. F., Egholm, M., and Lasken, R. S. (2002). Comprehensive human genome amplification using multiple displacement amplification. *Proc. Natl. Acad. Sci. USA* **99**, 5261–5266.

Dudoit, S., Gentleman, R. C., and Quackenbush, J. (2003). Open source software for the analysis of microarray data. *Biotechniques Suppl.* 45–51.

Eberwine, J., Yeh, H., Miyashiro, K., Cao, Y., Nair, S., Finnell, R., Zettel, M., and Coleman, P. (1992). Analysis of gene expression in single live neurons. *Proc. Natl. Acad. Sci. USA* **89**, 3010–3014.

Fare, T. L., Coffey, E. M., Dai, H., He, Y. D., Kessler, D. A., Kilian, K. A., Koch, J. E., LeProust, E., Marton, M. J., Meyer, M. R., Stoughton, R. B., Tokiwa, G. Y., and Wang, Y. (2003). Effects of atmospheric ozone on microarray data quality. *Anal. Chem.* **75**, 4672–4675.

Feinberg, A. P., and Vogelstein, B. (1983). A technique for radiolabeling DNA restriction endonuclease fragments to high specific activity. *Anal. Biochem.* **132**, 6–13.

Feinberg, A. P., and Vogelstein, B. (1984). A technique for radiolabeling DNA restriction endonuclease fragments to high specific activity: Addendum. *Anal. Biochem.* **137**, 266–267.

Golub, T. R., Slonim, D. K., Tamayo, P., Huard, C., Gaasenbeek, M., Mesirov, J. P., Coller, H., Loh, M. L., Downing, J. R., Caligiuri, M. A., Bloomfield, C. D., and Lander, E. S. (1999). Molecular classification of cancer: Class discovery and class prediction by gene expression monitoring. *Science* **286**, 531–537.

Gordon, P. M., and Sensen, C. W. (2004). Osprey: A comprehensive tool employing novel methods for the design of oligonucleotides for DNA sequencing and microarrays. *Nucleic Acids Res.* **32**, e133.

Greshock, J., Naylor, T. L., Margolin, A., Diskin, S., Cleaver, S. H., Futreal, P. A., deJong, P. J., Zhao, S., Liebman, M., and Weber, B. L. (2004). 1-Mb resolution array-based comparative genomic hybridization using a BAC clone set optimized for cancer gene analysis. *Genome Res.* **14,** 179–187.

Harbison, C. T., Gordon, D. B., Lee, T. I., Rinaldi, N. J., Macisaac, K. D., Danford, T. W., Hannett, N. M., Tagne, J. B., Reynolds, D. B., Yoo, J., Jennings, E. G., Zeitlinger, J., Pokholok, D. K., Kellis, M., Rolfe, P. A., Takusagawa, K. T., Lander, E. S., Gifford, D. K., Fraenkel, E., and Young, R. A. (2004). Transcriptional regulatory code of a eukaryotic genome. *Nature* **431,** 99–104.

Hodgson, G., Hager, J. H., Volik, S., Hariono, S., Wernick, M., Moore, D., Nowak, N., Albertson, D. G., Pinkel, D., Collins, C., Hanahan, D., and Gray, J. W. (2001). Genome scanning with array CGH delineates regional alterations in mouse islet carcinomas. *Nat. Genet.* **29,** 459–464.

Hosono, S., Faruqi, A. F., Dean, F. B., Du, Y., Sun, Z., Wu, X., Du, J., Kingsmore, S. F., Egholm, M., and Lasken, R. S. (2003). Unbiased whole-genome amplification directly from clinical samples. *Genome Res.* **13,** 954–964.

Hughes, T. R., Mao, M., Jones, A. R., Burchard, J., Marton, M. J., Shannon, K. W., Lefkowitz, S. M., Ziman, M., Schelter, J. M., Meyer, M. R., Kobayashi, S., Davis, C., Dai, H., He, Y. D., Stephaniants, S. B., Cavet, G., Walker, W. L., West, A., Coffey, E., Shoemaker, D. D., Stoughton, R., Blanchard, A. P., Friend, S. H., and Linsley, P. S. (2001). Expression profiling using microarrays fabricated by an ink-jet oligonucleotide synthesizer. *Nat. Biotechnol.* **19,** 342–347.

Jackson, A. L., Bartz, S. R., Schelter, J., Kobayashi, S. V., Burchard, J., Mao, M., Li, B., Cavet, G., and Linsley, P. S. (2003). Expression profiling reveals off-target gene regulation by RNAi. *Nat. Biotechnol.* **21,** 635–637.

Johnson, J. M., Castle, J., Garrett-Engele, P., Kan, Z., Loerch, P. M., Armour, C. D., Santos, R., Schadt, E. E., Stoughton, R., and Shoemaker, D. D. (2003). Genome-wide survey of human alternative pre-mRNA splicing with exon junction microarrays. *Science* **302,** 2141–2144.

Johnson, J. M., Edwards, S., Shoemaker, D., and Schadt, E. E. (2005). Dark matter in the genome: Evidence of widespread transcription detected by microarray tiling experiments. *Trends Genet.* **21,** 93–102.

Kronick, M. N. (2004). Creation of the whole human genome microarray. *Expert Rev. Proteom.* **1,** 19–28.

Kuhn, K. M., DeRisi, J. L., Brown, P. O., and Sarnow, P. (2001). Global and specific translational regulation in the genomic response of *Saccharomyces cerevisiae* to a rapid transfer from a fermentable to a nonfermentable carbon source. *Mol. Cell. Biol.* **21,** 916–927.

Marton, M. J., DeRisi, J. L., Bennett, H. A., Iyer, V. R., Meyer, M. R., Roberts, C. J., Stoughton, R., Burchard, J., Slade, D., Dai, H., Bassett, D. E., Jr., Hartwell, L. H., Brown, P. O., and Friend, S. H. (1998). Drug target validation and identification of secondary drug target effects using DNA microarrays. *Nat. Med.* **4,** 1293–1301.

Miki, R., Kadota, K., Bono, H., Mizuno, Y., Tomaru, Y., Carninci, P., Itoh, M., Shibata, K., Kawai, J., Konno, H., Watanabe, S., Sato, K., Tokusumi, Y., Kikuchi, N., Ishii, Y., Hamaguchi, Y., Nishizuka, I., Goto, H., Nitanda, H., Satomi, S., Yoshiki, A., Kusakabe, M., DeRisi, J. L., Eisen, M. B., Iyer, V. R., Brown, P. O., Muramatsu, M., Shimada, H., Okazaki, Y., and Hayashizaki, Y. (2001). Delineating developmental and metabolic pathways *in vivo* by expression profiling using the RIKEN set of 18,816 full-length enriched mouse cDNA arrays. *Proc. Natl. Acad. Sci. USA* **98,** 2199–2204.

Mukherjee, S., Berger, M. F., Jona, G., Wang, X. S., Muzzey, D., Snyder, M., Young, R. A., and Bulyk, M. L. (2004). Rapid analysis of the DNA-binding specificities of transcription factors with DNA microarrays. *Nat. Genet.* **36,** 1331–1339.

Nacht, M., Ferguson, A. T., Zhang, W., Petroziello, J. M., Cook, B. P., Gao, Y. H., Maguire, S., Riley, D., Coppola, G., Landes, G. M., Madden, S. L., and Sukumar, S. (1999). Combining serial analysis of gene expression and array technologies to identify genes differentially expressed in breast cancer. *Cancer Res.* **59**, 5464–5470.

Nordberg, E. K. (2005). YODA: Selecting signature oligonucleotides. *Bioinformatics* **21**, 1365–1370.

Odom, D. T., Zizlsperger, N., Gordon, D. B., Bell, G. W., Rinaldi, N. J., Murray, H. L., Volkert, T. L., Schreiber, J., Rolfe, P. A., Gifford, D. K., Fraenkel, E., Bell, G. I., and Young, R. A. (2004). Control of pancreas and liver gene expression by HNF transcription factors. *Science* **303**, 1378–1381.

Pinkel, D., Segraves, R., Sudar, D., Clark, S., Poole, I., Kowbel, D., Collins, C., Kuo, W. L., Chen, C., Zhai, Y., Dairkee, S. H., Ljung, B. M., Gray, J. W., and Albertson, D. G. (1998). High resolution analysis of DNA copy number variation using comparative genomic hybridization to microarrays. *Nat. Genet.* **20**, 207–211.

Pokholok, D. K., Harbison, C. T., Levine, S., Cole, M., Hannett, N. M., Lee, T. I., Bell, G. W., Walker, K., Rolfe, P. A., Herbolsheimer, E., Zeitlinger, J., Lewitter, F., Gifford, D. K., and Young, R. A. (2005). Genome-wide map of nucleosome acetylation and methylation in yeast. *Cell* **122**, 517–527.

Pollack, J. R., Perou, C. M., Alizadeh, A. A., Eisen, M. B., Pergamenschikov, A., Williams, C. F., Jeffrey, S. S., Botstein, D., and Brown, P. O. (1999). Genome-wide analysis of DNA copy-number changes using cDNA microarrays. *Nat. Genet.* **23**, 41–46.

Ramaswamy, S., Tamayo, P., Rifkin, R., Mukherjee, S., Yeang, C. H., Angelo, M., Ladd, C., Reich, M., Latulippe, E., Mesirov, J. P., Poggio, T., Gerald, W., Loda, M., Lander, E. S., and Golub, T. R. (2001). Multiclass cancer diagnosis using tumor gene expression signatures. *Proc. Natl. Acad. Sci. USA* **98**, 15149–15154.

Rimour, S., Hill, D., Militon, C., and Peyret, P. (2005). GoArrays: Highly dynamic and efficient microarray probe design. *Bioinformatics* **21**, 1094–1103.

Schadt, E. E., Edwards, S. W., GuhaThakurta, D., Holder, D., Ying, L., Svetnik, V., Leonardson, A., Hart, K. W., Russell, A., Li, G., Cavet, G., Castle, J., McDonagh, P., Kan, Z., Chen, R., Kasarskis, A., Margarint, M., Caceres, R. M., Johnson, J. M., Armour, C. D., Garrett-Engele, P. W., Tsinoremas, N. F., and Shoemaker, D. D. (2004). A comprehensive transcript index of the human genome generated using microarrays and computational approaches. *Genome Biol.* **5**, R73.

Segal, E., Friedman, N., Koller, D., and Regev, A. (2004). A module map showing conditional activity of expression modules in cancer. *Nat. Genet.* **36**, 1090–1098.

Shannon, K. W., Wolber, P. K., Delenstarr, G. C., Webb, P. G., and Kincaid, R. H. (2001). "Method for Evaluating Oligonucleotide Probe Sequences." Agilent Technologies Inc., Palo Alto, CA.

Sharp, A. J., Locke, D. P., McGrath, S. D., Cheng, Z., Bailey, J. A., Vallente, R. U., Pertz, L. M., Clark, R. A., Schwartz, S., Segraves, R., Oseroff, V. V., Albertson, D. G., Pinkel, D., and Eichler, E. E. (2005). Segmental duplications and copy-number variation in the human genome. *Am. J. Hum. Genet.* **77**, 78–88.

Shoemaker, D. D., Schadt, E. E., Armour, C. D., He, Y. D., Garrett-Engele, P., McDonagh, P. D., Loerch, P. M., Leonardson, A., Lum, P. Y., Cavet, G., Wu, L. F., Altschuler, S. J., Edwards, S., King, J., Tsang, J. S., Schimmack, G., Schelter, J. M., Koch, J., Ziman, M., Marton, M. J., Li, B., Cundiff, P., Ward, T., Castle, J., Krolewski, M., Meyer, M. R., Mao, M., Burchard, J., Kidd, M. J., Dai, H., Phillips, J. W., Linsley, P. S., Stoughton, R., Scherer, S., and Boguski, M. S. (2001). Experimental annotation of the human genome using microarray technology. *Nature* **409**, 922–927.

Singh, D., Febbo, P. G., Ross, K., Jackson, D. G., Manola, J., Ladd, C., Tamayo, P., Renshaw, A. A., D'Amico, A. V., Richie, J. P., Lander, E. S., Loda, M., Kantoff, P. W., Golub, T. R., and Sellers, W. R. (2002). Gene expression correlates of clinical prostate cancer behavior. *Cancer Cell* **1**, 203–209.

Snijders, A. M., Nowak, N., Segraves, R., Blackwood, S., Brown, N., Conroy, J., Hamilton, G., Hindle, A. K., Huey, B., Kimura, K., Law, S., Myambo, K., Palmer, J., Ylstra, B., Yue, J. P., Gray, J. W., Jain, A. N., Pinkel, D., and Albertson, D. G. (2001). Assembly of microarrays for genome-wide measurement of DNA copy number. *Nat. Genet.* **29**, 263–264.

van de Vijver, M. J., He, Y. D., van't Veer, L. J., Dai, H., Hart, A. A., Voskuil, D. W., Schreiber, G. J., Peterse, J. L., Roberts, C., Marton, M. J., Parrish, M., Atsma, D., Witteveen, A., Glas, A., Delahaye, L., van der Velde, T., Bartelink, H., Rodenhuis, S., Rutgers, E. T., Friend, S. H., and Bernards, R. (2002). A gene-expression signature as a predictor of survival in breast cancer. *N. Engl. J. Med.* **347**, 1999–2009.

Van Gelder, R. N., von Zastrow, M. E., Yool, A., Dement, W. C., Barchas, J. D., and Eberwine, J. H. (1990). Amplified RNA synthesized from limited quantities of heterogeneous cDNA. *Proc. Natl. Acad. Sci. USA* **87**, 1663–1667.

van't Veer, L. J., Dai, H., van de Vijver, M. J., He, Y. D., Hart, A. A., Bernards, R., and Friend, S. H. (2003). Expression profiling predicts outcome in breast cancer. *Breast Cancer Res.* **5**, 57–58.

van't Veer, L. J., Paik, S., and Hayes, D. F. (2005). Gene expression profiling of breast cancer: A new tumor marker. *J. Clin. Oncol.* **23**, 1631–1635.

Wang, D., Coscoy, L., Zylberberg, M., Avila, P. C., Boushey, H. A., Ganem, D., and DeRisi, J. L. (2002). Microarray-based detection and genotyping of viral pathogens. *Proc. Natl. Acad. Sci. USA* **99**, 15687–15692.

Waring, J. F., Ciurlionis, R., Jolly, R. A., Heindel, M., and Ulrich, R. G. (2001a). Microarray analysis of hepatotoxins *in vitro* reveals a correlation between gene expression profiles and mechanisms of toxicity. *Toxicol. Lett.* **120**, 359–368.

Waring, J. F., Jolly, R. A., Ciurlionis, R., Lum, P. Y., Praestgaard, J. T., Morfitt, D. C., Buratto, B., Roberts, C., Schadt, E., and Ulrich, R. G. (2001b). Clustering of hepatotoxins based on mechanism of toxicity using gene expression profiles. *Toxicol. Appl. Pharmacol.* **175**, 28–42.

Wilson, M., DeRisi, J., Kristensen, H. H., Imboden, P., Rane, S., Brown, P. O., and Schoolnik, G. K. (1999). Exploring drug-induced alterations in gene expression in Mycobacterium tuberculosis by microarray hybridization. *Proc. Natl. Acad. Sci. USA* **96**, 12833–12838.

[3] Illumina Universal Bead Arrays

By Jian-Bing Fan, Kevin L. Gunderson, Marina Bibikova, Joanne M. Yeakley, Jing Chen, Eliza Wickham Garcia, Lori L. Lebruska, Marc Laurent, Richard Shen, and David Barker

Abstract

This chapter describes an accurate, scalable, and flexible microarray technology. It includes a miniaturized array platform where each individual feature is quality controlled and a versatile assay that can be adapted for various genetic analyses, such as single nucleotide polymorphism genotyping,

METHODS IN ENZYMOLOGY, VOL. 410   0076-6879/06 $35.00
                          DOI: 10.1016/S0076-6879(06)10003-8

DNA methylation detection, and gene expression profiling. This chapter describes the concept of the BeadArray technology, two different Array of Arrays formats, the assay scheme and protocol, the performance of the system, and its use in large-scale genetic, epigenetic, and expression studies.

Introduction

DNA polymorphisms and differences in gene expression provide the genetic basis for phenotypic variation. The systematic screening of an organism's genome for genetic variants and identification of environmental conditions that affect gene regulation and linking the former to a possible phenotype may advance our understanding of the underlying genetic or epigenetic mechanisms and lead to the identification of alleles (or haplotypes) and genes that contribute to phenotypic diversity. However, any comprehensive genome-wide association studies would need to genotype a large number of single nucleotide polymorphisms (SNPs) (or measure expression of a large number of genes) in large sample sets (Botstein and Risch, 2003), thus requiring a system that combines very high throughput and accuracy with very low cost per SNP (or gene) analysis (Kwok and Chen, 2003; Syvanen, 2005). The technology described in this chapter provides a solution to these challenges by integrating a miniaturized array platform with a high level of assay multiplexing and scalable automation. The system uses the Sentrix BeadArray technology in combination with a solid-phase, allele-specific extension, followed by polymerase chain reaction (PCR) amplification to achieve high multiplex levels. The system is able to maintain high assay multiplex levels and high sample throughput in an informatically integrated production environment. Gunderson et al. (2005) described a genome-wide scalable genotyping assay, the Infinium assay, which is also deployed on Sentrix arrays and allows simultaneous whole genome genotyping of over 100,000 SNP markers.

Material and Methods

BeadArray Technology

Conventional microarrays are manufactured by spotting or synthesizing probes onto two-dimensional substrates at known locations (Fodor et al., 1991; Holloway et al., 2002; Schena et al., 1995). In contrast, BeadArray technology is based on the random self-assembly of a bead pool onto a patterned substrate (Fan et al., 2003; Michael et al., 1998; Oliphant et al., 2002). As a part of the array manufacturing process, a decoding process is implemented to map the precise location of a specific bead type on the array (Gunderson *et al.*, 2004). The number of independent array features is dependent on the complexity of the assay pool. A high degree of miniaturization can be achieved by exploiting the intrinsic size of the beads and patterned substrate. For example, the density of a randomly assembled 300-nm-diameter bead array is ~40,000 times higher than a typical spotted microarray (Michael *et al.*, 1998).

In the past, bead assemblies were encoded in a number of ways. The simplest method involved impregnation or attachment of dyes to the bead. Initial attempts to determine the location and identity of beads using this approach were largely affected by the variability in quantitation, instability, and array-to-array variation of the dye, among other technical problems, which limited the usefulness of the approach (Michael *et al.*, 1998). A number of other schemes that aim to encode particles directly with combinatorial codes generated by mixtures or spatial arrangements of optical signaling molecules have similar issues (Braeckmans *et al.*, 2002; Chan *et al.*, 2002; Fulton *et al.*, 1997; Han *et al.*, 2001; Lockhart and Trulson, 2001; Nicewarner-Pena *et al.*, 2001).

We sidestepped the need for complex dye chemistries and painstaking labeling processes by devising a novel, highly efficient decoding, rather than encoding, strategy that employs the specificity and reversibility of DNA hybridization (Gunderson *et al.*, 2004). Quantitatively pooled bead libraries are randomly self-assembled into etched microwell substrates (Barker *et al.*, 2003) (Fig. 1). The physical location or "map" of the beads is determined by serial hybridization steps using fluorescently labeled complementary oligonucleotides. The advantage of this approach is that dense packing with unlimited scalability can be achieved concomitantly with increased reliability. Specifically, each bead is derivatized on its surface with approximately 700,000 covalently attached oligonucleotide probes of a unique sequence (Kuhn *et al.*, 2004). Dye-labeled oligonucleotides then hybridize to the address sequence of each probe type in the hybridization or "decoding" process. The physical location of each of the bead types is thus mapped by using a decoding algorithm (Gunderson *et al.*, 2004) that tracks oligonucleotides successfully hybridizing to a specific address. With this compiled information, the physical location of each bead is mapped. The decoding process also serves as a quality control step for each feature of each array.

The address sequences are designed computationally to have no significant homology to genomic sequences within the species queried. With these "universal arrays," the address sequences, called IllumiCodes, are used for decoding as well as analytical readout in genotyping or gene expression applications. Currently, up to 1624 unique bead types are represented in each of the universal arrays, with an average 30-fold bead type redundancy in each array (~50,000 beads).

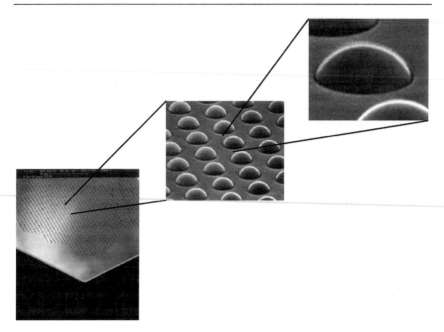

Fig. 1. BeadArray technology. Silica beads, each 3 μm in diameter, are assembled randomly into wells with 5-μm center-to-center spacing. Each bead is represented with an average 30-fold redundancy on each array.

Array of Arrays Format

Illumina has developed two different Array of Arrays formats: the Sentrix Array Matrix (SAM) and the Sentrix BeadChip (Fig. 2). The SAM is an 8 × 12 matrix consisting of 96 fiber bundles (i.e., 96 independent arrays) with ~50,000 light-conducting strands per bundle. Thus, one SAM can assay 96 separate samples simultaneously. The SAM is compatible with standard microtiter plates used to carry out sample preparation and enzymatic steps prior to exposure of the beads in the array to the sample. Therefore, the platform can be readily incorporated into automated routines using standard robotic equipment.

The second array format, termed the Sentrix BeadChip, is a silicon substrate with the dimensions 2.5 × 8.25 cm. This platform is highly flexible and will allow configurations that assay 1 to 16 samples at a time. For example, our Infinium Whole Genome Genotyping Assay (100 K genotypes) is performed on a single BeadChip. If the BeadChip is formatted differently, such as in the Human-6 BeadChip for whole genome gene expression analysis, 6 samples may be analyzed on one BeadChip. The

FIG. 2. Two different Array of Arrays formats: the 96-sample Sentrix Array Matrix (SAM) and the multisample Sentrix BeadChip. The SAM substrate (upper) is composed of 96 arrays. Each array is composed of ~50,000 fiber-optic strands chemically etched to create a well for a single bead. The BeadChip formats (lower) feature flexible configuration with dense packing ratios that can be used for 1 to 16 samples per BeadChip.

16-sample BeadChip has the same feature-to-feature spacing as standard multichannel pipettes.

GoldenGate Genotyping Assay

The GoldenGate genotyping assay (Fan *et al.*, 2003; Shen *et al.*, 2005) provides a flexible, accurate, and high-throughput SNP genotyping system for large-scale genetic analysis (Fig. 3). It allows for a high degree of locus multiplexing through highly specific extension and amplification steps. The multiplexed assay measures from 384 to 1536 SNPs in a single reaction. For example, a single SAM can be used to genotype 1536 SNPs across 96 samples, allowing the researcher to determine up to 150,000 genotypes simultaneously. Proprietary OligoDesigner software has been used to design hundreds of thousands of GoldenGate genotyping assays. It evaluates sequences flanking the targeted SNP, such as degree of homology to repeated sequences across the genome, palindromic sequences, GC&AT

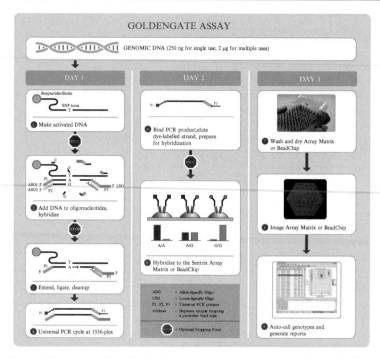

FIG. 3. Illustration of the GoldenGate genotyping assay process. (See color insert.)

content, and neighboring polymorphisms. The allele-specific sequence is designed at a T_m of 60° (57– 62°), whereas the locus-specific probe is designed at a T_m of 57° (54–60°). To assure high-quality data, various internal assay controls are used to assess the GoldenGate assay and array hybridization at various experimental steps, including gDNA/oligonucleotide annealing, PCR, array hybridization, and imaging. For example, controls include G/T or G/G mismatches; amplification balance of the two types of universal PCR primers; hybridization controls; and a double label control to estimate the optical balance of Cy3/Cy5 channels.

This section describes the standard GoldenGate genotyping assay procedures, although details of the protocol may vary depending on the specifics of the genotyping system used. All robotic processes were performed on a Tecan Genesis Workstation 150 (Tecan).

Biotinylation of Genomic DNA (Step 1). Genomic DNA (5 μl at 50 ng/μl) is mixed with 5 μl biotinylation reagent (MS1; Illumina) and incubated at 95° for 30 min. A precipitation reagent (PS1; Illumina; 5 μl) is then added, followed by the addition of 15 μl of 2-propanol. The samples are mixed thoroughly and spun at high speed (3000g). The pellets are resuspended in buffer (RS1; Illumina; 10 μl) and used in the subsequent assay step.

Annealing of Assay Oligonucleotides to Genomic DNA (Step 2). Genotyping assay oligonucleotides (corresponding to 384–1536 specific SNPs) and paramagnetic particles are then combined with the activated DNA in the oligonucleotide/target annealing step, in which the query oligonucleotides hybridize to the genomic DNA as it binds to paramagnetic particles. Two allele-specific oligonucleotides (ASOs) and one locus-specific oligonucleotide (LSO) are designed for each SNP (Fig. 3) (Fan *et al.*, 2003). All three oligonucleotide sequences contain regions of genomic complementarity and universal PCR primer sites; the LSO also contains a unique IllumiCode sequence complementary to a particular bead type. The IllumiCode sequence hybridizes to the universal bead type probes in the last hybridization step of the genotyping assay (see Step 6). The IllumiCodes represented on the universal array comprises 1624 artificial sequences carefully selected to not cross-hybridize with each other or with sequences in the human genome. LSOs are synthesized with a 5′ phosphate to enable ligation. Annealing reagent (OB1; Illumina; 30 μl) and SNP-specific oligonucleotides (OPA; Illumina; 10 μl) are combined with DNA (10 μl) to a final volume of 50 μl. Annealing is carried out by ramping the temperature from 70 to 30° over ∼2 h and then holding at 30° until the next processing step. Up to 1536 SNPs may be assayed simultaneously in this manner.

Assay Oligonucleotides Extension and Ligation (Step 3). Following assay oligonucleotide hybridization, several wash steps are performed to remove excess and mishybridized oligonucleotides. In a single reaction, a DNA polymerase with high specificity for a 3′ match and no strand displacement or exonuclease activity is used to extend the ASO(s) that perfectly matches the target sequence at the SNP site and fill the gap between the ASO and the LSO (Step 3), and a DNA ligase is used to seal the nick between the extended ASO and the LSO to form PCR templates that can be amplified with universal PCR primers (Step 4). A master mix for extension and ligation (MEL; Illumina; 37 μl) is added to the beads. Extension and ligation are carried out at 45° for 15 min. High locus specificity is achieved by the requirement that both the ASO and the LSO oligonucleotides hybridize to the same target site; extension of the appropriate ASO and ligation of the extended product to the adjacent LSO join information about the genotype present at the SNP site to the address sequence on LSO.

PCR Amplification (Step 4). After extension and ligation, the beads are washed with universal buffer 1 (UB1; Illumina) to reduce noise by removing excess and mishybridized oligonucleotides. The beads are then resuspended in 35 μl elution buffer (IP1; Illumina) and heated at 95° for 1 min to release the ligated products. The supernatant is then used in a 60-μl PCR reaction. PCR reactions are thermocycled as follows: 10 min at 37°; 34 cycles (35 s at 95°, 35 s at 56°, 2 min at 72°); 10 min at 72°; and cooled to 4° for 5 min. The three universal PCR primers (P1, P2, and P3) are 5′ labeled with Cy3, Cy5, and biotin, respectively. If necessary, uracil DNA glycosylase (UDG) can be added to the PCR master mix to help prevent PCR product contamination.

PCR Product Preparation (Step 5). Double-stranded PCR products are immobilized onto paramagnetic particles by adding 20 μl of paramagnetic particle b reagent (MPB; Illumina) to each 60-μl PCR reaction and then incubated at room temperature for a minimum of 60 min. Bound PCR products are washed with universal buffer 2 (UB2; Illumina) and denatured by adding 30 μl 0.1 N NaOH. After 1 min at room temperature, the released single-stranded (ss) DNAs are neutralized with 30 μl of hybridization reagent (MH1; Illumina) and hybridized to the universal array.

Array Hybridization (Step 6), Wash (Step 7), and Imaging (Step 8). The single-stranded, dye-labeled DNA products are hybridized to their complement bead type through their unique IllumiCode sequences (Step 6). The SAM or BeadChip is then washed (Step 7) and imaged (Step 8). Specifically, arrays are hydrated in UB2 for 3 min at room temperature and then preconditioned in 0.1 N NaOH for 30 s. Arrays are returned to the UB2 reagent for at least 1 min to neutralize the NaOH. The pretreated arrays are exposed to the labeled ssDNA samples described earlier. Hybridization is conducted under a temperature gradient program from 60 to 45° over ~12 h. The hybridization is held at 45° until the array is processed (Step 6). After hybridization, the arrays are first rinsed twice in UB2 and once with IS1 (IS1; Illumina) at room temperature with mild agitation (Step 7), dried for 20 min, and then imaged at a resolution of 0.8 μm using a BeadArray reader (Illumina; Barker *et al.*, 2003) (Step 8). PMT settings are optimized for dynamic range, channel balance, and signal-to-noise ratio. Cy3 and Cy5 dyes are excited by lasers emitting at 550 and 630 nm, respectively. A scan of the 96 arrays in a SAM takes approximately 90 min.

Automatic Genotype Scoring (Step 9). Based on the intensities detected from the two channels for the two respective alleles of each SNP, genotypes are called automatically using the proprietary genotyping software of Illumina (GenCall). GenCall software considers multiple factors; one of the first is the distribution of beads of the same type. Outliers are rejected to ensure genotyping accuracy.

A "quality" score, the GenCall score, is calculated for each genotype call, reflecting the degree of separation between homozygote and heterozygote clusters for that SNP and the placement of the individual call within a cluster. It ranges from 0 to 1 and has been shown to correlate with the overall accuracy of the genotyping call (Fan *et al.*, 2003). Figure 4 shows four examples of GenCall software-produced data plots. These are genotype clusters generated from 90 individual DNA samples from a set of families affected by bipolar disorder, assayed with an oligonucleotide pool corresponding to ~1400 SNPs. Separation between homozygote and heterozygote clusters is excellent and variation within each cluster is small.

Results and Discussion

Assay Flexibility

One key aspect of the GoldenGate genotyping assay design is incorporation of a universal IllumiCode sequence, allowing the assay products to be read out on a universal array (Chen *et al.*, 2000; Fan *et al.*, 2000; Gerry *et al.*, 1999; Iannone *et al.*, 2000). This approach offers substantial flexibility for assay content development. Any custom set of assays can be made on demand simply by building the complement of the IllumiCode into the SNP-specific assay oligonucleotides (see Fig. 3, step 2). In addition, use of a universal array greatly simplifies the manufacturing process and reduces the assay development time (because there is no need to create custom arrays) and overall genotyping costs.

Genotyping Throughput

The ability to use either the Sentrix BeadChip or the SAM (Fig. 2) allows the user to perform as few as 6000 genotype calls (384 loci multiplex on a 16-sample BeadChip) to as many as 300,000 genotype calls (1536 loci multiplex on two 96-sample array matrices) in a single day. Automating the GoldenGate assay with a few liquid handling robots and a laboratory information management system allows throughput of over 1.6 million genotype calls per day.

An international consortium of leading investigators (including Illumina scientists) was formed to define a haplotype map for the human genome at a 5-kb resolution. The International Haplotype Map (HapMap) project aims to genotype more than 1 million SNPs in 288 individuals (Caucasians, Africans, and Asians) and to identify "tag" SNPs that represent the common haplotypes in the general populations (The International HapMap Consortium, 2003). The GoldenGate genotyping platform of Illumina is used by the consortium to genotype ~65% of the SNP loci in this project. This effort, along with others (Hinds *et al.*, 2005), provides the tools and

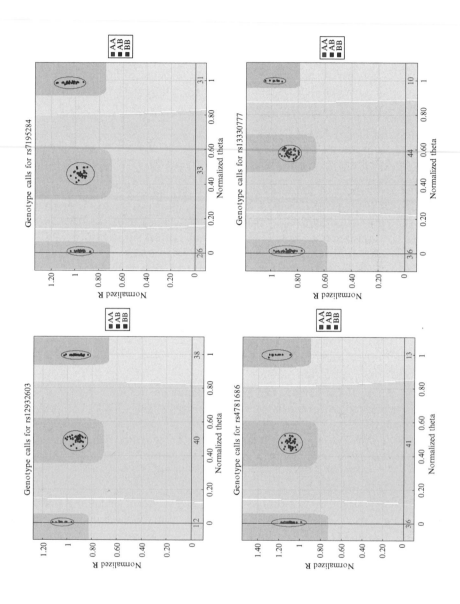

resources required for understanding the causal role of common human DNA variations in complex human traits.

Genotyping Performance

To date, over 500 million high-quality genotypes have been generated using the GoldenGate assay. About 1 million SNP loci have been designed and 88% of the attempted assays were successful. The assay development success rate is highly dependent on the SNPs selected. Of the DNAs processed, 97.8% were genotyped successfully and high-quality data were obtained. A primary reason for the failure to generate genotype calls successfully was insufficient DNA quantity. The genotyping result can be compromised significantly when less than 20% of the required input DNA (250 ng) is used. In general, the GoldenGate genotyping assay delivers a high call rate (99.8%) and call accuracy (>99.9%).

In addition, the GoldenGate assay can tolerate a certain degree of DNA degradation. We have obtained highly accurate genotyping results with whole genome-amplified DNA samples (Barker *et al.*, 2004) and DNA samples isolated from formalin-fixed, paraffin-embedded (FFPE) tissue samples (Fan *et al.*, unpublished data).

Implementation of the Methylation Profiling Assay on the SNP Genotyping Platform

We adapted the GoldenGate SNP genotyping system for high-throughput DNA methylation detection based on genotyping of bisulfite-converted genomic DNA. In this assay, nonmethylated cytosines (C) are converted to uracil (U) when treated with bisulfite, while methylated cytosines remain unchanged. The hybridization behavior of uracil is similar to that of thymine (T). Detection of the methylation status of a particular cytosine is carried out using a genotyping assay for a C/T polymorphism. The methylation status of the interrogated locus can be determined by

FIG. 4. GenCall software-produced plots of four randomly selected loci of 90 samples in polar coordinate representation. Each image is a graph of a single locus (assayed simultaneously with ~1400 other loci) with 90 "dots" representing individual DNA samples. The y axis is normalized intensity (sum of intensities of the two channels), and the x axis is the "theta" value ($\frac{2}{\pi} Tan^{-1}$ ($Cy5/Cy3$)). Theta values near 0 (left side of graph) are homozygotes for allele "A," and theta values near 1 (right side of graph) are homozygotes for allele "B." The GenCall software automatically grouped the 90 DNAs for each locus into two homozygote clusters (left and right) and the heterozygote cluster (middle). (See color insert.)

calculation of the ratio of the fluorescent signals from the "C" (methylated) and "T" (unmethylated) alleles.

We have characterized this assay system extensively. We established a standard process for probe design and data analysis, a standard bisulfite conversion protocol, and a set of internal controls and reference samples for assay development and calibration. We demonstrated that the assay is sufficient to detect changes in methylation status at 1536 different CpG sites simultaneously using 200 ng of human genomic DNA; it can detect as little as 2.5% of methylated DNA target in a complex sample and can distinguish a 17% methylation difference between samples with 95% confidence (Bibikova et al., unpublished data). We also observed high concordance between results obtained by our microarray-based methylation analysis and those by traditional methods such as methylation-specific PCR and bisulfite sequencing. This technology should prove useful for comprehensive methylation analyses in large populations.

High-Throughput Gene Expression Profiling on Universal Array Matrices

We have developed a flexible, sensitive, accurate, and cost-effective gene expression profiling assay, the DASL assay (cDNA-mediated annealing, selection, extension, and ligation) based on the same universal Sentrix arrays used for SNP genotyping (Fan et al., 2004). In this assay, RNA is first converted to cDNA by random priming; query oligonucleotides are designed for specific mRNA sequences in a way similar to the SNP assay design (Fig. 3, step 2) and bind to the corresponding cDNA and become extended and ligated enzymatically. The ligated products are then amplified and labeled fluorescently during PCR and are detected by hybridization to the universal array. The hybridization intensity is proportional to the original mRNA abundance in the sample and therefore represents the expression level of the targeted transcripts. The DASL assay protocol is identical to the GoldenGate assay described earlier (Fig. 3), except for the cDNA synthesis step. Specifically, a 20-μl reverse transcription reaction containing a reaction mix (MMC; Illumina) and total RNA (up to 1 μg) is incubated at room temperature for 10 min and then at 42° for 1 h.

The DASL assay multiplexes to over 1500 sequence targets, e.g., 500 genes at three probes per gene or 1500 genes at one probe per gene. It can measure gene expression level accurately across a dynamic range of 2.5–3 logs and resolve a less than twofold difference in expression between samples using 100 ng total RNA (Bibikova et al., 2004a; Fan et al., 2004). It fills the gap between two existing RNA profiling technologies: quantitative RT-PCR (high sample throughput/low gene content) and high-density oligonucleotide or cDNA microarray technology (high gene content/low sample throughput), with medium to high sample/gene throughput and high sensitivity.

Using the DASL assay, highly reproducible tissue- and cancer-specific gene expression profiles have been obtained with partially degraded RNAs isolated from formalin-fixed, paraffin-embedded tissues that had been stored from 1 to over 10 years (Bibikova *et al.*, 2004a,b) (Fig. 5). The method can tolerate RNA degradation in these fixed tissues because it uses a random priming procedure for cDNA synthesis. In general, RNA is extracted from four to five 5-μm tissue sections using the High Pure RNA Paraffin kit (Roche), yielding 0.3–15 μg total RNA. For each FFPE sample, 200 ng of total RNA is converted to cDNA and processed using the DASL assay.

Due to its high sequence specificity, the DASL assay can be used for splice variant detection (Yeakley *et al.*, 2002), as well as allele-specific expression profiling (i.e., measuring the relative abundance of allelic

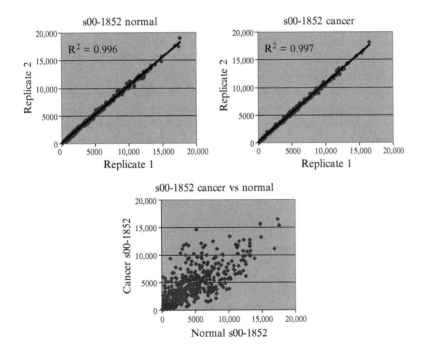

Fig. 5. Gene expression profiling in FFPE samples using the DASL assay. Highly reproducible gene expression profiles are obtained with RNAs isolated from normal and tumor FFPE tissue of the same patient (top panels). Differential expression (bottom panel) can be identified readily between normal and cancer samples.

transcripts) (Fan *et al.*, 2003). Differential allelic expression is relatively common, which occurs not only in imprinted genes, but also in nonimprinted autosomal genes (Knight, 2004; Pastinen and Hudson, 2004; Pastinen *et al.*, 2004; Yan *et al.*, 2002). We have developed genotyping assays for over 2000 SNPs derived from the coding region or 5'- and 3'-UTRs of more than 700 cancer-related genes (Fan *et al.*, unpublished data). Assay probes targeting specific SNPs are designed the same way as the standard GoldenGate SNP design, except that each assay target is chosen from the sense strand of mRNA and within one exon. Thus, they can be used to interrogate both genomic DNA (for genomic heterozygosity) and RNA (for relative abundance of allelic transcripts at the heterozygous loci). Imbalanced allelic expression is detected when the allele ratio in RNA (cDNA) differs from the corresponding ratio in genomic DNA (which is expected to be 1:1).

Conclusion

The GoldenGate genotyping platform combines a highly multiplexed genotyping assay with highly flexible Sentrix arrays to deliver high data quality and throughput. One of the most important features of the GoldenGate genotyping assay is that it genotypes directly on the genomic DNA and does not require prior PCR amplification of the genotyping target. It has been used in many genetic studies, including the International HapMap Project (The International HapMap Consortium, 2003), and has generated a large amount of genotyping data with an exceptional call accuracy and call rate. The scalability of the system allows the user to perform either small pilot studies or large-scale SNP genotyping association studies under the same system.

The DASL assay is particularly useful for profiling the expression of hundreds or thousands of genes in degraded RNAs, including RNAs from formalin-fixed, paraffin-embedded tissues. Formalin-fixed archival tissues represent an invaluable resource for genetic analysis and are the most widely available materials for which patient outcomes are already known. The ability to perform genetic analysis in these samples will enable both prospective and retrospective studies, facilitating research in correlating genetic profiles with clinical parameters.

In summary, BeadArray technology offers a complete solution to high-throughput genomic analysis, including SNP genotyping, LOH and genome amplification detection, DNA methylation profiling, and gene expression profiling, including allele-specific expression profiling and mRNA isoform profiling. It can be used to generate comprehensive genetic, epigenetic, and gene expression profiles for any biological system.

Acknowledgments

We thank John Stuelpnagel, Nicky Espinosa, and Melanie Smith for critical reading of this chapter and constructive suggestions. We are indebted to the many other scientists and engineers at Illumina whose dedicated work has created the platforms and assays reviewed here.

References

Barker, D. L., Hansen, M. S., Faruqi, A. F., Giannola, D., Irsula, O. R., Lasken, R. S., Latterich, M., Makarov, V., Oliphant, A., Pinter, J. H., Shen, R., Sleptsova, I., Ziehler, W., and Lai, E. (2004). Two methods of whole-genome amplification enable accurate genotyping across a 2320-SNP linkage panel. *Genome Res.* **14**, 901–907.

Barker, D. L., Theriault, D., Che, D., Dickinson, T., Shen, R., and Kain, R. (2003). Self-assembled random arrays: High-performance imaging and genomics applications on a high-density microarray platform. *Proc. SPIE* **4966**, 1–11.

Bibikova, M., Talantov, D., Chudin, E., Yeakley, J. M., Chen, J., Doucet, D., Wickham, E., Atkins, D., Barker, D., Chee, M., Wang, Y., and Fan, J. B. (2004a). Quantitative gene expression profiling in formalin-fixed, paraffin-embedded tissues using universal bead arrays. *Am. J. Pathol.* **165**, 1799–1807.

Bibikova, M., Yeakley, J. M., Chudin, E., Chen, J., Wickham, E., Wang-Rodriguez, J., and Fan, J. B. (2004b). Gene expression profiles in formalin-fixed, paraffin-embedded tissues obtained with a novel assay for microarray analysis. *Clin. Chem.* **50**, 2384–2386.

Botstein, D., and Risch, N. (2003). Discovering genotypes underlying human phenotypes: Past successes for mendelian disease, future approaches for complex disease. *Nat. Genet.* **33** (Suppl.), 228–237.

Braeckmans, K., De Smedt, S. C., Leblans, M., Pauwels, R., and Demeester, J. (2002). Encoding microcarriers: Present and future technologies. *Nat. Rev. Drug Discov.* **1**, 447–456.

Chan, W. C., Maxwell, D. J., Gao, X., Bailey, R. E., Han, M., and Nie, S. (2002). Luminescent quantum dots for multiplexed biological detection and imaging. *Curr. Opin. Biotechnol.* **13**, 40–46.

Chen, J., Iannone, M. A., Li, M. S., Taylor, J. D., Rivers, P., Nelsen, A. J., Slentz-Kesler, K. A., Roses, A., and Weiner, M. P. (2000). A microsphere-based assay for multiplexed single nucleotide polymorphism analysis using single base chain extension. *Genome Res.* **10**, 549–557.

Fan, J. B., Chen, X., Halushka, M. K., Berno, A., Huang, X., Ryder, T., Lipshutz, R. J., Lockhart, D. J., and Chakravarti, A. (2000). Parallel genotyping of human SNPs using generic high-density oligonucleotide tag arrays. *Genome Res.* **10**, 853–860.

Fan, J. B., Oliphant, A., Shen, R., Kermani, B. G., Garcia, F., Gunderson, K. L., Hansen, M., Steemers, F., Butler, S. L., Deloukas, P., Galver, L., Hunt, S., McBride, C., Bibikova, M., Rubano, T., Chen, J., Wickham, E., Doucet, D., Chang, W., Campbell, D., Zhang, B., Kruglyak, S., Bentley, D., Haas, J., Rigault, P., Zhou, L., Stuelpnagel, J., and Chee, M. S. (2003). Highly parallel SNP genotyping. *Cold Spring Harb. Symp. Quant. Biol.* **68**, 69–78.

Fan, J. B., Yeakley, J. M., Bibikova, M., Chudin, E., Wickham, E., Chen, J., Doucet, D., Rigault, P., Zhang, B., Shen, R., McBride, C., Li, H. R., Fu, X. D., Oliphant, A., Barker, D. L., and Chee, M. S. (2004). A versatile assay for high-throughput gene expression profiling on universal array matrices. *Genome Res.* **14**, 878–885.

Fodor, S. P., Read, J. L., Pirrung, M. C., Stryer, L., Lu, A. T., and Solas, D. (1991). Light-directed, spatially addressable parallel chemical synthesis. *Science* **251**, 767–773.

Fulton, R. J., McDade, R. L., Smith, P. L., Kienker, L. J., and Kettman, J. R., Jr. (1997). Advanced multiplexed analysis with the FlowMetrix system. *Clin. Chem.* **43**, 1749–1756.

Gerry, N. P., Witowski, N. E., Day, J., Hammer, R. P., Barany, G., and Barany, F. (1999). Universal DNA microarray method for multiplex detection of low abundance point mutations. *J. Mol. Biol.* **292**, 251–262.

Gunderson, K. L., Kruglyak, S., Graige, M. S., Garcia, F., Kermani, B. G., Zhao, C., Che, D., Dickinson, T., Wickham, E., Bierle, J., Doucet, D., Milewski, M., Yang, R., Siegmund, C., Haas, J., Zhou, L., Oliphant, A., Fan, J. B., Barnard, S., and Chee, M. S. (2004). Decoding randomly ordered DNA arrays. *Genome Res.* **14**, 870–877.

Gunderson, K. L., Steemers, F. J., Lee, G., Mendoza, L. G., and Chee, M. S. (2005). A genome-wide scalable SNP genotyping assay using microarray technology. *Nat. Genet.* **37**, 549–554.

Han, M., Gao, X., Su, J. Z., and Nie, S. (2001). Quantum-dot-tagged microbeads for multiplexed optical coding of biomolecules. *Nat. Biotechnol.* **19**, 631–635.

Hinds, D. A., Stuve, L. L., Nilsen, G. B., Halperin, E., Eskin, E., Ballinger, D. G., Frazer, K. A., and Cox, D. R. (2005). Whole-genome patterns of common DNA variation in three human populations. *Science* **307**, 1072–1079.

Holloway, A. J., van Laar, R. K., Tothill, R. W., and Bowtell, D. D. (2002). Options available—from start to finish—for obtaining data from DNA microarrays II. *Nat. Genet.* **32**(Suppl.), 481–489.

Iannone, M. A., Taylor, J. D., Chen, J., Li, M. S., Rivers, P., SlentzKesler, K. A., and Weiner, M. P. (2000). Multiplexed single nucleotide polymorphism genotyping by oligonucleotide ligation and flow cytometry. *Cytometry* **39**, 131–140.

Knight, J. C. (2004). Allele-specific gene expression uncovered. *Trends Genet.* **20**, 113–116.

Kuhn, K., Baker, S. C., Chudin, E., Lieu, M. H., Oeser, S., Bennett, H., Rigault, P., Barker, D., McDaniel, T. K., and Chee, M. S. (2004). A novel, high-performance random array platform for quantitative gene expression profiling. *Genome Res.* **14**, 2347–2356.

Kwok, P. Y., and Chen, X. (2003). Detection of single nucleotide polymorphisms. *Curr. Issues Mol. Biol.* **5**, 43–60.

Lockhart, D. J., and Trulson, M. O. (2001). Multiplex metallica. *Nat. Biotechnol.* **19**, 1122–1123.

Michael, K. L., Taylor, L. C., Schultz, S. L., and Walt, D. R. (1998). Randomly ordered addressable high-density optical sensor arrays. *Anal. Chem.* **70**, 1242–1248.

Nicewarner-Pena, S. R., Freeman, R. G., Reiss, B. D., He, L., Pena, D. J., Walton, I. D., Cromer, R., Keating, C. D., and Natan, M. J. (2001). Submicrometer metallic barcodes. *Science* **294**, 137–141.

Oliphant, A., Barker, D. L., Stuelpnagel, J. R., and Chee, M. S. (2002). BeadArray technology: Enabling an accurate, cost-effective approach to high-throughput genotyping. *Biotechniques* (Suppl.), 56–58, 60–61.

Pastinen, T., and Hudson, T. J. (2004). Cis-acting regulatory variation in the human genome. *Science* **306**, 647–650.

Pastinen, T., Sladek, R., Gurd, S., Sammak, A., Ge, B., Lepage, P., Lavergne, K., Villeneuve, A., Gaudin, T., Brandstrom, H., Beck, A., Verner, A., Kingsley, J., Harmsen, E., Labuda, D., Morgan, K., Vohl, M. C., Naumova, A. K., Sinnett, D., and Hudson, T. J. (2004). A survey of genetic and epigenetic variation affecting human gene expression. *Physiol. Genom.* **16**, 184–193.

Schena, M., Shalon, D., Davis, R. W., and Brown, P. O. (1995). Quantitative monitoring of gene expression patterns with a complementary DNA microarray. *Science* **270**, 467–470.

Shen, R., Fan, J. B., Campbell, D., Chang, W., Chen, J., Doucet, D., Yeakley, J., Bibikova, M., Wickham Garcia, E., McBride, C., Steemers, F., Garcia, F., Kermani, B. G., Gunderson, K., and Oliphant, A. (2005). High-throughput SNP genotyping on universal bead arrays. *Mutat. Res.* **573**, 70–82.

Syvanen, A. C. (2005). Toward genome-wide SNP genotyping. *Nat. Genet.* **37**(Suppl.), 5–10.

The International HapMap Consortium (2003). The International HapMap Project. *Nature* **426**, 789–796.

Yan, H., Yuan, W., Velculescu, V. E., Vogelstein, B., and Kinzler, K. W. (2002). Allelic variation in human gene expression. *Science* **297**, 1143.

Yeakley, J. M., Fan, J. B., Doucet, D., Luo, L., Wickham, E., Ye, Z., Chee, M. S., and Fu, X. D. (2002). Profiling alternative splicing on fiber-optic arrays. *Nat. Biotechnol.* **20**, 353–358.

[4] Microarray Oligonucleotide Probes

By DAVID P. KREIL, ROSLIN R. RUSSELL, and STEVEN RUSSELL

Abstract

Oligonucleotide probes are increasingly the method of choice for many modern DNA microarray applications. They provide higher target specificity, probe selection gives improved experimental control of hybridization properties, and targeting of specific gene subsequences allows better discrimination of highly similar targets such as splice variants or gene families. Only recently has there been substantial progress in dealing with the complexities of probe set design and probe-specific signal interpretation. After a discussion of advantages and disadvantages of oligonucleotide probes in comparison to amplicons, this chapter focuses on recent advances and remaining key challenges in probe design and computational data analysis for spotted and *in situ*-synthesized oligonucleotide microarray technologies. Both experimental questions and computational aspects are addressed. Experimental issues discussed include the choice of an optimal number of probes per target and probe lengths and their influence on bias and random measurement noise, effects of different probe or substrate modifications, and laboratory protocols on signal specificity and sensitivity. Computational topics include practical considerations and a case study in probe sequence design, the exploitation of probing multiple target regions, and the modeling of probe sequence-specific signals. The current state of the art of the field is examined, and principled thermodynamic probe design criteria are proposed that are based on the free energy of the probe–target complex at the hybridization temperature rather than its melting temperature. Finally, this chapter notes and discusses an emerging trend in recent computational work toward a focus on signal interpretation rather than probe sequence design.

METHODS IN ENZYMOLOGY, VOL. 410   0076-6879/06 $35.00
DOI: 10.1016/S0076-6879(06)10004-X

DNA microarray technology is pervading many aspects of the life sciences. From humble beginnings, detecting the expression of a few tens of genes, entire eukaryotic genomes can now be interrogated (Bertone *et al.*, 2004; Schena *et al.*, 1995). While the technology can be used for a variety of applications (Hanlon and Lieb, 2004; MacAlpine and Bell, 2005; Pinkel and Albertson, 2005), its main use is still in gene transcript expression profiling. Often, microarrays are used to screen for genes involved in a particular biological process of interest; however, larger data sets of comprehensive transcript coverage measured under a variety of conditions have considerable potential for much wider, systems-level analysis, for example, via the detection of coregulated groups of genes (Ihmels *et al.*, 2002, 2004; Lee and Batzoglou, 2003; Saidi *et al.*, 2004). Sensitive pattern detection tools require particularly accurate data so that biologically meaningful signatures can be distinguished from confounding experimental effects, which can only partly be removed at the analysis stage (Kreil and Russell, 2005). At present, unfortunately, hybridization signal levels measured are not easily related to absolute quantities of target transcripts. This chapter outlines the advantages and challenges of using oligonucleotide probes for transcript expression profiling, discusses typical considerations and practical aspects in probe sequence design, and high-lights recent developments in the modeling of hybridization behavior that are of relevance for probe design and the interpretation of hybridiza-tion signals. While recognizing that there are many sources of bias and noise in microarray data, developing an understanding of probe hybridiza-tion behavior will be instrumental in achieving a quantitative view of the transcriptome.

The Case for Oligonucleotide Probes

There are two types of common DNA microarray probes: oligonucleo-tides and double-stranded amplicons (Johnston *et al.*, 2004; Schena *et al.*, 1996). Amplicon probes have particularly high sensitivity and, for some applications, their relatively large tolerance to small transcript sequence variations can be helpful, for example, transparently tolerating naturally occurring polymorphisms. This same property, however, makes amplicon probes less well suited for the discrimination of very similar targets, such as alternative splicing variants, or families of paralogous genes. With all amplicon-based probes, moreover, the technical problems associated with polymerase chain reaction amplification of thousands of clones are not easily overcome (Hegde *et al.*, 2000, and Burr *et al.*, this volume). Conse-quently, some laboratories reported that only 66–79% of probes were not contaminated and matched their respective targets (Hager, this volume).

Nevertheless, probing species without a fully sequenced genome, comparing highly related strains, or exploiting specialized cDNA libraries indicate the use of amplicon arrays (Diatchenko et al., 1996; Suchyta et al., 2003). With the increased experimental control available with oligonucleotide probes and because of the challenges of manufacturing amplicon probes of uniform and validated quality, many modern microarray applications use synthesized oligonucleotide probes. Either multiple shorter probes per target are employed, as with Affymetrix chips (Lockhart et al., 1996), or longer oligonucleotide probes are used, typically 35- to 70-mers (Hughes et al., 2001; Kane et al., 2000; Nuwaysir et al., 2002). Oligonucleotide probes overcome many of the difficulties of amplicons and show increased target sequence discrimination (Duggan et al., 1999; Relogio et al., 2002). Moreover, one can ensure uniform probe concentrations, hybridization affinities, and minimal cross-hybridization. Consequently, very clean arrays can be achieved.

Considerations for Oligonucleotide Probe Design

Many issues affect probe design, no matter whether arrays are to be produced commercially or in-house. The number of probes per target and their lengths must be chosen first, and the following section discusses tradeoffs that must be made due to technical limitations of production platforms. Additional complexities, deferred to in later sections, include the discrimination of multiple splice variants of a transcript, issues of target secondary structure, and the detection of RNA degradation. Even without considering these, thorough studies regarding the optimal choice of probe length or the number of probes per target are difficult, and few systematic comparisons exist.

Number of Probes per Target and Probe Lengths

There are two properties of the microarray measurement process that one wishes to *maximize* for transcript expression profiling: (1) the *sensitivity*, a measure of how little is lost of the signal reflecting specific hybridization between the probe and its target; and (2) the *specificity*, a measure of how little nonspecific hybridization there is of the probe with molecules other than its target. At the same time, one aims to *minimize* two other properties of the measurement process: (1) the random signal variation or *noise*, often expressed as a coefficient of variation (CV), the standard deviation divided by the mean of multiple measurements; and (2) the *bias*, the systematic deviation of the measurement from the true signal due to probe-specific or other confounding technical effects.

Sensitivity generally increases with probe length, as the binding energy for longer probe–target hybrid complexes is typically higher; e.g., 60-mers detect targets with eightfold higher sensitivity than 25-mers (Chou *et al.*, 2004). The specificity of very short probes decreases with diminishing probe length because of the increasing chance of a random match to nontarget sequences. However, the specificity of very long probes decreases with growing probe length because the chance that a fragment of the probe matches an unwanted target increases with probe length. The fact that biological nucleotide sequences are not random further contributes to this because different targets can share domains of high sequence similarity.

Noise can be reduced by increased binding energy via greater probe length and by making multiple measurements per target—be that through replicate or multiple probes. A bias of individual probes should ideally be measured or modeled and removed. The bias of a set of probes for a given target can also be reduced through the combination of multiple probes of random bias, where the average bias decreases with the number of probes.

Although experimental comparisons are difficult because optimal hybridization protocols differ for probes of different lengths and binding affinities, qualitative trends observed for 25- to 500-mers (Fig. 1) suggested that, depending on how one weighs optimization criteria, probe lengths between 50 and 150 may yield a good compromise (Fig. 2). For the shorter probes, replicates from different arrays or multiple probes should be used to reduce noise, and probe bias compensated, for example, by the computational means discussed later.

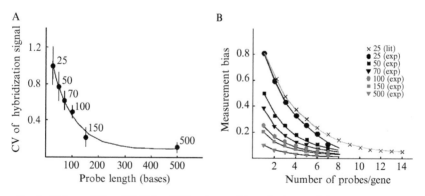

Fig. 1. Noise and bias. (A) Noise (CV) reduces with probe length. (B) Bias decreases with increased probe lengths (see symbols in legend) and with larger numbers of probes per gene. Bias was assessed as the average deviation between the robust means of the signals of random probe subsets and the robust mean of the signals of the full probe set (Chou *et al.*, 2004).

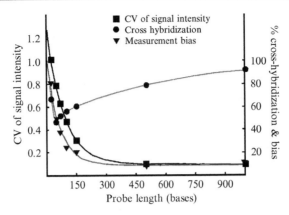

FIG. 2. Optimization by multiple criteria: three of the four criteria discussed are shown. The fourth, sensitivity, increases monotonically with probe length over the studied range (Chou *et al.*, 2004).

A very different type of "specificity" measure, indicative of detection performance for single nucleotide polymorphisms (SNPs), is obtained in studies comparing hybridization signals from "perfect match" probes and "mismatch" probes featuring a single base mismatch (Relogio *et al.*, 2002). In transcript expression profiling, however, sensitivity to SNPs is detrimental because of natural sequence polymorphisms in samples.

Microarray Production and Hybridization Protocols

The hybridization behavior of probes, measurement sensitivity, and specificity strongly depend on microarray production and hybridization protocols. The steric hindrance of the solid support to which probes are attached, for example, reduces hybridization more than twofold and, together with electrostatic effects, may render the terminal bases close to the support effectively invisible (Shchepinov *et al.*, 1997). Terminal uncharged amphiphilic spacer groups with 30–60 carbon–carbon bonds provide a remedy. A length of 40 seems optimal, as longer spacers reduce the effective concentration of the probe by allowing diffusion. Thus, creating sufficient space between substrate and probes to mitigate sterical and electrostatic effects on hybridization, yet also limiting probe diffusion, such optimal spacer groups could achieve a 150-fold increase in hybridization sensitivity while making the entire probe available for specific hybridization (Shchepinov *et al.*, 1997). Slide substrate coatings with a "gel-like" spacing effect provide an alternative, for example, GE Healthcare Code-Link or FullMoonBioscience PowerMatrix substrates (Le Berre *et al.*, 2003; Ramakrishnan *et al.*, 2002).

Relevant hybridization protocol parameters include hybridization temperature and duration, hybridization buffer composition and additives (such as formamide), and the stringency of washes. In the manufacture of spotted arrays, other protocol parameters of relevance include spotting buffer and microarray slide substrate chemistries, probe concentrations, and ambient temperature and humidity (Auburn *et al.*, 2005; Kreil *et al.*, 2003). The protocols employed should be validated for sensitivity and specificity, for example, by spike-in experiments, which are also easily assessed visually, an efficiency factor in protocol parameter screens. Typical results reflect the confounding effects of experimental conditions on hybridization behavior: False-color hybridization images of a spotted probe (see Supplement) showed clear detection of a difference in transcript expression, yet a change in just the microarray manufacturing chemistry made the same probe hybridize nonspecifically in both channels. While not suitable for immediate visual assessment, an optimization of the detection sensitivity for differential expression provides an alternative quantitative assay less dependent on a set of spike probes being representative of an array. Multiple replicate hybridizations and the use of dye swaps for multichannel systems ensure that differential signals cannot occur by chance and that a maximization of the number of difference calls corresponds to optimal hybridization conditions, once not-detected spots are accounted for correctly (Kreil *et al.*, in preparation).

In order for the measurement process to be both sensitive and specific for, ideally, *all* the probes on the array, one requires probes of uniform hybridization properties *at the reaction conditions employed*. This is the objective of probe sequence design.

Practical Considerations in Probe Sequence Design, a Case Study

Probe sequence design is complex for a variety of reasons, ranging from trivial technical nuisances to difficult theoretical problems that are the subject of active research. The prediction of actively transcribed genome regions, for instance, is still far from exhaustive, and hence the mixture of target transcripts that need to be discriminated in a sample is not fully known. Already a prediction of the properties of a candidate probe under competitive hybridization with a known mixture of different transcripts is a difficult thermodynamic modeling challenge. As a consequence, more or less crude approximations and heuristics are often employed, with modeling replaced by sequence similarity searches and alignments, and ad hoc rules of thumb regarding probe sequence complexity and secondary structure. These simplified approaches, however, are typically not sufficiently reliable on their own and usually require experimental validation of candidate probes in a final probe selection step. Approaches exploiting the different hybridization kinetics of specific and nonspecific binding to detect nonspecific probes are only applicable to probes for sufficiently highly expressed targets (Dai *et al.*, 2002). Comprehensive experimental validation is hence rarely performed for reasons of complexity and cost.

While a plethora of probe design software is available (cf. Supplement), it is clear that designs that are less crude approximations to proper thermodynamic modeling are better predictors of oligonucleotide hybridization performance (Luebke *et al.*, 2003). The common limitations of readily available software can roughly be classified into the following categories.

1. Limited validity of heuristics and approximations in lieu of proper thermodynamic modeling; for example, sequence similarity or "consecutive matches" to probe complements as indicative of cross-hybridization; or sequence palindromes as indicative of hairpin secondary structure instead of proper thermodynamic hybridization and folding models.
2. Replacement of parameter optimization by (a) acceptance of any solution with parameters meeting specified thresholds or (b) requiring users to fix the parameter for design runs; for example, accepting any probe with binding energy in a given range; or requiring users to specify both probe length and energy thresholds rather than inferring one from the other.
3. Lack of support for dealing with similarities in biological sequences and the consequential difficulty of designing specific probes for groups of similar sequences, for example, alternative splice forms or paralogous genes.
4. Technical issues such as malfunctions or undocumented software requirements.

This section does not attempt a comprehensive discussion but instead highlights *typical* consequences of problems encountered using the design of a genome-scale transcript array for *Drosophila melanogaster* (Kreil *et al.*, in preparation) as a case study, covering:

- Consequences of the choice of probe design software
- Construction of a set of target transcript sequences
- Choosing design parameters
- Searching for "optimal" probes
- Postprocessing, for example, to account for underprediction of transcripts from the genomic sequence

Although critical at all phases of employment, technical issues are collected in the Supplement because of their often-transient nature.

Choice of Software and Construction of Target Transcript Set

OligoArray 2.1 ('OA2', http://berry.engin.umich.edu/oligoarray2_1/) uses relatively few heuristic shortcuts (Rouillard *et al.*, 2003): cross-hybridization and self-folding are assessed using full two-state thermodynamic models employing the mfold algorithm, allowing for mismatches, bulges, loops, and hairpins (SantaLucia, 1998; Zuker, 2003). Otherwise, OA2 is comparable to many other probe design tools in using a BLAST sequence similarity search (Altschul *et al.*, 1997) together with heuristics to screen nontarget transcripts for potential cross-hybridization. Access to the software source code (unpublished) allowed for verification of the implementation and was most valuable in dealing with technical issues as they emerged. Sources of a revised version are published elsewhere (J.-M. Rouillard, personal communication, 2006).

As OA2 has no concept of "related" sequences and treats all predicted stable hybridizations to nontarget transcripts equally, duplicate and very similar sequences had to be removed in the construction of a "nonredundant" set of target transcript sequences using tools such as nrdb90 or CD-HI (Holm and Sander, 1998; Li *et al.*, 2001). For compatibility to common labeling methods, design was restricted to the 1500 base 3' regions of targets and sense probes had to be built for the labeled (antisense) targets derived by reverse transcription from the (sense) mRNAs in samples (Marko *et al.*, 2005).

Choice of Design Parameters, the Search for "Optimal" Probes

OA2 execution parameters provide thresholds for the acceptance of probe candidates. The probe candidate closest to the 3'-terminal of the target sequence that passes all criteria is selected: probe length and probe–target melting temperature T_m within given ranges, no stable probe secondary structure (self-folding), GC content in range (which we did not restrict), no tandem repeats, and a minimal number of predicted stably hybridizing nontarget transcripts. Accepted probe lengths were set to 65–69, as pilot experiments had demonstrated a good compromise between sensitivity and specificity with the protocols employed in our laboratory.

OA2 default parameters for 45- to 47-mers permit $85° = T_m = 90°$, tolerate stable cross-hybridization only for $T_x < 65°$, and stable probe secondary structure for $T_s < 65°$. Examining the T_m values of all candidate probes in the 1500 base 3' regions of target sequences yielded a set of T_m

values per sequence. Some target sequences had extreme probe candidate T_m distributions, with $\min[Q3(T_m)] = 76.3$ and $\max[Q1(T_m)] = 97$; with Q1/3 denoting the first and third quartiles, respectively. However, most targets had melting temperatures in a common range, with (Q1, median, Q3)[median(T_m)] = (87.0,89.0,90.5). This was well matched to the suggested tolerated T_m interval of 85–90°: More than 90% of target sequences were covered with at least 25% of candidate probes per target having a T_m in this interval. For our target set, the optimal 5° range maximizing coverage for 45- to 47-mers was 86.6–91.6°.

In contrast, for 65- to 69-mers, the extremes were $\min[Q3(T_m)] = 81.5$ and $\max[Q1(T_m)] = 100.6$, while (Q1, median,Q3)[median(T_m)] = (91.7,93.3,94.7). Less than half the target sequences, however, were covered with at least 25% of candidate probes having a T_m in the default interval 85–90°, severely reducing the number of probe candidates that could be considered. Shifting the 5° window to 90.6–95.6° (a 5.6° offset to OA2 defaults), however, could achieve coverage of 94% of all target sequences with at least 25% of candidate probes in range (Fig. 3). Thus, for most target sequences, a large number of probe candidates could be considered, increasing the likelihood that a specific probe with no cross-hybridization could be found. For a small number of target sequences (6%), however, probe design meeting these parameter thresholds was difficult.

The T_x and T_s thresholds were conservatively adjusted by 4° from 65 to 69°, leaving a margin of 1.6° for the effect of T_m overestimation at large temperatures (Rouillard et al., 2003), also matching the observed shift of the "optimal" T_m window. Targets with no satisfactory probes were rerun with increasingly relaxed parameters.

Employment and Postprocessing

The design runs for the parameter sets considered were executed on a distributed collection of computers using Grid Engine (http://gridengine. sunsource.net/). Accounting for possible underprediction of transcripts from the genomic sequence, all OA2-selected probes were screened to exclude predicted stable hybridizations to any genomic DNA sequence (BLAST search of both genomic sequence strands plus standard OA2 heuristics and mfold thermodynamic calculation). To partly compensate for the lack of support for transcript groups, when no specific probe was found, probes only predicted to cross-hybridize to alternative splice forms of the target gene were chosen over probes with predicted cross-hybridization to transcripts of different genes.

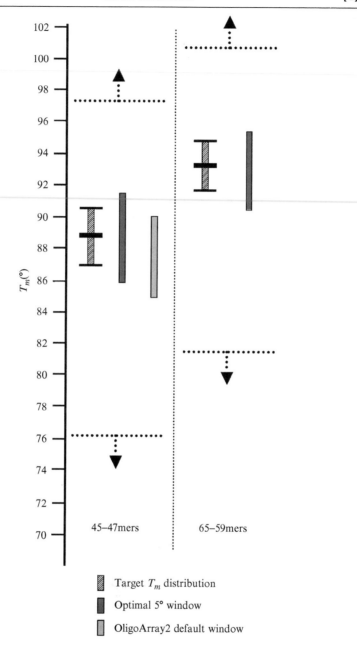

FIG. 3. Choice of parameter thresholds for oligonucleotide probe design. A design parameter selection well matched to the parameter distribution in the probe candidates extends the search space of acceptable candidates, increasing the likelihood that specific

Conclusion

It is noteworthy that there are no tools presently capable of automatically selecting a uniform range of thermodynamic properties allowing high specificity for most probes and delivering a probe set appropriately dealing with families of paralogous genes and alternative splicing variants. Combining results from multiple OA2 runs with carefully selected parameters, however, a "state-of-the-art" probe set could be obtained, with probes for more than 90% of all targets meeting all design criteria. For about 4% of targets, however, probes were predicted to cross-hybridize with transcripts from nontarget genes, most likely orthologues. Support for orthologues and splice forms and more automated parameter selection could simplify employment considerably. Finally, while OA2 gives very little control of probe placement, this property is a particular strength of oligonucleotide arrays.

Locations of Probe Sequence Target Regions: Discrimination of
    Highly Similar Targets

The ability of probing specific target regions can be exploited, e.g., to test for RNA integrity. Probe location-dependent trends in signals from multiple probes for an abundant transcript indicate RNA degradation. While bacterial RNA is degraded in a 3'-to-5' direction, eukaryotic RNA is degraded by exonuclease digestion from the 5' end (Brown, 2002). The transcript secondary structure may, however, hide particular target regions and hence affect the probe hybridization signal (Ratushna et al., 2005), a complication that should be considered (but usually is not) in probe design and signal interpretation.

In contrast to the 3' bias of many commonly used labeling methods, modern protocols can provide labeled full-length targets (Castle et al., 2003; Johnson et al., 2003). Calculated placement of probes then allows the discrimination of highly similar targets, such as families of paralogous

---

probes can be found. The 65- to 69-mers are compared to the 45- to 47-mer probes (the OA2 default probe lengths). Distribution of the median T_m of all possible candidate probes per target is shown. The median of these for the set of target transcripts is marked by a black bar; whiskers indicate first and third quartiles. The OA2 default T_m window is reasonable for 45- to 47-mers but needs adjustment for different probe lengths. While a 5° window can be found to suit most targets, there typically are targets with unusual properties requiring reruns with different parameter sets. This is illustrated for the most extreme cases: 75% of probe candidates for these transcripts have a T_m beyond the values indicated by the dotted lines. (Predicted T_m values as calculated by OA2.)

genes or alternative splice forms. The latter are of particular interest in the quest for understanding complex eukaryotes. Alternative splice forms are predicted for ∼50% of all human genes, comprising a complex variety of transcriptional constructs hard to distinguish with microarrays (Lander et al., 2001; Shai, 2004). While exon junction-spanning probes (Kane et al., 2000) can improve discrimination (Fig. 4), they can also cross-hybridize with alternative splice forms due to construction constraints. Moreover, 5% of human splice acceptor sites have NAGNAG motifs (N being a SNP), parts of which are received by some splice forms (Hiller et al., 2004). Probes for expression analysis must tolerate both this alternative motif inclusion and the motif degeneracy while being highly specific to the target splice form queried.

Clearly, signal interpretation constitutes a critical and challenging aspect of these microarray applications, which is reflected in a wide range of approaches. Adding gene structure-specific effects per splice form in a linear model of effects (Li and Wong, 2001), specific splice forms could be discriminated for genes of known structure (Wang et al., 2003). GenASAP could deduce splicing events from exon and exon-junction probe data by fitting a Bayesian generative linear model for single-cassette exon inclusion/exclusion using structured variational expectation-maximization (Shai, 2004). Comparing samples against their mixture and introducing (unknown) probe- and splice-form-specific affinities in a linear model of effects (Li and Wong, 2001), differences in splicing patterns between samples could be detected (Le et al., 2004). Present approaches to analyzing such complex data sets do not explicitly model cross-hybridization. With the severity of such probe-level effects, however, further progress is expected from including individual probe characteristics into the modeling process. The subsequent sections give an overview of our current state-of-the-art understanding of probe behavior.

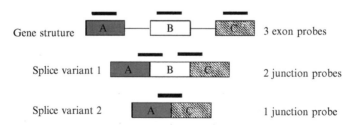

FIG. 4. Exon and exon junction probes. Black bars indicate probe locations. Direct measurement of splice variant 2 requires exon junction probes.

In Situ Synthesis vs Deposition of Presynthesized Oligonucleotides

Both robotic deposition of presynthesized oligonucleotides on arrays (Auburn et al., 2005) and in situ synthesis of probes each have their advantages and disadvantages. Using fixed mask lithography, an approach pioneered by Affymetrix (Lockhart et al., 1996), oligonucleotide synthesis is achieved by repeated cycles of base additions with different masks for light-directed deprotection of terminal hydroxyl groups (Pirrung, 2002). The typical coupling efficiency of only 92–94% per step (McGall et al., 1997), however, limits the technology to short probes (Fig. 5), although improved photosensitive groups exist (Pirrung, 2000). Typically, 11–14 probes of 25 bases are used per target for transcript expression profiling. Fixed mask lithography produces ~1,300,000 probes/chip, making this the technology of choice for extremely high numbers of probes. However,

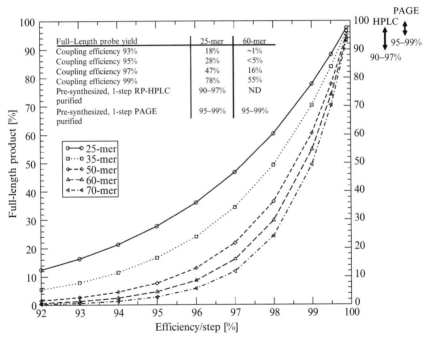

FIG. 5. Response of synthesis yields to varying coupling efficiencies for oligonucleotides of different lengths. Photolithographic and inkjet synthesis typically achieve efficiencies of 94–96% and ~97% per step, respectively. Presynthesized nucleotides are made with efficiencies >99%, and subsequent purification steps are feasible, of particular relevance for longer oligonucleotide probes. A single round of RP-HPLC obtains 90–97% of full-length product; PAGE yields 95–99% purity.

while well suited for industrial production of standard arrays, making small numbers of specialized arrays is uneconomical.

Very high density arrays can flexibly be produced by *in situ* synthesis via digital micromirror device (DMD) lithography, yielding ~400,000 features/ array. Improvements in photosensitive deprotection efficiencies, up from 95% to 98%, giving stepwise synthesis yields of up to 96% (Buhler *et al.*, 2004; Nuwaysir *et al.*, 2002; Singh-Gasson *et al.*, 1999), means that arrays for transcript expression profiling are available with 60-mer probes that typically employ five or more probes per target (cf. Nimblegen arrays). *In situ* synthesis by inkjet deposition can flexibly produce high density arrays of ~40,000 spots of excellent spot morphologies. Coupling efficiencies of up to 98% allow a higher yield synthesis of 60-mer probes (Hughes *et al.*, 2001; Lausted *et al.*, 2004). Typically one probe is used per target for transcript expression profiling.

As an alternative to *in situ* synthesis, *presynthesized* oligonucleotide probes can be spotted at high density, giving arrays of ~40,000 probes. Compared to *in situ* synthesis, presynthesized probes can be produced at much higher purity and yield. A coupling efficiency of >99% can be achieved in synthesis, and purification of the final product is possible by one or multiple rounds of reversed-phase high-performance liquid chromatography (RP-HPLC), which works well for shorter oligonucleotides, and/ or polyacrylamide gel electrophoresis (PAGE). Typically, 50- to 70-mers are used for transcript expression profiling, with one probe per target. Spotted arrays also allow more complex designs in which probes for multiple targets are spotted as composite probes for multiplexed target measurements or normalization purposes (Shmulevich *et al.*, 2003; Yang *et al.*, 2002).

Because probes containing mixtures of prematurely terminated oligonucleotides reduce measurement specificity at optimal hybridization conditions (Jobs *et al.*, 2002) and because purification steps are expensive, many laboratories spot probes with 5′-terminal amino groups onto aldehyde substrates. Only full-length probes bind to the substrate covalently while prematurely terminated oligonucleotides are washed off. Increased probe purity extremely simplifies thermodynamic modeling.

Thermodynamic Modeling of Microarray Probe Hybridization

Microarray Specific Effects

While the thermodynamics of nucleic acid hybridization in solution has long been an area of extensive research (Dimitrov and Zuker, 2004; SantaLucia and Hicks, 2004), only the recent popularization of microarrays has brought the more convoluted issue of hybridization behavior of oligonucleotides tethered to a solid support into the focus of current research. The solid support can interfere with target molecule binding sterically and chemically. Even with gel-like substrate coatings or spacers attached to probes reducing this effect, it was surprising that models for hybridization behavior in solution could directly be applied for presynthesized probes attached to a gel substrate once a linear correction was applied to thermodynamic parameters (Table I); this was unaffected by fluorescent end labels (Fotin et al., 1998).

The situation for probes from manufacturing processes giving mixtures of prematurely terminated oligonucleotides is more complicated. For a long time, therefore, probe sequence-specific variation in signal intensity from such arrays was not understood. Sequence-specific probe bias, particularly strong for short sequences, was reduced by combining measurements from multiple probes, yet without exploiting probe sequence information (Bolstad et al., 2003; Li and Wong, 2001). However, empirical models of sequence-specific binding with position-specific weights have been introduced: The predicted contributions of probe regions to the overall binding strength are attenuated depending on their positions along the probe sequence.

TABLE I
LINEAR CORRECTIONS TO THERMODYNAMIC PARAMETERS FOR OLIGONUCLEOTIDE
PROBES ATTACHED TO A SOLID SUPPORT[a]

Thermodynamic parameter	Linear correction for microarrays
ΔH°	$\Delta H^0_{array} = \Delta H^0_{solution} - 24$
ΔD°	$\Delta D^0_{array} = \Delta S^0_{solution} - 70$
ΔG°, original paper, slope constrained $= 1$	$\Delta G^0_{array} = \Delta G^0_{solution} - 3.2$
ΔG°, original paper, slope unconstrained	$\Delta G^0_{array} = 1.1 \Delta G^0_{solution} - 3.2$
ΔG°, recalculated, slope unconstrained	$\Delta G^0_{array} = 0.78 \Delta G^0_{solution} - 1.0$
ΔG°, HyTher	$\Delta G^0_{array} = 0.85 \Delta G^0_{solution} - 2.33$

[a] The first alternative formula for ΔG^0 gives the relationship published in the original paper by Fotin et al. (1998), where the slope has been assumed to be 1. The next line shows regression results without this constraint, as published. This does not, however, fit data in Fotin et al. (J. SantaLucia, Jr., personal communication, 2005). The formula labeled "recalculated" was obtained by linear least-squares regression from original table data (Fotin et al., 1998), while the last line shows the correction suggested by HyTher (http://ozone2.chem.wayne.edu/).

For data from Affymetrix chips, Zhang *et al.* (2003) successfully fit the signal intensities of a particular probe i for a target j as sum of contributions from specific and nonspecific binding to the probe plus a global background constant B:

$$I_{ij} = \frac{N_j}{1 + \exp(E_{ij})} + \frac{N^*}{1 + \exp(E_{ij}^*)} + B.$$

where N_j is the number or target molecules and N^* is the number of molecules binding nonspecifically to (all) probes. For a probe sequence $(b_1, b_2, \ldots, b_k, \ldots, b_{25})$, the free energy terms for specific and nonspecific binding, $E_{ij} = \sum_{k=1}^{25} \omega_k \epsilon(b_k, b_{k+1})$ and $E_{ij}^* = \sum_{k=1}^{25} \omega_k^* \epsilon^*(b_k, b_{k+1})$, are parameterized by empirical base-pair stacking energies $\varepsilon/\varepsilon^*$ and position-dependent weights ω_k/ω_k^*. This simple model fitted probe signal levels well, removing probe sequence-specific bias, apparently of particular relevance for low-intensity signals. The probe center gave the largest contribution to binding (Fig. 6). The empirical base-pair stacking energies, however, can vary considerably between different chip designs (data from http://odin.mdacc.tmc. edu/~zhangli/PerfectMatch/), reflecting the empirical nature of the model.

Naef and Magnasco (2003) use position-dependent affinities A_k in modeling probe-specific signal intensities for Affymetrix chips,

$$\ln\left(\frac{I_{ij}}{\underset{i}{median}\ I_{ij}}\right) = \sum_{k=1}^{25} A_k(b_k),$$

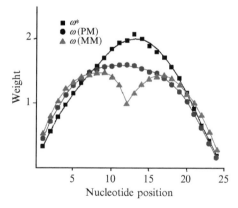

FIG. 6. Position-specific weights in a position-dependent, nearest-neighbor model. The center part of an Affymetrix probe gives the strongest contribution to binding. The curve for the mismatch probes (MM) reflects destabilization from the central mismatch base. Redrawn after Zhang *et al.* (2003.)

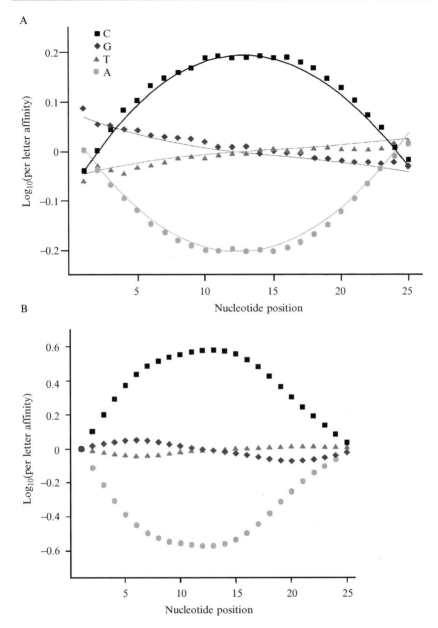

FIG. 7. Position-specific affinities. (A) Position-specific affinities for each of the four bases from the model of Naef and Magnasco (2003). A/T and C/G asymmetries are due to labeled pyrimidines U/C impeding binding for A/G. Positions are in synthesis order, with 1 denoting the 3′-terminal attached to the chip. Redrawn from Naef and Magnasco (2003) . (B) The same model parameters but as obtained by Wu *et al.* (2003). Note the differences for G/T in comparison with A. Redrawn from Wu *et al.* (2003.)

giving position-dependent scores for each of the four bases. Figure 7A shows the distinct base-specific profiles. The destabilizing effects of in-sequence labels indicate possible advantages of labeling target sequences outside the probe-binding regions. Overall, probe centers contributed most to overall binding.

GC-RMA (www.bioconductor.org) adopted the Naef and Magnasco model and in combination with data from nonspecific hybridization predicts probe signals corrected for background and bias. Affinities obtained for G and T (Fig. 7B) showed somewhat different behavior to that observed earlier, as can be expected for an empirical model, yet the predominant contribution to binding was again from the centers of the probes (Naef and Magnasco, 2003; Wu et al., 2003).

Common to all these approaches is the apparent attenuated influence of terminal probe regions. For the improvement of microarray manufacture and/or signal modeling, one wonders what could be its physical cause. At the 5' end, one may well see the result of diminishing synthesis yield through premature termination (Naef and Magnasco, 2003), while the reduced effect of bases in the 3'-terminal region could be due to steric hindrance of the solid support or overly dense population by short oligonucleotides (J. SantaLucia, Jr., personal communication, 2005).

Models for Hybridization in Solution

Even predicting hybridization in solution is a very complex modeling problem that is an area of active research (Dimitrov and Zuker, 2004; Santa-Lucia and Hicks, 2004). A hybridized complex or a folded structure actually assembles cooperatively in three-dimensional space, interacting dynamically with multiple other nucleic acid molecules and smaller molecules in solution as well as the solvent itself. In dependence on the temperature, what nucleic acids are present and at what concentrations, and the concentrations of salt ions and other buffer components (such as formamide), the nucleic acids can form a variety of heterogeneous complexes while at the same time folding within themselves. Therefore, to infer the concentration of a particular target transcript from microarray probe hybridization intensity, a fairly detailed understanding of the binding behavior of the probe and its potential binding partners is required. To make modeling tractable, several approximations are necessary. A focus on secondary structure elements is justified because the tertiary structure is a much weaker, second-order effect. The strong Watson–Crick interactions further allow the "discrete pairing approximation": positions in a sequence are either paired or not, rendering structure prediction suitable for dynamic programming algorithms, which have brought structure prediction for nucleic acids of up to 10,000 bases within reach for modern desktop computers (SantaLucia and Hicks, 2004).

The most common additional approximation in predicting microarray probe hybridization is looking at only one or two molecules at a time. Calculations for the hybridization of two molecules are typically simplified much further by assuming a "two-state model," where the two molecules are either in a "bound state" or not. To model the properties of the binding process under the two-state approximation, only differences of thermodynamic parameters between the two states need to be calculated. For such computations, corresponding rules have been derived from the measurement of thermodynamic properties of selected nucleotides with purposefully designed sequences and structures, which contained basic reoccurring motifs (SantaLucia, 1998). An important part of this rule set is formed by the unified Watson–Crick base-pair, nearest-neighbor parameters obtained by multiple-linear regression of measurements from several laboratories (SantaLucia, 1998) used by most microarray probedesign tools. State-of-the-art algorithms for the prediction of folding or hybridization structures of minimal and near-minimal energy use these parameters together with the corresponding rule set for more complicated structural motifs such as mismatched pairs, bulges, hairpins and various loops, and dangling ends (SantaLucia, 1998). Tools such as mfold (Zuker, 2003), HyTher (http://ozone2.chem.wayne.edu/), and ViennaRNA (Hofacker, 2003) can more accurately assess regions of nontarget transcripts that are suspected of nonspecific hybridization to a probe. Traditionally, these regions are selected by sequence similarity and heuristics; however, the development of tools that can identify regions in a longer target DNA that will hybridize with a shorter probe by direct thermodynamic calculation (SantaLucia and Hicks, 2004) will soon make this inaccurate heuristic approximation unnecessary (M. Zuker, personal communication, 2004).

Importantly, the most recent advances in thermodynamic computation now go beyond two-state models in the prediction of hybridization behavior (Fig. 8).

Again, the same thermodynamic rule set is used, but care has to be taken in order to avoid overcounting microstates: Although the experimental setup for the determination of the rule set has been designed to minimize this effect, the parameters measured for the two-state model are for two *effective* states ("bound" and "unbound"), each of which is actually a combination of multiple microstates. DNA Software's commercial OMP products account for this (SantaLucia and Hicks, 2004) and can provide correct multistate modeling allowing multiple folding and binding events to be considered, including multiple simultaneous interactions per molecule. The improvements achievable by moving beyond two-state models can also be seen in DINAMelt, which for two molecules

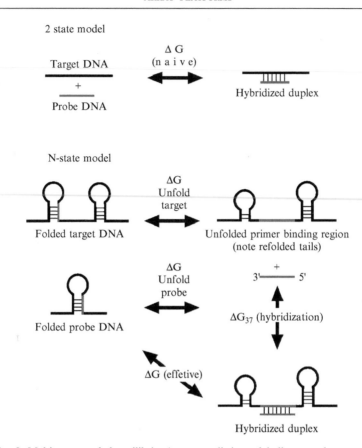

FIG. 8. Multistate-coupled equilibria. A more realistic model allows much more accurate predictions of hybridization behavior (Dimitrov and Zuker, 2004; Markham and Zuker, 2005; SantaLucia and Hicks, 2004). Redrawn after SantaLucia and Hicks (2004.)

(A and B) models: self-folding (A_{self} and B_{self}), self-binding (A-A and B-B), and heteroduplex formation (A-B) (Dimitrov and Zuker, 2004; Markham and Zuker, 2005). DINAMelt calculates full partition sums (i.e., accounting for all possible microstates), also taking care to avoid overcounting (N. Markham, personal communication, 2005). The multiple folding and binding events are modeled in competition to one another, giving temperature-dependent yields for each effective state. While these methods are currently too slow to be used as primary screens of oligonucleotide probe candidates during microarray design, they allow much more sophisticated evaluations of probe sets.

Thermodynamic Probe Design Criteria

When aiming for uniform probe characteristics across a microarray, many probe designs aim for uniform melting temperatures T_m. These alone, however, only give information about the behavior of the probes at their respective melting temperatures. Probes with the same T_m can behave quite differently at a reaction temperature $T_{hyb} < T_m$. For a given reaction temperature T_{hyb}, aiming for similar free energies at T_{hyb} would hence actually result in more uniform hybridization (J. SantaLucia, Jr., personal communication, 2005). This can be improved on even further by accounting for competitive hybridization and actually calculating, for a target transcript, what proportion of molecules will be bound to its probe at T_{hyb}, aiming for uniformity across probes.

In screening probes, designs typically aim to avoid secondary structure. Clearly, a strong secondary structure may render a probe inaccessible for its target. However, exploiting *competitive* hybridization, the secondary structure can contribute much to the specific recognition of a probe's target. This is actually exploited by other experimental techniques, such as molecular beacons (Bonnet *et al.*, 1999). Using thermodynamic models for competitive hybridization, one can actually employ probe as a secondary structure to adjust the level of specificity in target binding to that required (M. Zuker, personal communication, 2005), for example, highest for the discrimination of SNPs and highly similar targets and lower for transcript profiling transparently allowing for polymorphisms.

Outlook

With the increasing understanding of hybridization on microarrays, for many future microarray applications, the issue of probe design will yield to the task of probe signal interpretation. Increasingly, modern methods leave little freedom in probe selection because probes have to target a very well-defined region, e.g., in probing particular gene regions to elucidate regulatory binding or splicing events. Many of these probes will show cross-hybridization or strong secondary structure, and probe sets will display a wide spectrum of thermodynamic properties. To make the most of such data, a combination of experimental advances and sophisticated modeling will be instrumental. Repeated measurements under different hybridization conditions can, for example, discriminate specific from nonspecific signal by exploiting hybridization kinetics (Dai *et al.*, 2002).

A further advance in quantitative microarray analysis has come with algorithms directly motivated by physical models. Application of the most elementary representation of surface adsorption, the Langmuir isotherm (Atkins and de Paula, 2004), could account for the nonlinearities observed at high signal intensity due to saturation of the probe with target molecules (Hekstra *et al.*, 2003)—not to be confused with saturation effects in the scanning of fluorescent images. Combination of such a Langmuir adsorption model with thermodynamic free-energy calculations was very successful; however, despite the significant improvements seen, systematic variation was still detectable in data, highlighting the need for further studies (Held *et al.*, 2003).

The measurement process on microarrays is, over time, increasingly better understood and hence modeled. This correspondingly gives data that better reflect the true abundances of transcripts in samples, giving better detection characteristics in screens of samples for biological differences and providing a prerequisite for more sophisticated work in computational biology. While, overall, a lot of progress has been achieved, quantitative microarray analysis remains a challenging and active field of research.

Supplement

Further information is available at www.flychip.org.uk/MethEnz2005/.

Acknowledgments

We are grateful to Michael Zuker and John SantaLucia, Jr., for helpful discussions and advice. We also thank Richard Auburn, Nicholas Markham, Lisa Meadows, and Andrew Thompson for helpful discussions and Jean-Marie Rouillard for the kind provision of the OligoArray 2.1 source code. The group of D. Kreil is funded by the Vienna Science and Technology Fund (WWTF), the Austrian Centre of Biopharmaceutical Technology (ACBT), Austrian Research Centres (ARC) Seibersdorf, and Baxter AG. The laboratory of S. Russell is funded by the Biotechnology and Biological Sciences Research Council, the UK Medical Research Council, and the Wellcome Trust.

References

Altschul, S. F., Madden, T. L., Schaeffer, A. A., Zhang, J., Zhang, Z., Miller, W., and Lipman, D. J. (1997). Gapped BLAST and PSI-BLAST: A new generation of protein database search programs. *Nucleic Acids Res.* **25,** 3389–3402.

Atkins, P., and de Paula, J. (2004). "Atkins' Physical Chemistry." OUP, Oxford.

Auburn, R. P., Kreil, D. P., Meadows, L. A., Fischer, B., Matilla, S. S., and Russell, S. (2005). Robotic spotting of cDNA and oligonucleotide microarrays. *Trends Biotechnol.* **23,** 374–379.

Bertone, P., Stolc, V., Royce, T. E., Rozowsky, J. S., Urban, A. E., Zhu, X., Rinn, J. L., Tongprasit, W., Samanta, M., Weissman, S., Gerstein, M., and Snyder, M. (2004). Global identification of human transcribed sequences with genome tiling arrays. *Science* **306,** 2242–2246.

Bolstad, B. M., Irizarry, R. A., Astrand, M., and Speed, T. P. (2003). A comparison of normalization methods for high density oligonucleotide array data based on variance and bias. *Bioinformatics* **19**, 185–193.

Bonnet, G., Tyagi, S., Libchaber, A., and Kramer, F. R. (1999). Thermodynamic basis of the enhanced specificity of structured DNA probes. *Proc. Natl. Acad. Sci. USA* **96**, 6171–6176.

Brown, T. A. (2002). "Genomes." BIOS Scientific, Oxford.

Buhler, S., Lagoja, I., Giegrich, H., Stengele, K. P., and Pfleiderer, W. (2004). New types of very efficient photolabile protecting groups based upon the [2-(2-nitrophenyl)propoxy] carbonyl (NPPOC) moiety. *Helv. Chim. Acta* **87**, 620–659.

Castle, J., Garrett-Engele, P., Armour, C. D., Duenwald, S. J., Loerch, P. M., Meyer, M. R., Schadt, E. E., Stoughton, R., Parrish, M. L., Shoemaker, D. D., and Johnson, J. M. (2003). Optimization of oligonucleotide arrays and RNA amplification protocols for analysis of transcript structure and alternative splicing. *Genome Biol.* **4**, R66.

Chou, C. C., Chen, C. H., Lee, T. T., and Peck, K. (2004). Optimization of probe length and the number of probes per gene for optimal microarray analysis of gene expression. *Nucleic Acids Res.* **32**, e99.

Dai, H., Meyer, M., Stepaniants, S., Ziman, M., and Stoughton, R. (2002). Use of hybridization kinetics for differentiating specific from non-specific binding to oligonucleotide microarrays. *Nucleic Acids Res.* **30**, e86.

Diatchenko, L., Lau, Y. F., Campbell, A. P., Chenchik, A., Moqadam, F., Huang, B., Lukyanov, S., Lukyanov, K., Gurskaya, N., Sverdlov, E. D., and Siebert, P. D. (1996). Suppression subtractive hybridization: A method for generating differentially regulated or tissue-specific cDNA probes and libraries. *Proc. Natl. Acad. Sci. USA* **93**, 6025–6030.

Dimitrov, R. A., and Zuker, M. (2004). Prediction of hybridization and melting for double-stranded nucleic acids. *Biophys. J.* **87**, 215–226.

Duggan, D. J., Bittner, M., Chen, Y., Meltzer, P., and Trent, J. M. (1999). Expression profiling using cDNA microarrays. *Nat. Genet.* **21**, 10–14.

Fotin, A. V., Drobyshev, A. L., Proudnikov, D. Y., Perov, A. N., and Mirzabekov, A. D. (1998). Parallel thermodynamic analysis of duplexes on oligodeoxyribonucleotide microchips. *Nucleic Acids Res.* **26**, 1515–1521.

Hanlon, S. E., and Lieb, J. D. (2004). Progress and challenges in profiling the dynamics of chromatin and transcription factor binding with DNA microarrays. *Curr. Opin. Genet. Dev.* **14**, 697–705.

Hegde, P., Qi, R., Abernathy, K., Gay, C., Dharap, S., Gaspard, R., Hughes, J. E., Snesrud, E., Lee, N., and Quackenbush, J. (2000). A concise guide to cDNA microarray analysis. *Biotechniques* **29**, 548–556.

Hekstra, D., Taussig, A. R., Magnasco, M., and Naef, F. (2003). Absolute mRNA concentrations from sequence-specific calibration of oligonucleotide arrays. *Nucleic Acids Res.* **31**, 1962–1968.

Held, G. A., Grinstein, G., and Tu, Y. (2003). Modeling of DNA microarray data by using physical properties of hybridization. *Proc. Natl. Acad. Sci. USA* **100**, 7575–7580.

Hiller, M., Huse, K., Szafranski, K., Jahn, N., Hampe, J., Schreiber, S., Backofen, R., and Platzer, M. (2004). Widespread occurrence of alternative splicing at NAGNAG acceptors contributes to proteome plasticity. *Nat. Genet.* **36**, 1255–1257.

Hofacker, I. L. (2003). Vienna RNA secondary structure server. *Nucleic Acids Res.* **31**, 3429–3431.

Holm, L., and Sander, C. (1998). Removing near-neighbour redundancy from large protein sequence collections. *Bioinformatics* **14**, 423–429.

Hughes, T. R., Mao, M., Jones, A. R., Burchard, J., Marton, M. J., Shannon, K. W., Lefkowitz, S. M., Ziman, M., Schelter, J. M., Meyer, M. R., Kobayashi, S., Davis, C., Dai, H., He, Y. D.,

Stephaniants, R., Cavet, G., Walker, W. L., West, A., Coffey, E., Shoemaker, D. D., Stoughton, R., Blanchard, A. P., Friend, S. H., and Linsley, P. S. (2001). Expression profiling using microarrays fabricated by an ink-jet oligonucleotide synthesizer. *Nat. Biotechnol.* **19,** 342–347.

Ihmels, J., Bergmann, S., and Barkai, N. (2004). Defining transcription modules using large-scale gene expression data. *Bioinformatics* **20,** 1993–2003.

Ihmels, J., Friedlander, G., Bergmann, S., Sarig, O., Ziv, Y., and Barkai, N. (2002). Revealing modular organization in the yeast transcriptional network. *Nat. Genet.* **31,** 370–378.

Jobs, M., Fredriksson, S., Brookes, A. J., and Landergren, U. (2002). Effect of oligonucleotide truncation on single-nucleotide distinction by solid-phase hybridization. *Anal. Chem.* **74,** 199–202.

Johnson, J. M., Castle, J., Garrett-Engele, P., Kan, Z., Loerch, P. M., Armour, C. D., Santos, R., Schadt, E. E., Stoughton, R., and Shoemaker, D. D. (2003). Genome-wide survey of human alternative pre-mRNA splicing with exon junction microarrays. *Science* **302,** 2141–2144.

Johnston, R., Wang, B., Nuttall, R., Doctolero, M., Edwards, P., Lu, J., Vainer, M., Yue, H., Wang, X., Minor, J., Chan, C., Lash, A., Goralski, T., Parisi, M., Oliver, B., and Eastman, S. (2004). FlyGEM, a full transcriptome array platform for the Drosophila community. *Genome Biol.* **5.**

Kane, M. D., Jatkoe, T. A., Stumpf, C. R., Lu, J., Thomas, J. D., and Madore, S. J. (2000). Assessment of the sensitivity and specificity of oligonucleotide (50mer) microarrays. *Nucleic Acids Res.* **28,** 4552–4557.

Kreil, D. P., Auburn, R. P., Meadows, L., Russell, S., and Micklem, G. (2003). Quantitative microarray spot profile optimization: A systematic evaluation of buffer/slide combinations. *In* "German Conference in Bioinformatics." Munich, Germany.

Kreil, D. P., and Russell, R. R. (2005). There is no silver bullet: A guide to low-level data transforms and normalisation methods for microarray data. *Brief Bioinform.* **6,** 86–97.

Lander, E. S., Linton, L. M., Birren, B., Nusbaum, C., Zody, M. C., Baldwin, J., Devon, K., Dewar, K., Doyle, M., FitzHugh, W., Funke, R., Gage, D., Harris, K., Heaford, A., Howland, J., Kann, L., Lehoczky, J., LeVine, R., McEwan, P., McKernan, K., Meldrim, J., Mesirov, J. P., Miranda, C., Morris, W., Naylor, J., Raymond, C., Rosetti, M., Santos, R., Sheridan, A., Sougnez, C., Stange-Thomann, N., Stojanovic, N., Subramanian, A., Wyman, D., Rogers, J., Sulston, J., Ainscough, R., Beck, S., Bentley, D., Burton, J., Clee, C., Carter, N., Coulson, A., Deadman, R., Deloukas, P., Dunham, A., Dunham, I., Durbin, R., French, L., Grafham, D., Gregory, S., Hubbard, T., Humphray, S., Hunt, A., Jones, M., Lloyd, C., McMurray, A., Matthews, L., Mercer, S., Milne, S., Mullikin, J. C., Mungall, A., Plumb, R., Ross, M., Shownkeen, R., Sims, S., Waterston, R. H., Wilson, R. K., Hillier, L. W., McPherson, J. D., Marra, M. A., Mardis, E. R., Fulton, L. A., Chinwalla, A. T., Pepin, K. H., Gish, W. R., Chissoe, S. L., Wendl, M. C., Delehaunty, K. D., Miner, T. L., Delehaunty, A., Kramer, J. B., Cook, L. L., Fulton, R. S., Johnson, D. L., Minx, P. J., Clifton, S. W., Hawkins, T., Branscomb, E., Predki, P., Richardson, P., Wenning, S., Slezak, T., Doggett, N., Cheng, J. F., Olsen, A., Lucas, S., Elkin, C., Uberbacher, E., Frazier, M., *et al.* (2001). Initial sequencing and analysis of the human genome. *Nature* **409,** 860–921.

Lausted, C., Dahl, T., Warren, C., King, K., Smith, K., Johnson, M., Saleem, R., Aitchison, J., Hood, L., and Lasky, S. R. (2004). POSaM: A fast, flexible, open-source, inkjet oligonucleotide synthesizer and microarrayer. *Genome Biol.* **5,** R58.

Le, K., Mitsouras, K., Roy, M., Wang, Q., Xu, Q., Nelson, S. F., and Lee, C. (2004). Detecting tissue-specific regulation of alternative splicing as a qualitative change in microarray data. *Nucleic Acids Res.* **32,** e180.

Le Berre, V., Trevisiol, E., Dagkessamanskaia, A., Sokol, S., Caminade, A. M., Majoral, J. P., Meunier, B., and Francois, J. (2003). Dendrimeric coating of glass slides for sensitive DNA microarrays analysis. *Nucleic Acids Res.* **31,** e88.

Lee, S. I., and Batzoglou, S. (2003). Application of independent component analysis to microarrays. *Genome Biol.* **4,** R76.

Li, C., and Wong, W. H. (2001). Model-based analysis of oligonucleotide arrays: Expression index computation and outlier detection. *Proc. Natl. Acad. Sci. USA* **98,** 31–36.

Li, W., Jaroszewski, L., and Godzik, A. (2001). Clustering of highly homologous sequences to reduce the size of large protein databases. *Bioinformatics* **17,** 282–283.

Lockhart, D. J., Dong, H., Byrne, M. C., Follettie, M. T., Gallo, M. V., Chee, M. S., Mittmann, M., Wang, C., Kobayashi, M., Horton, H., and Brown, E. L. (1996). Expression monitoring by hybridization to high-density oligonucleotide arrays. *Nat. Biotechnol.* **14,** 1675–1680.

Luebke, K. J., Balog, R. P., and Garner, H. R. (2003). Prioritized selection of oligodeoxyribonucleotide probes for efficient hybridization to RNA transcripts. *Nucleic Acids Res.* **31,** 750–758.

MacAlpine, D. M., and Bell, S. P. (2005). A genomic view of eukaryotic DNA replication. *Chromosome Res.* **13,** 309–326.

Markham, N. R., and Zuker, M. (2005). DINAMelt web server for nucleic acid melting prediction. *Nucleic Acids Res.* **33,** W577–W581.

Marko, N. F., Frank, B., Quackenbush, J., and Lee, N. H. (2005). A robust method for the amplification of RNA in the sense orientation. *BMC Genom.* **6,** 27.

McGall, G. H., Barone, A. D., Diggelmann, M., Fedor, S. P. A., Gentalen, E., and Ngo, N. (1997). The efficiency of light-directed synthesis of DNA arrays on glass substrates. *J. Am. Chem. Soc.* **119,** 5081–5090.

Naef, F., and Magnasco, M. O. (2003). Solving the riddle of the bright mismatches: Labeling and effective binding in oligonucleotide arrays. *Phys. Rev. E Stat. Nonlin. Soft Matter Phys.* **68,** 011906.

Nuwaysir, E. F., Huang, W., Albert, T. J., Singh, J., Nuwaysir, K., Pitas, A., Richmond, T., Gorski, T., Berg, J. P., Ballin, J., McCormick, M., Norton, J., Pollock, T., Sumwalt, T., Butcher, L., Porter, D., Molla, M., Hall, C., Blattner, F., Sussman, M. R., Wallace, R. L., Cerrina, F., and Green, R. D. (2002). Gene expression analysis using oligonucleotide arrays produced by maskless photolithography. *Genome Res.* **12,** 1749–1755.

Pinkel, D., and Albertson, D. G. (2005). Comparative genomic hybridization. *Annu. Rev. Genom. Hum. Genet.* **6,** 331–354.

Pirrung, M. C. (2000). Production by quantitative photolithographic synthesis of individually quality checked DNA microarrays. *Chemtracts* **13,** 487–490.

Pirrung, M. C. (2002). How to make a DNA chip. *Angew. Chem. Int. Edn.* **41,** 1276–1289.

Ramakrishnan, R., Dorris, D., Lublinsky, A., Nguyen, A., Domanus, M., Prokhorova, A., Gieser, L., Touma, E., Lockner, R., Tata, M., Zhu, X., Patterson, M., Shippy, R., Sendera, T. J., and Mazumder, A. (2002). An assessment of Motorola CodeLink microarray performance for gene expression profiling applications. *Nucleic Acids Res.* **30,** e30.

Ratushna, V. G., Weller, J. W., and Gibas, C. J. (2005). Secondary structure in the target as a confounding factor in synthetic oligomer microarray design. *BMC Genom.* **6,** 31.

Relogio, A., Schwager, C., Richter, A., Ansorge, W., and Valcarcel, J. (2002). Optimization of oligonucleotide-based DNA microarrays. *Nucleic Acids Res.* **30,** e51.

Rouillard, J. M., Zuker, M., and Gulari, E. (2003). OligoArray 2.0: Design of oligonucleotide probes for DNA microarrays using a thermodynamic approach. *Nucleic Acids Res.* **31,** 3057–3062.

Saidi, S. A., Holland, C. M., Kreil, D. P., MacKay, D. J., Charnock-Jones, D. S., Print, C. G., and Smith, S. K. (2004). Independent component analysis of microarray data in the study of endometrial cancer. *Oncogene* **23**, 6677–6683.

SantaLucia, J. (1998). A unified view of polymer, dumbbell, and oligonucleotide DNA nearest-neighbor thermodynamics. *Proc. Natl. Acad. Sci. USA* **95**, 1460–1465.

SantaLucia, J., Jr., and Hicks, D. (2004). The thermodynamics of DNA structural motifs. *Annu. Rev. Biophys. Biomol. Struct.* **33**, 415–440.

Schena, M., Shalon, D., Davis, R. W., and Brown, P. O. (1995). Quantitative monitoring of gene expression patterns with a complementary DNA microarray. *Science* **270**, 467–470.

Schena, M., Shalon, D., Heller, R., Chai, A., Brown, P. O., and Davis, R. W. (1996). Parallel human genome analysis: Microarray-based expression monitoring of 1000 genes. *Proc. Natl. Acad. Sci. USA* **93**, 10614–10619.

Shai, O. F. B. J., Morris, Q. D., Pan, Q., Misquitta, C., and Blencowe, B. J. (2004). Probabilistic inference of alternative splicing events in microarray data. *In* "18th Annual Conference on Neural Information Processing Systems" (L. K. Saul, Y. Weiss, and L. Bottou, eds.) Vol. 17. Neural Information Processing Systems Foundation.

Shchepinov, M. S., Case-Green, S. C., and Southern, E. M. (1997). Steric factors influencing hybridisation of nucleic acids to oligonucleotide arrays. *Nucleic Acids Res.* **25**, 1155–1161.

Shmulevich, I., Astola, J., Cogdell, D., Hamilton, S. R., and Zhang, W. (2003). Data extraction from composite oligonucleotide microarrays. *Nucleic Acids Res.* **31**, e36.

Singh-Gasson, S., Green, R. D., Yue, Y., Nelson, C., Blattner, F., Sussman, M. R., and Cerrina, F. (1999). Maskless fabrication of light-directed oligonucleotide microarrays using a digital micromirror array. *Nat. Biotechnol.* **17**, 974–978.

Suchyta, S. P., Sipkovsky, S., Halgren, R. G., Kruska, R., Elftman, M., Weber-Nielsen, M., Vandehaar, M. J., Xiao, L., Tempelman, R. J., and Coussens, P. M. (2003). Bovine mammary gene expression profiling using a cDNA microarray enhanced for mammary-specific transcripts. *Physiol. Genom.* **16**, 8–18.

Wang, H., Hubbell, E., Hu, J. S., Mei, G., Cline, M., Lu, G., Clark, T., Siani-Rose, M. A., Ares, M., Kulp, D. C., and Haussler, D. (2003). Gene structure-based splice variant deconvolution using a microarray platform. *Bioinformatics* **19**(Suppl. 1), i315–i322.

Wu, Z., Irizarry, R. A., Gentleman, R., Murillo, F. M., and Spencer, F. (2003). A model based background adjustment for oligonucleotide expression arrays. *In* "Department of Biostatistics Working Papers." John Hopkins University, Baltimore, MD.

Yang, Y. H., Dudoit, S., Luu, P., Lin, D. M., Peng, V., Ngai, J., and Speed, T. P. (2002). Normalization for cDNA microarray data: A robust composite method addressing single and multiple slide systematic variation. *Nucleic Acids Res.* **30**, e15.

Zhang, L., Miles, M. F., and Aldape, K. D. (2003). A model of molecular interactions on short oligonucleotide microarrays. *Nat. Biotechnol.* **21**, 818–821.

Zuker, M. (2003). Mfold web server for nucleic acid folding and hybridization prediction. *Nucleic Acids Res.* **31**, 3406–3415.

[5] Automated Liquid Handling and High-Throughput Preparation of Polymerase Chain Reaction-Amplified DNA for Microarray Fabrication

By ANGELA BURR, KEVIN BOGART, JASON CONATY, and JUSTEN ANDREWS

Abstract

Genome-wide studies of gene expression and transcription factor-binding sites using DNA microarrays are leading to new systems level insights. The massively parallel nature of microarrays presents technical challenges: fabricating high-quality microarrays at the front end and data analysis and interpretation downstream. A principal challenge in fabricating microarrays is preparation of the DNA samples. This is particularly the case for polymerase chain reaction-amplified DNA samples. The challenge is to scale up efficiently to high-throughput preparation of tens of thousands of DNA samples while ensuring a uniform high quality. This chapter outlines strategic considerations, including automated liquid handling and workflow development to maximize efficiency, and quality control (QC) measures to ensure uniform quality. The protocols are presented with commentary to illustrate their logic and specific techniques. These principles and techniques are extensible to other high-throughput molecular biological applications.

Introduction

Studies of gene regulation using DNA microarrays are helping to unravel the regulatory networks that underlie the determination of, and the transition between, cell states. Transcriptome microarrays, fabricated with DNA corresponding to transcribed sequences, are used to interrogate gene expression. Genome microarrays, fabricated with DNA corresponding to all or part of a genome, including sliding windows, tiling paths, or regularly spaced sequences, are generally used for discovering previously unannotated transcribed sequences or the location of transcription factor-binding sites occupied *in vivo* (Ren *et al.*, 2000; Weinmann, 2004). The choice between platforms is dictated primarily by the experimental question or system being studied. Irrespective of the intended end use, DNA microarrays are created with either chemically synthesized oligonucleotides or DNA fragments that are customarily amplified by the polymerase chain reaction (PCR). Although oligonucleotides are a simple and robust alternative, amplified

METHODS IN ENZYMOLOGY, VOL. 410   0076-6879/06 $35.00
DOI: 10.1016/S0076-6879(06)10005-1

DNA platforms—from cDNA for instance—have certain advantages, especially in cases where limited genome sequence data are available (Holloway *et al.*, 2002; Stowe-Evans *et al.*, 2004). This chapter provides a practical guide to the high-throughput preparation of such probes. It describes a workflow implementation that is robust and flexible enough for adoption by any laboratory undertaking high-throughput techniques. Special attention is given to the issues of cost, scalability, verifiability, and management in creating these systems. In general, solutions to these challenges are suitable for a broad variety of problems in modern biology that demand scaling up of previously low-throughput techniques.

Overview

Preparing PCR-amplified DNA for microarrays entails PCR amplification, purification, quality control gel electrophoresis, concentration, and transfer to 384-well plates with spotting buffer in preparation for printing (Bowtell and Sambrook, 2002; Hedge *et al.*, 2000). The preparation of individual PCR-amplified DNA samples is so common as to be considered facile and is thoroughly documented elsewhere (Dieffenbach and Dveksler, 1993; Sambrook and Russell, 2001). However, what may be a trivial matter for one sample is an entirely different enterprise when applied to tens of thousands of samples. The act of scaling up introduces challenges and problems that are a consequence of the scale. The complete process, termed a *workflow*, must be developed with consideration given to cost, efficiency, quality control, and information management. In other words, it encompasses aspects of automation and operations management that are usually found in industrial settings but are not traditionally a part of academic research laboratories. This section describes workflows used for the production of DNA probe microarrays. For those readers using similar instruments and protocols, we provide a detailed description of the methods and have made our liquid handling data management scripts available for download off the World Wide Web (http://cgb.indiana.edu/downloads). However, because genomics laboratories probably have unique sets of instruments, production levels, and reagents, most workflows are tailored to local conditions. We therefore begin with an overview of high-throughput workflows with the aim of exploring the logic and mechanics of these solutions.

General Considerations

By scaling up a molecular biology protocol to high throughput, investigators must ensure that quality is not compromised while simultaneously achieving increases in time and cost efficiencies afforded by economies of scale. Efficiency can be increased by using robots (discussed later) and by processing multiple samples in parallel, which generally saves time and reduces the consumption of plasticware.

The first level of parallel processing is achieved by using multiwell plates. Multiwell plates are available in 96-, 384-, and 1536-well formats with industry-wide standards for the dimensions of footprint and well-to-well spacing. There are also a wide range of accessories designed for working with multiwell plates. These include seals (adhesive seals or heat seals), filter plates, vacuum manifolds, multichannel micropipettes, multiwell solid pin replicators (disposable and reusable), thermal cyclers, centrifuges, Speed-Vacs, spectrophotometers, and gel electrophoresis systems. Using a manual multiwell format apparatus, a skilled operator can produce 96 PCR-amplified DNA samples in the same time it takes to produce 1 sample. Indeed there are a number of genomics laboratories that have and are still processing several thousand PCR samples manually.

The second level of parallel processing is achieved by processing plates in batches. Again, a skilled operator can process two plates simultaneously in significantly less time than it would take to process the same two plates in series. To state the obvious, parallel processing of plates reduces the redundant steps common to each batch, such as preparing a master reaction mix and even accessing source plates in the freezer. Hardware constraints, such as the number of thermal cyclers available for use or the number of plates a liquid handling robot can accommodate, will usually determine the batch size.

The third level of parallel processing is to process replicates in parallel, that is, two or more of the same plate side by side. This reduces redundancy common to each plate. For instance, one visit with clean pipette tips can be used to dispense to two duplicate PCR plates. If duplicates were processed in series it may be necessary to use two racks of clean tips.

Reusing plasticware without compromising quality can increase cost efficiency further. Plasticware can be reused in the following ways. (i) One set of tips can be used to aliquot master reaction mixes to several destination plates. (ii) Some liquid handling robots allow disposable pipette tips to be reracked (shucked back into the original tip box) and used again. This option allows tips to be reused either at a subsequent step in the workflow or for a subsequent replicate of that particular plate. (iii) Some plasticware, such as deep-well plates for culturing bacteria, can be washed, sterilized, and reused. (iv) Some plasticware can be reused for unintended purposes. For example, inverted tip box lids are serviceable reagent reservoirs.

Automation

Laboratory automation is an important aspect of increasing both the throughput and the accuracy of the workflow. While manual sample processing is feasible, it must be done with a *zealous* attention to detail to avoid simple errors such as transferring a sample to an incorrect destination. Furthermore, when processing thousands of samples, fatigue, boredom, and distraction increase the likelihood for errors dramatically. In contrast, robots naturally do exactly what they are programmed to do, time and time again, thus reducing error rates significantly.

A wide number of options are available for automating sample preparation processes, such as PCR amplification, plasmid preparation, and sequencing reactions. These options range from fully automated systems— often custom systems composed of an articulated arm on a linear rail that can transfer plates between automated stations and custom software to integrate the instruments—to small benchtop liquid handling robots. Most academic genomic facilities opt for the middle ground with semiautomated workflows designed around one or more midsized liquid handling robots. These robots are generally constructed from three orthogonal linear rails (X,Y,Z axes) in a cantilever or gantry configuration that move a *head* or *pod* above a stationary *deck* where labware (plates, tip boxes, etc.) are held (Fig. 1). The head is generally equipped with liquid handling probes (fixed tips or mandrels for disposable tips) and/or a *gripper hand* to grip and move labware between stations on the deck. There are two major configurations for the liquid handling probes. In the first configuration, there are between 1 and 8 separate probes, each capable of independent movement in the Z axis. Pipetting with individual probes is generally controlled by a syringe pump and valve dedicated to each probe. Robots with this configuration are quite flexible and can transfer individually specified volumes between any two individual locations on the deck. This is particularly useful for *cherry picking* or *rearraying* or for normalizing concentrations across an entire plate. In the second configuration, the head is equipped with a fixed array of probes, usually 96 or 384. With arrayed probes, pipetting is generally controlled by a mechanical pipette incorporated in the head. This format is significantly faster than individual probes and is suited to transferring identical volumes between every well in plates. Various accessories can be incorporated on or in the deck, including static plate positions, shakers, incubators, thermal cyclers, barcode readers, vacuum manifolds, plate stacking carousels (hotels), tip washing stations, and disposal chutes. The final component of liquid handling robots is the software, which on most commercial robots has a graphical user interface that allows programs to be developed without the need to write code. We use

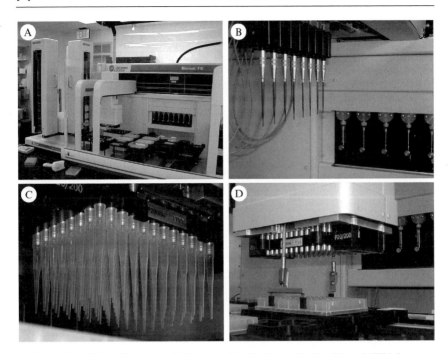

Fig. 1. Liquid handling robot: (A) image of the Beckman Coulter Biomek FX laboratory automation workstation, (B) individual probes, (C) 96-arrayed probes, and (D) gripper hand. (See color insert.)

a Beckman Coulter Biomek FX Laboratory Automation Workstation configured with one arm equipped with a 96-probe head and a gripper hand and a second arm equipped with 8 individual probes (Fig. 1). It is designed for flexibility, with a deck that can be reconfigured very easily and features user-friendly software, the importance of which should not be underestimated.

When developing liquid handling programs the following principles should be considered. (i) The process of loading the plates onto the deck is a step at which errors may occur. This could result in errors related to sample processing or even a physical crash. A visual map of the deck is a common feature of liquid handling robot software. We physically label each position on the deck, and the programs will not execute until the operator actively confirms that the deck is loaded as specified by the program. (ii) Pipetting parameters should be set to avoid drips and aerosols. This includes aspirating from just below the liquid surface (avoids liquid adhering to the outer surface), drawing a small air gap after aspirating (avoids the sample dripping), including a tip touch (tips gently touching the upper wall of the well) after aspirating or dispensing (avoids droplets adhering to the tip), ensuring that the tip is above the bottom of the well when dispensing (avoids sudden pressure release), using a small volume of air and low speed if a blow-out step is used when dispensing (avoids bubbles), and mixing at slow speeds (avoids bubbles). (iii) The high spatial accuracy of some liquid handling robots allows liquid to be dispensed to a destination plate without becoming contaminated by the sample in the destination plate. The technique of dispensing the first sample to the bottom of the wells and then dispensing the second sample to the upper portion of the well wall allows one set of tips to be reused to dispense multiple aliquots from one source plate to multiple destination plates without cross contamination. These steps help assure high quality.

Contamination

While parallel processing in multiwell plates increases efficiency, it also greatly increases the risk of cross contamination between samples. It is essential to design the workflow with this in mind and to conduct quality assurance tests. Special attention should be paid to the following. (i) Plates must be sealed effectively. When using adhesive seals, it is critical to ensure that they are applied without wrinkles and that sufficient pressure is exerted to form an impermeable seal. This can be done by rubbing the surface with a roller or a wad of tissue paper. If using heat seals, it is essential that the sealing tool be adjusted correctly. (ii) Sealing tape must be removed without creating airborne droplets. This is less of a concern if care is taken to not agitate the plates and to keep them in the horizontal position. With frozen plates, seals are removed as soon as the plates are out of the freezer. With liquid samples, seals are removed after brief centrifugation. We have tested both methods and found no cross contamination. (iii) Particulate contaminants must be avoided whenever plates are not covered by a lid, such as on the deck of a liquid handling robot. The simplest and most effective precaution is to keep the workspace clean. When designing a program on the liquid handling robot, moving labware or pipette tips over destination plates that may be contaminated by falling particles or drops must be avoided. Carefully planning the deck layout and programming the precise coordinate path that the robot takes can achieve this. (iv) Drips and aerosols must be avoided when pipetting (see earlier discussion). Having taken all these precautions, it is prudent to conduct tests designed to detect potential cross contamination. This is done by seeding a test plate with a positive control (bacterial culture or primer pairs) and a negative control (culture media or PCR buffer) in a checkerboard pattern so that each well containing a positive control is surrounded by eight wells containing negative controls. Then process the plate through all the steps of the workflow. The absence of product (growth or DNA) in the negative control wells confirms that cross contamination is not occurring.

Purity and Quality Control

Downstream steps in microarray fabrication—spotting, DNA attachment, and hybridization—can be affected dramatically by impurities in the DNA samples. First, particulate impurities can block microarray printing pins. Second, impurities that affect the hydrophobicity of the spotting solution can affect the morphology of the spots. Third, impurities may interfere with the attachment chemistry. Finally, impurities may fluoresce at wavelengths that overlap with the emission wavelengths of the detection dyes that are used. For these reasons, the following precautions are suggested: (i) Avoid *enhanced* PCR buffers containing various additives unless they are validated to not affect downstream steps. (ii) Use powder-free nitrile gloves and avoid latex gloves that may leach a fluorescent contaminant. (iii) Follow a purification protocol that has been validated with the specific slide chemistry that is used. Using standardized reagents and protocols is the easiest way to ensure high-quality and reproducible data.

The first step to ensuring a uniform high-quality microarray is to control as many variables as possible. The major variables are procedures and reagents. Procedures can best be controlled with well-documented standard operating practices and training to ensure uniformity among personnel. Reagents can best be controlled by sourcing reagents from large reputable suppliers and purchasing large volumes from the same batch. Major variables, such as operator, date, batch, instrument, and protocol version number, should also be recorded so that problems or variation can be traced rapidly. Even apparently trivial changes in reagent, such as a new batch of enzyme, can have a major impact on overall data quality.

In most high-throughput applications, some failure rate is unavoidable. The purpose of QC is to detect systematic deficiencies that may affect the quality of the end product, which in this case are fluorescence intensity values. If the deficiency is due to an error in the process (e.g., a critical component is defective or missing), then remedial action can be taken. This is only possible if QC data are monitored in real time. High-throughput production can result in either high-throughput success or high-throughput failure!

The advantage of thorough documentation and monitoring is that both systematic and episodic failures can be identified and their scope assessed at each stage along the process.

For PCR-amplified DNA, QC tests need to confirm that (i) there is a single product, (ii) the product has the correct sequence, (iii) there is no significant contamination with nonspecific amplified DNA, and (iv) there is sufficient yield. All except the sequence can be assessed by agarose gel electrophoresis. The ultimate QC test of amplification is to actually sequence the product. However, this approach is impractical due to financial and time constraints. Consequently, agarose gel electrophoresis is the most common QC test of PCR products, and each product should be electrophoresed and visualized on a gel. Performing the QC test after the purification step allows this step to be monitored as well. If the expected sizes are known, then concordance between the observed and expected molecular weights of the PCR products can be used as a measure of confidence that the sequence is correct and to flag products suspected to be incorrect. This can be combined with the sequencing of a subset of samples to estimate the rate of misidentified sequences.

Because all multiwell plates look identical and symmetrical at first glance, mistakes resulting from confusing or rotating plates are a pernicious problem. If such an error occurs at a step preceding the QC of the PCR products, it can be detected by inspecting the fragment sizes. However, if such mistakes occur subsequent to the QC test, it may go entirely undetected. For this reason it is advisable to include a single sample, referred to as a *bar code element* (a control DNA sample placed at unique locations within each sample plate to encode the identity of each sample plate unambiguously), in coded locations that are unique in each different sample plate. Once the samples are printed to microarrays, hybridization with a labeled complement to the bar code element can be used to confirm that all plates were in the correct order and in the correct orientation.

Information Management

The physical samples and data associated with each sample must be tracked throughout the workflow. It is essential that this be approached systematically. For instance, with a set of 15,000 samples and a workflow with eight physical steps and four data records there are 180,000 entities to be tracked. This can be achieved using a laboratory information management system (LIMS), a custom database, spreadsheets, or even flat text files. The essential requirement for all systems is that each entity, physical and data, is associated with a unique identifier (ID). While a database can function using arbitrary numerical IDs, finding and processing physical samples in the wet laboratory are facilitated greatly by using alphanumeric IDs that have a logical meaning. This is best achieved by devising a hierarchical syntax that identifies each of the following: (i) the project or platform (e.g., mouse array version 3), (ii) sample plate (e.g., plate containing 96 unique primer pairs or clones), (iii) the step in the workflow (e.g., plate containing PCR reaction or plate containing purified DNA), (iv) the batch number (e.g., batch 1), and (v) the well identity within the plate (e.g., A01 - H12). In the wet laboratory, each plate in the workflow is labeled. For example, DGRC_053_CLN_0605 designates: Project = DGRC, sample plate number = 053, workflow step = CLN (purified "clean" DNA), batch = 0605 (June 2005). Using such a scheme any operator who understands the logic of the system is able to unambiguously identify any sample in the process. We extend this system to label boxes of disposable liquid handling robot pipette tips that are to be reused with particular plates or processes. Adhesive thermal transfer labels printed by a barcode printer are resistant to most solvents and cold temperatures, and these print-plate IDs provide a robust human and machine-readable format for production, archiving, and data management. Regardless of the labeling method that is used, a working system must have a well-documented process and history in order to make full use of quality control data.

To track the progress of samples through the workflow, we maintain a log in the laboratory. The log is effectively an interface between QC and assessment on the one hand and process management and documentation on the other. Each step in the process can be vetted, and changes or problems, which are inevitable, can be managed or incorporated in subsequent analysis. For instance, to manage agarose gel quality control data from PCR amplifications, we use custom Excel macros to parse and analyze the flat files. The output of this process is a single file with QC data recorded against each sample so that (for example) failed reactions can be flagged. We then import that data into a customized implementation of our LIMS and analysis pipeline package, which is built on open source software called BioArray Software Environment (BASE). BASE is a web-based database solution to store, view, and analyze data generated by microarray experiments (Saal *et al.*, 2002). BASE has LIMS capabilities to record experimental information, including materials, protocols, and annotation for microarray production and hybridizations.

Methods

Workflow

We have designed a workflow that accommodates three parameters (Fig. 2). First, because our thermal cycler capacity is eight plates, we process PCR reactions in *batches* of eight plates. Second, we purify two 100-μl PCR reactions sequentially using the same filter plate. Thus, we simultaneously process *duplicates* of each plate (2 × 96-well plates with the identical set of samples) through the cleanup protocol. Third, our liquid handling robot has two vacuum manifold stations. Therefore, we simultaneously process the

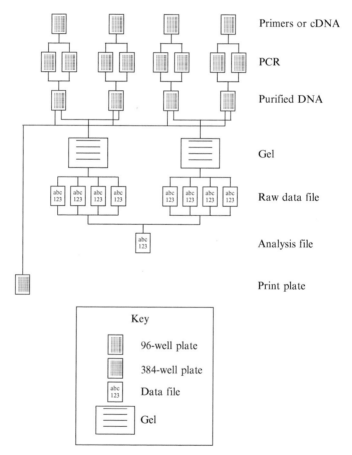

Primers or cDNA

PCR

Purified DNA

Gel

Raw data file

Analysis file

Print plate

Key

96-well plate

384-well plate

Data file

Gel

FIG. 2. Workflow. The schematic chart depicts the progression of samples through the workflow. The stage in the process is indicated on the right and a key is shown below the chart.

two pairs of duplicate plates during each robot program. This workflow can obviously be adjusted according to available equipment.

PCR Amplification

PCR amplifications of microarray-quality DNA must meet the following criteria: (i) maximal specificity, (ii) maximal yield, and (iii) no contaminants that affect hydrophobicity or attachment to microarray substrates or which autofluoresce in wavelengths used for labels.

PCR reactions are performed in skirted, 96-well, thin-walled polypropylene microwell plates in a total volume of 100 μl. Reactions to amplify from genomic DNA template contain 50 mM KCl, 10 mM Tris–HCl, pH 8.3, 1.5 mM MgCl$_2$, 200 μM each dNTP, 0.6 ng/μl genomic DNA, 200 nM each primer, and 0.025 U/μl Taq polymerase (Table I) and are cycled as follows: 94° for 2 min, 35 cycles of 94° for 1 min, 55° for 1 min, 72° for 1 min, 1 cycle of 72° for 2 min, and then held at 4°. Reactions to amplify from cDNA template are seeded with 10 μl of diluted (10/100 μl) and heat-killed overnight bacterial culture as described in Hedge $et\ al.$ (2000). The reactions contain 50 mM KCl, 30 mM Tris–HCl, pH 8.0, 2 mM MgCl$_2$, 100 μM each

TABLE I
MATERIALS AND EQUIPMENT

Material	Source
Taq DNA polymerase	Brinkmann-Eppendorf
dNTP set (100 mM)	Brinkmann-Eppendorf
Ultrapure distilled water	Gibco-Invitrogen
100-bp ladder	New England BioLabs
Salts and buffers	Sigma, Fisher
Skirted PCR plates	Marsh Bio Products
96-well plates	NUNC
384-well plates	Genetix
Reagent reservoir	Marsh Bio Products
Sealing tape	Corning-Costar
FX tips	Beckman-Coulter
Multiscreen-PCR 96	Millipore
Electrophoresis unit	Owl Separation Systems, Model A6
Peltier thermal cycler	MJ Research DNA Engine Tetrad, PTC-225
Speed-Vac	Thermosavant SpeedVac Plus SC210A
Filtration manifold	Millipore
Liquid handling robot	Beckman-Coulter Biomek FX
Digital imaging	KODAK Image Station 440CF and Kodak 1D Image Analysis Software, version 3.5

dNTP, 200 nM each primer, and 0.02 U/μl *Taq* polymerase and are cycled as follows: 94° for 3 min followed by 40 cycles of 94° for 45 s, 56° for 1 min, and 72° for 1 min with a final extension of 10 min at 72°.

Steps in the amplicon workflow are as follows. (i) Prepare a PCR master mix sufficient for a batch of plates. For the standard 8-plate batch, we prepare an 800-reaction master mix, allowing for up to 4% pipetting error. The master mix can be stored for up to 2 h at 4°. (ii) Retrieve 4 × 96-well primer plates from −80° storage, each containing 96 unique forward/reverse primers pairs (2.5 μM each primer). Thaw and centrifuge at 2000 rcf for 1 min. (iii) Load the deck of the liquid handling robot with reagents and plasticware (Fig. 3). (iv) Run the Biomek FX Protocol *PCR Setup* (modified from Beckman Coulter eLabNotebook: Automated Methods for PCR Setup on the Biomek FX, http://www.beckmancoulter.com). The protocol aliquots 92 μl of master mix and then dispenses 8 μl of primer mix to each well. The robot then mixes the reagents in place (Fig. 3). (v) Seal the

A

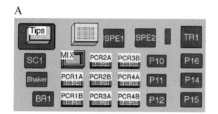

1. LOAD: TIP
2. TRANSFER: 92 μl MIX to PCR1A
3. Repeat Step 2 to add master-mix to PCR1B – PCR4A
4. UNLOAD: TIP
5. Manually transfer master-mix to final plate PCR4B

B

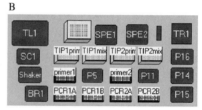

6. LOAD: Tip1prim
7. MIX: Primer1
8. TRANSFER: 8 μl PRIMER1 to PCR1A
9. TRANSFER: 8 μl PRIMER1 to PCR1B
10. UNLOAD: TIP1prim
11. Repeat 6–10 using Tip2prim, primer2, PCR2A and PCR2B
12. Check that primers are mixed with master mix
13. LOAD: Tip1mix
14. MIX: PCR1A
15. MIX: PCR1B
16. UNLOAD: Tip1mix
17. Repeat 13–16 using Tip2mix, PCR2A and PCR2B
18. Repeat 6–17 using PCR3A-PCR4B, PRIMER3–4, Tip3prim-Tip4prim, and Tip3mix-Tip4mix

Fig. 3. Automated liquid handling protocol PCR setup. Maps of the liquid handling robot deck are shown above the program steps. (A) Dispensing master mix. TIPS, P200 filter tips; MIX, reagent reservoir with master mix; PCR1A-PCR4B, empty PCR plates. (B) Adding primers. TIP1prim-TIP2mix, P200 tips; PRIMER1–2, plates of primers; PCR1A-PCR2B, PCR plates containing master mix. Notes: The mix at step 7 is 30 μl at 100% speed repeated 10 times. At steps 8 and 9, dispensing is made above the liquid level with a tip touch to avoid contaminating the tips with the master mix. Step 12 is used to ensure that the hanging droplet of primers has flowed down the wall of the well to mix with the master mix. At steps 14 and 15 the mix is 30 μl at 50% speed repeated five times. (See color insert.)

plates when the automated protocol is complete. (vi) Place the plates in the Peltier thermal cycler and run the cycling program. (vii) When the reactions are complete, store the plates at $-20°$ or proceed immediately with cleanup.

This workflow takes approximately 30 min to set up 8×96-well plates (duplicates of 4 different plates), and the thermal cycling takes approximately 2.5 h. Thus one operator can comfortably perform 16–24 96-well PCR amplifications per day.

Purification of PCR Product

After testing several alternatives we adopted size-exclusion filtration for purifying PCR products. The process is compatible with downstream microarray applications (Hedge *et al.*, 2000), and the format is more automation friendly than the alternative bind–wash–elute methods. By using filtration the sample is added and recovered from the top of the plate. Therefore, there is no need to assemble filtration and collection plates in a vacuum manifold, and aerosols in the vacuum manifold will not cause cross contamination. Furthermore, two 100-μl PCR reactions can be *serially* purified over a single plate without appreciable reduction in purity or yield. After the first sample is purified and eluted, the second sample must be immediately added to the plate and purified *before* the membrane dries out. After a plate is used once it cannot be stored and used at a later date. We tested two methods for resuspending the retentate—shaking and pipetting—and found that automated pipetting is the most efficient method.

Steps involved in the purification workflow are as follows. (i) Retrieve $8 \times$ 96-well PCR plates from $-20°$ storage (each plate contains 96 unique PCR products). Thaw and centrifuge the plates at 2000 rcf for 1 min. (iii) Load the deck with labware and reagents as shown in Fig. 4. Note that tips used previously to mix the PCR mixture are reused to transfer the reactions to the filtration plate. (iv) Execute the Biomek FX program *Cleanup* (modified from Beckman Coulter eLabNotebook: Increasing Throughput and Quality of Sequencing by Automation of MultiScreen Filter Technology on the Biomek FX, http://www.beckmancoulter.com). This program is semiautomated; the robot pauses once to prompt the operator to add reagents and labware to the deck. The steps in the program are illustrated in Fig. 4. The robot transfers the PCR product to a Multiscreen plate and applies a 7400-mm Hg vacuum for 4 min to the Multiscreen plate via the filtration manifold automated labware position (ALP). Then ultrapure water (100 μl) is added to each well in the Multiscreen plate. The retentate is pipetted up and down 25 times (*in situ*) and is then transferred to a multiwell collection plate. When the automated protocol is complete, the plates are sealed and stored at $-20°$.

A

B

1. Manually turn on vacuum pump
2. LOAD: TIPuse 1
3. TRANSFER: 100 μl PCR1A to filter1
4. UNLOAD: TIPuse1
5. Repeat 2–4 using TIPuse2, PCR2A, and Filter2
6. Vacuum 4 minutes
7. MOVE: Tipwater to position TL1
8. MOVE: Filter1 to position P4
9. MOVE: Filter2 to position P7

10. LOAD: TIPwater
11. TRANSFER: 100 μl Water to Filter1
12. TRANSFER: 100 μl Water to Filter2
13. UNLOAD: Tipwater
14. LOAD: TIPcln1
15. MIX: Filter1
16. TRANSFER: Filter1 to Dest1
17. UNLOAD: TIPcln1
18. Repeat 14–17 with TIPcln2, Filter2, and Dest2
19. Repeat 2–18 with PCR1B and PCR2B

FIG. 4. Automated liquid handling protocol cleanup. Maps of the liquid handling robot deck are shown above the program steps. (A) Filter purification. Filter1–2, filter plates; Dest1–2, empty storage plates; PCR1A-PCR2B, PCR plates with PCR product; TIPuse1–2, P200 tips (can be reused from PCR setup protocol); TIPcln1–2, P200 tips; TIPwater, P200 tips. Note: The pipetting head is used to gently press on the filter plates to ensure that a seal forms under vacuum. (B) Recovery of retentate. Millipore, Millipore collar for filter plates; Filter1–2, purification filter plates; Dest1–2, empty storage plates; Water, reservoir containing water; PCR1A-PCR2B, PCR plates with PCR product; TIPuse1–2, P200 tips; TIPcln1–2.P200 tips; TIPwater, P200 tips. Notes: Mix at step 15 is 50 μl at 50% speed, repeated 25 times. The aspiration at step 16 is done at 50% speed while the tips circle the bottom of the well to recover all of the liquid. (See color insert.)

This workflow can process 8 × 96-well plates per run. Each run takes approximately 45 min. Therefore, an operator can comfortably process 24–32 plates in a day.

Visualization of Purified PCR Products by Electrophoresis

We check every sample by agarose gel electrophoresis. Automated image analysis can distinguish between samples with one band, no band, a faint band, multiple bands, and a smear or a band whose electrophoretic mobility differs from its expected size. After testing most options, our laboratory has adopted a protocol for manually dry loading large-format gels using multichannel micropipettes. Manual loading is robust, and large-format gels provide sufficient size resolution to evaluate the expected sizes.

We use a gel format that is compatible with multiwell plates. The dimensions of the gels are 23 × 25 cm with four evenly spaced rows of 50 wells. Each row is loaded with 48 DNA samples and includes size standards in two positions (Fig. 5A). Thus, samples from two 96-well plates are accommodated by a single gel. The well-to-well spacing in the gel (4.5 mm) is exactly half the well-to-well spacing of a standard 96-well plate (9 mm). Therefore, when loading a gel from a 96-well plate using an eight-channel pipette, adjacent samples from one column in the plate are loaded into every other well in the gel. Repeating the process with an offset of one well interleaves the samples from the next row as follows: A1, A2, B1, B2, C1, C2, . . . , G1, G2, H1, H2. The sample order is subsequently deconvoluted at the data analysis step.

We prepare 1.5% agarose gels with 0.5× TBE buffer and then add 0.2 μg/ml of ethidium bromide (EtBr) to a cooling gel. The warm agarose is poured within the gel-casting trays (Sambrook and Russell, 2001; caution EtBr is a mutagen). We dry load 10 μl of a 100-bp standard (50 ng/μl 100-bp ladder, 1× gel-loading dye) in the first and last wells of each row and then immediately dry load 5 μl of each PCR sample with eight-channel pipettes as described earlier. The size standard has fragments of 100–1500 bp to measure the molecular weight and bands of variable and known amounts of DNA ranging from 18 to 97 ng to estimate the yield of our purified PCR products.

We carefully submerge the loaded gel within the electrophoresis unit containing 0.5× TBE, and a field of 3.75 V/cm is applied for ~1 h or until the leading band of the standard is nearing the next row of wells. The gel is then cut in half above the third row of wells and marked at the upper left corner of each half as a way to ensure proper orientation. Each half is visualized under ultraviolet light and recorded using a digital imaging station.

Although the following protocol is specifically designed for Kodak ID Image Analysis software, the process of converting a large number of images into data for downstream analysis is general. Scaling up begins by using predefined gel analysis templates and settings in the "Preferences" directory that can be applied to all images. Images are captured using identical camera settings. Following image capture, a template with pre-defined lanes is imported, which specifies the correct position of loaded samples in the gel. Two templates are saved for such use, one for each half of the gel image. The position of the first template is adjusted to exclude the wells from the template area, and lanes 1 and 50 are identified as standards by using predefined settings that specify the molecular weight and relative DNA concentration of each band of the ladder. Standards are positioned at both ends of a row of 48 samples to take advantage of software functions that perform point-to-point linear interpolation

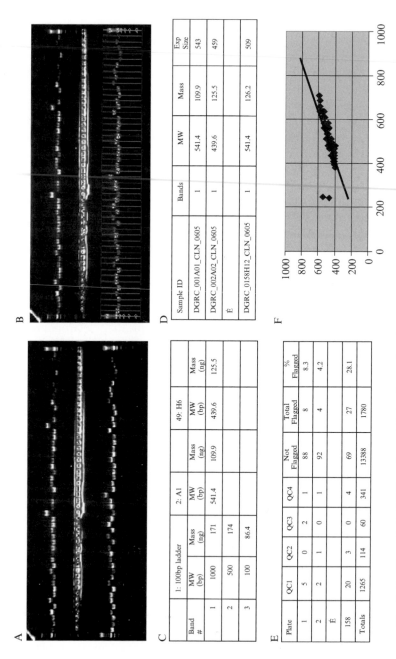

C

Band #	1: 100bp ladder		2: A1		49: H6	
	MW (bp)	Mass (ng)	MW (bp)	Mass (ng)	MW (bp)	Mass (ng)
1	1000	171	541.4	109.9	439.6	125.5
2	500	174				
3	100	86.4				

D

Sample ID	Bands	MW	Mass	Exp Size
DGRC_001A01_CLN_0605	1	541.4	109.9	543
DGRC_002A02_CLN_0605	1	439.6	125.5	459
Ė				
DGRC_0158H12_CLN_0605	1	541.4	126.2	509

E

Plate	QC1	QC2	QC3	QC4	Not Flagged	Total Flagged	% Flagged
1	5	0	2	1	88	8	8.3
2	2	1	0	1	92	4	4.2
Ė							
158	20	3	0	4	69	27	28.1
Totals	1265	114	60	341	13388	1780	

FIG. 5. Quality control of PCR products. (A) An electrophoresis gel image for one 96-well plate of purified PCR products. (This is half of the gel.) (B) The same image as A with an overlay by Kodak 1D software, identifying bands on each lane of one row on the gel.

between each standard and will therefore accurately adjust for occasional localized mobility variation.

Software functions called "Find Bands" and "Adjust Bands" automatically mark bands on the image (Fig. 5B). Genuine bands are distinguished from artifacts according to sensitivity and band-width settings, which are also optimally predefined if the protocol is followed consistently throughout. Finally, a function called "Display" calculates and prints predefined measurements of interest, including molecular weight, mass, and relative intensity for each sample band based on their position and fluorescence relative to the standards. Data are exported as a tab-delimited text file and saved for the quality control tests. The same procedure applies to the remaining bottom half of the image and is repeated for each plate.

Given that electrophoresis takes roughly 2 h and that we typically process batches of 8 plates, a single person can routinely process 12–24 plates in a day. By optimizing and using predefined settings for the image templates, the standards, the sensitivity parameters, and the measurements of interest, images can be processed within a reasonable time frame.

Quality Control of PCR Products

The challenges associated with scaling up laboratory procedures to minimize costs and time are not limited to processing reagents. Data collected for the purpose of assessing the quality of products during the manufacturing of DNA probes can also be massive and unwieldy. The twin challenges of data are (1) collection and storage and (2) downstream analysis and use of data. Without proper attention to data management

---

(C) An excerpt from tab-delimited data generated by Kodak 1D analysis. The first row identifies the lane of the gel. The molecular weight and mass of each band in each lane are recorded under the lane headings. (D) An excerpt of consolidated quality control data stored as a tab-delimited file. Included in the file is sample identification, number of bands (PCR products), molecular weight, mass, and expected size. The molecular weight and mass of only the dominant PCR product are included. (E) An example of quality control summary information. For each 96-well plate of amplicons, the number of samples failing quality control is listed categorically. They are identified as missing products (QC1), weak products (QC2), multiple products (QC3), and products of unexpected size (QC4). From these quality control statistics, the following calculations are recorded: the total number of samples passing quality control, the number of samples flagged during quality control, and the percentage of samples flagged. (F) An example of a scatter plot generated during quality control analysis. For one 96-well plate, the expected molecular weight of the amplicons (*Y* axis) is plotted against the observed molecular weight (*X* axis) of the dominant PCR product. Regression analysis is used to identify samples that deviate from their expected molecular weight. (See color insert.)

issues, the overall process of fabricating microarrays can be compromised by delays in discovering technical problems, by mistakes in interpreting data, and by creating a bottleneck in the production timeline. For these reasons, we have created a semiautomated *pipeline* of helpful computer programs that (i) combines and reorganizes data files, (ii) identifies and flags samples that do not meet predefined quality control criteria, and (iii) verifies that the 96-well plates and their orientations are labeled correctly. This pipeline tool involves three programs that run as Microsoft Excel macros (Plate_QC, Batch_QC, Flag_QC). Their summary files allow for a rapid assessment of individual samples and provide an overall assurance on the quality of each plate and of each batch.

Plate_QC. For each 96-well plate of purified PCR products that is gel electrophoresed, two tab-delimited text files are created from the Kodak 1D software. These files contain information on the number of bands recorded for each sample, their estimated DNA concentration, and their molecular weight (Fig. 5C). The first program of our pipeline is designed to merge the information from each pair of files derived from individual 96-well plates into single files in a readable format while sequestering additional information linked to the workflow, including the identification of personnel and dates when procedures were completed. Specifically, the script requests the location of the Kodak 1D output files and sample/experimental identification parameters, including plate ID and batch ID. The program then opens the two sister files derived from individual plates, copies the content from one file to the other, sorts each sample (lane in electrophoresis gel) by band mass to identify the dominant product, and prints the sample ID and number of bands for each sample. The molecular weight, mass, and percent relative intensity for the dominant band are also recorded. Finally, Plate_QC sorts the output by sample ID, which is saved in a tab-delimited text format (Fig. 5D).

Batch_QC. This second program of our pipeline combines all 158 newly created plate files into a single composite file arranged by sample ID. Specifically, the script requests the location of the output files from Plate_QC, identifies and opens the first plate file, and copies its content and pastes into a new output file. The program recursively processes all files in numerical order that are located in the source directory.

Flag_QC. The final program in the pipeline performs analysis and summary of data. PCR products can fail QC tests in four ways: (i) no band, (ii) weak band, (iii) multiple bands or a smear, and (iv) a molecular weight that deviates from the expected amplicon size. Given that samples are processed in 96-well formats, problems linked to plate sample tracking would be made obvious. Therefore, Flag_QC lists the samples in each plate that failed the PCR, lists the samples with multiple PCR products, identifies samples with deviant molecular weights, and identifies plates with molecular weight profiles that do not match expected profiles (Fig. 5E). Mismatched molecular weights and plate profiles are identified by a regression analysis of expected versus observed weights across all 96 samples within each plate and are measured by deviation of a sample from the regression line (Fig. 5F). Specifically, this final script requires a user-defined value limit with which to evaluate the deviation of samples from the regression line and the location of the file created by the Batch_QC program and requires a tab-delimited text file containing the sample ID and the expected molecular weights of the amplicons. The program then combines data from both files and, for each plate, creates a scatter plot and regression line of the observed versus expected molecular weights within Microsoft Excel. This first plot offers a visual representation of data to identify systematic errors. The formula of the regression line is parsed to calculate whether samples deviate within the user-defined percentage range from the line. A summary table of flagged elements is then created, identifying elements that have failed or passed each of the four QC tests. A second scatter plot is finally created that differentiates flagged and nonflagged samples using a color code.

Processing of data using pipelines for the purpose of quality control is an in-house solution to the laboratory's requirement of managing a large number of samples and for finding problems. Therefore, this pipeline is unique for our laboratory.

Consolidation

Following the quality assurance tests of the purified PCR products, the remaining steps of the workflow prepare the DNA for deposition onto the glass-slide microarrays. Consolidation is the process of lyophilizing, reformatting to 384-well plates, and reconstituting the DNA in spotting solution. Although these steps require multiple days to complete, the consolidation of samples into a higher density plate format permits the use of large number of pins within microarray print heads, diminishes evaporation of the samples in plates during a print run by reducing the surface area of the spotting solution, and conserves space.

Our workflow is designed to simultaneously process four 96-well plates to accommodate the number of positions available on the decks of our laboratory orbital shaker and liquid handling robot. As a result, the protocols minimize direct supervision at each step. The workflow includes drying the purified PCR products, resuspending the DNA in water, thus ensuring that all samples have equal and appropriate volumes for transfer, reformatting the samples from 96-well plates into 384-well plates, lyophilizing the samples in the new format, and resuspending the samples in spotting solution for printing. The details of these procedures are as follows. (i) First we retrieve 28 × 96-well purified DNA plates from −20° storage. The samples are thawed and then centrifuged at 2000 rcf for 1 min. The seals from each plate are removed, and the samples are placed in a laminar flow hood for 24–48 h to dry. Alternatively, the samples are dried under vacuum in a Speed-Vac for 3 h with heat or overnight without heating. The dried plates are retrieved in groups of four and placed on the deck of the liquid handling robot. The Biomek FX program *Resusupend-1* is then executed. (ii) *Resusupend-1* is a protocol that resuspends four 96-well plates containing the dried DNA. The deck of the robot holds four racks of P20 tips (TipsA–TipsD), the four 96-well DNA plates (SrcA–SrcD), and one reservoir of ultrapure water (Water). For each plate, the corresponding tips are loaded onto the 96-channel pipetting pod to transfer 20 μl of water to each sample. The labeled tip boxes are saved for the next liquid handling step, *Consolidation*. The plates are sealed and stored at −20° for over 24 h. (iii) We next retrieve and thaw the plates in groups of four, shake at 800 rpm for 1 h, and then centrifuge at 2000 rcf for 1 min. The Biomek FX program *Consolidation* is then executed. This program transfers the resuspended DNA from four 96-well plates (SrcA–ScrD) into a single virgin 384-well plate (Dest). For specifying the correct addresses for each transfer of 96 samples, the 384-well plate is divided into four quadrants (A, B, C, D). The deck of the robot also holds the same four racks of P20 tips (TipsA–TipsD) that were used during the previously executed *Resuspend-1* protocol. For each Src plate, the corresponding tips are loaded to transfer 10 μl of sample to each quadrant of the 384 well (Dest) in sequential order. (iv) When the transfer is complete, the samples are dried either in a laminar flow hood for 24–48 h or under vacuum in a Speed-Vac. (v) The lyophilized DNA samples are finally reconstituted in spotting buffer containing 3× SSC and 1.5 M betaine in groups of two plates using the Biomek FX program *Resuspend-2*. This program uses the 96-channel pipetting pod to transfer 8 μl of spotting solution from a reagent reservoir (Soln) to each sample, one quadrant at a time (specified as 384A-D and 384E-H). The deck of the robot therefore holds eight sets of virgin P20 tips (TipsA–TipsH), and for all quadrants of both plates, the corresponding tips are loaded for the transfers. When this last automated protocol is complete, the plates are sealed and stored at 4° for at least 16 h, shaken at 800 rpm for 1 h, and finally stored at 4° before they are used for printing.

This last stage of the workflow contains many rate-limiting drying steps. However, plates can be processed continually starting at different stages of the workflow. Our use of a liquid handling robot for unsupervised processing at many stages permits a single person to complete the task of consolidating 158 × 96-well plates within 2 weeks.

Summary

We have developed and utilized these protocols in fabricating spotted DNA microarrays for three experimental organisms: *Drosophila* (http://dgrc.cgb.indiana.edu/), *Daphnia* (John Colbourne, Brian Eads, Elizabeth Bohuski, and Justen Andrews, unpublished; http://daphnia.cgb.indiana.edu/), and *Helianthus* (sunflower) (Lai *et al.*, 2006). These microarrays differ in important respects. The *Drosophila* arrays are designed beginning with a well-developed genome annotation of a fully sequenced genome. As a result, the probes are produced using gene-specific primers to PCR amplify DNA from genomic template. In contrast, the *Daphnia* and sunflower arrays are created using universal primers to PCR amplify cloned cDNA from heat-killed bacterial cultures (low/medium copy or high copy) or purified plasmid DNA. Despite these differences, the same workflow was used successfully with minor modifications for each project. The success rates ranged between 91 and 96% (Table II). The dominant mode of failure for amplifications from genomic DNA using gene-specific primers was obtaining no amplified product, probably reflecting a failure in primer design or synthesis. However, the dominant mode of failure when using universal primers to amplify cDNA clones varied between libraries. The average yield from 100-μl PCR reactions seeded with cDNA clones was generally in the range of 6–10 μg with a low to medium copy plasmid giving a significantly lower yield of 2 μg.

TABLE II
PCR Statistics

Organism	Template	Single band	No band	Multiple bands	Average yield[a]
Drosophila	Genomic DNA	92%	7%	1%	9.6 μg
Daphnia	Bacterial culture	91%	6%	3%	9.6 μg
Daphnia	Plasmid DNA (high copy)	96%	1%	3%	6.3 μg
Helianthus	Bacterial culture (low/medium copy)	91%	7%	2%	1.9 μg
Helianthus	Bacterial culture (high copy)	91%	1%	8%	6.6 μg

[a] Yield per 100-μl reaction.

The protocols and workflows described here represent one solution among many possibilities for high-throughput production of DNA samples for microarrays. In the absence of budgetary constraints, highly sophisticated custom installations of laboratory robotics can achieve complete automation. Where budgetary constraints exist, most facilities are operating with semiautomated workflows. The protocols presented here can either be implemented directly or used as a starting point to guide investigators in developing custom solutions for high-throughput challenges.

Acknowledgments

We thank the many members of the Drosophila Genomics Resource Center and the Center for Genomics and Bioinformatics who contributed in many ways to the protocols described here. We gratefully acknowledge the invaluable assistance of John Colborne, Brian Eads, Elizabeth Bohuski, Zhao Lai, James Costello, and Monica Sentmenant in preparing this manuscript. This work was supported in part by the Indiana Genomics Initiative and by NIH Grant 1 P40 RR017093 to J.A.

References

Bowtell, D., and Sambrook, J. (2002). "DNA Microarrays: A Molecular Cloning Manual." Cold Spring Harbor Laboratory Press, Cold Spring Harbor, NY.

Dieffenbach, C. W., and Dveksler, G. S. (1993). Setting up a PCR laboratory. *PCR Methods Appl.* **3,** S2–S7.

Hedge, P., Qi, R., Abernathy, K., Gay, C., Dharap, S., Gaspard, R., Hughes, J. E., Snesrud, E., Lee, N., and Quackenbush, J. (2000). A concise guide to cDNA microarray analysis. *Biotechniques* **29,** 548–550, 552–554, 556 passim.

Holloway, A. J., van Laar, R. K., Tothill, R. W., and Bowtell, D. D. (2002). Options available—from start to finish—for obtaining data from DNA microarrays II. *Nat. Genet.* **32**(Suppl.), 481–489.

Lai, Z., Gross, B. L., Zou, Y., Andrews, J., and Rieseberg, L. H. (2006). Microarray analysis reveals differential gene expression in hybrid sunflower species. *Mol. Ecol.* **15,** 1213–1227.

Ren, B., Robert, F., Wyrick, J. J., Aparicio, O., Jennings, E. G., Simon, I., Zeitlinger, J., Schreiber, J., Hannett, N., Kanin, E., *et al.* (2000). Genome-wide location and function of DNA binding proteins. *Science* **290,** 2306–2309.

Saal, L. H., Troein, C., Vallon-Christersson, J., Gruvberger, S., Borg, A., and Peterson, C. (2002). BioArray Software Environment (BASE): A platform for comprehensive management and analysis of microarray data. *Genome Biol.* **3,** SOFTWARE0003.

Sambrook, J., and Russell, D. W. (2001). "Molecular Cloning: A Laboratory Manual." Cold Springs Harbor Laboratory Press, Cold Spring Harbor, NY.

Stowe-Evans, E. L., Ford, J., and Kehoe, D. M. (2004). Genomic DNA microarray analysis: Identification of new genes regulated by light color in the cyanobacterium *Fremyella diplosiphon. J. Bacteriol.* **186,** 4338–4349.

Weinmann, A. S. (2004). Novel ChIP-based strategies to uncover transcription factor target genes in the immune system. *Nat. Rev. Immunol.* **4,** 381–386.

[6] The Printing Process: Tips on Tips

By REED A. GEORGE

Abstract

While the microarray printing process consists of a few simple operations, many variables can affect the final quality of printed arrays. As in most high-throughput processes, the ultimate goal is to reduce and control variability. This chapter describes how to minimize variation in the printing process through proper selection and installation of printing tips, printing buffers, and implementation of quality control procedures.

Introduction

This chapter addresses the equipment, reagents, and methods required to successfully set up and operate a microarray printing system. Much of the discussion focuses on the variables that affect contact printing and reviews the tools that are available for monitoring quality control in the printing process.

The purpose of this chapter is to provide some basic troubleshooting and quality control tools to aid in producing high-quality printed microarrays on a consistent basis. Discussion of the variables that most affect print quality should aid in the troubleshooting of existing printing systems as well. In this chapter DNA samples that are deposited on the slide are referred to as probes, whereas the labeled species that are hybridized to the probe are referred to as targets. The surfaces to which the probes are attached, typically coated microscope slides, are referred to as substrates. While the focus is to provide information applicable to printing of both cDNA and oligonucleotide probes, the controlling variables and methods are applicable to protein printing as well.

Because there are many commercial or custom robotic systems available for printing microarrays, this chapter does not address specific system software. While many of the variables discussed are related to processes performed prior to and in preparation for microarray printing, the quality control methods described focus on controlling the actual printing process. The scope is also limited to printing pins and does not include noncontact or ink jet systems (Lausted *et al.*, 2006).

Printing pin selection and installation on the arraying system are discussed. Further, the variables that affect print quality and process

METHODS IN ENZYMOLOGY, VOL. 410   0076-6879/06 $35.00
                                                       DOI: 10.1016/S0076-6879(06)10006-3

troubleshooting methods are described, along with ongoing quality control procedures.

The Printing Process and Equipment

The microarray printing process is described in Fig. 1.

Printing is performed through the use of a simple robotic system, most commonly a three-axis (X,Y,Z) Cartesian configuration. The X and Y motions are typically accomplished with fairly low accuracy, using long travel ball screw or lead screw axes. However, many systems now employ high-speed, high-accuracy linear motors. Linear motors are substantially more expensive than traditional screw systems, but provide significant advantages in terms of speed, acceleration/deceleration, and accuracy. The vertical or Z axis is typically a high-precision lead screw, with fairly small travel length requirements. Microarray substrates are typically fixtured to a large platter, sometimes referred to as a platen, which is attached to either or both of the X and Y axes. Platters are usually large enough to hold between 20 and 300 substrates at a time. Large platters are a mixed blessing in that they provide for higher throughput and better utilization of probe reagents, but are also typically less consistent in flatness and mechanical tolerances, requiring careful robot system adjustment.

The print head, which contains the printing pins and serves to attach them to the Z axis, is usually designed to arrange the pins on 4.5-mm spacing, allowing the use of SBS format 384-well plates for probe aspiration. Print heads with pins set on 1536 (2.25 mm) spacings have recently become available. Most commonly, 16, 32, 48, or 64 pins are loaded into a print head. As the printing tips must contact the microarray substrate during the printing process, print heads are designed to provide some compliance to avoid damaging the tips. The pins must also return to their original position after contact is made. This is accomplished through spring-loading, gravity, or compliant foam or other materials exerting a downward force on the pins.

In addition to the robot axes, substrate platter, and print head, the microarrayer must include fixtures to secure the source plates that will supply probes for aspiration by the pins and facilities for washing and drying the tips between probe aspirations. Washing may be accomplished through the use of an ultrasonic or recirculating water bath. Drying is typically accomplished by a vacuum fixture, which pulls excess liquid from the tips.

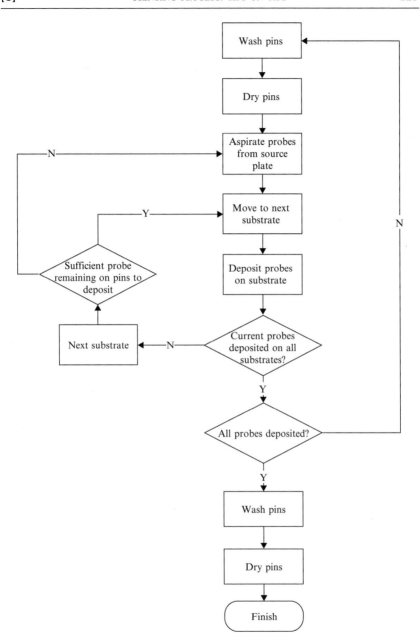

Fig. 1. Generic microarray printing process.

Printing Pin Technology Options

Currently, there are several commercial suppliers of microarray printing pins. The competing technologies are differentiated by design, manufacturing precision, and material. Contact printing pins take advantage of the surface tension properties of the liquid in which the probe is suspended to first aspirate the probe onto the pin tip, and subsequently to deposit the probe, or a fraction of it, onto the microarray substrate. Table I lists commercial suppliers and their advertised pin technologies.

Solid Pins

The most simple tip design consists of a solid metal (typically stainless steel or tungsten) cylindrical pin, machined to a fine point. Solid pins only pick up a small volume of liquid during the probe aspiration step (typically hundreds of picoliters), which is essentially completely transferred to a single substrate in one printing step. This necessitates an aspiration step for each array to be printed, increasing printing cycle times greatly. However, solid pins typically provide for highly repeatable spot morphologies, and the small uptake volume may be useful in cases where probe materials are limiting or extremely expensive. Solid pins are also the easiest to clean in the tip washing step and minimize carryover and clogging issues associated with other tip designs.

Closed Capillary Pins

Closed, cylindrical capillary pins have also been employed for microarray printing. The first type is based on precision ceramic capillaries from the microelectronics industry (Kulicke and Soffa, Pennsylvania) that have been adapted as drop-in replacements for stainless steel pins (George *et al.*, 2001). However, there is no commercial supplier of the tooling to

TABLE I
COMMERCIALLY AVAILABLE PRINTING PINS

Supplier	Tip types	Material
Majer Precision (www.majerprecision.com)	Open capillary	Metal
Telechem International (www.arrayit.com)	Open capillary, solid pin	Metal
Point Technologies (www.pointtech.com)	Open capillary, solid pin	Metal
Genetix (www.genetix.com)	Open capillary, solid pin	Metal
Kulicke and Soffa (http://www.kns.com)	Closed capillary	Ceramic
Labnext (www.labnext.com)	Closed capillary	Ceramic
Parallel Synthesis Technologies (www.parallel-synthesis.com)	Open capillary	Silicon

adapt these pins for microarray printing. A commercial closed capillary tip (Xtend Printing Pin) is available from LabNext, (Glenview, IL). Another group is using pulled quartz glass capillary tubes for microarray printing (Hamilton *et al.*, 2000).

Open Capillary Pins

The split pin or open capillary design is the most widely used in the field. For metal pins, the split tip design is typically manufactured from a pointed metal cylindrical pin, with a narrow channel machined into the pin tip, similar to a fountain pin. The channel is typically machined by electrical discharge machining and is on the order of 25 μm in width. The channel serves as a reservoir for aspirating larger volumes of probe (on the order of hundreds of nanoliters), allowing hundreds or thousands of substrates to be printed from a single aspiration step. This increases the throughput of the printing system greatly by minimizing the number of aspiration steps in a print run. The trade-off again is in the increased difficulty in achieving effective tip cleaning between samples and the possibility of tip clogging.

In addition, metal pin designs that rely on a sharp point to control the printed spot size have the associated issues of tip wear and deformation. As the tips are used in the printing process, making contact with the hard surface of substrates many thousands of times, the tips begin to deform and flatten, gradually increasing the size of the deposited spot. More drastic tip deformation can result from incorrect print head set up that causes pins to contact substrates with excessive force. One design that minimizes these issues (Telechem International, Sunnyvale, CA) incorporates a small flat area at the tip of the metal pin, which is less susceptible to deformation.

Micromachined Open Capillary Pins

In recent years, the techniques of photolithography-based micromachining have been applied to print pin technology in an effort to reduce print variability due to tip machining tolerances. Micromachined stainless steel tips have been fabricated and tested, but have not made a significant entry into the commercial market (Reese *et al.*, 2003).

Silicon microarray technology from Parallel Synthesis Technologies (Santa Clara, CA) uses tips micromachined from silicon. Because silicon is much harder than metal, the silicon microarray tips do not deform over time, thereby maintaining their high precision and repeatability. We have found that the micromachining techniques allow for significantly increased levels of precision and increased consistency in printing performance. Figure 2 shows the silicon microarray pins and print head.

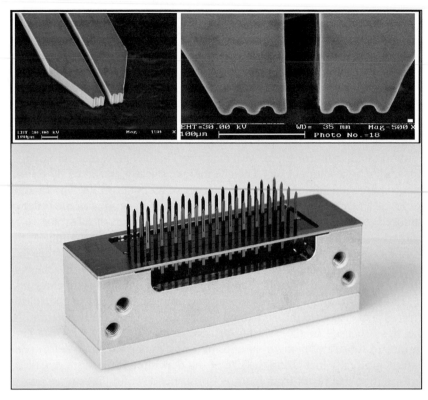

FIG. 2. Silicon microarray printing technology. (Top) Scanning electron microscope images of a silicon print tip, with a tip area of 200 × 200 μm. (Bottom) A silicon microarray print head with 48 tips installed. Images courtesy of Parallel Synthesis Technologies, Inc. (Santa Clara, CA). (See color insert.)

Critical Parameters and Troubleshooting

The microarray printing procedure is the point where all the processes leading to uniform probe spotting capability come together. A number of parameters must be optimized in the printing process itself, as well as in probe and substrate preparation. Printing parameters and influencing factors are discussed in detail in this section. Parameters related to the probe and substrate preparation are discussed briefly in order to aid in process troubleshooting.

Printing System Optimization

The robotic microarray printing system must be adjusted carefully to provide for the most robust printing possible. The robot system itself should be leveled accurately during installation, using a bubble level and adjusting or shimming the legs at the base of the system. If a commercial system is being used, then any adjustments or shimming of the robotic axes should be performed by the manufacturer at the time of installation. The level of the slide platter should be verified once the system table level is established, and the platter should be shimmed as required. Next, the coplanarity of the X and Y axes should be verified. The procedure will vary slightly from system to system, depending on the configuration of the robot. The best procedure is to use a dial indicator (e.g., a Starrett Model 701) on a stand, temporarily connected to the moving portion of the axis being tested. This configuration is shown in Fig. 3.

If any moving part on the axis is magnetic, a magnetic base for the indicator (such as the Starrett Model 657AA) will be helpful. The indicator can be zeroed with the indicator preloaded approximately 500 μm (.020 in.) against the platter. Next, with the power to the system disconnected, the axis is pushed manually so that the dial indicator measures any deviation in distance between the axis and the slide platter or another flat, level surface such as the system table. This procedure should be repeated for both the X and Y axes in several places spanning the travel length of each axis across the platter. Two criteria should be met: (1) minimal random

Fig. 3. Dial gauge as installed for axis coplanarity verification.

deviation (no consistently high or low side of the robotic axis relative to the platter) and (2) maximum deviation should be less than approximately 500 μm (0.020 in.). If these criteria are not met, the axes must be adjusted or shimmed until satisfactory measurements are obtained. Again, these specifications may vary between systems, depending on length of axis travel (shorter axes should have lower deviations) and platter flatness. Depending on the robot configuration, the measurements may have to be taken from different flat surfaces. For example, if the slide platter is attached to the X axis, height variation may be measured with respect to the system table.

Next, the print head mounting should be verified in similar fashion. This time, the dial indicator base should be attached to the slide platter, and measurements should be taken from the bottom edge of the print head, while moving the X and Y axes independently. Because the length of the print head is much smaller than the length of the axis, criteria for maximum deviation should be more on the order of 100 μm. Again, the print head mount should be adjusted or shimmed to ensure that this specification is met.

After the system is adjusted and verified mechanically, it is sometimes necessary to map the height variation over the slide platter so that each individual slide print height is optimally set in the printing software. This may be performed manually by using the system software to move the print head until the pins contact the slide and recording the Z axis coordinates or may be performed automatically with the appropriate software and sensor (George and Spellman, 2000). Many microarray printing facilities utilize the Arraymaker software, developed by Dr. Joseph DeRisi. A new feature for automated slide height mapping has been incorporated in ArrayMaker v. 2.6 (http://derisilab.ucsf.edu/arraymaker.shtml). Robotic axis adjustment and slide height mapping should only need to be performed at the startup of a new robotic system or if the system is moved or otherwise disrupted.

During installation of a new print head, the system must be adjusted properly such that the print, probe aspiration, tip cleaning, and drying locations match the modified mechanical arrangement. This adjustment operation should be performed with a few dummy pins in the print head to avoid damaging new pins. The dimensions provided here are tailored for the use of Parallel Synthesis Technologies' silicon microarray tips and may vary slightly from other pin technologies.

The printing height at the substrate should be adjusted such that all pins in the head contact the substrate, with as little overtravel (travel in the Z axis beyond where the tips contact the substrate) as practical, ensuring that all pins contact the substrate on each print cycle. This overtravel is typically set at 25–100 μm (.001–.004 in.). If the slide platter heights are mapped correctly, adjustment of the print height at one slide location will be sufficient.

The probe aspiration location should be adjusted such that the pins are centered in the source plate wells in the X, Y plane. The aspiration depth should be adjusted such that the pins are within approximately 100 μm (0.004 in.) of the well bottom. This allows for aspiration of virtually the entire probe volume, thereby minimizing waste.

The tip washing location should be adjusted such that the tips are immersed to a level deeper than the maximum probe level in the source plates. In some ultrasonic cleaning systems, cleaning efficiency is dependent on immersion depth, and the optimum height must be derived empirically from carryover testing (described later). Immersion depths that cause resonance (large amplitude vibration) of the system should be avoided. Further, the tip drying location should be tested empirically to ensure complete drying in the least amount of time.

The printing system should be installed in a humidity-controlled area. For silicon microarray printing tips, 65% relative humidity is ideal. At less than 45%, printing capability is seriously affected. The entire microarray printing facility may be humidity controlled or the arrayer may be enclosed in an environmentally controlled chamber. An enclosure can have the added benefit of protecting the microarrays from particulate matter during printing without the need for a large HEPA-filtered air source.

Printing Process Troubleshooting

Symptoms of an uncontrolled printing process are variability or out-of-specification spot diameters and/or missing spots on the microarrays. The causes can usually be traced to one of four sources: the robotic system, printing pins, reagents, or the printing environment. The best way to identify between them is to observe the symptoms themselves carefully.

Robotic System Issues

Robotic system problems are usually manifested in repeatable patterns. For example, if out-of-specification spots are detected repeatedly at the same substrate location on the platter, print height adjustment is likely to be the cause. If the spot diameter varies from one side of the print head to the other, the print head level is likely to be out of adjustment. One example of an exception to this repeatability is inaccurate adjustment of the pin drying location. Most printing pin and substrate combinations will not allow the printing of water, which is left in the tip if drying is not performed effectively. Therefore, incomplete drying may lead to missing spots. If the adjustment is such that the pins dry to varying degrees at each cycle, the symptom may appear as random pins not printing, with the problem pin locations changing from cycle to cycle.

Printing Pin Issues

Problems associated with printing pins are also typically systematic, in that particular pins fail repeatedly to print correctly. One effective test is to rearrange the pins in the print head and determine if the problem follows particular pins, or locations in the print head. One problem that will be specific to print head location is pin binding. In this case, the printing pin does not slide smoothly in the head, and the tip does not return to its original height after each printing step. This can result in the tip not contacting the substrate in the next print step, not being washed or dried effectively, or not aspirating a sample on the next cycle. This can usually be remedied by cleaning the printing pin and its location in the print head. Variable spot sizes associated with particular pins may be a result of print tip wear or damage from repeated contact with the hard substrate surface during normal printing operations, especially with metal pins. Damaged pins may produce larger spots if the tip is flattened or smaller spots if the open capillary is partially blocked by damage. Ceramic and silicon pins are less susceptible to this type of problem. Contamination of print tips with hydrophobic materials such as silicone lubricants can also cause pin-related failure. The contaminated tips will usually stop printing altogether in this case.

Reagent-Related Failures

Reagents most critical to uniform printing performance are the probe samples and the microarray substrates. Printed spot diameter and presence depend on consistent probe concentration and volume in the source plates, as well as appropriate selection and preparation of printing buffer. Low concentration probe samples will exhibit variable printing characteristics. In this case, printing from the same source plates multiple times should indicate problems with specific probe wells. Probe source plates should also be selected to ensure flatness for accurate aspiration in all tip locations.

Substrates may be purchased precoated or prepared in the laboratory. In either case, uniformity of the coating is critical to printing performance. Substrate coating uniformity may be assessed by using a standard probe preparation, such as the herring sperm DNA described in the quality control section to print multiple slides. Poor substrate surfaces will typically manifest themselves as random out-of-specification spots and will not be associated with pin location, specific probe samples, or locations on the robotic system. Most substrate coatings have a maximum shelf life that should not be exceeded.

In terms of printing buffer selection, most facilities using poly-L-lysine-coated or similar substrates print probe samples suspended in 3× sodium chloride and sodium citrate solution (SSC). However, many users also print from 50% dimethyl sulfoxide buffer, which reportedly denatures probe DNA, as well as exhibits a lower evaporation rate than SSC, reducing issues associated with storing probe plates uncovered for extended periods of time (Hegde *et al.*, 2000). Many substrate vendors also offer proprietary spotting buffers, which must be evaluated empirically by the user.

Environmental Issues

As mentioned earlier, the temperature and humidity in the microarray printing facility can affect overall print quality. High temperature and low humidity can cause premature drying of the probe on the print tips during printing and usually result in gradual printing degradation as the print head moves across the platter (early arrays are acceptable, later arrays are not). In general, facility conditions of approximately 70°F and 45–65% relative humidity are ideal.

Quality Control Testing

The most important factors influencing quality of the printing process are spot presence, consistency in size and morphology, and control of carryover or probe-to-probe contamination. Recommended quality control practices are provided in this section.

Printed Spot Presence and Uniformity

Printed spot consistency is best evaluated using the purpose-built fluorescent microarray scanner. Most scanners provide analysis packages that provide spot diameter as part of their standard output. In order to visualize the spots, a fluorescent-labeled random oligonucleotide target may be used, as it will hybridize to every spot on the array with a reasonable measure of uniformity (www.pangloss.com/seidel/Protocols/9-mer HybProtocol.pdf). Random 9-mers with appropriate fluorescent label modifications are available from several suppliers (e.g., Cy3-labeled 9-mers, Operon Biotechnologies, Inc., Huntsville, AL). In addition to gathering data on the morphology of the printed spots, the fluorescent intensity from the hybridization assay may be used to evaluate the consistency of DNA deposition, providing quality control of probe concentrations, as well as printing capability. The method is quite simple and takes very little time to perform. Because the method does not require any special conditions during the microarray printing process, it may be used to evaluate a representative sample from every print run.

Alternatively, if quality control of the printing process itself is the focus of the experiment, microarrays may be printed with sheared herring sperm DNA (Promega, Inc.), diluted to approximately 200 ng/μl concentration. It is important that the DNA be sheared (to approximately 100–3000 bp) in order to avoid tip clogging. This provides an ideal test array, as the hybridization to each spot, and therefore the fluorescent intensity across the entire array, should be identical.

The procedure is as follows. Microarrays are printed and postprocessed according to standard protocols (e.g., http://derisilab.ucsf.edu/microarray/protocols.html). A mock hybridization target, prepared according to standard protocols, but substituting 150 pmol of labeled random primer for the labeled nucleic acid target is denatured at 90° for 1 min, allowed to cool to room temperature, hybridized to the microarray using a hybridization chamber, and incubated at room temperature for 3–5 min. The target is then discarded, and the microarray is washed by standard protocols and scanned on a microarray scanner. The resulting image may be inspected visually to quantify the number of missing spots, and the diameter of each spot can be determined using the scanner software. A useful measure of repeatability is the coefficient of variation (CV), which is calculated as the standard deviation of the spot diameter, divided by the mean spot diameter, and is typically expressed as a percentage. With silicon microarray tips, we have been able to routinely achieve a CV of 6% for 32 tips, and less than 6% spot-to-spot CV with a single tip ($n = 300$–400 spots per pin, data not shown).

Probe Carryover

Evaluation of the amount of probe remaining on the printing tips after cleaning and drying, and therefore potentially contaminating the next probe in the printing sequence, is best performed in an experiment that represents as closely as possible a normal production print run. This can be achieved by printing probes representing ubiquitously expressed genes and then hybridizing with a labeled target generated from typical RNA samples. These housekeeping gene probes may be printed as controls by including them in the source plates for every print run or custom arrays may be printed specifically for testing probe carryover.

The source plates should be configured such that a representative number of pins will print first with standard printing buffer (containing no probe), followed by the housekeeping gene probe, and finally by at least two rounds of printing buffer only. Of course, these quality control arrays should be printed using the normal printing process, with the standard tip wash and dry cycles between each probe. The resulting microarrays can then be processed by the standard protocols and hybridized to a representative labeled target from the appropriate organism and/or tissue. Upon fluorescent scanning, the initial buffer spots can be used to establish a baseline nonspecific signal. The housekeeping gene spots will represent the signal to be expected for an undiluted probe. The sequential buffer spots printed after the gene spots are indicative of the percentage of probe that remains on the tips after each cycle of cleaning and drying. Carryover should be well below 1%, and typical values for silicon microarray pins are less than 0.5% (DeRisi and Carroll, unpublished data).

Alternatively, to assay carryover levels more quickly, it is possible to print a single oligonucleotide probe species in a similar procedure as described earlier, alternating with spotting buffer, and hybridize with the appropriate labeled antisense oligonucleotide. Because the complexity of the target is extremely low, the antisense probe can be hybridized in standard hybridization solution at room temperature for 5–10 min. Hybridized arrays may be scanned and evaluated immediately after performing the standard wash procedures. Exact hybridization and wash conditions will need to be determined empirically.

In order to get an even more rapid assessment of probe carryover, a fluorescent-labeled probe may be printed directly, in combination with spotting buffer-only wells as described earlier. However, this test is the least representative of the actual printing process and also puts fluorescent molecules in direct contact with printing pins. Gross carryover of the labeled probes could cause illegitimate signal in arrays that are subsequently printed with the same pins.

Future Developments

An interesting new area of research and development in printing technology is based on taking further advantage of silicon micromachining technology to produce an integrated print head with hundreds or thousands of essentially identical printing tips, which will be capable of extremely high-throughput printing. Figure 4 shows a device currently being developed by Parallel Synthesis Technologies under funding from the National Human Genome Research Institute of the National Institutes of Health. The device shown incorporates 100 print tips, on 2.25-mm spacing into a single silicon piece. Preliminary data have demonstrated spot diameter CV levels of less than 5% across all 100 tips.

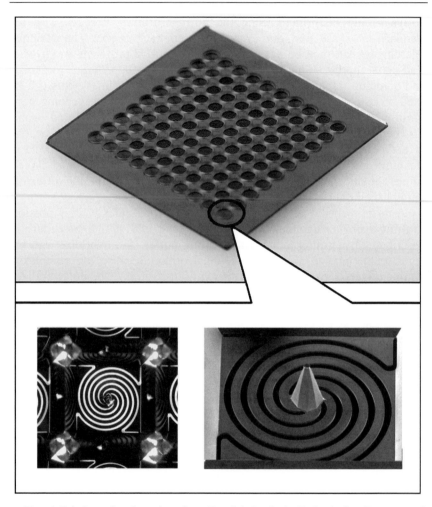

Fig. 4. Printing microtiter plate from Parallel Synthesis Technologies, Incorporated. Photograph and electron micrographs of a printing microtiter plate with 100 print tips on an 1536 SBS format. Images courtesy of Parallel Synthesis Technologies, Inc. (Santa Clara, CA). (See color insert.)

Acknowledgment

The author acknowledges a business relationship with Parallel Synthesis Technologies, Incorporated, whose products are described in this work.

References

George, R. A., and Spellman, P. T. (2000). Microarray technology improvements for increased density and repeatability. Poster presentation. Genome Sequencing and Analysis Conference, Miami, FL.

George, R. A., Woolley, J. P., and Spellman, P. T. (2001). Ceramic capillaries for use in microarray fabrication. *Genome Res.* **11,** 1780–1783.

Hamilton, G., Jones, A., Uber, D., Davy, D., Hansen, T., Nordmeyer, R., Wilson, D., Jaklevic, J., Gray, J., Albertson, D., and Pinkel, D. (2000). Technical approaches for producing and analyzing DNA microarrays. ISAC XX Abstracts 2000.

Hegde, P., Qi, R., Abernathy, K., Gay, C., Dharap, S., Gaspard, R., Hughes, J. E., Snesrud, E., Lee, N., and Quackenbush, J. (2000). A concise guide to cDNA microarray analysis. *BioTechniques* **29,** 548–562.

Lausted, C. G., Warren, C. B., Hood, L. E., and Lasky, S. (2006). Printing your own inkjet microarrays. *Methods Enzymol.* **410** (this volume), 168–189.

Reese, M. O., van Dam, R. M., Scherer, A., and Quake, S. R. (2003). Microfabricated fountain pens for high-density DNA arrays. *Genome Res.* **13,** 2348–2352.

[7] Making and Using Spotted DNA Microarrays in an Academic Core Laboratory

By Janet Hager

Abstract

Since its major launch into academia in the mid-1990s, spotted DNA microarray technology has expanded and matured into an important mainstream tool for genomic-scale gene expression studies across many species with many applications. Based on the principles of enzymatic nucleic acid labeling and DNA hybridization, the basic techniques were initially developed and disseminated by Patrick Brown's laboratory at Stanford and by others using "open source" approaches to techniques and instrumentation. Accessibility of microarrays has now become an important component of institutional research support. Indeed, the challenge facing many investigators when designing genome-scale experiments is to choose an appropriate platform and method from among the many microarray options available to them, both commercial and academic. The combination of microarray instrumentation and methods used for gene expression studies vary tremendously at different institutions and yet together function equally well as a whole. Instead of presenting a definitive set of instrumentation and methods, this chapter describes one such functional solution. It describes the specific implementation of instrumentation, standard operating procedures, and approaches for microarray fabrication

METHODS IN ENZYMOLOGY, VOL. 410   0076-6879/06 $35.00
          DOI: 10.1016/S0076-6879(06)10007-5

and gene expression studies that are used routinely at the Microarray Resource within the W. M. Keck Biotechnology Resource Laboratory at Yale. The procedures have evolved through 6 years of operation and have resulted in at least 50 publications acknowledging the use of microarray slides and/or services provided by the Resource. The protocols that are presented for array fabrication, quality control, labeling, and hybridization utilize both "home-brew" and commercially available products to achieve an optimized set of cost-effective tools. The aim is to provide a compendium of approaches and protocols to aid those starting out on the core laboratory path and to provide insight into the types of microarray services and studies that are undertaken in this particular academic core laboratory.

Introduction

Spotted microarrays have revolutionized the way in which biomedical researchers approach comparisons of cellular gene expression by allowing the interrogation of thousands of genes in a single experiment. The goal of the W. M. Keck Biotechnology Microarray Resource at Yale is to provide microarray tools and services in the most efficient and cost-effective way possible in a nonprofit setting. By its very nature, microarray technology is a relatively expensive platform to establish in terms of instrumentation, maintenance contracts, reagent costs, and personnel required for implementation. The Resource made the decision early on to focus on fabricating arrays at Yale and thus it is one of very few academic facilities nationwide with clean-room capabilities for printing microarrays. The costs associated with such a facility are of course more than that of a "benchtop" approach. In 1999 when the facility was established, there were very few academic core microarray labs or commercial vendors in existence. However, during the 6 years since its inception there has been a tremendous focus on microarray technology by both NIH funding and commercial vendors. The result is a plethora of options today for investigators wishing to use microarrays in their research, as duly noted in Gershon (2004). Along the way, a number of commercial companies have ceased to offer microarrays as products, for example, Incyte and Qiagen, perhaps indicating that the "market" for arrays is high in cost and low on return. With this in mind, it is quite difficult to keep abreast of all other options available commercially and to make comparisons based on microarrays available from other academic sites. The relative cost of microarray slides and services at various academic institutions is more often a reflection of the level of institutional and grant support rather than any difference in the *actual* cost incurred for fabrication and services. Also note that many academic microarray cores work on a collaborative basis *only,* whereas most of the work done in the Microarray Resource at Yale is noncollaborative, nonprofit, cost recovery. The glass-platform microarray slides distributed through the Microarray Resource range in cost to Yale investigators from ~$70 to $310 (without discount) depending on the number of probe features and the nature of the probes. The smallest arrays have 2000 features or less and the largest to date are 35,000 oligonucleotide probes per slide. At current printing density, Keck glass slides have two arrays per slide (i.e., two dual channel experiments per slide) for up to 9216 probes and a single array per slide above that level. It is estimated that the cost for standard labeling and hybridization, when done in an investigator's own laboratory without labor cost, is approximately $125. The total experiment cost without labor, for a pair of RNA samples using the human 28K array, would therefore be $435. This compares favorably with commercial array costs, for example, CodeLink Human Single Color Whole Genome 55KArrays from Amersham at $595 per slide, with reagents (no enzymes), and CodeLink Human Whole Genome 20KArrays at $495 per slide, with reagents (no enzymes). One advantage of using an academic core array, for which it is hard to assign a dollar value, is the availability of unlimited advice, assistance with labeling, training, and consultation on site. Additionally, a major strength and value of an academic core such as this is the printing capability for the development of new approaches and for custom arrays using print materials provided by individual investigators. Custom arrays may be offered cost effectively at ~$70 per array. Commercially designed and implemented custom array services are generally outside the reach of the budgets of most investigators. It is the arena of custom microarrays and the application of new approaches such as protein arrays, single nucleotide polymorphism (SNP) arrays, comparative genome hybridization (CGH), and diagnostic arrays that is anticipated as a growth area in usage over the next few years.

Technologies and Services Provided by the Keck Microarray Resource

The Keck Microarray Resource (http://keck.med.yale.edu/dnaarrays) was founded in 1999 as part of the W. M. Keck Biotechnology Resource Laboratory at Yale. The goal of the Resource is to enable its users to rapidly and effectively apply spotted microarray technology to answer important biological questions across a wide spectrum of biomedical research, while at the same time monitoring emerging technologies and implementing new approaches. During its first 16 months of operation, the facility processed >35,000 cDNA clones from bacterial stocks through plasmid isolation and polymerase chain reaction (PCR) amplification to arrayable cDNA. The Resource fabricates glass platform "generic," "targeted," and "custom" human, mouse, rat, and Arabidopsis microarrays under clean-room conditions for multiple applications using both cDNA and oligonucleotide probes. The Resource provides full-service projects—RNA QC, labeling, hybridization, and primary analysis—for investigators needing to obtain preliminary data for grant submission and feasibility studies, as well as for those clinical and nonclinical investigators without expertise, personnel, or time to undertake a study in their own laboratory and who need full-service microarray project completion. Custom array design and printing services for specialized, investigator-initiated, projects are also provided. The Resource currently provides an avenue for researchers to use any or all of these tools and services as they require, with the option of carrying out some procedures, such as labeling and hybridization, in their own laboratory. More than 259 researchers at Yale and 223 at other institutions have utilized slides and/or services of the Resource and a total of 51 publications have appeared to date in peer-reviewed journals (http://info.med.yale.edu/microarray/pub_serv.htm) acknowledging the Resource. Pilot and complete project services are provided using a variety of dual-color "dye"-labeling approaches, dependent on RNA quantity available, including aminoallyl, Genisphere 3DNA dendrimer, and Invitrogen/Genicon RLS gold and silver particles (see Brownstein, Chapter 11). We hybridize the labeled targets, scan the resulting slides, and acquire image and numerical data. Primary analysis of the resulting data is done in the Microarray Resource to generate expression ratios and scatter plots. We provide assistance in experimental design and grant writing and assess and implement emerging technologies. Examples of "generic" glass slide microarrays available include 15,000 NIA (National Institute Aging) mouse cDNAs; 25,000 mouse 70-mer oligonucleotides (Operon); 4800 rat 60-mer oligonucleotides (Compugen); 4608 human cDNAs (Research Genetics); 28,000 human 70-mer oligonucleotides (Operon); and 35,000 human 70-mer oligonucleotides (Operon, Fall 20005). Examples of focused, cost-effective "targeted" probe glass slide microarrays with two arrays per slide include human and mouse signal transduction and cancer arrays. A comprehensive web site with details of all these arrays is accessible through the Keck Biotechnology Resource Laboratory site (http://keck.med.yale.edu/dnaarrays) with links to additional information and publications of interest. All gene lists with GenBank Accession numbers and descriptions for each probe are posted at (http://keck.med.yale.edu/dnaarrays/genelists.htm).

A class 100 clean room with three bays for microarray printing and a class 1000 clean room for slide processing are part of the physical layout of the Microarray Resource suite. The clean rooms are under tight environmental control at 72° and 55% relative humidity and facilitate printing of high-quality microarrays. The following major instrumentation is used within the laboratory: genomic solutions, gene machines OmniGrid Arrayers (2); OmniGrid Server Arm 3; Perkin Elmer, ScanArray5000XL and molecular devices, Axon GenePix 4000A scanners; Invitrogen, Hi-Light reader for resonant light-scattering detection; Advalytix, ArrayBooster hybridization station; Matrix Technologies, Tango liquid handling system with plate stacker; Matrix Technologies, Hydra liquid handling system with TwisterH; Hitachi, GenSpec II spectrophotometer; Molecular Devices SpectraMax2 96-well plate reader; Qiagen, Biorobot 9600 (2) and biosafety cabinet.

Like most academic microarray core facilities at the outset, we fabricated glass slide arrays solely using PCR products from cDNA libraries and used protocols similar to those of Pat Brown's group at Stanford. In the face of increasing sequencing evidence that nonfidelity of cDNA targets is greater than 15% and as much as 30% when created by PCR from cDNA libraries (Taylor *et al.*, 2001) we made the transition in 2001 to primarily printing long oligonucleotides synthesized by commercial vendors. Oligonucleotides provide a relatively cost-effective, easily renewable resource for generic array probes and investigators obtain verifiable, accurate, reproducible, and meaningful data using this platform. We have faced and met the challenge of optimizing labeling and hybridization protocols to achieve the highest signal intensity, expression ratio fidelity, and least amount of variation in both qualitative and quantitative data from the arrays.

During the early years of operation of the Resource, a stumbling block that deterred many investigators from applying the power of whole genome or custom glass platform-spotted microarray approaches in their research was the fact that only very small quantities of total RNA (1 μg or less) or mRNA (nanogram amounts) of material for gene expression studies could be obtained in many cases. Driven by the fact that clinical samples are invariably handicapped by limited amounts of tissue, the need to focus on small areas of tissue, or the need to target cells of a specific type within a tissue, it has often not been possible to use glass platform microarray technology for these samples. To date, these problems have been addressed by employing a number of aRNA amplification methods. The expression patterns generated by these nucleic acid amplification approaches have been compared extensively to patterns resulting from nonamplified methods. Although there are many papers suggesting that the profile results of amplified vs unamplified RNA are very similar, a number of papers in the literature support the idea that expression patterns may be different due to skewing of the level of representation of some mRNAs in the final amplified labeled product, relative to the original representation of mRNA molecules in the cell. With these findings in mind, the Resource focused attention in 2004 on troubleshooting methods where RNA samples are nonamplified. First, we have optimized the Genisphere 3DNA dendrimer labeling approach without nucleic acid amplification and also with SenseAmp using only a single round of amplification. Second, we have added resonance light-scattering (RLS) technology as a means of signal detection for targets labeled with gold and silver nanoparticles using a specialized HiLight (white light source) reader and the GeniconRLS labeling system (Invitrogen). Estimates of the level of sensitivity of detection of low copy number RNA molecules in samples using standard fluorescently labeled microarrays do not appear to be well reported in the literature. Alexandre et al. (2001) showed that when using colorimetric gold–silver particles with multibiotinylated target DNA, the lower detection limit was 0.1 fmol with a dynamic range from 0.1 to 10 fmol and the authors suggest that fluorescent sensitivity may be in a similar range. Detection sensitivity for ABI (Applied Biosytems Inc.) chemiluminescent detection arrays is touted at 10–100 fmol, which is very comparable to that obtained with Cy3/Cy5 fluorescence. With the newer Q-dot fluorophores, 0.4–40 fmol detection has been reported for detection of miRNA on arrays (Liang et al., 2005) and even better detection is expected when optimum excitation and filter sets become available on standard microarray scanners. Au/Ag detection with white light resonant light scattering is in the 0.1- to 10-fmol range. In an operational sense, we can routinely detect a robust signal for Cy-3 and Cy-5 amino-allyl-labeled synthetic heterologous FX 174 cDNA fragments generated from 1 ng FX in vitro-transcribed RNA spiked into a labeling reaction with a 100-μg sample of total mammalian RNA (unpublished observation). However, we have not yet rigorously tested the LOD of cy-dye end-labeled FX174 probes by serial dilution. FX174 DNA has a molecular mass of 1.70×10^6 Da and is 5386 bases in length. The restriction fragments we use are 310, 604, 1072, and 1532 bases in length. Translating this into a detection sensitivity based on the number of molecules in the assay, each 1 ng spiked FX RNA (using the assumption that there is 1:1 conversion of FX RNA into single-stranded DNA during reverse transcription) represents for 310:9.9 fmol or 6.0×10^9 molecules; 604:5.1 fmol or 3.0×10^9 molecules; 1072:2.9 fmol or 1.7×10^9 molecules; and 1532:2.0 fmol or 1.2×10^9 molecules. This range of between 1 and 10 fmol is very much in line with published values and those reported by various vendors.

The GeniconRLS can be used for ultrasensitive detection of low copy number RNA molecules in expression studies. The system uses as little as 2 μg total RNA or 200 ng of mRNA without the need for amplification. Currently the technology is launched for nucleic acid analyses for both single and dual "color" microarrays. The Resource has successfully troubleshooted and tested the system with human cDNA and oligonucleotide microarrays made in the resource on poly-L-lysine-coated slides. GeniconRLS recommends commercially available slides and coatings, but we have shown that the system works as well for Keck arrays as these commercial surfaces with slight modifications during our standard slide postprocessing procedure. Having these two approaches for small RNA samples, we are able to make available a service that can accurately measure gene expression without amplification from well below 5 μg total RNA starting material. In combination with laser-capture techniques, these approaches bode well for future expansion of microarray expression studies at the level of small groups of cells and ultimately at the individual cell level.

Printing Spotted DNA Microarrays

Generic Glass Slide Microarray Printing

Arrays are printed in a class 100 clean room at 72°, 55% relative humidity, on TeleChem Superclean microscope slides coated in a class 1000 clean room with poly-L-lysine (http://www.microarrays.org/pdfs/Poly lysineSlides.pdf) using a Pat Brown protocol. cDNA probe arrays are composed of PCR products generated from cDNA clones. Plasmid DNA is isolated using Qiagen TurboPreps on QiaRobot 9600 robots. PCR products are generated in a 96-well plate format using PE GeneAmp thermocylers and are characterized on 1% agarose gels. After assessing global plate DNA concentration, the products are transferred to 384-well print plates using a Matrix Hydra with Twister H. PCR products are printed from 5 μl sample volume at between 200 and 400 ng/μl in print buffer (3× SSC, 0.005% sarcosyl) in Genetix X6004 plates. Duplicate adjacent spots are printed when possible using TeleChem Stealth SMP3 split pins on a GeneMachines Omnigrid-100 arrayer. Features are printed at 175 μm center-to-center distance using 16, 32, or 48 pins in a 4 × 4, 4 × 8, or 4 ×12 subarray pattern with up to 25 rows and 25 columns per subarray. Oligonucleotide probe arrays are composed of 65- to 70-mer Operon or Compugen oligonucleotides printed single spot, at 40 μM in print buffer (3× SSC, 0.01% sarcosyl) as described earlier. All arrays for standard fluorescence are outlined at the corners by scoring with a diamond pencil on the reverse side of the slide. Slides printed for Genicon resonance light-scattering applications are not scored. Slides are labeled with slide type and a unique number on Brothers white tape and are postprocessed in batches as needed according to a Pat Brown protocol (http://www.microarrays.org/pdfs/PostProcessing2001.pdf) by UV cross-linking and succinic anhydride blocking. Slides are stored at room temperature for up to 1 year in plastic slide boxes inside plexiglass cabinets containing desiccant.

Genomic Solutions OmniGrid 100 Microarrayer Print Settings

Because each printer is unique in its motor step distance, indeed as most X,Y,Z robotic arrayers age the motor step distance may change, it is not possible to provide hard and fast X,Y,Z settings for the arrayer. In general, when setting up a print with a 175-μm center-to-center spot distance with TeleChem Stealth SMP3 split pins, the following general guidelines may be followed.

1. Set up the print head with an appropriate number of pins for array size and layout. The main patterns used and maximum number of features per slide are 16 pin array: 4 × 4 blocks; 25 rows × 25 columns per block, total 10,000 features per array, two arrays per slide; 32 pin array: 4 × 8 blocks; 25 rows × 25 columns per block, total 20,000 features per array, one array per slide; 48 pin array: 4 × 12 blocks, 25 rows × 25 columns per block, total 30,000 features per array, one array per slide.

2. Prime the printing pins in the buffer anticipated for the print run (see later) by standing the pins in a 384-well plate with one index of buffer.

3. Set the print origin (usually 500 μm from top edge of slide, 500 μm from left edge of slide).

4. Set the Z value for pin contact by approaching the pin toward the slide with small motor steps until the pin contacts the slide as determined by eye. Although TeleChem recommends pulling the pin back off the slide by a motor step, we have found that this leads to a greater percentage of "drop-out" spots. Therefore we maintain contact of the pin with the slide. Theoretically, simply contacting the meniscus of the drop formed at the foot of the pin, and not the pin itself, to the slide should be sufficient to "pull" the drop out at each contact; however, this does not give good reproducibility under our conditions. By using this approach, we may cause slightly more rapid wear of the pin tips and consequently fewer total print spots per pin life when compared to using the "back-off" approach. This is something we have to balance in order to obtain printed slides with minimum drop-outs.

5. Set X and Y values based on spot size and number of features per array. X and Y values are generally the same. Most often we use a 175-μm center to center on both cDNA and oligonucleotide arrays.
6. Perform a test print to assure that all pins are both picking up and releasing sample optimally. If a pin is not performing well, examine the pin for lint or defects and either correct or replace.

Materials and General Settings for cDNA and Oligonucleotide Printing on In-House PLL-Coated Slides

Print plate: Genetix X6004 384-well low profile microplate, V bottom, with cover. Individually wrapped. (Many groups print in X7020.)

Sample volume per well: either 5 or 7.5 μl, cDNA (150–300 ng/μl); 5 μl, 40 μM (200 pmol per well) 70-mer oligonucleotide; or 7.5 μl, 40 μM (300 pmol per well) 70-mer oligonucleotide.

Slide type: in-house PLL-coated Telechem SuperClean slides.

Pin type: TeleChem Stealth SMP3 split pins.

Print buffer: oligonucleotide arrays: 3$\times$ SSC, 0.01% sarcosyl: 150 ml 20$\times$ SSC; 1 ml 10% sarcosyl; 849.0 ml ddH$_2$O. cDNA arrays: 3$\times$ SSC, 0.005% sarcosyl: 150 ml 20$\times$ SSC; 0.5 ml 10% sarcosyl; 849.5 ml ddH$_2$O. Filter sterilize using a 0.22-μm filter.

Spot spacing: 175 μm center to center.

Dip time: 6000 ms.

Sample redip: every 50 slides for single spot array; more frequently for multiple sample spots on a single array.

Dwell time (on slide): 0 ms.

Sample Blotting

The plain glass blot pad is cleaned with 70% ethanol prior to print runs and is cleaned at intervals during slide printing when full of spots, as queued by Omnigrid software. After each pin in the head picks up a sample, before printing on slides, spots are printed on the glass blot pad as follows.

For cDNA Samples. Number of blot spots/sample: 12; spot spacing: 250 μm center to center; contact time: 0 ms.

For Long Oligonucleotide Samples. Number of blot spots/sample: 6; spot spacing: 250 μm center to center; contact time: 0 ms.

Pin-Cleaning Procedure

The following procedure is applied before and after each print run and after every sample dip for both cDNA and oligonucleotide printing: pin wash: 8000 ms; pin vacuum dry: 5000 ms; repeat steps 1 and 2; pin wash: 3000 ms; pin vacuum dry: 12,000 ms.

Procedures After Printing

Genome scale printed slides are left on the Omnigrid 100 arrayer deck at least overnight; in some cases when printing is completed on Friday afternoon, the slides remain on the arrayer deck until Monday morning.

Quality Control Parameters for DNA Microarray Printing and Analysis Using Spotted Glass Slides

cDNA Print Material Q/C

PCR product gel images are generated for all cDNA printed. The "band" pattern is determined and the DNA concentration is estimated based on a lambda ladder. A subset of all cDNA plasmids and PCR products are resequenced, with additional sequencing of user-selected features. All of these data are stored in a Keck database and can be viewed by request.

Slide and Coating Q/C Begins with Use of TeleChem SuperClean Slides

All slides are inspected visually for defects and surface inconsistency before PLL coating in a class 1000 clean room. PLL slides are "cured" for 2 weeks and tested for hydrophobic properties before printing. All slides performing poorly (i.e., hydrophobicity test) or older than 8 weeks are discarded. Each PLL slide batch is prescanned to verify low background fluorescence and is inspected visually before placing on the arrayer deck.

Printing Q/C Begins with Printing in Class 100 Clean Rooms for Stringent Particle Control

A carryover test with Cy3/Cy5 end-labeled oligonucleotides is performed before each print run with new print head/pin xyz settings to assure successful pin washing and drying. Arrays are inspected visually at each plate change interval during printing. Temperature and humidity are maintained at 72°, 55% relative humidity and constantly electronically monitored in clean rooms. A spot morphology Q/C hybridization is carried out with Cy3/Cy5-end labeled, random 16-mers that highlight all features and allow establishment of "drop-out" spots where no DNA is printed. A spot drop-out level below 4% is currently considered acceptable for quality slide distribution. Slides with an above 4% spot drop-out level are considered subquality and are used for training purposes. Each slide batch undergoes a hybridization performance Q/C using standard amino-allyl labeling of RNA and hybridization. The S/N and dynamic range of signal is checked at this point. Since the year 2000, the Resource has printed 17,950 production microarrays with 22 different slide types, both custom and generic. Figure 1 shows the percentage of these slides, displayed for each slide type, that have passed our quality control parameters in terms of acceptable percentage of "drop-out" spots, even spot morphology and hybridization performance. The success rate varies with the type of material printed and its source. The overall success rate for slides printed at the Keck Resource is 72%.

First-Strand cDNA Synthesis Q/C

The efficiency of Cy3 and Cy5 dye conjugation to amino-allyl-labeled cDNA after nucleotide incorporation is estimated using the Hitachi Gene-Spec II spectrophotometer. The amount of nucleic acid is determined from

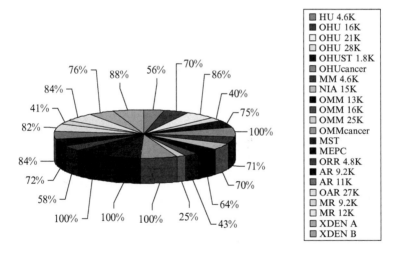

FIG. 1. Percentage printing success rate for all slide types with quality acceptable for distribution produced between 2000 and 2005 in the W. M. Keck Microarray Resource. A total of 17,950 microarray slides were printed and the success rate ranges from 25 to 100% depending on the slide type. The average success rate for all slides printed is 72%. Subquality sides are used for technical training in array use. HU, human cDNA; OHU, human operon 70-mer oligonucleotide; MM, mouse cDNA; NIA, Mouse National Institute of Aging cDNA; OMM, mouse operon 70-mer oligonucleotide; MST, mouse operon oligonucleotide; MEPC, Mouse Endocrine Pancreas Consortium cDNA; ORR, rat compugen oligonucleotide; AR, Arabidopsis MSU cDNA; OAR, Arabidopsis operon oligonucleotide; MR, mouse retinal cDNA; XDEN, rice operon oligonucleotide array. (See color insert.)

the OD_{260}, and using extinction coefficients of the dyes and readings at dye-specific wavelengths, the number of dye molecules per cDNA base is estimated. An acceptable range is one dye molecule per 20–50 bp. The pooled probe is inspected visually for color after unincorporated dye removal by column and should be purple (blue + red). The amount of labeled cDNA probe for each sample is adjusted to equivalence for hybridization.

Scanning Q/C

To optimize and increase uniformity of scanning on the Axon 4000A, adjust scanner PMT using the "histogram mode" to balance channel intensity. Balance the histogram at all points above background across signal intensity. Adjust laser power and PMT on "line scan mode" to balance channel signal intensity using the PE Scanarray 5000XL. Each scanner has a "reference grating" slide for wavelength checks. In addition, the Axon scanner constantly monitors itself and records diagnostics on the lasers.

Labeling and Hybridization Protocols Employed by the Keck Microarray Resource

A number of excellent reviews outline various labeling and hybridization approaches that may be referred to in addition to the Keck protocols described here (Cheung *et al.*, 1999; Hegde *et al.*, 2000). The following protocols have been adapted and developed through the efforts of Irina Tikhonova, Octavian Henegariu, and Janet Hager in the Keck microarray group. They include several steps that have proved to increase the quality of hybridization signals using Keck arrays in GeneMachines hybridization chambers (Genomic Solutions, Ann Arbor, MI) and in an Advalytix AG ArrayBooster (Olympus Life and Material Science Europa, GmbH) hybridization station. The procedures are based on protocols from Invitrogen, Genisphere, Qiagen, Amersham Biosciences, Current Protocols in Molecular Biology, and Dr. Geoff Childs' and Dr. Joseph DeRisi's groups. Table I shows reagents that are used routinely for the following protocols.

Tables II–VII provide detailed recipes for all stock solutions, buffers, and hybridization mixes required in the protocols: Table II, stock solutions; Table III, cDNA prehybridization solution; Table IV, cDNA hybridization mix; Table V, oligonucleotide hybridization mix; Table VI, hybridization chamber solution (store at room temperature); and Table VII, blocking solution (Childs) (store at $-20°$).

We routinely use and recommend the following labeling approaches that do *not* involve nucleic acid amplification steps:

TABLE I
REAGENTS USED IN RNA EXTRACTION, MICROARRAY LABELING AND HYBRIDIZATION PROTOCOLS

Item	Vendor	Catalog #
Microcon YM30 Filter	Millipore	42410
GFX Column	GE Healthcare	27–9606–01
CY3 dUTP Fluorolink	GE Healthcare	53022
CY5 dUTP Fluorolink	GE Healthcare	55022
CY3 Mono-Reactive Dye Pack	GE Healthcare	PA 23001
CY5 Mono-Reactive Dye Pack	GE Healthcare	PA 25001
CY2 Mono-Reactive Dye Pack	GE Healthcare	PA 22000
DMSO	Sigma-Aldrich	41638
RNase Free Water	GE Healthcare	US 70783
RNAse-Inhibitor	Roche	03335399001
pd(N)6 Random Hexamer	GE Healthcare	27–2166–01
Poly dA	GE Healthcare	27–7836–01
Human cot-1 DNA	Invitrogen	15279–011
Mouse cot-1 DNA	Invitrogen	18440–016
Yeast tRNA	Invitrogen	15401–029
Denhardt's solution 50X	Sigma-Aldrich	D2532–5ML
Oligo d(T) Cellulose Columns	Invitrogen	R545–01
TRIzol Reagent	Invitrogen	15596–026
SuperScript II RT	Invitrogen	18064–022
PBS 10X	Invitrogen	70011–044
Oligotex Columns	Qiagen	70022;70042;70061
RNeasy Midi Kit; RNeasy Mini Kit; RNeasy Micro Kit.	Qiagen	75142; 74104;74004
RNeasy MinElute Cleanup Kit	Qiagen	74204
QIAquick PCR Purification Kit	Qiagen	28104
Amino-allyl dUTP	Sigma Aldrich	A 0410
Ethanolamine	Sigma Aldrich	E 9508
Sephadex G-50 Spin Columns	Sigma Aldrich	55897
Salmon Sperm DNA	Stratagene	201190
Genicon RLS Two-color Kit	Invitrogen	40–1003–24
Mono-functional Biotin	Pierce	21329
Mono-functional Fluorescein	Invitrogen	F-2181
Array 900 3DNA Dendrimer	Genisphere	W500180
SenseAmp Plus Kit	Genisphere	RAMP110PLUS

1. Amino-allyl labeling and monofunctional Cy-dye conjugation: 50–100 μg total RNA; 5–10 μg mRNA.

2. Genisphere Dendrimer labeling—Cy3 and Cy5 dyes: 1 μg total RNA; 250 ng mRNA for arrays with 22 × 22-mm coverslip dimension; 2 μg total RNA; 500 ng mRNA for arrays with a 22 × 60-mm coverslip dimension.

TABLE II
STOCK SOLUTIONS

Stock solutions

33.3 mM dACG solution: Mix together 1 volume each of commercial nucleotide solutions (each 100 mM): 100 μl or 1 vol. dATP; 100 μl or 1 vol. dCTP; 100 μl or 1 vol. dGTP; 300 μl 33.3 mM dACG, then make 20 @ 15 μl aliquots and store at −20°.

25 mM dTTP solution: Mix 1 volume 100 mM dTTP and 3 vol. H_2O. Add 1M Tris, pH 7.5 to 10 mM final concentration.

20 mM aa-dUTP solution: Resuspend 1 mg of amino-allyl-dUTP in 95.5 μl 10 mM Tris pH 7.5. Store nucleotide solutions in small aliquots (20 μl) at −20°. After thawing, always spin 10 sec. in microfuge prior to use.

0.1 M NaHCO₃ pH 9.0: Dissolve 4.2 g of $NaHCO_3$ in 450 ml H_2O. Adjust the pH to 9.0 by adding approximately 25–30 ml of 0.1 M Na_2CO_3. Adjust volume to 500 ml and filter-sterilize the buffer.

TABLE III
PRE-HYBRIDIZATION SOLUTION FOR cDNA ARRAYS

Volume	Stock	Final concentration
60 μl	99.9% formamide	48% formamide
20 μl	20X SSPE	3.2X SSPE
5 μl	10% SDS	0.4% SDS
5 μl	50X Denhardt's	2X Denhardt's
2 μl	10 mg/ml ss salmon sperm DNA	0.177 mg/ml ss salmon sperm DNA
33 μl	ddH₂O	
Total Volume: 125 μl		

TABLE IV
cDNA ARRAY HYBRIDIZATION MIX

Volume	Stock	Final concentration before adding probe
78.5 μl	99.9% formamide	62.8% formamide
5 μl	20% SDS	0.8% SDS
10 μl	50X Denhardt's	4X Denhardt's
31.5 μl	20X SSPE	5X SSPE
Total Volume: 125 μl		

TABLE V
OLIGONUCLEOTIDE ARRAY HYBRIDIZATION MIX

Coverslip dimensions	22 × 22 mm	22 × 40 mm	22 × 50 mm	22 × 60 mm LifterSlip*
Hybridization mix volume	13 μl	24 μl	30 μl	55 μl
Labeled probe	8.3	16.2	20.6	38.5
2% SDS	1.3	2.4	3.0	5.5
20X SSC	1.7	3.1	3.8	7.0
poly dA (5 μg/μl)	0.7	1.3	1.6	3.0
Cot1 DNA (1 μg/μl)	1.0	1.0	1.0	1.0

*22 × 60 mm LifterSlip, Erie Scientific, Cat# 22X60I-2-4861.

TABLE VI
HYBRIDIZATION CHAMBER SOLUTION (STORE AT ROOM TEMPERATURE)

Volume	Stock	Final concentration
1220 μl	H_2O	-
320 μl	20 × SSC	3.2 × 23%
460 μl	99.9% formamide	
Total Volume: 2000 μl		

TABLE VII
BLOCKING SOLUTION (CHEUNG ET AL., 1999) (STORE AT −20°)

Volume	Stock	Final concentration after precipitation
20 μl	polydA (5 mg/ml)	2 mg/ml
20 μl	tRNA (10 mg/ml)	4 mg/ml
500 μl (1 vial)	human/mouse Cot-1 DNA (1 mg/ml)	10 mg/ml
Total Volume: 540 μl Ethanol precipitate the combined reagents; resuspend in **50 μl** filtered ddH$_2$O – **Final volume**		

3. Amino-allyl labeling with Genicon/RLS gold and silver particle conjugation: 5 μg total RNA; 500 ng mRNA. Only for total RNA quantities below 2 μg do we apply linear amplification methods mainly using SenseAmp Plus in combination with signal amplification using Genisphere 3DNA Dendrimer labeling.

General Considerations Regarding RNA Samples, Labeling, and Hybridization Strategies

RNA to be submitted to the Resource should be prepared and examined using the following assays.

1. For total RNA isolation from tissue samples use TRIzol (TRIZOL reagent, Invitrogen Corp., CA) followed by cleaning using the RNeasy kit [Qiagen Inc., CA; minikit (100 μg)]. Alternatively, total RNA can be isolated *directly* with RNeasy kits without using TRIzol. We recommend specialty kits for lipid-rich and fibrous tissues. When using mRNA it should be isolated from total RNA using Oligotex columns (Qiagen) or oligo(dT) cellulose columns (Invitrogen Corp.).

2. Measure the optical density (OD) of total RNA and mRNA samples, including OD_{260}, OD_{280}, and OD_{260}/OD_{280} and calculate the concentration (μg/μl) of your samples. Samples are submitted with all the values given earlier and the dilution factor used to take the readings.

The following amounts of starting material are required.

For each mRNA sample: Indirect aminoallyl labeling, 5 μg at a concentration of 0.33 μg/μl; Genisphere 3DNA Dendrimer labeling, 250 ng in greater than 5 μl aqueous solution; Genicon RLS labeling, 500 ng in greater than 5 μl aqueous solution.

For each *total* RNA sample: Indirect aminoallyl labeling, 50 μg at a concentration of 3.3 μg/μl; Genisphere 3DNA labeling, 1–2 μg in greater than 5 μl aqueous solution; Genicon RLS labeling, 1–5 μg in greater than 15 μl aqueous solution. Yale investigators bring their RNA samples to the laboratory on ice. Users outside of Yale campus send RNA samples in aqueous solution on dry ice.

3. Visualize each total RNA sample by running on an agarose gel containing 1.1% formaldehyde and photograph. Attach a photo of the results (please remember to clearly label each lane with the

sample label). RNA samples may be assayed for quality in the Microarray Resource after receipt using the Hitachi GeneSpec II instrument and/or an Agilent 2100 BioAnalyzer. If QC data are not provided with the sample, the Keck laboratory implements these two procedures before using the RNA for labeling and hybridization. The purpose of the QC is to establish that the quantity of RNA is as given by the investigator, as spectrophotometers from laboratory to laboratory vary in their calibration, and to establish that the RNA is not degraded significantly, as determined by visual inspection of intensity of the ribosomal RNA bands. The 28S rRNA:18S rRNA band intensity ratio should be very close to 2. RNA that is degraded appears smeared and rRNA bands are weak or not present. The RNA concentration is quantitated by UV spectrophotometry at an absorbance of 260 nm. An $A_{260/280}$ nm ratio is used as a measure of RNA purity. RNA with an $A_{260/280}$ nm ratio of 1.8–2.1 is considered acceptable for expression profiling. In our experience, because RNA of lower ratio values (1.5–1.7) may or may not give signal on an array, we therefore caution investigators not to proceed in this situation, but will do so if the sample is difficult to obtain or not easily reproducible. The cDNA yield after reverse transcription is often very much reduced for samples like this, perhaps due to enzymatic inhibition by impurities.

Hybridization Strategies

Two main approaches are used in the Resource: GeneMachines hybridization chambers, used for manual hybridizations of 4 × 4 arrays with small dimensions where an 22 × 18-mm coverslip is sufficient to cover the array, and Advalytix ArrayBooster Autohyb instrument, used for all 4 × 12 subarray slides where a 22 × 60-mm coverslip would be needed for manual hybridization. The ArrayBooster works on the principle of surface acoustic wave technology to actively mix the target liquid evenly over the array. Figure 2 shows a comparison of a hybridization signal between a manual hybridization of Genisphere-labeled plant cDNA interrogating an Arabidopsis 27,000 feature oligonucleotide array using a LifterSlip and the same target used in an AdvaCard hybridization in the ArrayBooster. There is a clear advantage in terms of signal intensity and evenness of hybridization signal when using the ArrayBooster. In general, cDNA arrays are hybridized at 42° with formamide buffer and oligonucleotide arrays at 55° or higher without formamide.

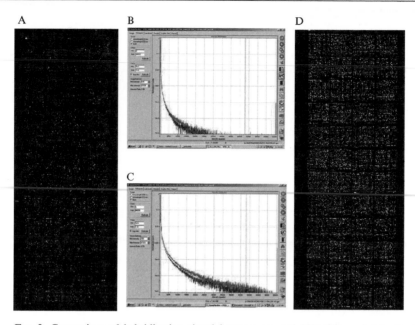

FIG. 2. Comparison of hybridization signal between manual LifterSlip and AdvaCard ArrayBooster hybridization of Genisphere-labeled plant cDNA interrogating an Arabidopsis 27,000 feature 70-mer oligonucleotide array. (A) Axon image of dual-color Genisphere hybridization using a manual LifterSlip with 5 μg total RNA starting material. (B) Axon GenePix Pro histogram image showing signal plot for A. (C) Axon GenePix Pro histogram image showing signal plot for D. (D) Axon image of dual-color Genisphere hybridization using an AdvaCard ArrayBooster with 2.5 μg total RNA starting material. Both hybridizations were done at 55°.

Indirect Amino-Allyl dUTP Target Labeling, Monofunctional Dye Conjugation, and cDNA and Oligonucleotide Array Hybridization Protocols

cDNA Array Method

Total RNA and mRNA Preparation. Total RNA is prepared using the Qiagen RNeasy midi or mini kit or micro kit. (Please follow the manufacturer's protocols for this step.) mRNA is prepared from total RNA using Oligotex columns (Qiagen) or oligo(dT) cellulose (GIBCO). (Please follow the manufacturer's protocols for this step.) We suggest using 50 μg of total RNA per labeling reaction when first beginning microarray experiments. Once technically proficient, this can most likely be reduced to 25 μg or less per labeling reaction. Prior to labeling, total RNA is concentrated to about 3.3 $\mu g/\mu l$ using either ethanol/isopropanol precipitation (1 volume RNA + 2 volumes ethanol:isopropanol 1:1) or microcon YM-30 columns.

Target Preparation and Labeling. Labeling reactions are performed using the metal block of a common thermocycler (PE GeneAmp 9700) in PCR tubes.

RNA Denaturing/Primer Annealing. Start with 15 μl of resuspended RNA per target (usually 50 μg total RNA or 4 μg mRNA). Add 4 μl oligo (dT) (1 $\mu g/\mu l$) + 1 μl random hexamers (5 $\mu g/\mu l$), giving a 20 μl total volume. Incubate for 3–5 min at 65–68° and place on ice when complete.

First Strand CDNA Synthesis and Labeling. Create labeling mix as follows: 8 μl 5× first strand buffer, 4 μl 0.1 M dithiothreitol (DTT), 0.6 μl 33.3 mM each dACG (final concentration is 500 μM), 0.32 μl 25 mM dTTP (final concentration is 200 μM), 4 μl 2 mM aa-dUTP (final concentration is 200 μM), and 1 μl RNAsin. Add 18 μl labeling mix to the 20 μl denatured RNA (see earlier). Add 2 μl SuperScript-II reverse transcriptase (Invitrogen, Carlsbad, CA). Incubate at 42° for 2 h and then heat inactivate at 94° for 2–3 min.

RNA Removal. Add 5 μl 0.5 M EDTA to each reaction tube and 10 μl 1 M NaOH. Heat 20 min at 60–65°. Add 6 μl 1 M HCl and 2 μl 1 M Tris, pH 7.5, to neutralize NaOH. Using HCl decreases the requirements for a large Tris volume. Target may be stored at −20° at this point. Otherwise, go directly to target cleanup.

CDNA Target Cleanup Using QIAquick PCR Purification Kit. Add 5 volumes of buffer PB to the 1 volume of cDNA synthesis as described earlier and mix. Apply the sample to the QIAquick column and centrifuge for 1 min. Transfer the QIAquick column to the empty 2-ml collection tube. Save the flow-through liquid until all cDNA products have been recovered from the column. Add 600 μl of 80% ethanol to each column and centrifuge at 13,000 rpm for 30 s. Remove the QIAquick column and discard the flow-through liquid. Repeat the washing step twice. After the final wash, discard the liquid and place each column back in the collection tube and spin additional 1 min to remove all traces of ethanol. Transfer each QIAquick column into a 1.5-ml tube and add 60 μl of 0.1 M NaHCO$_3$, pH 9.0, directly to the top of the membrane in each QIAquick column. Incubate the QIAquick column at room temperature for 5 min. Centrifuge at 13,000 rpm for 2 min to collect purified cDNA.

Target Labeling Using Aminoallyl dUTP and Monofunctional Dye Conjugation. Resuspend the monofunctional Cy3 and Cy5 dye aliquots purchased from Amersham/Molecular Probes (usually 0.2–0.3 mg dye/vial) in 40 μl dimethyl sulfoxide (DMSO), which can be stored for months at −20°. Use 2 μl for each labeling reaction. To avoid water condensation in the dye solution, let it warm for 5–10 min at room temperature each time before opening the vial with DMSO dye. Never spin this vial. Add 2 μl DMSO dye to each cDNA target. Mix with a pipette tip and incubate at room temperature in the dark for 60–90 min. After incubation, stop the reaction by adding 4 μl 2 M ethanolamine (Sigma), mix, and incubate 5 min at room temperature.

Target Purification After Dye Conjugation Using MinElute PCR Purification Kit (Qiagen). Add 5 volumes of buffer PB to the 1 volume of cDNA synthesis as described earlier and mix. Apply the sample to the QIAquick column and centrifuge for 1 min. Transfer the QIAquick column to the empty 2-ml collection tube. Save the flow-through liquid until all cDNA products have been recovered from the column. Add 600 μl of buffer PE to each column and centrifuge at 13,000 rpm for 30 s. Remove the QIAquick column and discard the flow-through liquid. Repeat the washing step twice. After the final wash, discard the liquid and place each column back in the collection tube and spin for an additional 1 min to remove all traces of ethanol. Transfer each QIAquick column into a 1.5-ml tube and add 12 μl of buffer EB directly to the top of the membrane in each QIAquick column. Incubate the QIAquick column at room temperature for 5 min. Centrifuge at 13,000 rpm for 2 min to collect purified cDNA.

Slide Denaturing and Prehybridization (While Preparing Target). Place prehybridization buffer, volume appropriate to the size of the array on a slide (see Table III), cover with a glass coverslip, and place the slide on the metal block of a thermocycler (or any surface heated at 76°). Before placing the slide on the block, spread a thin layer of water on the metal so that the glass and the metal have better adherence. Program the cycler for 2 min at 76°. After the temperature drops below 50°, place the slide in the GeneMachines hybridization chamber. Add 55 μl of chamber buffer in the grooves of the hybridization chamber for each slide. Incubate slide for 1 h in 50° water bath. Rinse off prehybridization buffer by placing the slide 2 min each in a glass jar or 50-ml tube consecutively with H_2O, 70% ethanol, and then 100% ethanol. Air dry the slide. Washing off the pre-hybridization buffer results in a better controlled, more even, and more reproducible hybridization.

Target Denaturation and Hybridization. In a tube, mix well all components of the hybridization mix with the volume appropriate to the size of the array. Heat tube at 90° for 3 min (usually done on the metal block of a thermocycler). Spin briefly at maximum speed in a microfuge and keep labeled target at 42°. As soon as the slides are prehybridized, pipette target to the array and cover with a coverslip. Place slide in the GeneMachines hybridization chamber and add 55 μl of chamber buffer in the grooves of the hybridization chamber for each slide. Hybridize overnight in a 42° water bath.

Array Wash. Remove the coverslip with a pair of fine forceps. Place slide in slide holder/glass dish or 50-ml tube consecutively with 200 ml 2× SSC/0.1% SDS at 42°, 200 ml 0.2× SSC/0.1% SDS at 42°, 200 ml 0.2× SSC shake gently for 10 min in each solution. Repeat the last step with fresh solution to make sure all residual SDS is removed. Place slide with label down in a 50-ml tube and spin in a swinging bucket rotor centrifuge for 5 min at 1000 rpm to dry slide. If washing is not stringent enough, the first wash can be done at 50°. Scan slides.

Oligonucleotide Array Method

RNA isolation, RNA denaturation, primer annealing, first strand synthesis, labeling, and target cleanup are the same as described earlier for cDNA arrays. The following protocols are specific for olionucleotide arrays. Note that there is no prehybridization step for oligonucleotide arrays.

Target Denaturation and Hybridization. Mix all components of the hybridization mix using the volume appropriate to the size of the array. Heat tube at 94° for 3 min (usually done on the metal block of a thermocycler). Spin briefly at maximum speed in a microfuge and keep tube at 55°. As soon as the slides are ready, pipette target onto the array and cover with a coverslip. Place slide in the hybridization chamber; add 55 μl of 3× SSC in grooves of the GeneMachines hybridization chamber for each slide. Hybridize overnight in a 55° water bath.

Array Wash. Remove the coverslip with a fine forceps. Place slide in slide holder/glass dish or 50-ml tube consecutively with 200 ml 2× SSC/0.1% SDS at 42°, 200 ml 0.2× SSC/0.1% SDS at 42°, 200 ml 0.2× SSC shake gently for 10 min in each solution. Repeat the last step with fresh solution to make sure all residual SDS is removed. Place slide with label down in a 50-ml tube and spin in a swinging bucket rotor centrifuge for 5 min at 1000 rpm to dry slide. If washing is not stringent enough, the first wash can be done at 50°. Scan slides.

Genisphere 3DNA Dendrimer Labeling and Oligonucleotide Array Hybridization

Protocol to Use with Genisphere Array 900 Kit and Erie Scientific LifterSlips

cDNA Synthesis from Total RNA. In a 0.25-ml PCR tube, prepare the RNA-RT primer mix: 1–5 μl total RNA (2.0 μg mammalian total RNA or 2.5 μg plant total RNA); 1 μl RT primer (vial 11, 5 pmol/μl);

add nuclease-free water (vial 10) to a final volume of 6 μl. Mix the RNA-RT primer mix and microfuge briefly to collect contents in the bottom of the tube. Heat to 80° for 5 min and immediately transfer to ice for 2–3 mins. Microfuge briefly to collect contents at the bottom of the tube and return to ice. Each cDNA synthesis requires 4.5 μl of reaction master mix. To reduce pipetting errors, the reaction master mix should contain at least 9 μl. Prepare a reaction master mix in a microtube on ice: 4 μl 5× Super-Script II first strand buffer; 2 μl 0.1 M DTT (supplied with enzyme); 1 μl Superase-in RNase inhibitor (vial 4); 1 μl dnTP mix (vial 3); 1 μl Super-script II enzyme, 200 units; 9 μl total volume. Microfuge briefly and keep on ice until ready to use. Add 4.5 μl of the reaction master mix to the 6 μl of RNA-RT primer mix (10.5 μl final volume). Gently mix (do not vortex) and incubate at 42° for 2 h. Stop the reaction by adding 1.0 μl of 1.0 M NaOH/100 mM EDTA. Incubate at 65° for 10 min to denature the DNA/RNA hybrids and degrade the RNA. Neutralize the reaction with 1.2 μl of 2 M Tris–HCl, pH 7.5. Proceed to hybridization of cDNA and 3DNA to microarray.

LifterSlip Blocking. Incubate LifterSlips in 40°, 2% bovine serum albumin, 0.2% SDS for 10 min. Rinse blocked LifterSlips with ddH$_2$O and dry with pressured air.

cDNA Hybridization. Thaw and resuspend the 2× hybridization buffer (vial 6) by heating to 70° for at least 10 min or until resuspended completely. Vortex to ensure that the components are resuspended evenly. If necessary, repeat heating and vortexing until all the material has been resuspended. Microfuge for 1 min. For each array, prepare the following cDNA hybridization mix for use with a 22 × 60-mm LifterSlip: 12.7 μl cDNA synthesis #1; 12.7 μl cDNA synthesis #2; 30 μl 2× hybridization buffer (vial 6); 1.0 μl CoT-1 DNA; 3.6 μl nuclease-free water (vial 10); 60.0 μl total volume. Vortex gently and microfuge the cDNA hybridization mix briefly. Incubate the hybridization mix first at 80° for 10 min and then at the hybridization temperature (55°) until loading the array. Prewarm the microarray slide and LifterSlip in the GeneMachine hybridization chamber on the top of a heat block to the hybridization temperature. Vortex gently and microfuge the cDNA hybridization mix briefly. Aspirate the cDNA hybridization mix from a tube. Slowly release the sample solution into the capillary gap between the microarray slide surface and the LifterSlip. Add 55 μl of 3× SSC to the grooves of the hybridization chamber. Hybridize overnight in a 55° water bath.

Post-cDNA Hybridization Wash. Prewarm the 2× SSC, 0.2% SDS wash buffer to 42°. Remove the LifterSlip with a fine forceps. Wash slides for 10–15 min in prewarmed 2× SSC, 0.2% SDS. Wash for 10–15 min in 2× SSC at room temperature. Wash for 10–15 min in 0.2× SSC at room temperature. Transfer the array to a dry 50-ml centrifuge tube, orienting the slide so that any label is down in the tube. Immediately centrifuge for 5 min at 1000 rpm to dry the slide.

3DNA Hybridization. Thaw the 3DNA Array 900 capture reagent (vial 1) in the dark at room temperature for 20 min. Vortex at maximum setting for 3 s and microfuge briefly. Incubate at 50° for 10 min. Vortex at maximum setting for 3–5 s and then spin the tube briefly to collect the contents at the bottom. Thaw and resuspend the 2× hybridization buffer (vial 6) by heating to 70° for at least 10 min. Vortex and microfuge for 1 min. For each array, prepare the following 3DNA hybridization mix: 2.5 μl Cy3-3DNA Array 900 capture reagent (vial 1); 2.5 μl Cy5-3DNA Array 900 capture reagent (vial 1); 1.0 μl Cot-1 DNA; 30 μl 2× hybridization buffer (vial 6); 24 μl nuclease-free water (vial 10); 60.0 μl total volume. Vortex gently and microfuge the 3DNA hybridization mix briefly. Incubate the 3DNA hybridization mix first at 80° for 10 min and then at the hybridization temperature (65°) until loading the array. Prewarm the microarrays slide and LifterSlip in the GeneMachine hybridization chamber on the top of a heat block to the hybridization temperature. Vortex gently and microfuge the 3DNA hybridization mix briefly.

Aspirate the 3DNA hybridization mix from a tube. Slowly release the sample solution into the capillary gap between the microarray slide surface and the LifterSlip. Add 55 μl of 3× SSC to the grooves of the hybridization chamber. Hybridize in a 65° water bath for 5 h.

Post-3DNA Hybridization Wash. After hybridization, wash slides several times to remove unbound 3DNA molecules. Perform these washes in the dark to avoid photobleaching and fading of fluorescent dyes. Prewarm the 2× SSC, 0.2% SDS wash buffer to 42°. Remove the LifterSlip with a fine forceps. Wash slides for 10–15 min in prewarmed 2× SSC, 0.2% SDS. Wash for 10–15 min in 2× SSC at room temperature. Wash for 10–15 min in 0.2× SSC at room temperature. Transfer the array to a dry 50-ml centrifuge tube, orienting the slide so that any label is down in the tube. Immediately centrifuge for 5 min at 1000 rpm to dry the slide.

Protocol for Use with Genisphere Array 900 Kit and Advalytix ArrayBooster DNA Microarray Hybridization Station

Proceed with cDNA synthesis as for the LifterSlip approach given earlier.

Array Prehybridization. Thaw and resuspend the 2× hybridization buffer (vial 6) by heating to 70° for at least 10 min or until resuspended completely. Vortex to ensure that the components are resuspended evenly. If necessary, repeat heating and vortexing until all the material has been resuspended. Microfuge for 1 min. For each array, prepare the following prehybridization mix for use with a 22 × 60-mm LifterSlip: 40.0 μl 2× hybridization buffer (vial 6); 1.0 μl Cot-1 DNA; 39.0 μl nuclease-free water (vial 10); 80.0 μl total volume. Vortex gently and microfuge the prehybridization mix briefly. Incubate the prehybridization mix first at 80° for 10 min and then at the prehybridization temperature (55°) until loading the array. Prewarm the microarrays slide and LifterSlip in the Gene-Machine hybridization chamber on the top of a heat block to the prehybridization temperature. Aspirate the prehybridization mix from a tube. Slowly release the sample solution into the capillary gap between the microarray slide surface and the LifterSlip. Add 55 μl of 3× SSC to the grooves of the hybridization chamber. Prehybridize for 1 h in a 55° water bath.

Postprehybridization Wash. Prewarm the 2× SSC, 0.2% SDS wash buffer to 42°. Remove the LifterSlip with a fine forceps. Wash slides for 10–15 min in prewarmed 2× SSC, 0.2% SDS. Wash for 10–15 min in 2× SSC at room temperature. Wash for 10–15 min in 0.2× SSC at room temperature. Transfer the array to a dry 50-ml centrifuge tube, orienting the slide so that any label is down in the tube. Immediately centrifuge for 5 min at 1000 rpm to dry the slide.

cDNA Hybridization. Thaw and resuspend the 2× hybridization buffer (vial 6) by heating to 70° for at least 10 min or until resuspended completely. Vortex to ensure that the components are resuspended evenly. If necessary, repeat heating and vortexing until all the material has been resuspended. Microfuge for 1 min. For each array, prepare the following cDNA hybridization mix for use with a 22 × 71-mm AdvaCard: 12.7 μl cDNA synthesis #1; 12.7 μl cDNA synthesis #2; 47.5 μl 2× hybridization buffer (vial 6); 1.0 μl CoT-1 DNA; 21.1 μl nuclease-free water (vial 10); 95.0 μl total volume. Vortex gently and microfuge the cDNA hybridization mix briefly. Incubate the hybridization mix first at 80° for 10 min and then at the hybridization temperature (55°) until loading the array. Prewarm the microarrays slide and AdvaCard to the hybridization temperature. Vortex gently and microfuge the cDNA hybridization mix briefly. Aspirate the cDNA hybridization mix from a tube. Insert the filled pipette tip firmly into the left-front hole of the AdvaCard and release the sample solution slowly until the liquid starts to enter the capillary gap. Eject the pipette tip and let the capillary effect draw the liquid into the gap between the microarray and the AdvaCard. Start the incubation.

Post-cDNA Hybridization Wash. Prewarm the 2× SSC, 0.2% SDS wash buffer to 42°. Submerse the slide/AdvaCard sandwich into a staining jar filled with prewarmed 2× SSC, 0.2% SDS wash buffer. By applying a small amount of pressure with a finger, separate the AdvaCard from the slide. Wash slides for 10–15 min in prewarmed 2× SSC, 0.2% SDS. Wash for 10–15 min in 2× SSC at room temperature. Wash for 10–15 min in 0.2× SSC at room temperature. Transfer the array to a dry 50-ml centrifuge tube, orienting the slide so that any label is down in the tube. Immediately centrifuge for 5 min at 1000 rpm to dry the slide.

AdvaCard Posthybridization Washing. Rinse the AdvaCard with ddH$_2$O and 0.5% SDS. Rinse with ddH$_2$O. Rinse with 95% ethanol. Rinse again with ddH$_2$O. Dry with a pressured air stream.

3DNA Hybridization and Post-3DNA Wash Proceeds as for LifterSlip Method Except Apply the AdvaCard Method for Sample Loading and AdvaCard/Slide Separation. For each array, prepare a larger volume of the following 3DNA hybridization mix for use with the AdvaCard: 2.5 μl Cy3 3DNA Array 900 capture reagent (vial 1); 2.5 μl Cy5 3DNA Array 900 capture reagent (vial 1); 1.0 μl Cot-1 DNA; 47.5 μl 2× hybridization buffer (vial 6); 41.5 μl nuclease-free water (vial 10); 95.0 μl total volume. Prewarm the microarray slide and AdvaCard to the hybridization temperature (65°). Vortex gently and microfuge the cDNA hybridization mix briefly. Aspirate the cDNA hybridization mix from the tube. Insert the filled pipette tip firmly into the left-front hole of the AdvaCard and slowly release the sample solution until the liquid starts to enter the capillary gap. Eject pipette tip and let the capillary effect draw the liquid into the gap between the microarray and the AdvaCard. Start the incubation. Perform post-3DNA hybridization wash and AdvaCard posthybridization washing (see earlier discussion).

Genicon/Invitrogen Resonance Light Scattering Using
  Gold and Silver Nanoparticle Labeling with HiLight
  Scanner Detection

We now provide microarray services to support the Genicon/Invitrogen ultrasensitive resonance light-scattering microarray platform for gold and silver particle labeling of as little as 1 μg total RNA without nucleic acid amplification. Our protocol utilizes amino-allyl labeling of first strand cDNA followed by the conjugation of monofunctional biotin and fluorescein and then conjugation with avidin-Ag and antifluorescein-Au particles. After archiving, the hybridized arrays are stable and can be stored and rescanned many times. The HiLight reader is a white light scanner that illuminates the array and captures the light scattered by the gold and silver particles using a CCD camera (Fig. 3). Data are stored as a .tiff image. We use ArrayVision software to grid and analyze data.

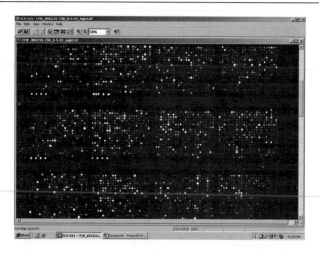

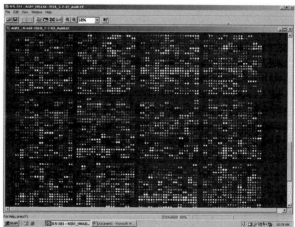

FIG. 3. The Resource has applied an amino-allyl approach for RLS labeling instead of directly incorporating the more expensive biotin or fluorescein d-UTP during reverse transcription. Five micrograms total HeLa and HL60 RNAs was amino-allyl labeled during cDNA creation using standard DeRisi protocols to provide less bias between samples than by direct incorporation. cDNAs were linked to monofunctional biotin or fluorescein and reacted with antibiotin- or antifluorescein-tagged gold or silver particles and hybridized against a human 21K 70-mer oligonucleotide array (top) and human 4.6K cDNA array (bottom). (See color insert.)

Genicon/Invitrogen RLS Indirect Target-Labeling Protocol

Target Preparation and Labeling. Follow RNA denaturing/primer annealing steps as for the amino-allyl labeling protocol. Start with 15 μl of resuspended RNA per probe (usually 5 μg total RNA). Add 5 μl oligo(dT)

(1 μg/μl) 20 μl total volume. Incubate 5 min at 68°. Place on ice. Follow first strand synthesis/labeling and RNA removal steps as for the amino-allyl labeling protocol. For cDNA target cleanup, use the CyScibe GFX purification kit (Amersham Biosciences). Add 500 μl of capture buffer to each GFX column. Transfer the unpurified aminoallyl-labeled cDNA products described earlier into each GFX column; mix cDNA by gently pipetting up and down five times. Centrifuge each column at 13,000 rpm for 30 s. Transfer the GFX column to the empty 2-ml tube. Save the flow-through liquid until all cDNA products have been recovered from the column. Add 600 μl of 80% ethanol to each column and centrifuge at 13,000 rpm for 30 s. Remove the GFX column and discard the flow-through liquid. Repeat the washing step twice. After the final wash, discard the liquid and place each column back in the collection tube and spin for an additional 30 s to remove all traces of ethanol. Transfer each GFX column into a 1.5-ml tube and add 60 μl of 0.1 M NaHCO$_3$, pH 9.0, directly to the top of the membrane in each GFX column. Incubate the GFX column at room temperature for 5 min. Centrifuge at 13,000 rpm for 2 min to collect purified cDNA.

Biotin and Fluorescein Target Labeling. Resuspend the 2 mg monofunctional conjugate aliquots in 68 μl DMSO. Use 2 μl for each labeling reaction. (Biotin can be purchased from Piece and fluorescein from Invitrogen.) Store for months at −20°. To avoid water condensation in the dye solution, let it warm for 5–10 min at room temperature each time before opening the vial with DMSO dye. Never spin this vial. Add 2 μl DMSO dye to each cDNA probe. Mix with a pipette tip and incubate at room temperature in the dark for 60–90 min. After incubation, stop the reaction by adding 4 μl 2 M ethanolamine (Sigma), mix, and incubate 5 min at room temperature.

Target Purification after Dye Conjugation Using CyScibe GFX Purification Kit. Add 500 μl of capture buffer to each GFX column. Transfer the unpurified labeled cDNA products as described earlier into each GFX column; mix cDNA by gently pipetting up and down five times. Centrifuge each column at 13,000 rpm for 30 s. Transfer the GFX column to the empty 2-ml tube. Save the flow-through liquid until all cDNA products have been recovered from the column. Add 600 μl of wash buffer to each column and centrifuge at 13,000 rpm for 30 s. Remove the GFX column and discard the flow-through liquid. Repeat the washing step twice. After the final wash, discard the liquid and place each column back in the collection tube and spin for an additional 30 s to remove all traces of ethanol. Transfer each GFX column into a 1.5-ml tube and add 60 μl of elution buffer directly to the top of the membrane in each GFX column. Incubate the GFX column at room temperature for 5 min. Centrifuge at 13,000 rpm for 2 min to collect purified labeled cDNA.

For hybridization and signal detection procedures, please refer to the GeniconRLS two-color DNA array kit and GeniconRLS GSD-501 detection and imaging instrument manuals. These manuals are shipped with these products and are available for download at www.invitrogen.com.

Scanning and Analyzing Spotted Microarrays

Scanning of Generic Glass Slides

After hybridization, slides are stored in a black box to exclude fluorescent laboratory light. Microarrays are scanned immediately after hybridization using one of three scanners, dependent upon the labeling. The Axon GenePix 4000A laser scanner is used for amino-allyl labeling and Genisphere dendrimer fluorescence at Cy3 and Cy5 wavelengths. Microarrays using fluorophores outside these wavelengths are scanned on the PE ScanArray 5000XL scanner with 16 fluorophore capability including external blue laser excitation and an autoloader. Genicon RLS experiments using gold and silver particles are scanned using a white light HiLight scanner instrument for dual-wavelength scattered light capture. All .tiff images are 16 bit and can be analyzed using most academic freeware (e.g., Scanalyze, TM-4) and commercially available analysis packages (e.g., Axon GenePix Pro).

Primary Data Analysis

Standard fluorescent amino-allyl and Genisphere dendrimer-labeled Cy3/Cy5 microarrays are gridded and analyzed using Axon GenePix Pro 3.0 or 5.0 software in conjunction with the appropriate .gal files, which provide spreadsheet data as .gpr files. Genicon RLS .tiff images generated by the HiLight scanner are analyzed using ArrayVision software (Imaging Research Inc., Ontario, Canada) providing spreadsheet data as a .txt file. Global normalization is used for standard fluorescent data, and linear normalization is applied to RLS data if requested by the investigator. Genespring 5.0 (Silicon Genetics) software is used for additional visualization and data mining. Analyses are offered as part of full-service projects or to complement projects done in individual investigators' laboratories. Training sessions are provided for all analysis software packages mentioned. At this point, it is suggested that data be examined by the staff in the Keck Biostatistics Resource under the direction of Hongyu Zhao (http://keck.med.yale.edu/biostats/) for further analyses using a variety of statistical methods and softwares.

Yale Microarray Database

Major challenges for any academic microarray core are data warehousing, distribution, query, and annotation. To address these issues,

development of the Yale Microaray Database (http://info.med.yale.edu/
microarray/) as a repository for .tiff images and numerical data
began in 2000 with collaboration between a group of individuals from
Yale Center for Medical Informatics (YCMI), departments within Yale
School of Medicine, and the Faculty of Arts and Science (Cheung et al.,
2002) (http://info.med.yale.edu/microarray/people.htm), as well as advisors
and contributors from other institutions. This Oracle-based database
continues to expand its data query potential and links to analysis tools
tailored to the needs of investigators (Cheung et al., 2004). User groups
in YMD are represented by participating laboratories (within Yale and
outside of Yale). Each laboratory has a designated principal investigator
(PI). There are five types of user roles in the following order: database
administration (DBA), laboratory administration (Lab Admin), level 1,
level 2, and guest. Functionally, the database stores image files and associ-
ates with them primary analysis data, either normalized or raw, with or
without application of data filters, all of which is arranged in an experimental hierarchy with MIAME compliant experimental descriptions.
Image files and primary data are uploaded via the YMD web-front.
Query functions allow users to define parameters for searching through
individual or multiple data sets. The query parameters may be saved
for repeated application and query results may be saved as text files or
HTML format with links to functional annotations. Modifications continue
to be made to two tools developed to assist users in creation of array lists
(MAC: at http://ymd.med.yale.edu/kei-cgi/kc_mac_dev8.pl) (Cheung et al.,
2002) and to query and annotate array features (KARMA: at http://biryani.
med.yale.edu/karma/cgi-bin/ugtest.pl). KARMA (Cheung et al., 2004),
Keck ARray Manager and Annotator, is a web-based annotation tool
(implemented using Perl CGI) that allows investigators to query the content of all microarray platforms available from the Resource (http://ymd.
med.yale.edu/karma/cgi-bin/karma.pl). The tool allows investigators to
display content overlap and unique features among arrays (either within
the same species or across different species), make decisions regarding
which array will best serve their project needs, and provides links to
annotation for all features on the arrays. The entry key identifier for
array features in the current version of KARMA is the NCBI GenBank
Accession Number from which subsequent annotation information is
retrieved via UniGene Cluster ID including LocusLink. With the imminent
retiring of LocusLink and the incomplete nature of the UniGene database,
plans are underway to use NCBI Entrez Gene instead of LocusLink as a
major source of annotation. Through LocusTag links can be made to,
for example, UCSC Genome Browser to show fine chromosome map
positions for CGH studies and Ensembl for transcriptome information.

The ability to sort and display KARMA output in HTML with an expanded number of annotation options will be valuable. The addition of links to GEO expression data for each feature will also give KARMA output a direct link to published expression data for each feature. The ultimate goal for KARMA is to generate a query output with a direct link to sequence information for each feature with the ability to query a gene list directly with sequence and output features carrying that sequence with full annotation. As mining of probes becomes more detailed, information at the sequence level is becoming even more of an imperative for all types of microarrays, -BAC arrays with large sequence probes for CGH, short exon-specific designed oligonucleotide probes, promoter arrays with both PCR products and oligonucleotides. This type of extension will most likely require high-performance computing to handle queries and such resources now available at Yale through the Center for High Performance Computation in Biology and Biomedicine (http://info.med. yale.edu/hpc/).

Major journals (e.g., *Nature, Cell*, and the *Lancet*) have made microarray data compliance with the MIAME (Minimum Information About a Microarray Experiment) standards a publication requirement for submission (http://www.biomedcentral.com/news/20021010/05/). A major repository for such published MIAME compliant data is the Gene Expression Omnibus database at NCBI. General information and links to this database are found at the Microarray Resource web site at http:// keck.med.yale.edu/dnaarrays/geo.htm. All microarrays distributed by the Resource are associated with a GEO platform accession number for use when data are submitted. To streamline the data submission process, development and refinement of code are underway that extract data automatically from YMD, generate platform, series, and sample data, and perform batch submission in SOFT format (http://www.ncbi.nlm.nih. gov/geo/info/soft2.html). GEO will also ultimately use the MAGE-ML format for data submission and so programming is under development to facilitate submission by this mechanism. To achieve a significant speedup of query performance, we have migrated primary data (image quantitative data) from the Oracle database server to a main memory database (MMDB) system that resides on a different server. While the experimental annotation is still stored in Oracle, a link is established between the experiments (stored in Oracle) and the corresponding primary data (stored in MMDB). A specialized query engine is built on top of the MMDB system to allow speedy query access to primary data. The MMDB approach is several orders of magnitude speedier than the Oracle approach.

Gene Expression Studies Using Keck Microarray Slides
and/or Services

The 51 papers published since the Microarray Resource began describe fruitful microarray studies, many of which provide direct insight into the molecular mechanisms of cancer, two of which are described. In a study published in 2004, Dr. Harriet Kluger's group used cDNA microarrays representing 4600 mouse genes to study differential mRNA expression in several mouse breast cancer cell lines (Kluger *et al.*, 2004). The aim was to determine which genes were differentially expressed when mouse cell lines were expressing either wild-type CSF1R or a version of the gene mutated at an autophosphorylation site. The study successfully identified genes that can be linked to invasive phenotypes or to tumorigenesis. Genes identified in the initial study provide a basis for more detailed studies of metastatic progression and local invasiveness and may also provide therapeutic targets for the development of new chemotherapies. In 2005, Kluger and colleagues continued their exploration of the phenomenon of metastasis, the primary cause of death, in breast cancer. Using a mouse xenograft model they identified genes potentially involved with metastasis, comparing expression in a poorly metastatic human breast cancer cell line and a highly metastatic variant. Microarray analyses of these isogenic variants were done using mouse 16K operon 70-mer oligonucleotide microarray slides. Differentially expressed genes were identified statistically and differences of >2.5-fold were found for 106 genes. Changes in protein or RNA expression for 10 of 12 genes that were most highly differentially expressed were confirmed. Three marker genes, heat shock protein 70 (HSP-70), chemokine (C-X-C motif) ligand 1 (CXCL-1), and secreted leukocyte protease inhibitor (SLPI), were studied further with breast cancer tissue microarrays using a novel method of automated quantitative analysis. Two variants of a human breast cancer cell line were used to identify genes expressed differentially in the more tumorigenic and metastatic variant. The protein expression for three of these differentially expressed genes was examined using human breast cancer specimens with associated clinical data. Two of the three (CXCL-1 and HSP-70) were associated with decreased survival, whereas all three were associated with lymph node metastases. Combined with the previous report using this model, four genes expressed differentially in the GI101A model of breast cancer progression were significantly associated with lymph node metastasis or shorter survival in breast cancer patients. Thus, cell line models can serve as a basis for further studies of the biology of breast cancer metastasis and provide models for further experimental analyses. The differentially expressed genes identified in this study might have an important role not only as prognostic markers, but also as the next generation of molecular therapeutic targets. The identification of genes associated with metastasis in experimental models may have clinical implications for the management of breast cancer because some of these are associated with lymph node metastasis and survival and might be useful as prognostic markers or molecular targets for novel therapies.

Dr. David Stern and his group used microarrays to explore the phenomenon that overexpression of ErbB2 and ErbB4 receptors in breast cancers may be accompanied by contrasting clinical outcomes (Amin *et al.*, 2004). To investigate the molecular mechanisms contributing to these differences, they undertook a comparative study of gene expression regulated by the two receptors using an oligonucleotide array. Agonistic antibodies were used to activate ErbB2 and ErbB4 in isolation from the other ErbBs in human breast cancer cells. It was discovered that, in the same cell line, ErbB2 and ErbB4 activation influence gene transcription differentially. Although there are genes that are regulated by signaling from both receptors, there are also receptor-specific targets that are preferentially regulated by each receptor. They also showed that two ligands acting via the same receptor homodimer may activate different subsets of genes. Many of the induced genes revealed by the microarray study were hitherto unidentified targets of ErbB signaling. These include ErbB4 targets EPS15R, GATA4, and RAB2 and ErbB2-activated HRY/HES1 and PPAP2A. The authors note that targets of ErbB2 homodimer signaling may be especially important as markers in breast cancer. In their newest publication (Amin *et al.*, 2005), Dr Stern's group continued to use microarrays to elucidate the molecular mechanisms underlying the observation that autocrine production of neuregulin (NRG), a growth factor that activates members of the epidermal growth factor receptor/ErbB family of protooncogenes, is sufficient for breast tumor initiation and progression. They undertook a global analysis of genes regulated by NRG in luminal mammary epithelial cell lines. Gene expression profiling of estrogen receptor-positive T47D cells exposed to NRG-1 revealed both previously identified and novel targets of NRG activation. Profiling of other estrogen receptor-positive breast cancer cell lines, MCF7 and SUM44, yielded a group of 21 genes whose transcripts are upregulated by NRG in all three lines tested. Because NRG activation of these cells induces resistance to antihormonal therapy, the identified genes may provide clues to molecular events regulating mammary tumor progression and hormone independence.

The applicability of microarray-based, genomic scale, massive analysis is often limited by the need for large amounts of high-quality RNA. Dr. Gabriel Capella and his group at the University of Barcelona, Spain, have developed a method for analyzing small clinical samples using Keck microarrays (Grau *et al.,* 2005). They used RNA arbitrarily primed PCR (RAP-PCR) as an unbiased fingerprinting PCR technique to reduce both the amount of initial material needed and the complexity of the transcriptome for microarray experiments. The aim of their study was to evaluate the feasibility of using hybridization of RAP-PCR products as transcriptome representations to analyze differential gene expression in colon cancer samples using a microarray platform. RAP-PCR products obtained from samples with limited availability of biological material, such as experimental metastases, were hybridized to conventional cDNA microarrays. To gain insight into the molecular basis of the metastatic process, they used orthotopic implantation of human primary tumors in nude mice. Implanted xenografts resemble, in their early stages, primary tumors and reproduce, in part, their dissemination pattern, allowing completion of the metastatic process in a short time. From their observations, they designed a protocol to successfully analyze paired xenograft-metastases samples. Using this approach, they found two genes, HER2 and MMP7, that may be downregulated during distal metastasis of colorectal tumors and that RAP-PCR glass array hybridization can be used for transcriptome analysis of small clinical samples.

Acknowledgments

The author thanks Irina Tikhonova, Lesa Moemeka (Clarke), and the Resource staff for their ongoing expert technical assistance and organization, especially during the establishment of the Resource and during development and troubleshooting of protocols. Thanks also to Kei Cheung for his leadership on YMD and to Ken Williams for his dedication to the Resource goals. The Keck Microarray Resource is supported in part by the Anna and Argall Hull Fund from the Yale Comprehensive Cancer Center; the Yale Microarray Center for Research on the Nervous System, NINDS-NIMH Grant 1U24NS051869–01 (PI Shrikant Mane); and the Yale/NIDA Neuroproteomics Research Center, Grant 5P30DA018343–02 (PI Kenneth Williams).

References

Alexandre, I., Hamels, S., Dufour, S., Collet, J., Zammatteo, N., Longueville, F. D., Gala, J.-L., and Remacle, J. (2001). Colorimetric silver detection of DNA microarrays. *Anal. Biochem.* **295,** 1–8.

Amin, D. N., Perkins, A. S., and Stern, D. F. (2004). Gene expression profiling of ErbB receptor and ligand-dependent transcription. *Oncogene* **23**(7), 1428–1438.

Amin, D. N., Tuck, D., and Stern, D. F. (2005). Neuregulin-regulated gene expression in mammary carcinoma cells. *Exp. Cell Res.* **309,** 12–23.

Cheung, K. H., de Knikker, R., Guo, Y., Zhong, G., Hager, J., Yip, K. Y., Kwan, A. K. H., Li, P., and Cheung, D. W. (2004). Biosphere: The interoperation of web services in microarray cluster analysis. *Appl. Bioinformat.* **3**(4), 253–256.

Cheung, K. H., Hager, J., Nelson, K., White, K., Li, Y., Snyder, M., Williams, K., and Miller, P. (2002). A dynamic approach to mapping coordinates between microplates and microarrays. *J. Biomed. Inform.* **35**(5–6), 306–312.

Cheung, K. H., White, K., Hager, J., Gerstein, M., Reinke, V., Nelson, K., Masiar, P., Srivastava, P., Li, Y., Li, J., Li, J. M., Allison, D. B., Snyder, M., Miller, P., and Williams, K. (2002). YMD: A microarray database for large-scale gene expression analysis. *Proc. Am. Med. Informat. Assoc. 2002 Annu. Symp.* 140–144.

Cheung, V. G., Morley, M., Aguilar, F., Massimi, A., Kucherlapati, R., and Childs, G. (1999). Making and reading microarrays. *Nat. Genet.* **21**(1 Suppl.), 15–19.

Gershon, D. (2004). Microarrays go mainstream. *Nat. Methods* **1**, 263–270.

Grau, M., Sole, X., Obrador, A, Tarafa, G., Vandrell, E., Valls, J., Moreno, V., Peinado, M. A., and Capella, G. (2005). Validation of RNA arbitrarily primed PCR probes hybridized to glass cDNA microarrays: Application to the analysis of limited samples. *Clin. Chem.* **51**(1), 93–101.

Hegde, P., Qi, R., Abernathy, K., Gay, C., Dharap, S., Gaspard, R., Hughes, J. E., Snesrud, E., Lee, N., and Quackenbush, J. (2000). A concise guide to cDNA microarray analysis. *Biotechniques* **29**(3), 548–550, 552–554, 556 passim.

Kluger, H. M., Kluger, Y., Gilmore-Hebert, M., Chang, J., Rodov, S., Mironenko, O., Kacinski, B., Perkins, A., and Eva Sapi, A. (2004). cDNA microarray analysis of invasive and tumorigenic phenotypes in a breast cancer model. *Lab. Invest.* **84**(3), 320–331.

Kluger, H. M., Lev, D. C., Kluger, Y, McCarthy, M. M., Kiriakova, G., Camp, R. L., Rimm, D. L., and Price, J. E. (2005). Using a xenograft model of human breast cancer metastasis to find genes associated with clinically aggressive disease. *Cancer Res.* **65**(13), 5578–5587.

Liang, R. Q., Li, W., Li, Y., Tan, C. Y., Li, J. X., Jin, Y. X., and Ruan, K. C. (2005). An oligonucleotide microarray for microRNA expression analysis based on labeling RNA with quantum dot and nanogold probe. *Nucleic Acids Res.* **33**(2), e17.

Taylor, E., Cogdell, K., Coombes, L., Hu, L., Ramdas, L., Tabor, A., Hamilton, S., and Zhang, W. (2001). Sequence verification as quality control step for production of cDNA microarrays. *BioTechniques* **31**, 62–65.

[8] Printing Your Own Inkjet Microarrays

By Christopher G. Lausted, Charles B. Warren,
Leroy E. Hood, and Stephen R. Lasky

Abstract

DNA arrays are now the tools of choice for high-throughput DNA/RNA analysis. While many technologies exist for mass-producing arrays, there are just a few ways to economically produce small batches of custom oligonucleotide arrays for prototyping experiments and specialized applications. Inkjet printing, adapted from the world of office electronics to the world of molecular biology, is one such method. With programmable oligonucleotide synthesizers, scientists can prototype DNA array assays quickly and inexpensively. A benchtop inkjet arrayer—nicknamed POSAM—can be built by most skilled molecular biology laboratories. Inkjet arrays can fulfill the changing needs of those studying the complex network of relationships in systems biology.

METHODS IN ENZYMOLOGY, VOL. 410
Copyright 2006, Elsevier Inc. All rights reserved.

0076-6879/06 $35.00
DOI: 10.1016/S0076-6879(06)10008-7

Introduction

Background

The DNA array is a powerful tool for the high-throughput identification and quantification of nucleic acids. Among the numerous uses, array analysis has become a standard technique for monitoring gene expression. Arrays can be made by depositing ("spotting") presynthesized DNAs [such as cDNA, polymerase chain reaction (PCR) products, or oligonucleotides] onto glass or membrane or by the synthesis of oligonucleotides directly on a suitable solid substrate. As the sequences for *de novo* oligo array synthesis are stored in data files instead of frozen DNA libraries, the costs and the potential for errors in amplification, storage, and retrieval are reduced greatly.

Oligonucleotide Arrays

Affymetrix (Santa Clara, CA) pioneered the *de novo* synthesis of oligonucleotide arrays using the photolithographic approach of the semiconductor industry to define microarray features. However, because the photomasks needed are numerous and expensive, the creation of new array designs is relatively slow and expensive. To address this problem, alternative systems have been developed that offer programmable synthesis. The Institute for Systems Biology's POSAM (Lausted *et al.*, 2004) and Agilent's (Palo Alto, CA) SurePrint system (Wolber *et al.*, 2006) utilize inkjet synthesis. The Nimblegen (Madison, WI) and Xeotron/Invitrogen (Carlsbad, CA) systems utilize dynamic micromirror device photolithography (LeProust *et al.*, 2001; Singh-Gasson *et al.*, 1999). The Combimatrix (Mukilteo, WA) CustomArray system utilizes electrochemical base deprotection (Tesfu *et al.*, 2004). Aside from the POSAM, these instruments were developed by groups that were, or were to become, companies dedicated to selling array chips. As such, those microarray synthesizers are not available to scientists who might wish to operate or modify their own machine. However, the schematics, assembly instructions, and user manual for a do-it-yourself POSAM inkjet arrayer are freely available online (http://techdev.systemsbiology.net/posam/).

Development of Inkjet Arrays

Most efforts to apply inkjet print heads to microarray production have involved depositing presynthesized DNA onto a reactive substrate. Commercial piezoelectric pipetting robots are available for noncontact microarray printing (Arrayjet, Dalkeith, UK; Perkin-Elmer, Wellesley, MA;

GeSiM, Großerkmannsdorf, Germany). Like pin-spotting microarrayers, they are intended for spotting libraries of amplified DNAs. With only four to eight piezoelectric jets, they are not easily convertible to high-throughput, *de novo* oligonucleotide array synthesis. Some of these printers are based on commercial "bubble jet" print heads containing tiny heating elements that rapidly vaporize a water-based solution in a capillary to eject a droplet (Okamoto *et al.*, 2000; Roda *et al.*, 2000; Stimpson *et al.*, 1998) onto a slide or membrane. These printers are difficult to clean and reload and, being water based, are not suitable for organic synthesis of oligonucleotides.

Inkjet arrays from Agilent, the largest commercial producer of inkjet arrays, have been used successfully for expression profiling, validating predicted mRNA transcripts, and detecting alternative splicing (Hughes *et al.*, 2001; Johnson *et al.*, 2003; Schadt *et al.*, 2004; Shoemaker and Linsley, 2002). Tiling arrays were constructed with 60-mers spanning 50,000 known and predicted transcripts and used to detect 28,000 transcripts. Splice-site arrays were constructed with 36-mers spanning the exon–exon junctions of 10,000 genes and used to find alternative splicing in 74% of the multiexon genes. These arrays range in density from 11,000 to 44,000 features per slide.

The POSAM Arrayer

The POSAM was designed to be a reasonably compact machine utilizing as many off-the-shelf components as possible. It utilizes a low-cost piezoelectric print head with six fluid channels each feeding 32 piezoelectric jets. Four channels (128 jets) deliver phosphoramidite precursors, one channel (32 jets) is used to deliver the activator (ethylthiotetrazole) onto the array, leaving one channel (32 jets) available for an optional linker or modified phosphoramidite base. A second print head may also be attached, although this is not yet supported by the software.

Piezoelectric jets can accurately deliver a wide range of nonvolatile solvents in volumes as low as 4 pl. Piezoelectric print heads, high-quality motion controllers, and standard phosphoramidite oligonucleotide synthesis chemistry allow users to synthesize arrays of any nucleic acid sequence at specific, closely spaced spots (features) on suitable substrates. The current configuration of the POSAM platform produces multiple unique microarrays, each with 9800 different reporter sequences, on modified silicon or glass microscope slides. Synthesis is rapid and relatively inexpensive (phosphoramidite usage is practically negligible), and as many as eight new arrays of 50-mers can be ready for hybridization within 1 day of their design. As these arrays can be designed using genome sequence information alone, they are suitable for a wide range of biological investigations, including the study of gene expression, alternative splicing, *in vivo* protein binding using ChIP-on-Chip (Ren *et al.*, 2000) or DamID (van Steensel and Henikoff, 2000), and the detection of single nucleotide polymorphisms. Additionally, covalent linkage of the oligonucleotides to the substrate is sufficiently robust so as to allow for the reuse of individual oligoarrays with little change in the signal-to-noise ratio.

Inside the POSAM

The POSAM is based on a three-axis servo positioning system that can move the print head over a wide range with 5 μm repeatability. The Epson F057020 piezoelectric print head and five reagent dispensing valves are mounted to the X- and Z-axis stages. Array substrates (glass or silicon, 25 × 76 mm) are vacuum chucked into a slide holder mounted on the Y-axis stage. The robotics is enclosed within a sealed acrylic cover to maintain a dry, inert atmosphere. A latched access door is opened to insert and remove materials, while a glove panel can be used to make adjustments inside the enclosure during synthesis runs. Nitrogen flow into the enclosure also powers the air amplifier—a device that recirculates the internal atmosphere through desiccant and activated charcoal filters. The motion controller, servo amplifiers, power supplies, and other circuits are packaged in a standard 19-in. vertical rack. The arraying process is directed by a PC running Pogo, our open-source control software.

Piezojet performance is checked before every printing event using laser droplet detection and after every printing event using camera-based machine vision. Jets that fail to trigger the laser detector are taken off-line during synthesis. Pogo then calculates a printing strategy to account for the failing jets. As partially occluded jets can fire off-target, a machine-vision system checks for the correct placement and mixing of each droplet on the substrate. The system illuminates the surface with collimated light and collects images with a 1024 × 768 pixel IEEE-1394 CCD camera. Droplet diameter and position are automatically calculated by Pogo. If desired, pictures of every jetting cycle may be saved using the lossless PNG format.

Array substrates may be either patterned or unpatterned. Patterned, silicon slides are hydrophobic surfaces containing a predefined array of reactive, hydrophilic wells and are available from Lumera, Inc. (Bothell, WA). These slides are preferred as they produce arrays with perfectly round, precisely positioned features (see Fig. 1). Alternatively, unpatterned glass slides can be prepared in the laboratory. These substrates are uniformly hydrophobic and rely on precise jetting to define microarray

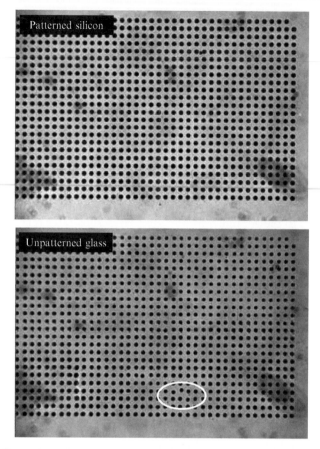

FIG. 1. Comparison of spot morphology and placement for Lumera-patterned silicon and unpatterned glass slides. A patterned silicon slide produces an array of perfectly round droplets with a fixed diameter. Unpatterned glass substrates are unable to prevent positioning errors. These errors (white circle) were detected automatically during inkjet synthesis by the imaging system of the POSAM.

feature position and size. The glass is derivatized using a mixture of two silanes: 3-triethoxysilylpropyl-4-hydroxybutyramide (3TSP4HB) provides reactive hydroxyl groups for DNA synthesis and a polydimethlysiloxane increases the contact angle so that jetted droplets remain discrete. These substrates are slightly easier to use because they do not need to be aligned with the print head.

How Long Should an Oligonucleotide Be?

Currently, oligoarrays are available with short (20 to 25 nucleotides) oligonucleotides or longer oligonucleotides (60 to 70 nucleotides) depending on the method used to synthesize the arrays. Affymetrix (Dalma-Weiszhausz et al., 2006), which entered early and dominates the field, produces multiple (as many as 20), short (20 to 25 nucleotides) oligonucleotides to identify each gene. Their use of photolithographic masks makes it more economical to produce a larger number of shorter sequences. Other in situ array manufacturers can make longer sequences more affordably through the use of programmable printing devices (e.g., inkjets and dynamic micromirrors). Both long and short oligonucleotides have advantages and disadvantages. Relogio et al. (2002) compared the specificity and sensitivity of 60-mers to 25-mers for array hybridization with RNA. The 60-mers were found to have, on average, sevenfold more sensitivity but with fourfold poorer specificity. Kane et al. (2000) showed that single, carefully selected oligonucleotides of 50 nucleotides in length reliably detect the presence of gene transcripts in complex biological samples, reporting two-color fluorescence ratios closely correlated to corresponding qPCR results. These results indicate that 50- to 60-mers are better suited to applications such as gene expression profiling, whereas shorter oligonucleotides are better suited to applications such as alternative splice detection and genotyping.

Designing Oligonucleotides

The first step in making an inkjet microarray is designing the oligonucleotide reporter sequences (Kreil et al., 2006). A number of open-source software packages are freely available for downloading. Most packages follow the same basic strategy. The user provides a file with all gene sequences of interest in FASTA format. Ranges for oligonucleotide length and melting temperature (T_m) are specified. Oligonucleotide candidates are chosen from a sliding window moving from 3' to 5' through each gene. Heuristics involving sequence complexity, GC content, and repeats are used to eliminate troublesome candidates. Candidate oligonucleotides are then checked for potential cross hybridization using a sequence-similarity search such as BLAST.

Biosap was developed by our group to design whole-genome oligonucleotide arrays. It runs on a Linux server and is accessed using any computer with a web browser as a client. The user provides Biosap with two FASTA files: a target file contains the genes for which oligonucleotides are to be chosen and the whole-genome file to be used for BLAST analysis. Other parameters include the number of oligonucleotides per gene, their length, the desired T_m, and the usual heuristics. Candidate oligonucleotides and their BLAST results are saved to an SQL database. The C++ and Java source code is available at http://techdev.systemsbiology.net/biosap.

Two similar programs available online are OligoWiz (Nielsen *et al.*, 2003) and OligoArray (Rouillard *et al.*, 2003). OligoWiz, from the Technical University of Denmark, is a java client-server application. The user provides a single FASTA file. For each gene, OligoWiz produces a line graph showing scores for homology T_m (closeness to desired melting temperature), complexity, and a weighted total. Scores are plotted on the y axis against the input sequence position on the x axis. OligoWiz recommends a sequence with the best total score and highlights it in color. It can be found at http://www.cbs.dtu.dk/services/OligoWiz2. OligoArray 2.1 is an open-source, java program that runs from the command line. It takes a FASTA file, masks out the low-complexity bits, uses BLAST to create a similarity matrix for each position of each input sequence, and moves a window from 3′ to 5′ to study candidate oligonucleotides. After the T_m is checked, the secondary structure and the potential cross hybridization with similar sequences are analyzed using OligoArrayAux, an MFOLD (Zuker, 2003) replacement. It can be found at http://berry.engin.umich.edu/oligoarray2_1.

Two other packages, SEPON (Hornshoj *et al.*, 2004) and ProbeSelect (Li and Stormo, 2001), are available from their developers upon request. SEPON uses a similar approach to the packages described earlier. It runs on Linux and uses BLAST, MFOLD, and Melting (Le Novere, 2001). ProbeSelect takes a different approach. While other methods save the slow BLAST step until the end, ProbeSelect looks for low-frequency sequences first. The algorithm uses sequence suffix arrays and landscapes efficiently create a list of oligonucleotide candidates.

All of the packages require some trial and error to design a complete set of oligonucleotides. Often, it is difficult to choose unique probe oligonucleotides for all genes in a genome. Kane *et al.* (2000) showed that cross hybridization is a problem in 50-mers with more than 75–80% homology or with stretches exceeding 15 bp. Selection parameters usually need to be adjusted several times until an acceptable oligonucleotide set is found.

Array Reusability

Arrays are generally considered to be single-use products. When oligonucleotides are covalently attached, however, they should be reusable as long as the background can be kept low. For reuse, arrays can be denatured by heat or by base-pH treatment. Consolandi *et al.* (2002) functionalized microarray slides using either amino or epoxy silane and found them to be reusable at least once after boiling. Signal was reduced 50% after 25 cycles of thermocycling between 50 and 95°. Dolan *et al.* (2001) functionalized microarray slides using *p*-aminophenyl trimethoxysilane and diazotization chemistry and showed them to be reusable. While signal declined steadily, the arrays did retain two-thirds of their initial signal after three rounds of stripping and rehybridization. Their stripping method used several 30-s immersions into 95° neutral-pH buffer. Beier and Hoheisel (1999) functionalized microarray slides with a dendrimeric linker by repeated cycles of acetylation and polyamination. These arrays stripped and rehybridized in the same manner and were found to maintain their initial signal levels for at least seven rounds of use. The arrays would also tolerate 30 cycles of thermocycling between 25 and 95° without a significant signal loss.

POSAM inkjet arrays synthesized on epoxy silane slides were also found to maintain two-thirds of their initial signal after seven rounds of hybridization and stripping (Lausted *et al.*, 2004). These arrays were stripped by immersion in 20 m*M* NaOH at room temperature for 2 min. The current glass substrates also show a level of reusability (see Fig. 2) .

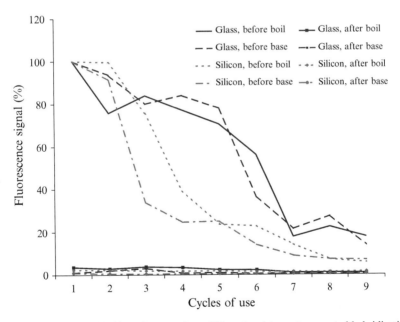

FIG. 2. Two types of inkjet substrates show different resistance to repeated hybridization and stripping. Glass slides derivatized with 3-triethoxysilylpropyl-4-hydroxybutyramide maintain two-thirds of their signal for five cycles. Silicon slides from Lumera maintain their signal for two or three cycles. Stripping with 20 m*M* sodium hydroxide or by boiling removes greater than 96% of the signal.

Unpatterned glass slides, derivatized with 3TSP4HB, maintain two-thirds of their signal after five rounds of hybridization and stripping. Patterned silicon slides were found to be reusable only one time. The arrays were either boiled 5 min in deionized water or immersed in NaOH solution for 5 min.

Modified Oligonucleotides

Because the POSAM uses standard phosphoramidite chemistry, arrayed oligonucleotides may contain a wide variety of modifications, such as fluorescent labels, spacers, RNA units, and locked nucleic acid (LNA) units. The LNA ribonucleoside contains a methylene linkage between the 2'-O and the 4'-C atoms (Petersen and Wengel, 2003). LNA oligonucleotides exhibit unusually high thermal stability toward complementary DNA and RNA that allows for increased microarray sensitivity and specificity. LNA array expression profiling has been demonstrated with a *C. elegans* toxicity chip. In that experiment, a set of 120 oligonucleotides containing 33% LNA nucleotides was synthesized and photocoupled to plastic substrates (Tolstrup *et al.*, 2003). We have evaluated the *in situ* synthesis of LNA microarrays. We synthesized an array containing variations of the sequence 3'-N_{12}-G*T*ACACGA*T*GAAG*T*GG*T*TAC where from zero to six DNA bases were replaced with LNA-T bases. (No more than one mismatch was allowed.) When hybridized with a complementary, Cy3-labeled, DNA 20-mer, the signal was found to increase with LNA-T content (see Fig. 3). The array feature containing six LNA-Ts produced a signal fourfold stronger than the unmodified DNA feature. Additionally, specificity was enhanced. While mismatches involving DNA-T reduced the signal 14–34%, mismatches involving LNA-T reduced the signal 40–43%.

Another type of modified microarray is the double-stranded (ds) DNA array. Such DNA arrays enable the genome-wide identification of putative transcription factor-binding sites. Binding can be measured by applying protein samples to dsDNA microarrays and labeling with a fluorescent primary/secondary antibody system (Bulyk, 2006; Bulyk *et al.*, 1999, 2001). Inkjet arrays of single-stranded DNA can be converted easily to double-stranded DNA by two methods: for short sequences, "snapback" oligonucleotides can be designed that self-hybridize into long hairpins; for longer sequences, dsDNA can be made by primer extension using the Klenow fragment of DNA-polymerase. Inkjet sequences are designed containing a common 12 base priming sequence near the 3' end of the oligonucleotide. Preliminary experiments have shown that the two methods produce comparable results for NF-κB binding (see Fig. 4).

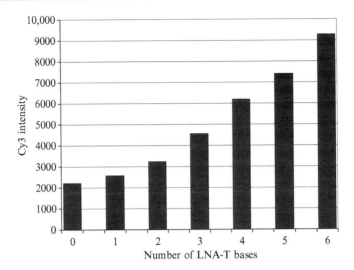

FIG. 3. Hybridization signal increases with increasing LNA-T content. An inkjet array of 32-mers was constructed containing from zero to six LNA-T substitutions and was hybridized with a complementary Cy3-labeled 20-mer.

Description of Methods

Materials

Substrate Preparation

- Sylgard 184 silicone elastomer kit (Dow Corning, Midland, MI)
- Hydrophobic, masked, silicon substrates (Lumera, Bothell, WA)
- Glass slides, "Fisher's Finest," 1 × 3; glass slide staining dishes; sodium hydroxide (NaOH); hydrochloric acid (HCl); Citranox acid detergent (Fisher Scientific, Hampton, NH)
- "mBox" polypropylene slide staining dishes (Erie Scientific, Portsmouth, NH)
- 3-Triethoxysilylpropyl-4-hydroxybutyramide (3TSP4HB) (Gelest, Morrisville, PA)
- Silanol-terminated polydimethylsiloxanes (STPDMS), 750 cS viscosity (United Chemical Technologies, Bristol, PA)
- Ethanol (Aaper Alcohol and Chemical, Shelbyville, KY)
- Toluene; acetone (Sigma, St. Louis, MO)
- Nanostrip glass cleaning solution (Cyantek, Fremont, CA)

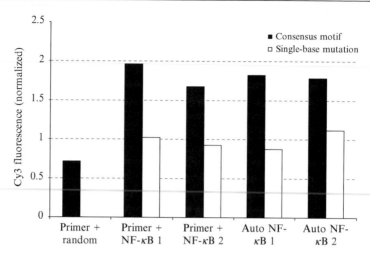

FIG. 4. On-chip detection of NF-κB protein binding to dsDNA by indirect immunolabeling. A mouse anti-NF-κB primary antibody and a Cy3-conjugated goat antimouse secondary antibody were used for labeling. Fluorescence was measured with a ScanArray 5000 microarray scanner. Double-stranded DNA was created using primer extension ("Primer") or self-complementary snapback oligonucleotides ("Auto"). The two methods produce comparable results. Both strand orientations ("1" and "2") are shown. The recombinant NF-κB p50 protein bound best to the consensus motif 5′-GGGACTTTCC.

Inkjet Synthesis

- Nucleoside phosphoramidites (bzdAMP, AcdCMP, ibudGMP, dTMP, LNA dTMP); 5-ethylthio-1H-tetrazole; oxidizer (0.02 M iodine in pyridine/tetrahydrofuran/water); detritylation acid (2.5% dichloroacetic acid in dichloromethane) (Glen Research, Sterling, VA)
- Methylglutaronitrile (MGN); 3-methoxypropionitrile (3MP) (Sigma)
- Acetonitrile, "optidry" grade; ammonia, aqueous; methylamine, aqueous; molecular sieves, grade 564; syringes, 1- and 5-ml polypropylene; needles, 26 gauge; "Drierite" desiccant with indicator (Fisher Scientific)
- Nitrogen gas, 99.998% pure
- Activated charcoal (Aquarium Pharmaceuticals, Chalfont, PA).

Hybridization and/or Primer Extension

- DIG Easy Hyb buffer (Roche Molecular Biochemicals, Indianapolis, IN)
- Tris–Cl, pH 7.5; magnesium chloride (MgCl₂); dithiothreitol (DTT); phosphate-buffered saline (PBS); saline sodium phosphate EDTA

(SSPE); sodium dodecyl sulfate (SDS); Tween 20; Triton X-100 (Fisher Scientific)
- DNA polymerase Klenow fragment; nucleotide mix (dNTPs) (Fermentas, Hanover, MD)
- DNA primer (5′-ACTGGCTTAGAC) (Integrated DNA Technologies, Coralville, IA)
- Hybridization chambers (Corning Life Sciences, Acton, MA)
- "LifterSlip" coverslips, 25 × 60 mm (Erie Scientific).

Setup and Synthesis

Slide Preparation. Either masked silicon Lumera slides or unmasked glass slides may be used. Glass slides need to be prepared several days in advance of synthesis to allow time for curing. A silicone barrier can be applied to both Lumera and glass slides to mask off the edges. This barrier reduces reagent usage and helps direct used reagents into the waste collection channel of the slide holder.

1. Slide cleaning.
   a. In a fume hood, load one or more slide staining dishes with 18 slides and fill the dishes two-thirds full with Nanostrip. Gently agitate the slides using the wire handle of the dish to dislodge any bubbles. Remove the wire handle. Incubate at room temperature for 1 h. (Caution: Nanostrip is a concentrated acid mixture containing hydrogen peroxide. It is extremely aggressive against organic material and should be considered very hazardous. Handle it in accordance with the manufacturer's instructions.)
   b. Carefully remove each rack from the staining dishes and rinse thoroughly under running deionized water. Immerse each rack in a staining dish filled with deionized water and incubate for 20 min at room temperature. Repeat this step three times.
   c. Repeat steps a and b, replacing Nanostrip with 10% (w/v) NaOH solution.
   d. Repeat steps a and b, replacing Nanostrip with 1% HCl solution and reducing the cleaning time from 1 h to 1 min.
   e. In a chemical fume hood, fill each staining dish two-thirds full with methanol and gently lower the glass racks into the solvent, agitating up and down several times. Incubate at room temperature for 5 min.
   f. Remove slide racks from the methanol and blow excess methanol clear while holding the slides down in the rack with one hand. Transfer slides to mBox holders. Allow slides to air dry in the hood for 1 h. Silanize immediately or store in a desiccator.

2. Slide silanization.
   a. Prepare a 500-ml silanization solution of 5% 3TSP4HB and 0.1% (v/v) STPDMS in toluene.
   b. Fill three polypropylene dishes: one with silanization solution, one with toluene, and one with ethanol.
   c. Place the 20-slide mBox holder in the silanization solution and stir for 10 min at room temperature.
   d. Wash the slides in toluene for 30 s with stirring. Wash the slides in ethanol for 60 s with stirring.
   e. Remove the slides from the ethanol wash and immediately spin them dry in a centrifuge. Let the slides stand for 30 min in a fume hood.
   f. Store slides in a desiccating cabinet at room temperature at least 2 days before use to allow surfaces to cure.
3. Edge-masking barriers.
   a. Prepare 5–10 g Sylgard 184 according to the manufacturer's instructions. This is enough to mask at least 20 slides.
   b. Spread the Sylgard across the slide surface to make barriers approximately 3 to 5 mm wide across the full 25-mm slide width. Locate one barrier 13 mm from the slide top and a second barrier 13 mm from the slide bottom (see Fig. 5).
   c. Bake the slides 15 min at 80°.

FIG. 5. Barriers of silicone elastomer can be spread on the slide edges to confine reagent to the center of the array. The liquid cures in 15 min at 80°. The silicone is removed after DNA synthesis.

Solvent Preparation. Dry solvent is critical to achieving high coupling efficiency during synthesis. Additionally, tetrazole and phosphoramidites will precipitate from wet solvent and cause nozzle failures during ink jetting. The solvent mixture must be prepared several days in advance of synthesis and left on washed molecular sieves until use.

1. Wash graduated cylinder, bottles, caps, and septa with 1% Citranox in warm water, rinsing thoroughly with deionized water. Rinse with acetone, invert, and allow to dry fully.
2. Place bottles and graduated cylinder in a convection oven at 250° for 30 min.
3. Remove from oven and allow to cool sufficiently for safe handling. Cap graduated cylinder with aluminum foil and set aside. Add approximately 25 g of molecular sieves to each amber bottle. Cap immediately.
4. Set oven to 300°. Wrap a Pyrex dish in aluminum foil. Remove caps, place bottles containing molecular sieves in the Pyrex dish, and bake in oven for 16–24 h (overnight).
5. Carefully remove bottles from oven (300° is very hot; use appropriate thermal gloves and caution when handling hot materials) and immediately flush each bottle with dry gas.
6. Place the Pyrex dish with bottles in desiccator. Pull maximum vacuum to evacuate air from the desiccator. Direct a stream of dry gas into the valve as vacuum is released to flood the desiccator with dry gas.
7. Repeat the vacuum/dry gas flush four times. Leave bottles under vacuum for 1 h to cool.
8. Using the graduated cylinder prepared as described earlier, add 50 ml of 3MP and MGN to each bottle. This should fill the bottles almost completely (almost no head space).
9. Cap bottles immediately with septa caps. Carefully seal the cap to the bottle with parafilm, being careful to cover the entire septa and the joint between the cap and the neck of the bottle.
10. Fill a 5-ml syringe with dry gas and inject gas into bottle through the septum. This provides positive pressure in the bottle and prevents the entry of atmospheric moisture during storage.
11. Shake bottle vigorously for 30 s and place on orbital shaker overnight.
12. Store bottles inverted clamped on a ring stand. Allow bottles to sit undisturbed for 2 to 3 days so particulates can settle. Solvent solution should be very clear after settling. Cloudy solvent is an indication of moisture and should be discarded.

Preparing Pogo with Sequence Data

13. Array sequences must be formatted into a file readable by Pogo, the POSaM control program. This file, called a lineup file, can be a comma-separated variable (CSV) spreadsheet file. CSV is the simplest format—intuitive and visual. The sequences can be arranged graphically in a block of 70 columns and 140 rows (or less). Almost any spreadsheet (OpenOffice, Excel, Gnumeric) will export CSV format. Another option is the four-column, tabular file. The space-delimited columns must contain row number, column number, description, and sequence, respectively. "Gridding" oligonucleotides are placed around the perimeter of the array. A complementary, fluorescence-labeled control 20-mer is either prehybridized to the array or included in the subsequent hybridization experiment. The gridding spots, usually numbering more than 416 on a full-size array, allow the user to check for correct spot morphology, to validate synthesis efficiently, and to register the image for the spot-finding operation.

Oligoarray Synthesis with the POSAM

14. Preparing tetrazole and phosphoramidites: Nucleoside phosphoramidites and (^{bz}dAMP, ^{Ac}dCMP, ^{ibu}dGMP, dTMP), 5-ethylthio-1H-tetrazole are dissolved at 0.2 M in the jetting solvent. Degas the solutions by drawing a vacuum of at least one-half atmosphere for 1 h.
15. Clean the inkjet print head: The inkjet print head is fed by six 500-μl vials with conical bottoms. Using a 5-ml polypropylene syringe of dry solvent fitted with a 26-gauge needle, flush 1 ml through each capped vial and out the piezo nozzles. Using a 50-ml polypropylene syringe of dry nitrogen gas, dry each fluid channel with 25 ml of the inert gas.
16. Loading reagents in inkjet: Using 1-ml polypropylene syringes fitted with 26-gauge needles, fill four fluid channels with the phosphoramidite solutions; fill one or two channels with tetrazole solution. For each channel, insert the needles of the reagent and of the nitrogen gas syringes. Also insert a third needle as a vent. Push 100 to 400 μl into each vial. Next, using slow and steady pressure on the nitrogen syringe, push reagent from the vials out the piezo nozzles. Remove the syringes. Wipe droplets from the piezo nozzles. Wipe from front to back, avoiding contamination of the neighboring nozzles. Remove the vent needles (see Fig. 6).

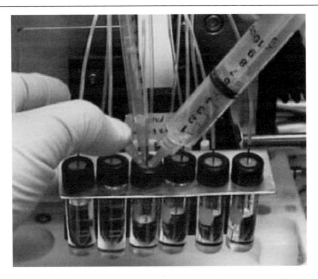

FIG. 6. Loading the inkjet print head. A vent needle, a 1-ml syringe with phosphoramidite solution, and a 5- to 10-ml syringe of nitrogen gas are used to load the supply vial for bank #3. (See color insert.)

17. Loading acetonitrile, oxidizer, and acid: Fill the acetonitrile, oxidizer, and acid bottles with enough reagent for the length of oligonucleotides to be synthesized. This is typically 1 ml per slide per base for oxidizer and acid. The acetonitrile is consumed at a rate six times higher. These bottles are kept under 3 psi pressure. Purge bubbles and old reagents out of the system by opening each solenoid for 6 s.

18. Drying down the box: Place approximately 500 g fresh desiccant in the POSaM recirculation basket. Flood the enclosure over the POSaM arrayer with nitrogen for 45 min. Check that the hygrometer reaches a steady-state minimum reading. Once synthesis begins, the hygrometer reading will be affected by the dichloromethane in the acid solution. A spike in apparent humidity each time oxidizer and acid are added is normal.

19. The synthesis cycle: Each cycle of synthesis includes (1) printing, (2) washing, (3) oxidization, (4) washing, (5) detritylation, and (6) washing. The first step of the reaction consists of phosphoramidite monomer printing followed by tetrazole printing, followed by a 2-min incubation. This first 3' base is reprinted to ensure a high density of oligonucleotide synthesis. Oxidization and detritylation steps are carried out for 30 s each.

20. Deprotection with AMA: After reporter synthesis is completed, remove the microarray slides from their slide holder and rinse with acetonitrile and then 95% ethanol. Remove the base protecting groups by a 105-min incubation in an aqueous solution of 13% ammonium hydroxide and 20% methylamine in a sealed container at room temperature. Rinse the deprotected slides five times with deionized water, dry, and store in a desiccator.

There are several common problems that can easily ruin a synthesis run. Careful attention must be paid to the nitrogen gas and liquid reagent levels. Estimate the reagent consumption before beginning—it is proportional to the number of slides and the length of the oligonucleotides arrayed. Check that the waste line is flowing freely. If waste is allowed to accumulate in the slide holder, the enclosure atmosphere will become wet and reduce synthesis efficiency. The print head must be aligned correctly. If it is not level and perpendicular to the direction of travel, droplets will miss their targets. Ensure that 95–100% of the nozzles are firing. Failing nozzles often indicate a wet phosphoramidite solution. Nozzles that fire intermittently are also less likely to be firing on target.

Double-Stranded DNA by Primer Extension

1. First blocking.
   a. Apply 65 μl of blocking solution (10 mg/ml bovine serum albumin in PBS) to each inkjet array and cover with 24 × 60-mm coverslips. Incubate for 1 h at room temperature.
   b. Warm a beaker containing 150 ml deionized water to 80°. Remove the coverslip by immersing each array in the beaker. Keep the array in the water for 1 min. Dry the array with a stream of nitrogen.
2. Hybridization and extension.
   a. Prepare 2× Klenow buffer solution (20 mM Tris–Cl, 10 mM MgCl$_2$, 15 mM DTT in deionized water).
   b. Prepare extension mix (1× Klenow buffer solution, 0.4 mM dNTPs, 0.5 μM DNA primer, 0.133 U/μl Klenow).
   c. Apply 65 μl of extension mix to each inkjet array and cover with 24 × 60-mm coverslips. Place the arrays in hybridization chambers. Incubate for 2 h at room temperature, followed by an additional 2 h at 30°.
   d. Remove the arrays from the hybridization chamber and dip each five times into a small beaker containing 6× SSPE and 0.005% Triton X-100.

    e. Dip each array five times into a small beaker containing 0.2×
       SSPE. Dry the arrays with a stream of nitrogen.
3. Second blocking.
    a. Place the arrays in a sealed container containing 2% skim milk in
       PBS. Rock the container gently for 1 h at room temperature.
    b. Wash the arrays by dipping five times in a small beaker
       containing 0.1% Tween 20 in PBS.
    c. Dip the arrays five times in another beaker containing 0.005%
       Triton X-100 in PBS. Dry the arrays with a stream of nitrogen.

Hybridization

The arrays are prehybridized with a control oligonucleotide to verify efficient synthesis and to check spot morphology. The control oligonucleotide contains at least 20 bases complementary to the border features and is labeled at the 5′ end. If the array is to be used later with Cy3- and Cy5-labeled samples, the control oligonucleotide may be labeled with a third fluorophore such as fluorescein. As many array scanners do not have a 488-nm laser, this is not always possible. In that case, Cy3 or Cy5 may be used. If a ScanArray 5000 confocal laser scanner is used, the signal from the border features is expected to nearly saturate the photomultiplier tube (PMT) when using 80% laser power and 80% PMT gain.

4. Apply 65 μl of hybridization solution (control hybridization solution
    contains 1 nM control oligonucleotide in 1× DIG Easy Hyb buffer)
    to each inkjet array and cover with 25 × 60-mm coverslips.
5. Place the arrays in hybridization chambers. Add 10 μl of water to
    the two dimples in the chambers and close the chambers. For test
    hybridization, incubate 30 min in a 37° water bath. Complex samples
    can incubate overnight.
6. Remove the arrays from the hybridization chamber. Remove the
    coverslips by dipping the slides in wash buffer 1 (2× SSPE, 0.1%
    SDS). Place the slides in a separate, sealed container of wash buffer
    1. Agitate gently for 15 min.
7. Wash the arrays in wash buffer 2 (0.1× SSPE) for 1 min, agitating
    gently.
8. Dry the arrays with a stream of nitrogen. Scan the arrays promptly.

Two-color competitive hybridizations are done in a similar manner, also using DIG Easy Hyb buffer. Hybridization temperature varies with the length of probe, with 37–45° being the usual range. Hybridizations are typically allowed to proceed overnight. Coverslips are removed and slides are washed in the same manner as described previously. The dry slides are inserted into a microarray reader and scanned with 10-μm resolution. The images are saved in a 16-bit uncompressed format. Microarray slides can be stripped by washing in 20 mM sodium hydroxide for 2 min, followed by five washes in deionized water.

Data Analysis

Inkjet array data are analyzed in the same manner as spotted oligonucleotide arrays. Fluorescence images are processed by spot-finding software, the two separate color channels are normalized, and data are stored in a Minimum Information About a Microarray Experiment (MIAME)-compliant database. We save our laser scans as 16-bit TIFF files and extract feature information using Dapple http://www.cs.wustl.edu/~jbuhler/research/dapple/ or TIGR Spotfinder http://www.tm4.org/spotfinder.html software. Both tools are freely available online. Our MIAME-compliant database, SBEAMS, is a modular system designed to allow the storage of both microarray and proteomic data. Microarray data can be tested for differential expression using variability and error assessment (VERA) and significance of array measurement (SAM). VERA and SAM are programs that apply maximum-likelihood analysis to generate a confidence value, lambda, indicating the likelihood an array feature is detecting a significant change in signal (Ideker *et al.*, 2000). SBEAMS, VERA and SAM are all available at http://www.sbeams.org/project_description.php.

Concluding Remarks

The POSAM was not developed for the sole purpose of expression profiling—the commercial sector provides a good selection of gene arrays. The POSAM is more appropriate for applications such as generating tiling arrays that cover the noncoding regions of genomic DNA. Such arrays are needed for the analysis of protein-binding sites in native chromatin by ChIP-on-chip and DamID (Kreil *et al.*, 2006; Negre *et al.*, 2006) methods. Few commercial microarrays are available containing these regions because of the cost associated with synthesizing or cloning every noncoding region in a given genome. For instance, we are synthesizing POSAM dsDNA arrays that tile across regions where expression profiling has suggested the presence of regulatory elements. Elements of *cis*-regulatory networks will be found by treating cells with different stimuli and applying the nuclear extracts to dsDNA tiling arrays. Protein–DNA binding is currently detected using specific antibodies and fluorescence; however, we are developing inkjettable surfaces that are also compatible with a new SPR array scanner and with available MALDI-TOF mass spectrometers.

SPR allows us to observe changes in dsDNA-binding activity, while subsequent MALDI-TOF analysis may identify peptides from the specific transcription factors bound to arrayed sequences.

The open-source inkjet arrayer is one of the most flexible ways for scientists to meet their array prototyping needs. The arrays themselves are flexible: sequences are stored in data files rather than in physical libraries, they can be designed and redesigned easily, and they can contain the wide range of modifications available from conventional phosphoramidite synthesis. Additionally, the arrayer is flexible: the hardware and control software is freely available online. The Pogo software is licensed under the terms of the General Public License. Users are both allowed and encouraged to improve and extend the code, providing they keep the code open. Similarly, it is hoped that users will improve and extend POSAM hardware and make those enhancements freely available.

References

Beier, M., and Hoheisel, J. D. (1999). Versatile derivatisation of solid support media for covalent bonding on DNA-microchips. *Nucleic Acids Res.* **27**, 1970–1977.

Bulyk, M. L. (2006). Analysis of sequence specificities of DNA-binding proteins with protein binding microarray. *Methods Enzymol.* **410** (this volume), 279–299.

Bulyk, M. L., Gentalen, E., Lockhart, D. J., and Church, G. M. (1999). Quantifying DNA-protein interactions by double-stranded DNA arrays. *Nat. Biotechnol.* **17**, 573–577.

Bulyk, M. L., Huang, X., Choo, Y., and Church, G. M. (2001). Exploring the DNA-binding specificities of zinc fingers with DNA microarrays. *Proc. Natl. Acad. Sci. USA* **98**, 7158–7163.

Consolandi, C., Castiglioni, B., Bordoni, R., Busti, E., Battaglia, C., Bernardi, L. R., and De Bellis, G. (2002). Two efficient polymeric chemical platforms for oligonucleotide microarray preparation. *Nucleosides Nucleotides Nucleic Acids* **21**, 561–580.

Dalma-Weiszhausz, D. D., Warrington, J., Tanimoto, E. Y., and Garrett Miyada, C. (2006). The Affymetrix GeneChip platform: An overview. *Methods Enzymol.* **410** (this volume), 3–28.

Dolan, P. L., Wu, Y., Ista, L. K., Metzenberg, R. L., Nelson, M. A., and Lopez, G. P. (2001). Robust and efficient synthetic method for forming DNA microarrays. *Nucleic Acids Res.* **29**, E107–E107.

Hornshoj, H., Stengaard, H., Panitz, F., and Bendixen, C. (2004). SEPON, a selection and evaluation pipeline for oligonucleotides based on ESTs with a non-target Tm algorithm for reducing cross-hybridization in microarray gene expression experiments. *Bioinformatics* **20**, 428–429.

Hughes, T. R., Mao, M., Jones, A. R., Burchard, J., Marton, M. J., Shannon, K. W., Lefkowitz, S. M., Ziman, M., Schelter, J. M., Meyer, M. R., Kobayashi, S., Davis, C., Dai, H., He, Y. D., Stephaniants, S. B., Cavet, G., Walker, W. L., West, A., Coffey, E., Shoemaker, D. D., Stoughton, R., Blanchard, A. P., Friend, S. H., and Linsley, P. S. (2001). Expression profiling using microarrays fabricated by an ink-jet oligonucleotide synthesizer. *Nat. Biotechnol.* **19**, 342–347.

Ideker, T., Thorsson, V., Siegel, A. F., and Hood, L. E. (2000). Testing for differentially-expressed genes by maximum-likelihood analysis of microarray data. *J. Comput. Biol.* **7**, 805–817.

Johnson, J. M., Castle, J., Garrett-Engele, P., Kan, Z., Loerch, P. M., Armour, C. D., Santos, R., Schadt, E. E., Stoughton, R., and Shoemaker, D. D. (2003). Genome-wide survey of human alternative pre-mRNA splicing with exon junction microarrays. *Science* **302,** 2141–2144.

Kane, M. D., Jatkoe, T. A., Stumpf, C. R., Lu, J., Thomas, J. D., and Madore, S. J. (2000). Assessment of the sensitivity and specificity of oligonucleotide (50mer) microarrays. *Nucleic Acids Res.* **28,** 4552–4557.

Kreil, D. P., Russell, R. R., and Russell, S. (2006). Microarray oligonucleotide probes. *Methods Enzymol.* **410** (this volume), 73–98.

Lausted, C., Dahl, T., Warren, C., King, K., Smith, K., Johnson, M., Saleem, R., Aitchison, J., Hood, L., and Lasky, S. R. (2004). POSAM: A fast, flexible, open-source, inkjet oligonucleotide synthesizer and microarrayer. *Genome Biol.* **5,** R58.

Le Novere, N. (2001). MELTING, computing the melting temperature of nucleic acid duplex. *Bioinformatics* **17,** 1226–1227.

LeProust, E., Zhang, H., Yu, P., Zhou, X., and Gao, X. (2001). Characterization of oligodeoxyribonucleotide synthesis on glass plates. *Nucleic Acids Res.* **29,** 2171–2180.

Li, F., and Stormo, G. D. (2001). Selection of optimal DNA oligos for gene expression arrays. *Bioinformatics* **17,** 1067–1076.

Negre, N., Sergey Lavrov, S., Jerome Hennetin, J., Michel Bellis, M., and Giacomo Cavalli, G. (2006). Mapping the distribution of chromatin proteins by ChIP on chip. *Methods Enzymol.* **410** (this volume), 316–341.

Nielsen, H. B., Wernersson, R., and Knudsen, S. (2003). Design of oligonucleotides for microarrays and perspectives for design of multi-transcriptome arrays. *Nucleic Acids Res.* **31,** 3491–3496.

Okamoto, T., Suzuki, T., and Yamamoto, N. (2000). Microarray fabrication with covalent attachment of DNA using bubble jet technology. *Nat. Biotechnol.* **18,** 438–441.

Petersen, M., and Wengel, J. (2003). LNA: A versatile tool for therapeutics and genomics. *Trends Biotechnol.* **21,** 74–81.

Relogio, A., Schwager, C., Richter, A., Ansorge, W., and Valcarcel, J. (2002). Optimization of oligonucleotide-based DNA microarrays. *Nucleic Acids Res.* **30,** e51.

Ren, B., Robert, F., Wyrick, J. J., Aparicio, O., Jennings, E. G., Simon, I., Zeitlinger, J., Schreiber, J., Hannett, N., Kanin, E., Volkert, T. L., Wilson, C. J., Bell, S. P., and Young, R. A. (2000). Genome-wide location and function of DNA binding proteins. *Science* **290,** 2306–2309.

Roda, A., Guardigli, M., Russo, C., Pasini, P., and Baraldini, M. (2000). Protein microdeposition using a conventional ink-jet printer. *Biotechniques* **28,** 492–496.

Rouillard, J. M., Zuker, M., and Gulari, E. (2003). OligoArray 2.0: Design of oligonucleotide probes for DNA microarrays using a thermodynamic approach. *Nucleic Acids Res.* **31,** 3057–3062.

Schadt, E. E., Edwards, S. W., GuhaThakurta, D., Holder, D., Ying, L., Svetnik, V., Leonardson, A., Hart, K. W., Russell, A., Li, G., Cavet, G., Castle, J., McDonagh, P., Kan, Z., Chen, R., Kasarskis, A., Margarint, M., Caceres, R. M., Johnson, J. M., Armour, C. D., Garrett-Engele, P. W., Tsinoremas, N. F., and Shoemaker, D. D. (2004). A comprehensive transcript index of the human genome generated using microarrays and computational approaches. *Genome Biol.* **5,** R73.

Shoemaker, D. D., and Linsley, P. S. (2002). Recent developments in DNA microarrays. *Curr. Opin. Microbiol.* **5,** 334–337.

Singh-Gasson, S., Green, R. D., Yue, Y., Nelson, C., Blattner, F., Sussman, M. R., and Cerrina, F. (1999). Maskless fabrication of light-directed oligonucleotide microarrays using a digital micromirror array. *Nat. Biotechnol.* **17,** 974–978.

Stimpson, D. I., Cooley, P. W., Knepper, S. M., and Wallace, D. B. (1998). Parallel production of oligonucleotide arrays using membranes and reagent jet printing. *Biotechniques* **25**, 886–890.

Tesfu, E., Maurer, K., Ragsdale, S. R., and Moeller, K. D. (2004). Building addressable libraries: The use of electrochemistry for generating reactive Pd(II) reagents at preselected sites on a chip. *J. Am. Chem. Soc.* **126**, 6212–6213.

Tolstrup, N., Nielsen, P. S., Kolberg, J. G., Frankel, A. M., Vissing, H., and Kauppinen, S. (2003). OligoDesign: Optimal design of LNA (locked nucleic acid) oligonucleotide capture probes for gene expression profiling. *Nucleic Acids Res.* **31**, 3758–3762.

van Steensel, B., and Henikoff, S. (2000). Identification of *in vivo* DNA targets of chromatin proteins using tethered dam methyltransferase. *Nat. Biotechnol.* **18**, 424–428.

Wolber, P. K., Patrick, J., Collins, P. J., Anne, B., Lucas, A. B., Anniek De Witte, A., Karen, W., and Shannon, K. W. (2006). The Agilent *in situ*-synthesized microarray platform. *Methods Enzymol.* **410** (this volume), 28–57.

Zuker, M. (2003). Mfold web server for nucleic acid folding and hybridization prediction. *Nucleic Acids Res.* **31**, 3406–3415.

[9]   Peptide Nucleic Acid Microarrays Made with (S,S)-*trans*-Cyclopentane-Constrained Peptide Nucleic Acids

By Mark A. Witschi, Jonathan K. Pokorski, and Daniel H. Appella

Abstract

Procedures to attach *trans*-cyclopentane-modified peptide nucleic acid oligomers to a glass slide are described. Peptide nucleic acids can offer distinct advantages to DNA detection compared to typically used DNA microarrays, especially with regard to chemical stability and quality of data. The *trans*-cyclopentane modification incorporated into the peptide nucleic acid is, in some cases, essential for successful DNA detection. It is hoped that these peptide nucleic acid-bearing glass slides will find application in several areas of DNA microarray technology.

Introduction

Genomic analysis can provide key diagnostic information about diseases and pathogens, and it is now an important component of biochemical and biomedical research (Hofman, 2005). The accuracy of data from such studies depends on the reliability of diagnostic techniques to signal the

METHODS IN ENZYMOLOGY, VOL. 410   0076-6879/06 $35.00
                  DOI: 10.1016/S0076-6879(06)10009-9

presence or absence of specific DNA sequences (Chaudhuri, 2005). Several technological advances have shown that synthetic DNA oligomers can be coupled to highly sensitive assays to signal binding of a complementary DNA sequence (Ivnitski et al., 2003; Meldrum, 2000) These devices are typically rated according to their ability to detect low levels of DNA, discriminate between very similar sequences of DNA (i.e., identify single nucleotide polymorphisms), and provide useful information in a reasonable timeframe. However, the reliability of these assays can be limited by the molecular recognition properties of DNA. Replacing DNA as the target capture strand with a synthetic oligomer designed to have better DNA recognition properties could improve the overall performance of these assays.

Replacing DNA probes with a peptide nucleic acid (PNA) analog has the potential to improve DNA detection technology (Demidov and Frank-Kamenetskii, 2004). PNAs bind to complementary DNA sequences with significantly higher affinity than the corresponding DNA sequences (Ratilainen et al., 2000), which improves sensitivity. This improvement in stability is very general with regard to nucleobase sequence. PNA is also significantly more stable on surfaces than DNA, yielding detection devices with longer shelf-lives (Kröger et al., 2002). Furthermore, PNA-based probes can bind to complementary DNA at low salt concentrations in solution because PNAs consist of a neutral polyamide backbone (Weiler et al., 1997). Therefore, one can create conditions that favor formation of DNA–PNA duplexes over DNA–DNA duplexes and tertiary structures by destabilizing DNA structures under low salt concentrations. This situation should give PNA-based detection probes a distinct advantage over DNA probes because the chance for false negatives should be reduced.

Production of PNA arrays via automated, robotic systems has been reviewed by Jacob et al. (2004). In one approach, an elegant system has been developed in which the PNAs for an array can be made using parallel synthesis on resin-filled microwell plates. After synthesis, the PNAs are cleaved, and then only the amine-reactive ends of the full-length PNAs are attached to a glass slide. This technique has been applied to the construction of a PNA microarray to probe the presence of a known point mutation in the *BRCA1* gene associated with breast cancer. Despite the development of technology to make PNA arrays, the use of PNA probes lags far behind that of DNA probes. One reason we suspect that PNA sees less use than DNA in array development is because the most commonly employed PNAs (derived from an aminoethylglycine backbone) do not have sufficiently improved oligonucleotide-binding properties compared to DNA to warrant the effort or funds necessary to synthesize or purchase these oligomers.

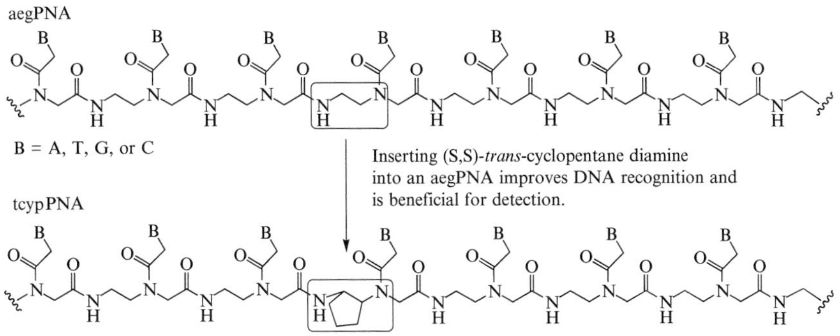

aegPNA

B = A, T, G, or C

tcypPNA

Inserting (S,S)-*trans*-cyclopentane diamine into an aegPNA improves DNA recognition and is beneficial for detection.

FIG. 1. General difference between aegPNA and tcypPNA. Note that cyclopentane stereochemistry is (S,S).

We have developed a new class of *trans*-cyclopentane-derived PNAs (*t*cypPNAs) in which *trans*-cyclopentane diamine has been incorporated into several positions, and in varying number, within PNA backbones of mixed base sequences (Fig. 1). Compared to unmodified PNA, *t*cypPNA displays improved binding affinity (i.e., higher T_m values) and sequence specificity to complementary DNA, indicating that these modified PNAs possess favorable properties for use in DNA diagnostics (Pokorski *et al.*, 2004). Furthermore, introduction of cyclopentanes into the PNA backbone affords PNAs with general and additive improvements in binding affinity to complementary DNA. It has been shown that *t*cypPNA is essential to improve a nanoparticle-based DNA detection system. In this system, the addition of one cyclopentane was crucial to achieve DNA detection (Pokorski *et al.*, 2005). This chapter reports details for the synthesis of *t*cypPNA monomer with a thymine base (*t*cypPNA-T), the synthesis of PNA oligomers, and the attachment of the PNA oligomer to a glass surface. It is hoped that this chapter will stimulate others to incorporate *t*cypPNA into detection systems for DNA.

Experimental Section

The synthesis of *t*cypPNA-T monomer (compound **6**) with the correct protecting groups for solid-phase PNA synthesis is outlined in Scheme 1. The detailed synthetic procedures to make **6** are presented later. The synthesis of **1** was accomplished following the procedure of LePlae *et al.* (2001). Following the synthesis of **6**, procedures for solid-phase synthesis of PNAs containing **6** with a linker for attachment to surfaces are described.

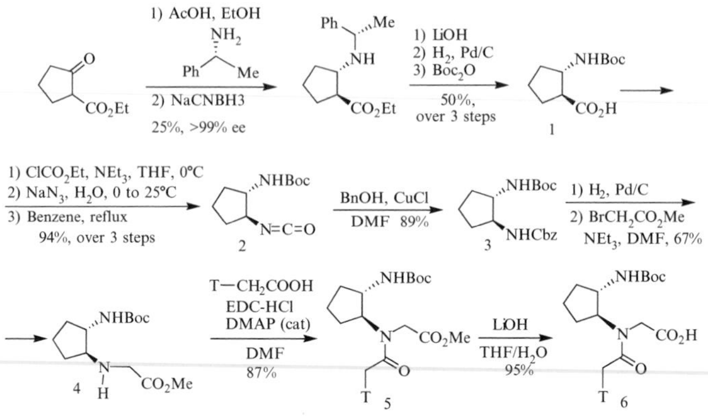

SCHEME. 1.

Next, procedures for purification of PNAs and attachment to an amine-active surface are presented.

General

Melting points (m.p.) are obtained on a Thomas Hoover capillary melting point apparatus and are uncorrected. Optical rotations ($[\alpha]_D$) are measured on a Perkin-Elmer 241 Polarimeter using sodium light (D line, 589.3 nm) and are reported in degrees; concentrations (c) are reported as g/100 ml. Infrared spectra (IR) are obtained on a Bio-Rad FTS60 FTIR. IR spectra are collected either as a thin film on a KBr disc or by preparing a pressed KBr pellet containing the title compound. Proton nuclear magnetic resonances (^{1}H NMR) are recorded in deuterated solvents on an iNOVA 500 (500 MHz) spectrometer. Chemical shifts are reported in parts per million (ppm, δ) relative to tetramethylsilane (δ 0.00). If tetramethylsilane is not present, the residual protio solvent is referenced [CDCl$_3$, δ 7.27; dimethylsulfoxide-d_6 (DMSO), δ 2.50]. ^{1}H NMR splitting patterns are designated as singlet (s), doublet (d), or quartet (q). Splitting patterns that could not be interpreted or visualized easily are designated as multiplet (m) or broad (br). Coupling constants are reported in hertz. Proton-decoupled (^{13}C NMR) spectra are obtained on an iNOVA 500 (125 MHz) spectrometer and are reported in ppm using the solvent as an internal standard (CDCl$_3$, δ 77.23; DMSO, δ 39.52). Low resolution mass spectra (LRMS) are obtained using a Micromass Quattro II Triple Quadrupole HPLC/MS/MS mass spectrometer. High resolution mass spectra (HRMS) are obtained on a Micromass Q-Tof Ultima at the University of Illinois at the Urbana–Champaign Mass Spectrometry Center. Elemental analysis data are collected by Atlantic Microlab, Inc. Analytical thin-layer chromatography (TLC) is carried out on Sorbent Technologies TLC plates precoated with silica gel (250 μm layer thickness). Flash column chromatography is performed on EM Science silica gel 60 (230–400 mesh). Solvent mixtures used for TLC and flash column chromatography are reported in v/v ratios. Unless noted otherwise, all commercially available reagents and solvents are from Sigma-Aldrich and used without further purification. Tetrahydrofuran (THF) is distilled from sodium and benzophenone prior to use. Dimethylformamide (DMF) is purified by passage through a bed of activated alumina. 1-(3-Dimethylaminopropyl)-3-ethylcarbodiimide hydrochloride (EDC), O-(7-azabenzotriazol-1-yl)-N,N,N', N'-tetramethyluronium hexafluorophosphate (HATU), and Boc-Lys-(2-Cl-Z)-OH are from Advanced ChemTech. All aeg PNA monomers are from Applied Biosystems. Kaiser test solutions are from Fluka. Boc-8-amino-3,6-dioctanoic acid (Boc-mini-PEG) is from Peptides International. CodeLink amine-active slides are from Amersham Biosciences.

(1S,2S)-2-(tert-Butoxycarbonylamino)-cyclopentyl Isocyanate (2)

β-amino acid **1** (2.3 g, 10.1 mmol) is added to an oven-dried 250-ml round-bottomed flask (RBF) under N_2 (g) atmosphere and dissolved in dry THF (50 ml) with stirring. The solution is cooled to 0°, and triethylamine (2.8 ml, 20.2 mmol, 2.0 equivalent) and ethyl chloroformate (1.9 ml, 20.2 mmol, 2.0 equivalent) are added dropwise via a syringe, forming a white precipitate. The heterogeneous mixture is stirred at 0° for 20 min. Sodium azide (2.0 g, 30.3 mmol, 3.0 equivalent) is dissolved in H_2O (35 ml) and added to the reaction mixture. The biphasic mixture is stirred vigorously at 0° for 20 min. The ice bath is removed and the mixture is allowed to warm to room temperature for 10 min. The mixture is diluted with H_2O (30 ml) and is extracted with ethyl acetate (3 × 30 ml). The combined organic phases are dried over Na_2SO_4 and concentrated (not to dryness!). The residue is taken up in benzene (50 ml) and refluxed for 30 min. The solution is cooled, concentrated, and dried under vacuum to yield 2.15 g (94%) of **2** as a colorless, crystalline solid. m.p. = 73–75°; IR (KBr) 3700, 3379, 2981, 2279 (N=C=O stretch), 1687, 1517, 1458, 1366, 1345, 1314, 1290, 1250, 1164, 1047, 909, 780, 580 cm^{-1}; ^{1}H NMR (500 MHz, CDCl$_3$) δ 4.48 (br s, 1H, Boc-NH) 3.85 (br s, 1H, O=C=N-CH) 3.71 (br s, 1H, Boc-NH-CH) 2.19–2.11 (m, 1H, cyclopentane-H) 2.04–1.97 (m, 1H, cyclopentane-H) 1.81–1.65 (m, 3H, cyclopentane-H) 1.46 (s, 10H, t-butyl-CH_3 + cyclopentane-H).

tert-Butyl (1S,2S)-2-(benzyloxycarbonylamino)-cyclopentyl Carbamate (3)

Copper (I) chloride (0.9 g, 9.5 mmol) is added to an oven-dried 100-ml RBF under N_2 (g) atmosphere and suspended in dry DMF (35 ml) with stirring. Benzyl alcohol (0.9 ml, 9.5 mmol) is added dropwise via a syringe and allowed to stir until a bright yellow-green suspension has formed (approximately 5 min). At this point, **2** (2.1 g, 9.5 mmol) is added and stirred at room temperature for 45 min. The dark green suspension is poured into H_2O (70 ml) and extracted with ethyl acetate (150 ml). The organic phase is washed with saturated NaCl (aq) (2 × 50 ml), dried over Na_2SO_4, concentrated, and dried under vacuum to yield 2.83 g (89%) of **3** as a colorless solid. R_f=0.38 (2% MeOH/CH_2Cl_2); m.p. = 150–151°; $[\alpha]_D^{23}$ = −8.0° (c = 1.0, EtOH 100%); IR (film) 3341, 2971, 1680, 1528, 1304, 1270, 1233, 1169, 1041, 779, 741, 696, 639 cm^{-1}; ^{1}H NMR (500 MHz, CDCl$_3$) δ? 7.34 (m, 5H, Ph-*H*) 5.30 (br s, 1H, Cbz-N*H*), 5.09 (s, 2H, Ph-C*H*$_2$), 4.81 (br s, 1H, Boc-N*H*), 3.68 (m, 2H, carbamate-NH-C*H*), 2.17–2.09 (m, 2H, cyclopentane-*H*) 1.70 (m, 2H, cyclopentane-*H*) 1.43 (s, 11H, *t*-butyl-C*H*$_3$ + 2 cyclopentane-*H*); ^{13}C NMR (125 MHz, CDCl$_3$) δ?156.9, 156.5, 136.7, 128.5, 128.0, 79.5, 66.6, 58.4, 57.2, 30.1, 29.8, 28.4, 19.7; LRMS (ESI-MS *m/z*): Mass calc'd for $C_{18}H_{27}N_2O_4$ [M + H]$^+$, 335.20. Found 335.3. Anal. Calc'd for $C_{18}H_{26}N_2O_4$: C, 64.65; H, 7.84; N, 8.38. Found C, 64.45; H, 7.86; N, 8.29.

tert-Butyl (1S,2S)-2-[(methoxycarbonyl)methylamino)]-cyclopentyl Carbamate (4)

Compound **3** (1.5 g, 7.6 mmol) is dissolved in methanol (150 ml) and added to an oven-dried Parr flask containing 10% Pd/C (0.5 g). The flask is placed on a Parr apparatus under 55 psi of H_2 (g) pressure and is shaken for 12 h. TLC analysis reveals no starting material [R_f=0.38 (2% MeOH/CH_2Cl_2)]. The suspension is filtered through Celite and concentrated. The residue is dissolved in dry DMF with stirring. Triethylamine (1.1 ml, 7.6 mmol) and methyl bromoacetate (0.6 ml, 6.8 mmol, 0.9 equivalent) are added dropwise via a syringe. The reaction is allowed to stir at room temperature for 3 h. The reaction mixture is diluted with saturated NaHCO$_3$ (aq) (50 ml) and extracted with ethyl acetate (2 × 35 ml). The combined organic phases are washed with saturated NaCl (aq) (2 × 25 ml), dried over Na_2SO_4, concentrated, and dried under vacuum. The residue is purified by flash column chromatography [R_f=0.30 (EtOAc)] to yield 1.26 g (67%) of **4** as a colorless oil, which forms a colorless, crystalline solid upon sitting at room temperature. R_f=0.30 (EtOAc); m.p. = 49–52°; $[\alpha]_D^{23}$ = +5.0° (c = 1.0, EtOH 100%); IR (film) 3343, 2960, 2871, 1741, 1703, 1522, 1438, 1392, 1367, 1244, 1207, 1173, 1044, 1021, 996, 778 cm^{-1}; ^{1}H NMR (500 MHz, CDCl$_3$)

δ 4.56 (br s, 1H, Boc-N*H*) 3.73 (s, 3H, CO$_2$C*H*$_3$) 3.67 (br m, 1H, Boc-NH-C*H*) 3.48 (q, J = 13.5 Hz, 2H, NH-C*H*$_2$-CO$_2$Me) 2.85 (m, 1H, C*H*-NH) 2.13 (m, 1H, cyclopentane-*H*) 1.93 (m, 2H, cyclopentane-*H*) 1.77–1.68 (m, 1H, cyclopentane-*H*) 1.67–1.59 (m, 1H, cyclopentane-*H*) 1.45 (s, 9H, *t*-butyl-C*H*$_3$) 1.42–1.34 (m, 1H, cyclopentane-*H*); ^{13}C NMR (125 MHz, CDCl$_3$) δ 172.8, 155.5, 78.6, 64.7, 57.2, 51.4, 48.8, 31.1, 30.8, 28.1, 21.2; LRMS (ESI-MS *m/z*): Mass calc'd for C$_{13}$H$_{25}$N$_2$O$_4$ [M+H]$^+$, 273.18. Found 273.2. Anal. Calc'd for C$_{13}$H$_{24}$N$_2$O$_4$: C, 57.33; H, 8.88; N, 10.29. Found C, 57.20; H, 8.96; N, 10.22.

Methyl N-[(2S)-tert-Butoxycarbonylaminocyclopent-(1S)-yl]-N-[thymin-1-ylacetyl]-glycinate (5)

Compound **4** (0.95 g, 3.5 mmol) is added to an oven-dried 50-ml RBF and dissolved in dry DMF (15 ml) with stirring. The solution is cooled to 0°. Thymine acetic acid (0.96 g, 5.25 mmol) and dimethylamino-pyridine (100 mg, 0.9 mmol) are added. EDC (1.3 g, 7.0 mmol, 2.0 equivalent) is added, and the reaction mixture is allowed to stir at 0° for 10 min. The ice bath is removed, and the reaction is stirred for 36 h at room temperature. The solution is diluted with H$_2$O (90 ml) and extracted with ethyl acetate (3 × 60 ml). The combined organic phases are washed with saturated NaCl (aq) (4 × 75 ml), dried over Na$_2$SO$_4$, concentrated, and dried under vacuum to yield 1.33 g (87%) of ester **5** as a colorless solid. If necessary, the solid is purified by flash column chromatography: R_f = 0.38 (5% MeOH/ CH$_2$Cl$_2$); ^{1}H NMR (500 MHz, DMSO) δ major rotomer 11.29 (s, 1H, imide-N*H*) 7.17 (s, 1H, thymine-*H*) 6.96 (d, J = 8.0 Hz, 1H, Boc-N*H*) 4.72 (q, J = 16.5 Hz, 2H, thymine-C*H*$_2$) 4.0–3.75 (m, 4H, NH-C*H*$_2$-CO$_2$Me + C*H*-NH) 3.59 (s, 3H, CO$_2$C*H*$_3$) 1.9–1.4 (m, 6H, cyclopentane-*H*) 1.75 (s, 3H, thymine-C*H*$_3$) 1.36 (s, 9H, *t*-butyl-C*H*$_3$) minor rotomer 7.20 (s, 1H, thymine-*H*) 6.75 (d, J = 8.0 Hz, 1H, Boc-N*H*) 4.45 (q, J = 16.5 Hz, 2H, thymine-C*H*$_2$) 3.70 (s, 3H, CO$_2$C*H*$_3$); ^{13}C NMR (125 MHz, DMSO) δ 169.6, 167.3, 164.4, 155.4, 151.0, 141.7, 108.3, 77.9, 61.1, 52.7, 51.6, 47.9, 43.6, 28.2, 25.9, 18.8, 12.0; LRMS (ESI-MS *m/z*): Mass calc'd for C$_{20}$H$_{31}$N$_4$O$_7$ [M+H]$^+$, 439.22. Found 439.2.

N-[(2S)-tert-Butoxycarbonylaminocyclopent-(1S)-yl]-N-[thymin-1-ylacetyl]-glycine (6)

Ester **5** (0.93 g, 2.12 mmol) is dissolved in THF (33 ml) and cooled to 0°. Lithium hydroxide monohydrate (1.2 g, 28.4 mmol) is dissolved in H$_2$O (28 ml), and the solution is added to the reaction mixture for 5 min. The solution is allowed to warm to room temperature and is stirred for 5 h. The mixture is diluted with H$_2$O (45 ml) and extracted with ethyl ether

(3 × 50 ml). The aqueous layer is acidified with 3 N HCl (aq) to pH 1. The solution is extracted with ethyl acetate (5 × 70 ml). The combined organic phases are dried over Na_2SO_4, concentrated, and dried under vacuum to yield 0.87 g (96%) of PNA monomer **6** as a colorless solid. m.p. = Decomposition at 203°; $[\alpha]_D^{23} = -36.5°$ (c = 1.0, MeOH 100%); IR (KBr) 3351, 3178, 2976, 2611, 2531, 1682, 1530, 1455, 1365, 1244, 1167, 1045, 853, 782, 467 cm^{-1}; ^{1}H NMR (500 MHz, DMSO) δ major rotomer 12.41 (br s, 1H, CO$_2$H) 11.28 (s, 1H, imide-NH) 7.15 (s, 1H, thymine-H) 6.94 (d, J = 8.0 Hz, 1H, Boc-NH) 4.71 (q, J = 17.0 Hz, 2H, thymine-CH_2) 3.9–3.7 (m, 4H, NH-CH_2-CO$_2$Me + CH-NH) 1.9–1.4 (m, 6H, cyclopentane-H) 1.75 (s, 3H, thymine-CH_3) 1.36 (s, 9H, t-butyl-CH_3) minor rotomer 6.75 (d, J = 8.0 Hz, 1H, Boc-NH) 4.44 (q, J = 17.0 Hz, 2H, thymine-CH_2); ^{13}C NMR (125 MHz, DMSO) δ 177.5, 167.1, 164.4, 155.4, 151.0, 141.8, 108.2, 77.9, 61.2, 52.7, 47.9, 43.8, 28.2, 26.0, 18.9, 12.0; HRMS (ESI-MS m/z): Mass calc'd for $C_{29}H_{29}N_4O_7$ [M+H]$^+$, 425.2036. Found 425.2043.

General Procedure for Manual Solid-Phase Synthesis of Peptide Nucleic Acids

Note on resin washes: All washes used at least enough solution to cover the resin (~1.5 ml for 50 mg resin, and ~5 ml for 1 g resin). Values in parentheses describe the number of times the resin is washed with the indicated solution and the agitation time for each wash. For example, (4× 30 s) indicates four washes, with each having an agitation time of 30 s and each followed by draining under vacuum. If no time or number of repetitions is indicated, a single wash and/or a 5 s agitation period is used.

Downloading Resin

Methyl benzhydryl amine (MBHA) resin (1.0 g, 0.3 mmol active sites/g) is downloaded to 0.1 mmol/g with Boc-Lys-(2-Cl-Z)-OH. The resin is first swelled in dichloromethane (DCM) for between 1 and 12 h. The following solutions are prepared: 0.2 M Boc-Lys-(2-Cl-Z)-OH in 1-methyl-2-pyrrolidinone (NMP) (A), 0.2 M HATU in NMP (B), and 0.5 M N,N-diisopropylethyl amine (DIEA) in NMP (C). These solutions are then combined appropriately to give two additional solutions: 0.45 ml of A + 0.46 ml of C + 1.59 ml NMP (solution 1) and 0.55 ml of B + 1.95 ml NMP (solution 2). Solutions 1 and 2 are premixed for 1 min and then added to the resin. The resin is agitated with a mechanical shaker for 1 h and then drained under vacuum. The resin is subsequently washed with DMF (4×), DCM (4×), 5% DIEA in DCM (1× 30 s), and again with DCM (4×). The remaining active sites are then capped with a 1:2:2 solution of acetic anhydride (Ac$_2$O): NMP:pyridine for 1.5 h. This is followed by washes with DCM (2× 5 s) and a qualitative Kaiser test to confirm that no primary amines remain. Resin is then washed with DCM (2×) and allowed to dry under vacuum for 30–60 min. Downloaded resin is stored in a desiccator until further use.

aegPNA Synthesis

The following is a representative coupling cycle for one PNA monomer. Downloaded resin (50 mg) is swelled in DCM for 1 h. The solvent is drained under vacuum, and a solution of 5% m-cresol in trifluoroacetic acid (TFA) is added to the resin. The resin is shaken for 4 min, and the solution is removed under vacuum. (Note: TFA deprotection is repeated three times for deprotection of the lysine residue and then twice for every subsequent monomer.) This is followed by subsequent washes with DCM, DMF (1× 5 s, 1× 30 s, 1× 5 s), DCM (2× 5 s), and pyridine (2× 5 s). A Kaiser test is performed to confirm deprotection. Upon a positive Kaiser test, 150 μl of a 0.4 M solution of PNA monomer in NMP is premixed with 150 μl of 0.8 M N-methyldicyclohexylamine (MDCHA) in pyridine and 300 μl of 0.2 M O-(benzotriazol-1-yl)-N,N,N',N'-tetrameth-yluronium hexafluorophosphate (HBTU) in DMF for 1 min. This solution is then added to the resin and agitated for 30 min. Following coupling, the resin is drained under vacuum and washed with DMF, 5% DIEA/DCM (1×-30 s) and DCM (2×-5 s). Again, a qualitative Kaiser test is performed and, if negative, the resin is capped with a 1:25:25 mixture of Ac$_2$O:NMP: pyridine (2×-2 min). (Note: If the Kaiser test is positive, the coupling cycle is repeated, beginning with the pyridine washes.) The capping step is followed by washes with DCM, 20% piperidine/DMF, and finally DCM (1×-5 s, 1×-30 s, 1×-5 s). This cycle is then repeated iteratively until the oligomer is complete on the resin.

Cyclopentane PNA Synthesis

The following is a representative coupling cycle for one cyclopentane PNA monomer. The resin is washed with a solution of 5% m-cresol in TFA (3×-10 min) and removed under vacuum. This is followed by subsequent washes with DCM, DMF (1× 5 s, 1× 30 s, 1× 5 s), DCM (2× 5 s), and pyridine (2×-5 s). A Kaiser test is performed to confirm deprotection. It should be noted that the primary amines of deprotected cyclopentane monomers do not show the characteristic blue color of a positive Kaiser test, but vary in color between gray/purple and red. Upon a positive Kaiser test, 150 μl of a 0.4 M solution of the cyclopentane PNA monomer in NMP is premixed with 150 μl of 0.8 M MDCHA in pyridine and 300 μl of 0.2 M HBTU in DMF for 1 min. This solution is then added to the resin and agitated for 60 min. Following coupling, the resin is drained and washed with DMF, 5% DIEA/DCM (1× 30 s) and DCM (2× 5 s). Again, a qualitative Kaiser is performed and, if negative, the resin is capped with a 1:25:25 mixture of Ac$_2$O:NMP:pyridine (2×-2 min). (Note: If the Kaiser test is positive, the coupling cycle is repeated, beginning with the pyridine washes.) The capping step is followed by washes with DCM, 20% piperidine/DMF, and finally DCM (1× 5 s, 1× 30 s, 1× 5 s).

Linker Attachment

The following is a representative coupling cycle for attaching one Boc-mini-PEG linker. Additional mini-PEG linkers could be attached by repeating this procedure. A solution of 5% m-cresol in TFA is added to resin that has a Boc-protected PNA oligomer. The resin is shaken for 4 min, and the solution is removed under vacuum. The same deprotection conditions are repeated a second time, followed by subsequent washes with DCM, DMF (1× 5 s, 1× 30 s, 1× 5 s), DCM (2× 5 s), and pyridine (2×-5 s). A Kaiser test is performed to confirm deprotection. Upon a positive Kaiser test, 150 μl of a 0.4 M solution of Boc-mini-PEG in NMP is premixed with 150 μl of 0.8 M MDCHA in pyridine and 300 μl of 0.2 M HBTU in DMF for 1 min. This solution is then added to the resin and agitated for 30 min. Following coupling, the resin is drained under vacuum and washed with DMF, 5% DIEA/DCM (1× 30 s), and DCM (2× 5 s). Again, a qualitative Kaiser test is performed and, if negative, the resin is capped with a 1:25:25 mixture of Ac$_2$O:NMP:pyridine (2× 2 min). (Note: If the Kaiser test is positive, the coupling cycle is repeated, beginning with the pyridine washes.) The capping step is followed by washes with DCM, 20% piperidine/DMF, and finally DCM (1× 5 s, 1× 30 s, 1× 5 s).

Cleavage of PNA from Resin

Cleavage from the resin is accomplished under acidic conditions. First, the resin is washed with TFA (2× 4 min) and drained under vacuum. Next, 750 μl of a solution cooled to 0° consisting of 75 μl thioanisole, 75 μl m-cresol, 150 μl trifluoromethanesulfonic acid, and 450 μl TFA is added to the resin and agitated for 1 h. The resulting solution is collected using positive N$_2$ pressure to force liquid through the fritted vessel. Another 750-μl portion of cleavage solution is added. After an additional hour of agitation, cleavage solutions are combined in a glass vial and volatiles are removed by passing a stream of dry nitrogen over the product to afford a yellow/brown oil.

Crude PNA Isolation

The resulting oil is partitioned between five 2.0-ml microcentrifuge tubes using a micropipettor to typically yield 100–150 μl of the oil per tube. A 10-fold excess (by volume) of diethyl ether is added to each tube. The solutions are mixed by vortexing until the brown color no longer remains and a cloudy white precipitate forms. The solutions are then placed on dry ice for 10 min. This is followed by centrifugation (5 min at 7000 rpm) and removal of solvent via decanting or pipetting to yield a white solid as the crude PNA product. The ether precipitation cycle is repeated four times with the following dry ice incubation times: $2\times$ 5 min and $2\times$ 2 min. Repeating the cycle twice more with no dry ice incubation completes crude PNA isolation. After decanting the final ether wash, residual solvent is removed by passing a stream of N_2 over the crude PNA product.

PNA Purification and Characterization

All peptide nucleic acid oligomers are purified on reversed-phase HPLC with UV detection at 260 nm. VYDEK C18 (d = 10 mm, l = 250 mm, 5 μm) semiprep columns are utilized, eluting with 0.05% TFA in water (solution A) and 0.05% TFA in acetonitrile (solution B). An elution gradient of 100% A to 100% B over 60 min at a flow rate of 2.2 ml/min is used. PNAs are characterized by mass spectroscopy using a PerSeptive Biosystems Voyager DE MALDI-TOF system with a 2′,4′,6′-trihydroxyacetophenone monohydrate matrix. Mass spectra are acquired using a N_2 laser (337 nm wavelength, 5 ns pulse) with at least 100 shots per sample. All PNA oligomers give molecular ions consistent with the final product.

Procedure Used to Attach PNA to Surface

The $t$cypPNA with the mini-PEG linker is spotted onto a CodeLink amine-active slide (Amersham Biosciences) using a DNA microarrayer (GMS 417 Arrayer, Genetic Microsystems, Woburn, MA) using a spot diameter of 300 μm and a distance between spots of 700 μm. Following overnight immobilization, the slide is washed with 0.2% SDS at 50° for 10 min, washed with NANOpure water (18 MΩ), dried with a stream of N^2, and used immediately. No additional capping steps are performed prior to hybridization experiments.

References

Chaudhuri, J. D. (2005). Genes arrayed out for you: The amazing world of microarrays. *Med. Sci. Monit.* **11,** RA52–RA62.

Demidov, V. V., and Frank-Kamenetskii, M. D. (2004). Two sides of the coin: Affinity and specificity of nucleic acid interactions. *Trends Biochem. Sci.* **29**, 62–71.

Hofman, P. (2005). DNA microarrays. *Nephron. Physiol.* **99**, 85–89.

Ivnitski, D., O'Neil, D. J., Gattuso, A., Schlicht, R., Calidonna, M., and Fisher, R. (2003). Rapid detection and identification of infectious disease agents. *Biotechniques* **35**, 862–869.

Jacob, A., Brandt, O., Würtz, S., Stephan, A., Schnölzer, M., and Hoheisel, J. D. (2004). Production of PNA-arrays for nucleic acid detection. *In* "Peptide Nucleic Acids: Protocols and Applications" (P. E. Nielsen, ed.), 2nd Ed. pp. 261–279. Horizon Bioscience, Wymondham, United Kingdom.

Kröger, K., Jung, A., Reder, S., and Gauglitz, G. (2002). Versatile biosensor surface based on peptide nucleic acid with label free total internal reflection fluorescence detection for quantification of endocrine disruptors. *Anal. Chim. Acta* **469**, 37–48.

LePlae, P. R., Umezawa, N., Lee, H., and Gellman, H-S. (2001). An efficient route to either enantiomer of trans-2-aminocyclopentanecarboxylic acid. *J. Org. Chem.* **66**, 5629–5632.

Meldrum, D. (2000). Automation for genomics: Sequencers, microarrays, and future trends. *Genome Res.* **10**, 1288–1303.

Pokorski, J. K., Nam, J.-M., Vega, R. A., Mirkin, C. A., and Appella, D. H. (2005). Cyclopentane-modified PNA improves the sensitivity of nanoparticle-based scanometric DNA detection. *Chem. Commun.* 2101–2103.

Pokorski, J. K., Witschi, M. A., Purnell, B. L., and Appella, D. H. (2004). (S,S)-trans-Cyclopentane-constrained peptide nucleic acids: A general backbone modification that improves binding affinity and sequence specificity. *J. Am. Chem. Soc.* **126**, 15067–15073.

Ratilainen, T., Holmén, A., Tuite, E., Nielsen, P. E., and Nordén, B. (2000). Thermodynamics of sequence-specific binding of PNA to DNA. *Biochemistry* **39**, 7781–7791.

Weiler, J., Gausepohl, H., Hauser, N., Jensen, O. N., and Hoheisel, J. D. (1997). Hybridization based DNA screening on peptide nucleic acid (PNA) oligomer arrays. *Nucleic Acids Res.* **25**, 2792–2799.

Section II

Wet-Bench Protocols

[10]   Optimizing Experiment and Analysis Parameters for Spotted Microarrays

By SCOTT J. NEAL and J. TIMOTHY WESTWOOD

Abstract

This chapter outlines considerations and methods of experimental design for spotted microarray studies that contribute to the robustness of the acquired data. The chapter is divided into two principal sections: (1) a summary of factors that affect the quality of individual array results; and (2) a discussion of experimental design features and replication criteria that enable researchers to gain confidence in their findings. The goal of section (2) is to promote the careful design of microarray studies, with downstream statistical analysis in mind, such that a minimal amount of biological and/or technical replication is required to achieve reproducible results. Topics addressed include choosing appropriate samples to be compared, determining the optimal amount of sample to use, flagging uninformative data, establishing confidence limits in normalized data and the application of statistical tests to identify differentially expressed genes.

Introduction

Now into its second decade, experimentation with high-density spotted microarrays has become a widely used approach to genome level studies of gene expression. Researchers embarking on microarray studies for the first time often face a number of decisions e.g., What array platform should I use? What kind, how much, and how should I isolate the RNA? What samples should I be comparing on the array? How many replicates do I need to do? What labeling and hybridization protocols should I use and do these choices affect the reproducibility of results? Without careful *a priori* consideration of the experimental parameters such as those mentioned above, many results cannot be properly analyzed nor stand up to the rigors of statistical evaluation. With regards to analysis, other decisions have to be made- e.g. What data should be filtered out or flagged? Should I subtract background signals? Do I need to normalize the data and if so, what method should I use? What kind of statistical analyses can or should be done?

The aim of this chapter is to provide practical advice on some of the decisions that must be made in microarray experiment design, experimental

METHODS IN ENZYMOLOGY, VOL. 410
0076-6879/06 $35.00
DOI: 10.1016/S0076-6879(06)10010-5

methodology, and data analysis. The chapter is divided into two major sections: (1) experimental factors that affect the quality and reproducibility of individual array results; and (2) experimental design features and replication criteria that enable researchers to have confidence in their findings.

Factors Affecting the Quality of Experimental Data

Although not the main focus of this chapter, it is important to consider the many factors that can affect the quality of microarray data before any meaningful discussion of experimental design can begin. From choosing the appropriate samples, to generating high quality RNA, to conducting stringent hybridizations; there are many aspects of microarray studies for which there is no golden standard to follow. Many of these aspects have been addressed in detail elsewhere (refer to Nature Genetics "Chipping Forecast" series – Volume 21(1S), 1999; Volume 32(4S), 2002; Volume 37 (1S), 2005) but certain ones deserve highlighting.

Choice of Array Substrates and Printing Conditions

The physical quality of spotted microarrays can be a major factor in determining the reproducibility of any given result. The glass substrate must be optically flat and must not bind non-specifically to either unbound dye molecules or to fluorescently-labeled oligonucleotides as this will lead to elevated background signals. Printing conditions and spotting buffers must also be optimized to generate spots with consistent morphologies. In their thorough primer on DNA microarray experimentation, Hegde and colleagues (2000) considered the influence of DNA purification methods, spotting buffers, temperature and relative humidity on overall spot morphology and array quality. Using commercially available microarray substrates with stringent quality control metrics such as Corning UltraGAPS slides and following the manufacturer's prescribed chemical blocking procedures to reduce background fluorescence, many of these factors are easily controlled and high quality arrays may be consistently produced either in house or by core service facilities.

RNA Type and Isolation

When high-quality microarrays are available the most important factor contributing to a successful experiment is to begin with high quality RNA. Depending on the protocol, either total RNA or mRNA may be used. Total RNA may be isolated using a variety of methods but those most commonly employed rely on homogenization of the cells, tissue or organism in the presence of a highly denaturing guanidine isothiocyanate

(GITC)-containing buffer which immediately inactivates RNases to ensure isolation of intact RNA. Such buffers can be combined with phenol:chloroform for a single solution approach to the isolation and purification of total RNA as originally described by Chomczynski and Sacchi (1987) and several commercially available reagents such as Invitrogen's TRIzol are examples of this approach. Other commercially available RNA isolation kits combine GITC-containing buffers with purification columns or membranes such as Qiagen's RNeasy products. Our experience has led us to rely on TRIzol to extract RNA from most samples. Whereas different purification columns are required depending on the mass or type of sample tissue, the TRIzol method is robust, scalable and samples may also be homogenized directly in the reagent, which eliminates the need to use additional RNase-protective buffers that may interfere with the extraction. In some instances, the use of total RNA may lead to poor or no reverse transcription of the template mRNA and/or poor incorporation of modified nucleotides and/or fluorescent background on the hybridized array due to contaminants present in the RNA preparation. In such cases, it may be preferable to use mRNA and a variety of column- or bead- based oligo dT purification kits are commercially available. As mRNA purification generally introduces further costs, additional time requirements and extra steps during which samples may become contaminated or degraded we do not favor this approach unless the particular source of the RNA makes it untenable to use total RNA. It should also be considered that the addition of equal amount of mRNA to reciprocal labeling reactions might misrepresent their relative abundance *in vivo*, and thus introduce bias into the results (Pradet-Balade *et al.*, 2001).

Labeling and Hybridization Methods

For transcript analysis, either total RNA or mRNA may be used in the two common methods employed to generate fluorescent targets to be hybridized to a microarray, known as the "direct" and "indirect" labeling methods. In the direct approach, dye-coupled nucleotides are added to the cDNA synthesis reaction and are directly incorporated into the nascent molecules during reverse transcription. To achieve a two-color comparative hybridization, different dyes must be used to label the respective samples. This necessity has given rise to the concept of "dye bias" whereby one dye-coupled nucleotide may be more efficiently incorporated than the other by reverse transcriptase, giving rise to a biased pool of labeled cDNA. In order to avoid this potential bias the indirect labeling approach may be used. In this method the same modified nucleotide, for example, amino-allyl dUTP, is added to the labeling reactions for the respective samples to be compared and a different dye ester is conjugated to each sample in a secondary reaction. Thus, there should be no bias in the incorporation step since the conditions are identical, but again, the extra steps required for the indirect labeling method may lead to sample degradation or a bias of a different kind. Another approach to dealing with "dye bias" has been to utilize dye-swap experiments where the two samples of interest are reciprocally labeled and hybridized to separate arrays. The necessity for such experiments will be further discussed in section (2). While not universally true, many researchers have found minimal or no dye bias in terms of finding differentially regulated genes in a given experiment. Moreover, the direct labeling method is often preferred because for many experimenters it is less prone to failure. In the indirect labeling protocol, a buffer exchange is often required prior to the conjugation of the dye ester to the modified nucleotide. Inefficient conjugation, and thus a poorly labeled sample may result from incomplete buffer exchange and/or prolonged handling of the un-conjugated dye esters. The pre-conjugated dye-nucleotides used in direct labeling are generally more stable.

Finally, the labeled cDNA targets must be successfully hybridized to the microarray. Some protocols recommend a prehybridization step, but if the microarrays are chemically "blocked" after printing it is generally not required (Martinez et al., 2003). Including competitors such as salmon sperm DNA or yeast tRNA in the hybridization solution also serves to reduce non-specific probe-target interactions. One variable that must be carefully considered is the hybridization temperature (Evertsz et al., 2001). Although raising the hybridization temperature from 60° to 65° reduces the overall signal intensity, it does increase the stringency of the hybridization such that sequences with up to 80% homology can generally be discriminated (Evertsz et al., 2001). Other products, such as Roche's DIG EasyHyb solution, may be used to simulate these stringent conditions at lower hybridization temperatures. Using lower temperatures decreases the likelihood that the small volume of hybridization solution will evaporate resulting in artifacts on the arrays. Many microarray facilities have robotic hybridization stations that can provide precise liquid removal and replacement in a controlled temperature environment and such devices can help ensure the reproducibility of hybridization conditions. However, high quality results can be obtained without such devices. Special hybridization chambers, like those made by Corning, preserve the humid microenvironment for the hybridization and prevent evaporation at the edges of the array. These chambers are also submersible, allowing the hybridization to take place in a circulating water bath where the temperature is held constant as opposed to in an air incubator where it may fluctuate. Consideration should also be paid to the washing and drying of the arrays after hybridization. Excess volumes of particle-free washing solutions should be used, and contact of the arrays to one another should be avoided. Drying of the arrays should be done quickly using either clean compressed nitrogen or air from a large tank or by centrifugation to avoid drying of the last wash solution on the edges of the array. Small cans of compressed air should be avoided since the propellant (e.g. liquefied 1,1-difluoroethane) droplets can cause background fluorescence if they contact the array surface.

Factors to Consider in Experimental Design and Analysis

There are many possible schemes or designs that can be employed in microarray analyses. The simplest experimental design is one in which two states are directly compared on a single microarray, and this will be the focus of the concepts presented herein. More complex designs include factorial comparisons for time course studies and larger scale studies where a reference sample is used as a comparator for divergent biological samples that otherwise do not have a logical control sample in common. Such designs may or may not feature dye swap comparisons as well. Pros and cons of the various designs are discussed in detail elsewhere (Churchill, 2002; Kerr and Churchill, 2001). In this section we will present some information concerning the choice of samples to compare but most of the attention will be paid to analytical methods that may be used to test the robustness of a given design and generate confidence measures for the results.

Choice of Samples to Compare

As stated earlier, the direct comparison of two samples on a single microarray represents the most basic experimental design. Generally, one sample should represent the control or untreated state and the other should represent the experimentally modulated state. One example of such a design is the contrasting of a cell line that has been treated with a drug with another that has received placebo only. In this case the logical control is the placebo sample, and all changes in gene expression will be expressed relative to this sample. However, it is possible that no logical control exists, for instance in the comparison of two different cell lines (Neal *et al.*, 2003). In this case either sample could represent the control state, and for the purpose of analysis one must be chosen as such, even if the choice is made randomly.

It is through direct comparisons that the most meaningful results are obtained because the concept of biological relevance can be intuitively applied. In the second example above, where two different cell lines were compared, it would have been possible to compare the expression of each cell line to a common reference, such as a sample derived from a pool of RNA from each cell line. However, the reference sample simply introduces an intermediate step in the analysis and the sample itself holds no biological relevance (Kerr and Churchill, 2001). The choice to pool samples or not is an oft-faced issue. Generally, pooling should be avoided if possible as it reduces the dimensionality of the analysis. The comparison of pooled samples over multiple arrays will reduce the inter-array variability but again, the biological relevance is diluted because the expression pattern is no longer attributable to a single sample. Some experimenters take a hybrid approach, that is, by pooling the control/reference RNA samples while individually labeling the biological replicates of the experimental samples. Two advantages of this approach include that the same reference RNA sample are used in the entire study so that differences between any two arrays are more easily compared, and secondly, variation in gene expression between biological replicates will primarily reflect variation in the experimental replicates and not the control or reference samples.

Quantification and Analysis Parameters

Once a successful co-hybridization of relevant samples has been conducted the data must be collected. Generally, a confocal or parfocal laser scanner is used to excite the dyes in the hybridized target cDNAs and capture their fluorescence in the form of a Tagged Information File Format (TIFF) image. The expression values are then quantified from the image using the pixel intensities in the areas defined as spots during image analysis. This process, which is automated in many commercial software packages, represents a critical step in the overall analysis because it is generally only done once. Two such software packages that come with widely used scanners include GenePix Pro (that accompanies Axon scanners) and QuantArray or ScanArray Express (that accompanies Perkin Elmer scanners). Another quantification program that is available for free is the open source TIGR Spotfinder software (www.tm4.org/spotfinder.html). Important considerations for image analysis are presented elsewhere (Yang *et al.*, 2001), though it is worth mentioning here that one of the most important things that should be done prior to further analysis is array-wide normalization of overall signal intensities between the two channels being compared. This is typically done by adjusting either the laser power for a particular channel and/or the PMT/gain for that channel such that the total intensity for a given channel is approximately the same (i.e. within 10%) as the other channel. If this is not done, the downstream normalization algorithms will need to utilize larger correction factors and thus be prone to producing artifacts in the data, particularly for spots with low signals. Some software packages such as GenePix Pro have a utility that makes it relatively easy to compare the overall signal intensity between channels.

We acknowledge that the above advice is not suitable to all applications and we strongly caution against scanning arrays multiple times simply to match the channel intensities. Particularly, the repeated use of high laser power settings will cause photobleaching of the dyes, likely in an asymmetric manner, and thus have a detrimental effect on the primary data. A number of recent studies have investigated the effects of PMT and laser power settings on the linearity and dynamic range of microarray data (Bengtsson *et al.*, 2004; Lyng *et al.*, 2004; Shi *et al.*, 2005). The consensus of these studies is that the linear response of individual scanners is limited to only a small portion of the PMT gain range. Performing scans at subobtimal PMT settings gives rise to strong intensity-dependent biases in the results (see Timlin, 2006, for more detail).

The quantified image data is exported in a tabular format and is ready to be analyzed. One of the first decisions a researcher will have to make is with regards to how the background will be treated. Background values are tabulated by quantifying the fluorescence of areas around the hybridized spots, however, the areas and calculations used to derive the value are implemented differently in many commercial software packages. Most applications offer the option to have these values subtracted from the spot intensities, despite there being no clear consensus as to whether or not the background around a spot is representative of the background within the spot. Background signals around spots are often higher than the signals of DNA spots to which there is no hybridization and occasionally background subtraction yields negative spot intensities. In ratio-based analyses background subtraction may generate extreme values if the control or experimental spot intensities are already very low, thus increasing the scatter of the data. To avoid such complications, in most instances it is better not to subtract background signals, while accepting the caveat that the resulting ratio data is likely to be compressed compared to the true expression difference.

Regardless of background, low intensity spots on the array should be dealt with before the data is analyzed. In microarray analysis it is assumed that fluorescence intensity is approximately proportional to the transcript abundance of a given gene. Therefore, spots that exhibit low signal intensities are likely to represent genes with very low expression but may also reflect deficiencies in the spotted probes themselves. Furthermore, a recent report has identified a non-linear response to low intensity data in several commercial scanners (Lyng *et al.*, 2004). With the uncertain contribution of background signals being proportionally higher to these low signal spots they are likely to produce highly variable ratios and are best ignored during analysis. This may be accomplished by applying an arbitrary low expression threshold value, or better, an experimentally derived cutoff such as being greater than two standard deviations greater than the mean background intensity (Quackenbush, 2002). Although genes with low expression may be of interest to researchers, spotted microarray technology is not ideal for evaluating such expression profiles. The reproducible detection of rare targets is negatively affected by the variable deposition of probe substrate on spotted arrays and the relatively large feature size, which dilutes the target signal over many pixels. Also, it has come to light that the effective dynamic range of most scanners is approximately three orders of magnitude; substantially less than the expected five orders of magnitude they can theoretically achieve (Lyng *et al.*, 2004). Proper statistical analysis will, however, determine whether or not low-intensity results are consistent enough to be meaningful.

Before results are further analyzed, it is also important to flag other low quality features on the hybridized array. These may include surface defects such as scratches, the presence of dust particles or other spots affected by elevated non-specific background signals. Control spots containing spotting buffer only or DNA from other species should also be flagged if they have not already been filtered out by the low intensity cut-off because any small reproducible biases in their signals will otherwise contribute to any normalization method used and perhaps skew the normalization of the biologically relevant data. We have observed such effects in our own data (not shown) by employing special flags that enable us to selectively ignore different classes of features (e.g., spotting buffer, control genes, spike-control probes) during normalization. This effect may be the result of the small channel-specific biases described by Bengtsson and colleagues (Bengtsson *et al.*, 2004). Software such as QuantArray allows easy flagging of undesirable spots and these flags are maintained when the data is exported into other analytical software packages. It is important to note that flagging spots, as we describe here, does not amount to the loss of any primary data. Defective features on the array are simply highlighted in a manner that allows them to be selectively excluded from downstream analyses in compatible programs. The actual deletion of any primary data is not advised as this may cause incompatibilities with certain programs and may prohibit the user from effectively correcting certain systematic effects in their data.

Normalization is a complex topic on its own, and may be applied in a number of ways. The fundamental goal of all normalization procedures is to remove sources of experimental bias so that authentic biological differences in gene expression may be detected. Intensity- and ratio-based normalization schemes simply translate the data so that either the mean intensity or the mean ratio of the values from the two channels is equal. However, these do not correct for intensity dependent biases in the signals, which may be visualized in MA plots (Yang *et al.*, 2002). MA plots are a tool commonly used to visualize intensity-dependent effects on log-transformed ratios in microarray data. Regression or spline normalizations effectively correct for intensity-dependent effects, resulting in a symmetrical distribution of data about the horizontal line M = 0 (no change in gene expression) (Workman *et al.*, 2002). One such method is Lowess smoothing of the experimental data. Lowess normalization considers only a defined portion of the data and is relatively insensitive to extreme values (Workman *et al.*, 2002; Yang *et al.*, 2002). Due to this insensitivity to extreme values, the Lowess algorithm does not automatically ignore flagged values during its calculations, but we have determined through experience that when a significant number of spots are flagged the algorithm performs better when these are ignored. Generally, the failure to ignore flagged spots during normalization introduces a shift in the distribution of ratios and thus affects the magnitudes of the measured normalized fold changes for all spots on the array.

A number of software packages have the ability to perform post-scanning normalization as well as statistical analyses and clustering of genes with similar expression patterns. These include commercial software such as GeneTraffic and ArrayExpress (Iobion Informatics/Stratagene), GeneSpring and Rosetta Resolver (Agilent), DecisionSite (Spotfire), Acuity (Axon/Molecular Devices) as well as open-source software such as TIGR MIDAS and MeV (www.tm4.org), BASE (http://base.thep.lu.se/) and Bioconductor and R (www.bioconductor.org).

Of particular importance is to know when and when not to normalize one's data. In general, if very few changes in gene expression are expected between the experimental and reference samples then normalization should probably be done- usually Lowess normalization applied at the sub-array or block level. Examples of experiments in which normalization should be applied include comparisons of drug treatment vs. placebo and mutant vs. genetically-similar wild type. Examples of experiments where global or sub-array normalization should not be applied include developmental time course studies in which individual time points are compared to a pool of all stages of development, and any study in which the experimental sample is being compared to a "universal" (e.g. commercial) RNA reference. In these types of studies the overall signal intensity between the two channels in a given array or sub-array is not expected to be similar and as a result, applying ratio- or intensity-based normalization will usually skew the data to minimize differences between the two channels.

In such instances external spike-in controls or other means are required to calibrate the array signals.

Sensitivity

When any new array platform is used its sensitivity should be tested to identify the optimal amount of RNA to use for each labeling reaction and subsequent hybridization. This optimization is not only important to conserve potentially precious biological samples but also to maximize the signal to noise ratio for the experimental system. In a previous study (Neal *et al.*, 2003), we performed such testing to characterize the *Drosophila* 7k2 array platform. For this, we first defined a "valid spot" using a number of criteria with the implication that the data from such a spot would be suitable for robust analysis. Labeled cDNA was prepared from increasing amounts of RNA from the same pool and was hybridized to sequentially printed arrays. In this experiment we determined that using 80 μg of total RNA in the labeling reaction generated double the mean signal intensity for valid spots when compared to 40 μg RNA samples (Fig. 1A) and also gave rise to the greatest number of valid spot pairs on the hybridized microarray (Fig. 1B). Although 40 μg of total RNA yielded reasonable results, the lower mean signal intensity would lead to more variable ratios and would thus require more replication to achieve the same confidence in

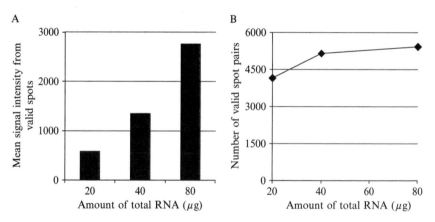

FIG. 1. The Sensitivity of the *Drosophila* 7k2 microarray platform was assessed by hybridizing different amounts of labeled cDNA to sequentially printed microarrays. Labeled cDNA was generated from increasing amounts (20, 40, and 80 μg) of total RNA from the same sample pool. Data were normalized with the Lowess algorithm in GeneTraffic and a low-intensity threshold was applied. The mean signal intensity from the valid spots was compared (A) between the three arrays. The total number of valid spot pairs from each array was also determined (B). Reproduced with permission from Neal *et al.* (2003). *Genome* **46**(5), 879–892. NRC Press, Canada.

the results. In general, if RNA is not limiting, use as high an amount of RNA that is practically possible. Note that the use of excessive RNA may be detrimental to the efficiency of the labeling reaction.

Reproducibility

Results must be reproducible in any experimental system for the researcher to be confident in their validity, and they must be further so to achieve statistical significance. Microarray experiments are no exception despite the early trend in this field of using arbitrary fold cutoffs to select differentially expressed genes. In the absence of rigorous statistical analysis, it is still possible to experimentally determine confidence limits for microarray data by performing self-self hybridizations. By co-hybridizing differentially labeled cDNA derived from the same RNA pool one can plot a histogram of the distribution of ratios in the replicated normalized data and determine the threshold values for a given confidence interval (Johnston *et al.*, 2004; Neal *et al.*, 2003). For the *Drosophila* 7k2 microarray, we previously determined (Neal *et al.*, 2003) that with triplicate data the threshold values for the 99% confidence interval were the equivalents of a ±1.3-fold change (Fig. 2); far less than the arbitrary 2-fold threshold

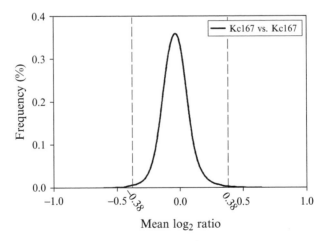

Mean log$_2$ ratio

FIG. 2. The variation in ratios in a series of homotypic hybridizations was assessed for the *Drosophila* 7k2 microarray platform. Three sequentially printed microarrays were hybridized with fluorescent cDNA probes generated from 80 μg of total RNA. Data were normalized and spot quality filters were applied as discussed in the original manuscript. For spots that passed the validity criteria at least 67% of the time (minimum 4 of 6 spots valid) the mean log$_2$-transformed ratios were plotted as a histogram (dotted line). The vertical bars illustrate the limits of the 99% confidence interval. Reproduced in a modified form with permission from Neal *et al.* (2003). *Genome* **46**(5), 879–892. NRC Press, Canada.

employed in many early microarray studies. In an independent study, Johnston and colleagues (2004) determined that approximately ± 1.5-fold was the threshold for the 99.5% confidence interval in quadruplicate data on their array platform. The significance of establishing these thresholds is that, in the case of the latter study, any gene in an experimental comparison whose expression changed more than 1.5-fold had only a 0.5% chance of being a false positive. The issue surrounding multiple hypothesis testing will be addressed later. Based on the above results, one can start to calculate the number of replicates required to support their conclusions. Although three independent replicates is a minimum for any statistical analysis, one could identify confidence thresholds for duplicate data, which would likely be sufficient to confirm a large fold change. When extending this approach to a large number of replicates it is conceivable that an extremely narrow confidence interval could be established, however, one must also consider the cost of the analysis and the potential biological relevance of smaller changes in gene expression.

Another common issue facing researchers wishing to conduct microarray experiments is the necessity of dye swap experiments. The purpose of these technical replicates is to normalize for biases in dye incorporation, as described earlier in the chapter. However, Lowess normalization effectively smoothes the distribution of results in an MA plot and reverses the intensity dependent dye bias, making these replicates potentially unnecessary. To test this directly we undertook the comparison of the divergent expression profiles of two *Drosophila* cell lines (Neal et al., 2003). Three independent RNA isolates were taken from each cell line and three pairwise comparisons were made between the samples. Technical replicate hybridizations were then conducted in which the same samples were labeled in the alternative dye configuration. These experiments were analyzed as a group ($n = 6$) and in separate groups ($n = 3$ each) for each dye configuration. In each case the major findings were recapitulated with only small changes in the ranking of the results (Table I). These results suggest that dye bias does not have a strong effect on Lowess normalized data. Despite this fact, it is important not to ignore the effects of dye incorporation differences in an experimental design. Thus, it is preferable to always label the control sample with the same dye if explicit dye swaps are not going to be conducted so that the outcome of normalization will be predictable from slide to slide even if the correction is not complete.

Statistical Analysis

Several statistical analyses may be applied to microarray data and the choice of method often relates to the experimental design. Simple

TABLE I
AN ASSESSMENT OF THE NECESSITY OF DYE-SWAP EXPERIMENTS[a]

	Experimental data used for analysis		
Criteria	SL2 vs. Kc167 (all) ($n = 6$ arrays)	SL2 vs. Kc167 (Cy5 vs. Cy3) ($n = 3$ arrays)	SL2 vs. Kc167 (Cy3 vs. Cy5) ($n = 3$ arrays)
Number of genes >2-fold up-regulated	241	249	267
Number of genes >1.8-fold up-regulated contained in the list of >2-fold up-regulated genes from all six experiments (%)	241 (100%)	226 (93.8%)	228 (94.6%)
Number of genes >2-fold down-regulated	261	300	255
Number of genes >1.8-fold down-regulated contained in the list of >2-fold down-regulated genes from all six experiments (%)	261 (100%)	257 (98.5%)	250 (95.8%)
Number of top 20 up-regulated genes maintaining this designation between the 3- and 6-array analyses	n/a	20	18[b]
Number of top 20 down-regulated genes maintaining this designation between the 3- and 6-array analyses	n/a	16	16

[a] The necessity of dye-swap experiments was conducted by comparing results from the independent analyses of a replicated dataset. The analysis of the entire set, in which reciprocal labeling reactions were performed, was compared to the analyses of two exclusive subsets of the data in which the control samples were consistently labeled with the same dye. Reproduced in a modified form with permission from Neal *et al.* (2003). *Genome* **46**(5), 879–892. NRC Press, Canada.

[b] One of the genes did not meet the general validation criteria in this group of experiments, effectively making this value out of 19 instead of 20.

pairwise comparisons can be evaluated using T-tests whereas timecourse and developmental studies are more likely to be evaluated using ANOVA. ANOVA offers the benefit of identifying the relative contributions of specific sources of variance and can be valuable when validating experimental protocols and designs (Johnston *et al.*, 2004). However, the implementation of statistical analysis to microarray results must also consider the number of times the test is conducted and address the errors associated with multiple hypothesis testing. If a common reference sample is not used, it may prove difficult to factor all of the experimental replicates into the analysis as well. Such issues have been the focus of a number of studies to date (Kerr and Churchill, 2001; Sherlock, 2001; Tusher *et al.*, 2001).

Cluster analysis, although not strictly statistical in nature, is favored in the microarray community. This approach classifies genes based on their expression values in a number of treatments, and groups them according to the similarity of their expression profiles. A number of different metrics may be used to perform clustering, such as hierarchical clustering, k-means clustering and self-organizing maps. Some of these algorithms were implemented into a basic user interface (Eisen *et al.*, 1998) that also outputs the results in graphical form. This approach has mainly been used to look for coordinate expression among families of genes and to attribute functions to previously uncharacterized genes. However, clustering does not measure the significance of the individual expression profiles in each group and thus the results of this approach are more qualitative in nature. The value of these results increases when genes with significant changes in expression are identified first, discussed below, and clustered after the fact.

A simple approach to assess the validity of reproduced experimental results is to consider the variance of the mean gene expression ratio. Standard deviation and coefficient of variance (COV) both measure this trait. Whereas the magnitude of the standard deviation is related to the magnitude of the mean, the COV is expressed as a percentage and is thus better suited to a large-scale analysis. Small COVs represent consistent results and larger COVs indicate that the data points contributing to the mean are more variable. Furthermore, as even a single data point approaches a $\log_2$ ratio of zero (no change in gene expression) the COV is greatly increased. We have previously used an arbitrary COV threshold of 100% to eliminate genes from consideration. This threshold is applied after measures of spot quality and minimum fold-change have already been implemented. In our experience this filter has helped identify several spurious results, such as when two clones purportedly from the same gene show divergent signals, and occasions where localized background was not properly flagged. Software such as GeneTraffic or ArrayExpress (Iobion/ Stratagene) is well suited for this type of analysis.

Several of the above concepts have been implemented in a widely used software tool known as the Statistical Analysis of Microarrays (SAM) program (Tusher *et al.*, 2001). SAM performs a modified T-test on the ratios of replicated data and models the randomized data to estimate the False Discovery Rate (FDR). FDR is a correction factor for multiple hypothesis testing and amounts to an estimate of how many results beyond the thresholds are likely to be present due to chance alone. The thresholds are adjusted by a parameter, δ, which allows the user to establish their own tolerance for false positives within the results. Routinely δ is adjusted such that less than one result is expected by chance, implying that all of the results are likely to be true positives.

We applied SAM analysis to our previously published experimental data (GEO accessions GSM6159-GSM6167) in order to contrast our *ad hoc* analysis approach of confidence intervals and COV thresholds with a true statistical analysis of the data. Since at least three data points are required for SAM to function, we combined the data from the duplicate spots on the array to increase the proportion of the results it would consider. We acknowledge that this amounts to technical-replication of the results and that it may artificially increase the perceived confidence in them. We first tested our control data set of self-self hybridizations with SAM and ran the analysis three times to determine what effects the random seeding had on the outcome. One important difference between the SAM analysis and our previous analysis is that SAM considers significant changes for individual spots whereas GeneTraffic operates at the level of "genes", where the expression values may be derived from multiple spots depending on their annotation. To refine the SAM gene list we filtered it to remove all elements with a mean ratio less than 1.3 fold and a COV greater than 100%; values derived from our *ad hoc* analysis. This reduced the list to only 6 features/genes that showed significant changes in the self-self hybridizations out of approximately 5000 features that were detected on the array. While we were not expecting to see any genes, examination of these genes revealed that 4 of them had signal intensities just above our arbitrary low expression threshold and two had legitimate signal differences that we cannot explain. Overall, we conclude that this validates our original analysis approach and furthermore supports the use of SAM for analysis of microarray data.

We used SAM to analyze the data from the comparison of expression profiles of the two *Drosophila* cell lines (Neal *et al.*, 2003). Each group of triplicate data in which the dyes were used in the same configuration was analyzed separately, again initializing SAM with three random integers. Using the fold-change and COV filters as above, SAM analysis identified 1347 (755 up and 592 down) significantly changed genes in the first data set and 2129 (1017 up and 1112 down) genes in the dye flip data set. There would appear to be a large discrepancy in the number of genes identified between the two data sets, which could represent the effects of dye bias or insufficient normalization. In this case a Lowess smoothing factor of 20% was used as opposed to the increased factor of 40% used in other studies (Yang *et al.*, 2002). We are of the opinion that the discrepancy is unlikely to be the result of dye bias, but rather it may relate to the fact that there was more sensitivity in the analysis of the dye flip experiments. The 2129 features identified in the latter analysis also compared favorably to the 2379 differentially expressed genes that we originally identified using our confidence interval approach.

Although SAM is a versatile tool, and the newly released version 2.0 supports the analysis of more complex experimental designs, it fails to consider the intensity of the spots contributing to the given ratios in its calculations. Because intensity is assumed to be representative of gene expression, one must consider the fact that a small fold change in a highly expressed gene may be more biologically significant than a larger fold change in a lowly expressed gene. This concept of "expression value" is central to Affymetrix array analysis but is difficult to implement in two-color analysis due to the lack of sufficient exogenous controls on most arrays. One approach that addresses this issue partially is to generate a volcano plot. This plot features the significance of a change (P-value) on the Y-axis versus the $\log_2$ ratio of gene expression on the X-axis. Genes with low expression levels are more likely to have less reproducible ratios and therefore higher P-values. The utility of these plots is to select differentially expressed genes based on a combination of both significance levels and fold-change. This approach is exemplified in a multivariate study of gene expression in *Drosophila* (Jin *et al.*, 2001).

Perspectives and Conclusions

In this chapter, we addressed some of the key aspects of: (1) experiment methodology that can affect the quality and reproducibility of microarray results; and (2) the issues and parameters that need to be considered in order to optimize the analysis of microarray data. A decision tree highlighting some of the key choices that are made in designing, performing and analyzing microarray experiments is shown in Fig. 3. For experiment methodology, some recommendations were made to help ensure high quality and reproducible data: use of commercial glass substrates with low background (e.g., Corning UltraGAPS); preparation of high quality total RNA using a scalable method (e.g., Invitrogen TRIzol); labeling with a robust and reproducible protocol (e.g., direct incorporation of Cy-dye labeled nucleotides); hybridization of samples using inexpensive commercial reagents and equipment (e.g., Roche DIG EasyHyb and Corning hybridization chambers).

Some recommendations were also made with regards to experiment design and analysis: hyphen;potential pooling of reference RNA so the same reference is used throughout a project; to flag and remove artifactual and other spots that do not contain information; not to subtract background fluorescence from spot signals to avoid highly variant ratios for low expressing genes; in most cases to normalize microarray data; to optimize RNA amounts used for maximum detection of low expressing genes; to establish confidence limits for fold-changes in the experimental

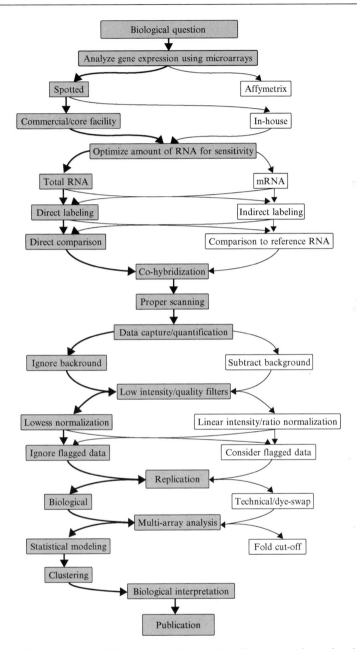

Fig. 3. Decision-tree highlighting some key choices that are made performing and analyzing microarray experiments. Shaded boxes and bold arrows indicate the authors' recommendations for the approach to take in a typical study. See text for details.

platform by performing a few self-self hybridizations of reference samples; and in general not perform dye-swap experiments using the same biological experiments but to label the reference RNA sample with the same dye. With regards to statistical analyses, we recommend methods that include some measure, either direct or indirect, of signal intensity for a given gene such as the filtering and COV method described above or the volcano plot approach, because they avoid some of the major pitfalls of methods that rely on strict fold-change cutoff that may include some non-significant genes (false positives) and exclude other significant values that are below the cut-off.

Few technologies offer the genome-wide expression screening potential of DNA microarrays for the same cost and effort. Unfortunately, the ease of performing these experiments has often led to poorly planned studies and unusable data. By understanding and implementing the basic concepts outlined in this chapter it is possible to exploit the true potential of spotted microarrays and generate robust and statistically-validated results.

Acknowledgments

We thank Neil Winegarden of the UHN Microarray Centre (Toronto) for advice regarding the printing of microarrays, experiment protocols and quantification of arrays and Jason Gonçalves (Iobion Informatics) for insights into microarray data analysis. We also thank past and present members of the Canadian *Drosophila* Microarray Centre and Westwood lab including Meredith Gibson, Jianming Pei, Martin Hyrcza, Sarah Gonsalves, Huafen Li, Tony So, Mandy Lam, Zak Razak and Ali Qureshi for their contributions towards this work. This work was supported by a NSERC Discovery grant to J.T.W. and the CDMC is supported by a multi-user maintenance and NET grant from CIHR.

References

Bengtsson, H., Jonsson, G., and Vallon-Christersson, J. (2004). Calibration and assessment of channel-specific biases in microarray data with extended dynamical range. *BMC Bioinformatics* **5**, 177.

Chomczynski, P., and Sacchi, N. (1987). Single-step method of RNA isolation by acid guanidinium thiocyanate-phenol-chloroform extraction. *Anal. Biochem.* **162**, 156–159.

Churchill, GA. (2002). Fundamentals of experimental design for cDNA microarrays. *Nat. Genet.* **32**(Suppl.), 490–495.

Eisen, M. B., Spellman, P. T., Brown, P. O., and Botstein, D. (1998). Cluster analysis and display of genome-wide expression patterns. *Proc. Natl. Acad. Sci. USA* **95**, 14863–14868.

Evertsz, E. M., Au-Young, J., Ruvolo, M. V., Lim, A. C., and Reynolds, M. A. (2001). Hybridization cross-reactivity within homologous gene families on glass cDNA micro-arrays. *Biotechniques* **31**, 1182, 1184, 1186 passim.

Hedge, P., Qi, R., Abernathy, K., Gay, C., Dharap, S., Gaspard, R., Hughes, J. E., Snesrud, E., Lee, N., and Quackenbush, J. (2000). A concise guide to cDNA microarray analysis. *Biotechniques* **29**, 548–550, 552–554, 556 passim.

Jin, W., Riley, R. M., Wolfinger, R. D., White, K. P., Passador-Gurgel, G., and Gibson, G. (2001). The contributions of sex, genotype and age to transcriptional variance in *Drosophila melanogaster*. *Nat. Genet.* **29**, 389–395.

Johnston, R., Wang, B., Nuttall, R., Doctolero, M., Edwards, P., Lu, J., Vainer, M., Yue, H., Wang, X., Minor, J., Chan, C., Lash, A., Goralski, T., Parisi, M., Oliver, B., and Eastman, S. (2004). FlyGEM, a full transcriptome array platform for the *Drosophila* community. *Genome. Biol.* **5**, R19.

Kerr, M. K., and Churchill, G. A. (2001). Statistical design and the analysis of gene expression microarray data. *Genet. Res.* **77**, 123–128.

Lyng, H., Badiee, A., Svendsrud, D. H., Hovig, E., Myklebost, O., and Stokke, T. (2004). Profound influence of microarray scanner characteristics on gene expression ratios: Analysis and procedure for correction. *BMC Genomics* **5**, 10.

Martinez, M. J., Aragon, A. D., Rodriguez, A. L., Weber, J. M., Timlin, J. A., Sinclair, M. B., Haaland, D. M., and Werner-Washburne, M. (2003). Identification and removal of contaminating fluorescence from commercial and in-house printed DNA microarrays. *Nucleic Acids Res.* **31**, e18.

Neal, S. J., Gibson, M. L., So, A. K., and Westwood, J. T. (2003). Construction of a cDNA-based microarray for *Drosophila melanogaster*: A comparison of gene transcription profiles from SL2 and Kc167 cells. *Genome* **46**, 879–892.

Pradet-Balade, B., Boulme, F., Mullner, E. W., and Garcia-Sanz, J. A. (2001). Reliability of mRNA profiling: Verification for samples with different complexities. *Biotechniques* **30**, 1352–1357.

Quackenbush, J. (2002). Microarray data normalization and transformation. *Nat. Genet.* **32** (Suppl.), 496–501.

Sherlock, G. (2001). Analysis of large-scale gene expression data. *Brief Bioinform.* **2**, 350–362.

Shi, L., Tong, W., Su, Z., Han, T., Han, J., Puri, R. K., Fang, H., Frueh, F. W., Goodsaid, F. M., Guo, L., Branham, W. S., Chen, J. J., Xu, Z. A., Harris, S. C., Hong, H., Xie, Q., Perkins, R. G., and Fuscoe, J. C. (2005). Microarray scanner calibration curves: Characteristics and implications. *BMC Bioinformatics* **6**(Suppl. 2), S11.

Timlin, J.A (2006). Scanning microarrays: Current methods and future directions. *Methods Enzymol.* **411**, 79–98.

Tusher, V. G., Tibshirani, R., and Chu, G. (2001). Significance analysis of microarrays applied to the ionizing radiation response. *Proc. Natl. Acad. Sci. USA* **98**, 5116–5121.

Workman, C., Jensen, L. J., Jarmer, H., Berka, R., Gautier, L., Nielser, H. B., Saxild, H. H., Nielsen, C., Brunak, S., and Knudsen, S. (2002). A new non-linear normalization method for reducing variability in DNA microarray experiments. *Genome. Biol.* **3**, research 0048.

Yang, Y. H., Buckley, M. J., and Speed, T. P. (2001). Analysis of cDNA microarray images. *Brief Bioinform.* **2**, 341–349.

Yang, Y. H., Dudoit, S., Luu, P., Lin, D. M., Peng, V., Ngai, J., and Speed, T. P. (2002). Normalization for cDNA microarray data: A robust composite method addressing single and multiple slide systematic variation. *Nucleic Acids Res.* **30**, e15.

[11] Sample Labeling: An Overview

By MICHAEL BROWNSTEIN

Abstract

It is not easy to write a critical review of the methods available for labeling RNA and DNA "extracts" for microarray studies. There are a number of reasons for this: Suppliers of the reagents and kits used for this purpose do research and development, quality control, and validation and then they provide a hard-wired, "optimized" product. They often give few details about the compositions of these products, are inclined to put the best face they can on what they sell and gloss over any deficiencies, and have no interest in paying for direct comparisons of their product to those of other companies. These comparisons can be expensive to perform, and there are few good examples in the literature. When comparative experiments have been done, it is not clear that each of the individual methods tested was executed with equal proficiency. Many experiments can be required to determine how best to hybridize any given labeled extract to a particular array and how to block, wash, and postprocess (e.g., stain) the array so that the signal-to-noise ratio is maximized. In addition, authors of comparative studies used different arrays, technical protocols (some of which are out of date), experimental designs, and analyses. Finally, some new techniques, which seem quite promising, have been employed so little that their strengths and shortcomings are difficult to assess.

Given these constraints, this chapter attempts to (1) outline the steps involved in labeling extracts, (2) touch on problems that remain to be solved, and (3) describe and evaluate a variety of labeling methods. Methods used for studying gene expression as opposed to genomic DNA are predominant in this chapter, but the latter are mentioned in passing.

Nucleic acids have to be extracted from cells or tissues before they can be labeled. Next, if there is too little template to label efficiently, the RNA or DNA must be amplified. Finally, the unamplified or unamplified products have to be tagged so that they can be detected. This last step can occur before the products are hybridized to elements on the array or afterwards, and techniques are available that allow the user to perform signal amplification as well as template amplification.

METHODS IN ENZYMOLOGY, VOL. 410   0076-6879/06 $35.00

Extraction of Nucleic Acids

RNA

There are two widely used sorts of methods for purifying RNA from cell and tissue samples. The first of these relies on guanidinium thiocyanate-phenol (Trizol) extraction and alcohol precipitation (Chomczynski and Sacchi, 1987). Several Trizol solutions are available that are optimized for use with mammalian tissues and cell lines, whole blood, or bacteria. Trizol extracts intact RNA efficiently from most eukaryotic cells and tissues. The high salt protocol (http://64.233.161.104/search?q=cache:9zVgz5MHi8wJ: www.invitrogen.com/content/sfs/manuals/15596026.pdf+trizol+high+salt +protocol&hl=en) can be used to reduce contamination by polysaccharides and proteoglycans. Tissues that are rich in RNase (e.g., pancreas) are diffi-cult, but not impossible, to process (Gill *et al.*, 1996).

Blood presents a special problem. Trizol BD yields high-quality RNA, which can be reverse transcribed well, but it is rich in globin mRNAs. In fact, globin messages can comprise as many as 70% of all message mole-cules extracted. Their abundance makes it difficult to detect white blood cell (WBC) RNAs in blood samples with arrays, and it is WBC transcripts that most workers want to study. The problem arises from the fact that humans, for example, have 5000–10,000 WBC per microliter of blood and 4 to 6 million red blood cells (RBCs), 0.5 to 2% of which are nucleated. While red cells rapidly degrade their RNA as soon as they expel their nuclei, there are as many immature, nucleated RBCs per microliter of blood as there are WBCs. Ambion has introduced a magnetic bead-based method for depleting globin mRNA from blood samples. Marketed as GLOBINclear-Human, it is described at http://www.ambion.com/techlib/ prot/fm_1980.pdf. It is worth noting that WBCs can be enriched using methods that depend on centrifugation, but these techniques do not re-move nucleated RBCs completely and take time and effort. One's clinical collaborators are sometimes unable to devote much effort to sample prep-aration, and in the hour that it may take to enrich WBCs, some transcripts are induced (e.g., those encoding cytokines) and others are lost.

Extraction of RNA from yeast requires much more aggressive proce-dures than extraction from mammalian cells. The RiboPure-Yeast RNA isolation method accomplishes the lysis of yeast cells with a phenol-based solution and vigorous vortexing in the presence of Zirconia beads—a so-called "bead beating" method (see http://www.ambion.com/techlib/prot/ fm_1926.pdf).

Preparing RNA from some bacteria can also be troublesome because they lyse poorly and are prone to degrade their RNA rapidly when their cell walls are disrupted. Whereas Trizol LS was designed specifically to improve yields of RNA from prokaryotes, it may not work well for all species. An extraction kit similar to the RiboPure-Yeast RNA isolation method (the RiboPure-Bacteria RNA isolation kit) appears to give good yields of intact RNA from a variety of prokaryotes and does not require chemical or enzymatic pretreatments that may affect expression profiles.

In addition to methods that depend completely or in part on organic extractions, there are techniques that are based on the lysis of cells in the presence of chaotropic agents and adsorption of the RNA on solid supports (e.g., glass fiber filters in spin columns). The RiboPure kits described earlier are hybrid methods in that they include organic extraction and subsequent purification on solid supports.

While organic extractions are quite reliable, they do not permit one to obtain RNA from small tissue samples (e.g., those obtained by means of laser microdissection or laser capture) very efficiently. To get around this problem, a number of companies, including Ambion, Arcturus, and Stratagene, have developed kits that nominally permit the user to isolate RNA from as few as 10 cells (50–100 pg of RNA), and on-column treatment with DNase can be used to eliminate DNA from the samples. We have been satisfied with these kits, but have only used them to purify products from hundreds of cells.

Spin columns are also used in conjunction with PAXgene tubes (Thach *et al.*, 2003). These are used for collecting blood samples and contain tetradecyltrimethylammonium oxalate (Catrimox-14; Dahle and Macfarlane, 1993), a cationic surfactant that lyses cells and simultaneously precipitates nucleic acids.

RNAlater is a reagent that is in common use because of the convenience that it affords. This low pH, concentrated ammonium sulfate solution can be used at room temperature to prevent RNA degradation in tissue and cell samples and allows them to be stored until it is convenient to process them.

It is worth commenting on formalin-fixed tissues specifically. Because there are so many of these available and because they represent such a valuable resource, it would be very useful to be able to extract nucleic acids from them, but this has proven difficult. Formalin is a very efficient crosslinker that reacts with proteins, DNA, and RNA. Consequently, the DNA and RNA that can be extracted from formalin-fixed tissues is a poor substrate for enzymes. In addition, the longer a sample was fixed and subsequently stored, the less useful template can be obtained from it. For example, to amplify microsatellite markers from DNA extracted from formalin-fixed, paraffin-embedded (FFPE) tissues, primers that yield short (<100 bp) amplicons have to be designed, and even with these there is a significant failure rate. Despite this, it appears that whole genome amplification can be accomplished successfully using DNA from FFPE sections (see http://www.sigma-origins.co.uk/pdfs/articles/1115138157.pdf). Working with RNA is harder of course. Arcturus offers an RNA extraction kit for FFPE samples. The information at their web site is a bit sketchy, but they say that 5000–10,000 cells have to be harvested to obtain a sufficient amount of RNA (5 ng minimum) to amplify and label. They indicate that the RNA obtained is often quite degraded: "FFPE samples are often too degraded to get reliable 28S/18S ratios using Agilent's bioanalyzer. At Arcturus, we depend on bioanalyzer profiles only to get an idea of the molecular weight distribution of RNA isolated from FFPE samples. A predominance of high molecular weight RNA fragments in the sample is indicative of a sample that is amplifiable. Further, we perform a quantitative reverse transcriptase qRT-PCR on β-actin using both a 3' and 5' (400 bases from the 3' end of the mRNA) probe to measure 3'/5' ratios. These ratios are based on Ct values and allow us to determine the level of degradation in our sample and the transcribability of our RNA. High 3'/5' ratios indicate a difference in amplification between the 5' end and 3' end which might arise as a result of degradation. As a rule of thumb, samples with a 3'/5' ratio below 20–40 are considered suitable." Clearly, the quality of the RNA extracted is not very good, and the oligo(dT)-primed amplification of this template, which is recommended by Arcturus, yields a very 3'-biased product. This would best be used with an equally biased array.

Xiang et al. (2004) described a novel reagent that can be used to fix, immunostain, and harvest specific cell populations from heterogeneous tissue sections (e.g., brain). The reagent, dithio-bis(succinimidyl propionate) (DSP) reacts with primary amines, but not RNA, and has an -S-S- bridge in the middle that can be reduced, reversing the cross-linking. Unfortunately, DSP-fixed tissue only retains about 15% of its RNA after it has been stained, but the RNA seems quite intact and it is readily reverse transcribed and amplified. It is clear that better fixatives are needed, but DSP might be fine if large numbers of cells could be dissected quickly. Microdissections currently require extensive user intervention, and it is difficult to harvest any more than 500 cells in a single session if they are not members of an abundant class. Perhaps this problem will be solved if RNA can be extracted from cells obtained by "expression microdissection" (Tangrea et al., 2004). To date the method has only been used successfully for studies of DNA and protein.

Whenever possible, it is important to analyze RNA extracts. Three instruments facilitate this: the NanoDrop ND-1000 spectrophotometer, the NanoDrop fluorospectrometer ND-3300, and the Agilent 2100 bioanalyzer.

The first two of these allow DNA and RNA to be assayed in a volume of 1 μl with no cuvette. As little as 10 pg of RNA can be detected with the fluorospectrometer using RiboGreen dye. The bioanalyzer permits one to determine the quality of subnanogram amounts of total RNA. Best experimental results are obtained with arrays when the amounts and qualities of the templates used are well matched.

DNA

DNA is much less prone to degradation than RNA, and extracting it is considerably easier. Many DNA extraction kits are available commercially; some permit the user to purify subnanogram amounts of product, but because DNA is not broken down as readily as RNA, it can be released from microdissected cells and processed for certain studies without being purified (Klein *et al.*, 1999).

Template Amplification

When microarrays first came into use, investigators were not unhappy to have to use as much as 200 μg of RNA in their labeling reactions, but this changed rapidly. More and more workers wanted to study defined cell populations or small biopsy samples.

RNA

The various labeling methods that are available today require 0.5 to 20 μg of total RNA template—the amount of RNA in roughly 100,000 to 5,000,000 mammalian cells. Thus, the development of methods for microdissecting cells and small, homogeneous regions of organs created a demand for techniques that would boost template levels from picograms to more than a microgram without altering the complexity or composition of the original mRNA pool in the process. Broadly speaking, two methods have been used for this: linear amplification with phage RNA polymerases (e.g., T7 RNA polymerase), techniques commonly referred to as *in vitro* transcription (IVT), or exponential amplification by means of the polymerase chain reaction. Van Gelder and colleagues (1990) were the first to develop a method for linear amplification and most of the other IVT protocols that have been described are derivatives of theirs. Briefly, reverse transcription of purified mRNA or, more commonly, the poly(A) plus component of total RNA is driven by an oligonucleotide with oligo(dT) on the 3′ end and a phage RNA polymerase promoter sequence on the 5′ end. After the cDNA is made, the RNA is replaced with a second DNA strand, and the product is used as a substrate for the production of antisense RNA (aRNA) by the phage polymerase. In the version of the technique adopted by Affymetrix (Dalma-Weiszhausz *et al.*, 2006), biotinylated nucleotides are incorporated into the aRNA. In other applications, the aRNA product is used as a template for the production of labeled DNA or in additional rounds of amplification. It should be clear that the method is potentially 3′ biased, especially if more than one round of amplification is used. Xiang and co-workers (2003b) described a less biased method based on random vs oligo(dT)-primed DNA synthesis in each round of amplification. They showed that this technique amplifies degraded RNA more faithfully than the original one.

IVT is not problem free. It may amplify templates that are rare, GC rich, or plagued by a secondary structure inefficiently. While it seems to increase the number of genes that can be detected in small samples, it has also been reported to decrease the number of differentially expressed transcripts found by 20 to 30% (Xiang *et al.*, 2003a). This is often an acceptable price to pay for being able to study more homogeneous and informative samples.

Perhaps it is worth commenting on IVT of RNA isolated from single cells. A mammalian cell contains 5–10 pg of RNA. It has been found that linear amplification gives quantitative results if a minimum of 0.1–1 ng of total RNA is used as template. In the absence of RNA, amplification reactions still generate DNA products. These appear to be concatenations of primers, and when they are labeled, they can give signals on arrays. Thus, when a few picograms of RNA are amplified and labeled, the signals detected on arrays are partly real, partly artifactual. The more template added, the less troublesome this problem is.

Polymerase chain reactions are much quicker and simpler to perform than linear amplifications, which has stimulated workers to try to develop RT-PCR-based methods to amplify RNA for array studies. Two different strategies have been used for the exponential amplification of RNA. Iscove and co-workers (2002) have used oligo(dT) to drive cDNA synthesis from RNA isolated from single cells. By using limiting concentrations of deoxyribonucleotide triphosphates (dNTPs) and brief reaction times, they produced short (several hundred base) cDNAs, to which they added poly(A) tails with terminal transferase. This allows them to PCR the resulting templates efficiently with a single primer [oligo(dT) with a more complex sequence on the 5′ end]. While the method has certainly proven useful for cataloging genes that are expressed in single cells and while it does seem to preserve sample-to-sample differences in selected genes, its value for global surveys of gene expression microarray is unproven. A 3′-biased microarray would surely have to be used to take advantage of this technique.

Wang *et al.* (2000) took a different approach to PCR-based amplification of RNA by taking advantage of the fact that reverse transcriptase is prone to add untemplated Cs to cDNAs when it reaches the end of an RNA template. This string of Cs can be used for "template switching." In this method, first-strand cDNA synthesis is primed using an oligonucleotide with oligo(dT) on the 3' end and a defined sequence on the 5' end. Following this, an oligonucleotide with three Gs on its 3' end (complementary to the Cs added by RT to complete reverse transcripts) is used to drive second-strand synthesis. The 5' end of this oligonucleotide also has a defined sequence, and the defined sequences on the ends of each cDNA allow PCR to be used to amplify them. When the number of PCR cycles is kept to a minimum, the method appears to work as well, or nearly as well, as IVT (Saghizadeh *et al.*, 2003; Wang *et al.*, 2003) even though PCR is not an "equal opportunity employer." Some templates work well; others do not. This is probably true of linear amplification methods too, but because they have not been employed to amplify or clone single products, we are probably less aware of their shortcomings. Underlying the adoption of either method for use in expression profiling is the assumption that no matter how well or how poorly a specific template is amplified, the biasing is constant from one reaction to the next. In this event, amplified samples can be compared to one another and differentially expressed genes can be detected. To a useful degree, this assumption appears to hold true.

There may be times when qualitative vs quantitative results are acceptable. For example, if one's goal is simply to detect pathogens in tissues or body fluids or to catalog transcripts that are too rare to be detected with conventional microarray methods (e.g., those encoding G protein-coupled receptors), using multiplex RT-PCR or PCR and a sufficient number of cycles to achieve signal saturation has proven useful. Analyzing the reaction products with arrays contributes to specificity and sensitivity (Qi, unpublished result; Hansen, unpublished result).

DNA

Array-based studies of genomic DNA have increased dramatically in the last few years. Arrays have been used for the detection of single nucleotide polymorphisms (SNPs; Gunderson *et al.*, 2006), analysis of DNA copy-number changes (Erickson and Spana, 2006), global surveys of chromatin accessibility, and DNA methylation. Specialized protocols have been used for template amplification or preparation in each of the aforementioned cases. While some workers have used PCR for whole genome preamplification (Cheung and Nelson, 1996; Grothues *et al.*,

1993; Liu *et al.*, 2004; Ludecke *et al.*, 1989; Telenius *et al.*, 1992; Zhang *et al.*, 1992), others have employed strand displacement amplification mediated by phage polymerase Phi29 in the presence of random hexamer primers (Dean *et al.*, 2002; Lage *et al.*, 2003; Lovmar *et al.*, 2003). Both methods appear to suffer from allelic amplification bias when they are used to amplify fewer than 100 genome equivalents of DNA (Paul and Apgar, 2005).

Labeling

Chemical Methods

Chemical methods for labeling extracts are not yet widely used, but they seem quite promising. The only reagents of this sort that have been commercialized to date are *cis*-platinum derivatives (van de Rijke *et al.*, 2003). Kreatech markets cyanine3 (Cy3)- and cyanine5 (Cy5)-labeled *cis*-platinum compounds. These react monofunctionally with electronegative moieties in nucleic acids, notably N7 of guanine residues. The reactions only require 30-min incubations at 85°. Optimal signal-to-noise ratios (30–40) are seen when approximately 1% of all nucleotides are labeled; self-quenching occurs at 2–3% (Raap *et al.*, 2004). Little information is available regarding the use of Cy–*cis*-platinum compounds for labeling RNA, but the Kreatech web site provides some experimental data. Company scientists used Eberwine's method to make aRNA from 5 μg of total RNA and then labeled 2 μg of the amplified product. The signals obtained on a printed oligonucleotide array seemed strong and even. Because microarrays often contain sense vs antisense elements, reverse transcription of total RNA (plus or minus linear amplification) is required to use *cis*-platinum labeling with these arrays. If RNA is not limiting, it may be more ideal to use total RNA in the labeling reaction, especially if one's goal is to study transcript processing (Hughes *et al.*, 2006), because mRNAs could be labeled from end to end. In this case, sense arrays would have to be printed. Such an array has already been produced by PerkinElmer for studies of miRNAs, which can be enriched by centrifugation of total RNA through an Amicon YM-100 spin column, and then labeled with a *cis*-platinum derivative (see http://las.perkinelmer.com/Content/Related-Materials/mRNAapp.pdf). In this case, the company used their MICRO-MAX ASAP reagents: fluorescein- and biotin-*cis*-platinum derivatives. The labeled miRNAs are hybridized to the array, which is washed and then stained. Fluorescein-labeled miRNAs are stained with an antifluorescein antibody conjugated to horseradish peroxidase (HRP), which catalyzes the release of Cy3 from Cy3 tyramine and results in the deposition of

Cy3 on the array. Biotin-labeled miRNAs are stained with streptavidin conjugated to HRP using Cy5 tyramine as the substrate. The sensitivity and specificity of the method seem to be excellent, and the expression levels seen were similar to those obtained by means of Northern blotting. This same method has been used to label total RNA for microarray studies (Gupta *et al.*, 2003). Ten micrograms of template was required to obtain optimal signals. Richter *et al.* (2002) studied enzymatic incorporation of haptens followed by antibody staining and found that "wide scattering of the nonregulated spots and high background levels mask the regulation of genes that show lower signal-to-noise ratios and/or less pronounced regulation."

Carbodiimide-based cyanine dyes, which react with unpaired thymine, guanine, or uracil bases in DNA or RNA, have been used to label cDNA reverse transcribed from 1 to 5 μg of mouse liver or brain total RNA using oligo(dT) priming (Kimura *et al.*, 2005). The labeled products were hybridized to printed oligonucleotide arrays (70-mers), and maximum signals (vs signal-to-noise ratios, which were not reported) were found when one dye molecule was incorporated per 30–50 bases. Differentially expressed genes could be detected and validated, and the method performed as well as conventional methods that were used in parallel. Because only 1–5 μg of RNA was used in the conventional direct and indirect labeling reactions, the results were surely suboptimal (see later), and the comparisons are hard to interpret. Carbodiimide-based dye labeling is fast and efficient, and it would be nice to see more work done with it.

One final group of reagents that have been used for covalent labeling of nucleic acids is the aryldiazomethanes (Laayoun *et al.*, 2003). Instead of attaching to bases as platinum and carbodiimide compounds do, aryldiazomethanes react with phosphate groups. The conditions used allow RNA and DNA to be labeled while they are being cleaved or they can be cleaved afterward. The method has not yet been used for expression profiling.

Huber *et al.* (2004) described a gold nanoparticle-based labeling technique. Total brain RNA was hybridized to printed (antisense strand) oligonucleotide arrays. Then 15-nm-diameter gold nanoparticles with oligo(dT$_{20}$) attached to their surfaces were hybridized to the poly(A) tails of RNAs bound to elements on the array. Finally, the arrays were stained with silver enhancer solutions (Sigma), and the scatter signal of the silver-enhanced gold nanoparticles was captured. While scatter analysis has been shown to be 1000 times as sensitive as fluorescence detection, the method described earlier only seems to be about 10 times as sensitive as conventional protocols, and rigorous comparisons of the techniques are still lacking. The reagents and reader required for gold nanoparticle detection will soon be available from Nanosphere, Inc. (see http://www.nanosphere-inc.

com/). In theory, it seems possible that a method that is virtually identical to this one could be used to label RNA with strepavidin quantum dots (see http://www.qdots.com/live/render/content.asp?id=32) after adding biotinylated oligo(dT) to their surfaces. Quantum dots—small semiconductor particles—fluoresce intensely with narrow and symmetrical emission spectra, are quite stable, and do not exhibit photobleaching. When optimal readers are produced to visualize them, they will be outstanding reagents.

Enzymatic Methods

The first method that was developed for labeling RNA involved the direct incorporation of nucleotides derivatized with Cy dyes into cDNA (Schena *et al.*, 1995). DNA was produced by reverse transcriptase in a reaction driven by oligo(dT). The most recent version of this technique (see http://cmgm.stanford.edu/pbrown/protocols/4_human_RNA.html) specifies that 50–100 μg of total RNA should be used per labeling. While some workers have used as little as 20 μg of template, results are indeed better when more RNA is used. For this reason and because of concerns about "dye bias"—dye-specific differences in the incorporation of derivatized nucleotides into cDNA products—this method is gradually falling out of favor. Similarly, while the use of ^{32}P- or ^{33}P-dNTP-labeled extracts to develop DNA arrays printed on nylon membranes or plastic (Cox, 2001; Jokhadze *et al.*, 2003; http://www.clontech.com/clontech/products/literature/pdf/brochures/AltasBR.pdf) may be convenient for some users because it depends on methods and instruments that are readily available in many molecular biology laboratories, this strategy has not been widely adopted.

To overcome the problem of dye bias, Clontech introduced aminoallyl-modified bases, which can be incorporated into DNA. Subsequently Cy3 or Cy5 dyes are coupled to the reactive amino groups in the DNA products (see http://www.clontech.com/clontech/archive/JAN02UPD/pdf/PowerScript.pdf). (It is worth noting that the differential stability of the cyanine dyes may still result in dye bias, and users should be aware of the problem and control for it; see Wolber *et al.*, 2006.) Initially, this "indirect" labeling method was used in conjunction with oligo(dT) priming of cDNA synthesis from total or mRNA samples and required no less template for reliable results. Subsequently, Xiang *et al.* (2002) showed that random priming of total RNA had no effect on background, but increased signal strength when indirect labeling was used. In addition, these authors demonstrated that adding an aminoallyl moiety to the base on the 5′ end of each primer increased signal strength even further. Indeed, prelabeling primers with Cy3 or Cy5 also gave good results (Johnston, 2004).

This should not be surprising. If the average incorporation of dye molecules is, optimally, about 1 molecule per 100 and if placing an additional molecule on the 5′ end of each product results in no quenching of the signal from dye stacking (Randolph and Waggonner, 1997), then the short cDNAs that are formed in random primed labeling reactions should indeed benefit from the addition of single additional dye molecules. Random priming has an additional positive feature. Products are formed along the entire transcript, which should facilitate studies of differential splicing.

Han *et al.* (2005) added four amino groups with two thymine spacers between each one to the 5′ end of an oligo(dT)^{18} primer and used this to drive the synthesis of cDNA. Just as Xiang *et al.* (2002) did, Han and colleagues (2005) incorporated aminoallyl-modified bases into the cDNA that was synthesized. Labeling 5 μg of RNA using the modified primer appeared to give signals as strong as those seen when 20 μg of RNA was reverse transcribed with an unmodified oligo(dT) primer.

Based on the aforementioned studies, it is logical to infer that the addition of many labeled molecules to the 5′ end of each cDNA produced from an RNA extract could give excellent signals. Based on the studies of Stears *et al.* (2000), the Genisphere 3DNA expression array detection kits were designed to do just this. The kits are available in two versions: one is designed to be used with 1–5 μg of total RNA and the other is designed to be used with 10–20 μg. In the case of the former, reverse transcription is driven by an oligo(dT) primer on the 5′ end of which is a specified "capture" sequence (see http://www.genisphere.com/pdf/array350hs_061603a.pdf). After the RNA is degraded and removed, the cDNA is hybridized to an array. Then a DNA dendrimer with a sequence complementary to that of the capture sequence is used to detect cDNAs that have bound to elements on the array. Because each dendrimer has about 375 dyes per molecule, the resulting signals are strong, but the technique seems to suffer from two problems. The dendrimers are large and they may hinder hybridization to the capture sequence. Even though the manufacturer only recommends a 12-h hybridization of dendrimers to the array, others have suggested that considerably more time may be required to maximize the signal strength (Livesey, 2003). While the large-scale labeling method was reported to work well in one study (Badiee *et al.*, 2003), the accuracy and sensitivity of the small-scale method have been questioned (Badiee *et al.*, 2003; Richter *et al.*, 2002). This is attributable, in part, to the compression of signals at both the high and the low ends of the spectrum (Manduchi *et al.*, 2002).

Users of Affymetrix short oligonucleotide (25-mer) arrays employ a labeling method that is unique to this platform (see http://www.affymetrix.com/support/technical/datasheets/sample_brochure.pdf). As noted earlier, it includes an IVT step mediated by the T7 RNA polymerase. Biotinylated ribonucleotides are incorporated into the aRNAs produced in this reaction, and after the RNAs are fragmented and hybridized to arrays, the latter are stained with a steptavidin–phycoerythrin conjugate. The labeling method is 3' biased, but the elements on Affymetrix arrays are designed to be in the region of 600 bases proximal to the 3' end of each transcript to compensate for this.

Because streptavidin-coated quantum dots that fluoresce at several different wavelengths are available (see earlier discussion), using them to label biotinylated RNA or DNA is feasible. Even without an optimal reader, preliminary experience in using them has been quite encouraging.

It should be obvious that variations of the labeling methods described can be used to label DNA and that these methods share the same strengths and weaknesses that the corresponding RNA-labeling protocols have.

Conclusions

While the majority of workers who use microarrays employ products that are made or endorsed by Affymetrix, a significant number of investigators use other reagents. It seems likely that most groups will migrate away from cDNA arrays. Their only obvious advantage may be that they permit one to study species for which commercial arrays or oligonucleotide collections are not available. For example, Mutsuga *et al.* (2004, 2005) used mouse cDNA arrays to study both rat and hibernating ground squirrel samples. However, problems with coverage, annotation, well-to-well contamination, and cross talk have been hard to correct.

Unfortunately, many of the comparative studies of amplification and labeling methods that have been done to date relied on cDNA arrays. Such studies should now focus on developing and validating methods for short and long oligonucleotide arrays instead.

The techniques used for RNA and DNA purification are well understood and validated. The choice of a method can be based safely on price and convenience, but using a good method does not guarantee obtaining a good product. Not all kits remove enzyme inhibitors that are present in crude extracts of blood, melanocytes, soil, and so on. Furthermore, no extraction method can create intact templates out of degraded ones, and it is important to measure the quantity and to examine the quality of RNA and DNA in advance of amplifying and/or labeling it.

To study small samples, template and/or signal amplification is essential. Template amplification in particular has been used successfully in many laboratories. Among the many methods used to amplify RNA, IVT by phage RNA polymerases (e.g., T7 RNA polymerase) seems to be the most robust, but it time-consuming and expensive to practice. PCR-based, exponential amplifications may introduce somewhat more bias than linear amplifications, but PCR is certainly faster and easier to perform.

A new generation of labeling methods will be marketed soon. The direct, chemical labeling of templates should prove very valuable. The steps involved are inexpensive, fast, and easy to execute and should give excellent results. Unfortunately, too few experiments have been done to permit one to evaluate the strengths and weaknesses of these techniques at this time.

All of the enzymatic labeling methods that have been introduced seem to work. That is, they generate products that give signals on arrays and allow differentially expressed genes to be detected, but some are better than others. Incorporation of aminoallyl-modified bases followed by covalent addition of dyes to the product ("indirect labeling") is less expensive and preferable to direct incorporation of dye-labeled bases because there is no dye bias. Adding primary amines to the 5′ ends of the primers used to drive cDNA synthesis increases the strength of the signals seen with indirect labeling, and as little as 1 to 5 μg of RNA instead of 20–50 μg gives reliable results. Attaching a dendrimer with hundreds of dye molecules to the 5′ end of each cDNA gives strong signals as well, but the amount of RNA required per labeling reaction does not drop dramatically and the quality of the results may not be as good. This also seems to be true when enzymes are used for signal amplification after fluorescein- or biotin-modified bases are used to label extracts. It is difficult to keep these reactions in their linear range. Using a steptavidin–phycoerythrin conjugate to detect biotin adducts, however, seems to be quite reliable.

Whether the methods currently in use for expression profiling will be superseded by new, highly parallel versions of serial analysis of gene expression or quantitative PCR remains to be seen. This is unlikely to occur in the near future.

References

Badiee, A., Eiken, H. G., Steen, V. M., and Lovlie, R. (2003). Evaluation of five different cDNA labeling methods for microarrays using spike controls. *BMC Biotechnol.* **3,** 23.

Cheung, V. G., and Nelson, S. F. (1996). Whole genome amplification using a degenerate oligonucleotide primer allows hundreds of genotypes to be performed on less than one nanogram of genomic DNA. *Proc. Natl. Acad. Sci. USA* **93,** 14676–14679.

Chomczynski, P., and Sacchi, N. (1987). Single-step method of RNA isolation by acid guanidinium thiocyanate-phenol-chloroform extraction. *Anal. Biochem.* **162,** 156–159.

Cox, J. M. (2001). Applications of nylon membrane arrays to gene expression analysis. *J. Immunol. Methods* **250,** 3–13.

Dahle, C. E., and Macfarlane, D. E. (1993). Isolation of RNA from cells in culture using Catrimox-14 cationic surfactant. *Biotechniques* **15**, 1102–1105.

Dalma-Weiszhausz, D. D., Warrington, J., Tanimoto, E. Y., and Miyada, C. G. (2006). The Affymetrix GeneChip platform: An overview. *Methods Enzymol.* **410** (this volume), 3–28.

Dean, F. B., Hosono, S., Fang, L., Wu, X., Faruqi, A. F., Bray-Ward, P., Sun, Z., Zong, Q., Du, Y., Du, J., Driscoll, M., Song, W., Kingsmore, S. F., Egholm, M., and Laske, R. S. (2002). Comprehensive human genome amplification using multiple displacement amplification. *Proc. Natl. Acad. Sci. USA* **99**, 5261–5266.

Erickson, J. N., and Spana, E. P. (2006). Mapping *Drosophila* genomic aberration breakpoints with comparative genome hybridization on microarrays. *Methods Enzymol.* **410** (this volume), 377–386.

Gill, S. S., Aubin, R. A., Bura, C. A., Curran, I. H., and Matula, T. I. (1996). Ensuring recovery of intact RNA from rat pancreas. *Mol. Biotechnol.* **6**, 359–362.

Grothues, D., Cantor, C. R., and Smith, C. L. (1993). PCR amplification of megabase DNA with tagged random primers (T-PCR). *Nucleic Acids Res.* **21**, 1321–1322.

Gunderson, K. L., Steemers, F. J., Ren, H., Ng, P., Zhou, L., Tsan, C., Chang, W., Bullis, D., Musmacker, J., King, C., Lebruska, L. L., Barker, D., Oliphant, A., Kuhn, K. M., and Shen, R. (2006). Whole genome genotyping. *Methods Enzymol.* **410** (this volume), 359–376.

Gupta, V., Cherkassky, A., Chatis, P., Joseph, R., Johnson, A. L., Broadbent, J., Erickson, T., and DiMeo, J. (2003). Directly labeled mRNA produces highly precise and unbiased differential gene expression data. *Nucleic Acids Res.* **31**, e13.

Han, J., Lee, H., Nguyen, N. Y., Beaucage, S. L., and Puri, R. K. (2005). Novel multiple 5'-amino-modified primer for DNA microarrays. *Genomics* **86**, 252–258.

Huber, M., Wei, T. F., Muller, U. R., Lefebvre, P. A., Marla, S. S., and Bao, Y. P. (2004). Gold nanoparticle probe-based gene expression analysis with unamplified total human RNA. *Nucleic Acids Res.* **32**, e137.

Hughes, T. R., Hiley, S. L., Saltzman, A. L., Babak, T., and Blencowe, B. J. (2006). Mircroarray analysis of RNA processing and modification. *Methods Enzymol.* **410** (this volume), 300–316.

Iscove, N. N., Barbara, M., Gu, M., Gibson, M., Modi, C., and Winegarden, N. (2002). Representation is faithfully preserved in global cDNA amplified exponentially from sub-picogram quantities of mRNA. *Nat. Biotechnol.* **20**, 940–943.

Johnston, R., Wang, B., Nuttall, R., Doctolero, M., Edwards, P., Lu, J., Vainer, M., Yue, H., Wang, X., Minor, J., Chan, C., Lash, A., Goralski, T., Parisi, M., Oliver, B., and Eastman, S. (2004). FlyGEM, a full transcriptome array platform for the Drosophila community. *Genome Biol.* **5**, R19.

Jokhadze, G., Chen, S., Granger, C., and Chenchik, A. (2003). Nylon cDNA expression arrays. *Methods Mol. Biol.* **224**, 9–29.

Kimura, N., Tamura, T. A., and Murakami, M. (2005). Evaluation of the performance of two carbodiimide-based cyanine dyes for detecting changes in mRNA expression with DNA microarrays. *Biotechniques* **38**, 797–806.

Klein, C. A., Schmidt-Kittler, O., Schardt, J. A., Pantel, K., Speicher, M. R., and Riethmuller, G. (1999). Comparative genomic hybridization, loss of heterozygosity, and DNA sequence analysis of single cells. *Proc. Natl. Acad. Sci. USA* **96**, 4494–4499.

Laayoun, A., Kotera, M., Sothier, I., Trevisiol, E., Bernal-Mendez, E., Bourget, C., Menou, L., Lhomme, J., and Troesch, A. (2003). Aryldiazomethanes for universal labeling of nucleic acids and analysis on DNA chips. *Bioconjug. Chem.* **14**, 1298–1306.

Lage, J. M., Leamon, J. H., Pejovic, T., Hamann, S., Lacey, M., Dillon, D., Segraves, R., Vossbrinck, B., Gonzalez, A., Pinkel, D., Albertson, D. G., Costa, J., and Lizardi, P. M. (2003). Whole genome analysis of genetic alterations in small DNA samples using hyperbranched strand displacement amplification and array-CGH. *Genome Res.* **13,** 294–307.

Liu, D., Liu, C., DeVries, S., Waldman, F., Cote, R. J., and Datar, R. H. (2004). LM-PCR permits highly representative whole genome amplification of DNA isolated from small number of cells and paraffin-embedded tumor tissue sections. *Diagn. Mol. Pathol.* **13,** 105–115.

Livesey, F. J. (2003). Strategies for microarray analysis of limiting amounts of RNA. *Brief Funct. Genom. Proteom.* **2,** 31–36.

Lovmar, L., Fredriksson, M., Sigurdsson, S., Liljedahl, U., and Syvanen, A. C. (2003). Quantitative evaluation by minisequencing and microarrays reveals accurate multiplexed SNP genotyping of whole genome amplified DNA. *Nucleic Acids Res.* **31,** e129.

Ludecke, H. J., Senger, G., Claussen, U., and Horsthemke, B. (1989). Cloning defined regions of the human genome by microdissection of banded chromosomes and enzymatic amplification. *Nature* **338,** 48–50.

Manduchi, E., Scearce, L. M., Brestelli, J. E., Grant, G. R., Kaestner, K. H., and Stoeckert, C. J., Jr. (2002). Comparison of different labeling methods for two-channel high-density microarray experiments. *Physiol. Genom.* **10,** 169–179.

Mutsuga, N., Shahar, T., Verbalis, J. G., Brownstein, M. J., Xiang, C. C., Bonner, R. F., and Gainer, H. (2004). Regulation of gene expression in magnocellular neurons in rat supraoptic nucleus during sustained hypoosmolality. *Endocrinology* **146,** 1254–1267.

Mutsuga, N., Shahar, T., Verbalis, J. G., Xiang, C. C., Brownstein, M. J., and Gainer, H. (2005). Selective gene expression in magnocellular neurons in rat supraoptic nucleus. *J. Neurosci.* **24,** 7174–7185.

Paul, P., and Apgar, J. (2005). Single-molecule dilution and multiple displacement amplification for molecular haplotyping. *Biotechniques* **38,** 553–554, 556, 558–559.

Raap, A. K., van der Burg, M. J., Knijnenburg, J., Meershoek, E., Rosenberg, C., Gray, J. W., Wiegant, J., Hodgson, J. G., and Tanke, H. J. (2004). Array comparative genomic hybridization with cyanin cis-platinum-labeled DNAs. *Biotechniques* **37,** 130–134.

Randolph, J. B., and Waggoner, A. S. (1997). Stability, specificity and fluorescence brightness of multiply-labeled fluorescent DNA probes. *Nucleic Acids Res.* **25,** 2923–2929.

Richter, A., Schwager, C., Hentze, S., Ansorge, W., Hentze, M. W., and Muckenthaler, M. (2002). Comparison of fluorescent tag DNA labeling methods used for expression analysis by DNA microarrays. *Biotechniques* **33,** 620–628, 630.

Saghizadeh, M., Brown, D. J., Tajbakhsh, J., Chen, Z., Kenney, M. C., Farber, D. B., and Nelson, S. F. (2003). Evaluation of techniques using amplified nucleic acid probes for gene expression profiling. *Biomol. Eng.* **20,** 97–106.

Schena, M., Shalon, D., Davis, R. W., and Brown, P. O. (1995). Quantitative monitoring of gene expression patterns with a complementary DNA microarray. *Science* **270,** 467–470.

Stears, R. L., Getts, R. C., and Gullans, S. R. (2000). A novel, sensitive detection system for high-density microarrays using dendrimer technology. *Physiol. Genom.* **3,** 93–99.

Tangrea, M. A., Chuaqui, R. F., Gillespie, J. W., Ahram, M., Gannot, G., Wallis, B. S., Best, C. J., Linehan, W. M., Liotta, L. A., Pohida, T. J., Bonner, R. F., and Emmert-Buck, M. R. (2004). Expression microdissection: Operator-independent retrieval of cells for molecular profiling. *Diagn. Mol. Pathol.* **13,** 207–212.

Telenius, H., Carter, N. P., Bebb, C. E., Nordenskjold, M., Ponder, B. A., and Tunnacliffe, A. (1992). Degenerate oligonucleotide-primed PCR: General amplification of target DNA by a single degenerate primer. *Genomics* **13,** 718–725.

Thach, D. C., Lin, B., Walter, E., Kruzelock, R., Rowley, R. K., Tibbetts, C., and Stenger, D. A. (2003). Assessment of two methods for handling blood in collection tubes with RNA stabilizing agent for surveillance of gene expression profiles with high density microarrays. *J. Immunol. Methods* **283**, 269–279.

van de Rijke, F. M., Heetebrij, R. J., Talman, EG., Tanke, H. J., and Raap, A. K. (2003). Fluorescence properties, thermal duplex stability, and kinetics of formation of cyanin platinum DNAs. *Anal. Biochem.* **321**, 71–78.

Van Gelder, R. N., von Zastrow, M. E., Yool, A., Dement, W. C., Barchas, J. D., and Eberwine, J. H. (1990). Amplified RNA synthesized from limited quantities of heterogeneous cDNA. *Proc. Natl. Acad. Sci. USA* **87**, 1663–1667.

Wang, E., Miller, L. D., Ohnmacht, G. A., Liu, E. T., and Marincola, F. M. (2000). High-fidelity mRNA amplification for gene profiling. *Nat. Biotechnol.* **18**, 457–459.

Wolber, P. K., Collins, P. J., Lucas, A. B., De Witte, A., and Shannon, K. W. (2006). The Agilent *in situ*-synthesized microarray platform. *Methods Enzymol.* **410** (this volume).

Xiang, C. C., Chen, M., Kozhich, O. A., Phan, Q. N., Inman, J. M., Chen, Y., and Brownstein, M. J. (2003a). Probe generation directly from small numbers of cells for DNA microarray studies. *Biotechniques* **34**, 386–388, 390, 392–393.

Xiang, C. C., Chen, M., Ma, L., Phan, Q. N., Inman, J. M., Kozhich, O. A., and Brownstein, M. J. (2003b). A new strategy to amplify degraded RNA from small tissue samples for microarray studies. *Nucleic Acids Res.* **31**, e53.

Xiang, C. C., Kozhich, O. A., Chen, M., Inman, J. M., Phan, Q. N., Chen, Y., and Brownstein, M. J. (2002). Amine-modified random primers to label probes for DNA microarrays. *Nat. Biotechnol.* **20**, 738–742.

Xiang, C. C., Mezey, E., Chen, M., Key, S., Ma, L., and Brownstein, M. J. (2004). Using DSP, a reversible cross-linker, to fix tissue sections for immunostaining, microdissection and expression profiling. *Nucleic Acids Res.* **32**, e185.Zhang, L., Cui, X., Schmitt, K., Hubert, R., Navidi, W., and Arnheim, N. (1992). Whole genome amplification from a single cell: Implications for genetic analysis. *Proc. Natl. Acad. Sci. USA* **89**, 5847–5851.

[12] Genomic DNA as a General Cohybridization Standard for Ratiometric Microarrays

By BRIAN A. WILLIAMS, RICHELE M. GWIRTZ, and BARBARA J. WOLD

Abstract

Feature variability on ratiometric microarrays is accommodated by simultaneous cohybridization of a labeled reference standard with a labeled experimental sample. An optimal reference standard would provide full and equal representation for all array features from a given genome so that it would function on any array, would represent all features with similar signal intensity, and would be highly reproducible—both technically and

METHODS IN ENZYMOLOGY, VOL. 410
Copyright 2006, Elsevier Inc. All rights reserved.

0076-6879/06 $35.00
DOI: 10.1016/S0076-6879(06)10012-9

biologically—from preparation to preparation and laboratory to laboratory. A low cost and a good shelf life are also highly desirable. Finally, providing for straightforward recovery of RNA prevalence information and for integration of data across multiple, initially unrelated studies would be significant advances over current methods. For virtually all ratiometric array studies published to date the reference standard has been some kind of RNA sample assembled from a number of different cell lines, tissues, or experimental time points. These RNA references fall short of the desired universality, uniformity, and reproducibility criteria, which then affect data quality and integration across studies. Also, the various mixed RNA standards cannot be used to derive RNA prevalence information from an experimental sample. In contrast, genomic DNA is a natural choice to meet all the criteria, although it has not yet been widely exploited for eukaryotic array experiments. Principal stumbling blocks have been achieving high enough absolute signals for large mammalian and plant genomes and finding a way to stabilize labeled DNA so that it can be stored and used with ease. This chapter describes two genomic DNA-labeling methods that make it possible to use genomic DNA as a universal microarray cohybridization standard. The indirect labeling method permits production of a large quantity of a stable genomic DNA standard that can then be quality tested and stored frozen. This optimizes experimental consistency and significantly improves ease of use. This chapter also shows that the genomic DNA reference standard can deliver RNA prevalence measurements from ratiometric array platforms.

Introduction

DNA microarrays made by mechanical deposition, inkjet, or microelectromechanical systems now comprise a major fraction of microarray studies because they offer great design flexibility, availability, and, sometimes, cost advantages over other platforms. However, the source of measurement inconsistencies between microarray experiments continues to receive close attention, resulting in a number of proposals for the standardization of experimental and data analysis techniques (Consortium, 2005; Irizarry et al., 2005; Larkin et al., 2005). Reliable quantification of RNA transcripts in microarray experiments must accommodate variation in array features within and between gene chips, as differences in probe density can affect the accuracy of measurement for a given labeled target (Peterson et al., 2001). This is done by making the measurements ratiometric (Eisen and Brown, 1999; Shalon et al., 1996). Thus, the pertinent measurement used for subsequent analysis is the ratio of the fluorescence signal for the experimental RNA sample ("numerator") relative to the fluorescence signal for the same feature from a reference RNA sample ("denominator"). These paired signals come from a simultaneous cohybridization reaction of samples with the microarray, in which each sample has been labeled with a different dye (usually Cy3 or Cy5) (Shalon *et al.*, 1996). Cohybridization of two differentially labeled RNA samples to the same array, as practiced in a "loop design," can reduce measurement variance and provides a method for statistically modeling the components of measurement variance (Churchill, 2002). For larger scale studies that require reproducible cohybridization standards with minimal lot-to-lot variation, "reference design" experiments are the norm. Reference design experiments cohybridize the same reference standard sample to all arrays compared; the ratio for each feature on an array is then computed and compared across all arrays in a study. In contrast, features on microarrays from Affymetrix are manufactured by a photolithographic process, which is sufficiently reproducible that they are commonly used without a cohybridization reference. As a result, Affymetrix RNA profiling has been based on "single color" hybridization of one RNA sample per array. However, even on Affymetrix arrays, it has been shown that two-color ratioing can deliver extra confidence and information or simply generate twice as much data in the form of two measurements per array (Hacia *et al.*, 1996, 1998).

Ratiometric standardization provides a conceptually elegant workaround for feature variation across and between arrays. It also creates problems in experimental design and data analysis that affect data quality, robustness, interpretation, and, importantly in our view, the ability to use data sets in larger scale combinations and meta-analyses. Much of the difficulty arises from characteristics of the widely used reference standards. Properties wanted in an ideal standard include (1) sequence comprehensivity for all genes or other elements on an array; (2) no biological or technical variability across preparations; (3) capacity to deliver RNA prevalence information; and (4) potential to combine data from diverse studies. Minimally, for comparison of multiple experimental results between laboratories or over long time frames, a viable cohybridization standard should be invariant in sequence content. It should also be sequence comprehensive so that every feature on the array, when reacted with the standard, produces a signal well within the linear range of detection of the laser scanner. Finally, it should be readily available so that it can be adopted easily. If not available commercially, it should be straightforward to prepare with minimal laboratory-to-laboratory variation and should be relatively inexpensive. These minimal criteria can be met by using genomic DNA, labeled in a manner that provides sufficient signal for application to large mammalian genomes. Beyond the minimal criteria, a reference standard that represents each gene at a uniform and known concentration should preserve RNA prevalence information in the experimental sample. Also, a standard that enables the integration of data from multiple unrelated studies could significantly elevate the value of data for meta-analysis. These additional functions of the genomic DNA standard are addressed in this chapter, along with basic labeling and hybridization protocols.

Current Cohybridization Standards: RNA

Since the advent of microarrays considerable experimental design energy has gone into selecting reference standards for major studies (Zhu *et al.*, 2004) and small experiments alike, and the choices have mainly been pragmatic ones that fall short of the criteria outlined earlier. One prominent strategy is to get broader and more even array feature coverage by combining RNA samples from multiple cell lines (Perou *et al.*, 2000; Ross *et al.*, 2000) or multiple experimental time points or tissue samples specific to a given study (Bergstrom *et al.*, 2002; Spellman *et al.*, 1998). Currently, a mixture of 11 cell lines with broad representation of known expressed sequences in mouse is available as the Universal Mouse Reference RNA (UMRR) standard (Stratagene). A similar Universal Human Reference RNA (UHRR) standard is also available (Stratagene). Inherent biological variability in the cell line starting material, and the fact that feature coverage is inevitably partial, sets limits on these approaches, although they are generally more comprehensive than a single source RNA reference.

DNA Standards

In contrast with the varied standards made from natural RNAs, there is a simple and rather compelling logic for using genomic DNA as a universal cohybridization standard for ratiometric arrays. Recent successes with microarray-based comparative genome hybridization (CGH) in which two human genomic DNA samples are cohybridized to cDNA microarrays suggest that genomic DNA signals could be made reliable enough to act as a universal denominator for expression measurements (Pollack *et al.*, 1999). Genomic DNA offers complete sequence representation, sequence stability over time without preparation-to-preparation variation, uniform prevalence for the vast majority of genes that are single copy, and low cost. In principle, it should apply universally to arrays made from standardized collections of oligonucleotides, home-made custom libraries of expressed sequences, micro-RNA sequences, splice isoform probes, or even intergenic DNAs for ChIP and transcript mapping experiments. The challenge for genomic DNA in most of these uses has been that in large genomes, most of the DNA mass is not exonic. This raised the questions of whether the genomic DNA signal can be made robust enough and whether background from the rest of the genomic DNA in the reaction would be low enough. It has been shown that the technical issues can be overcome, even for large mammalian genomes that present the greatest challenge in genome size and composition (Williams *et al.*, 2004).

Evaluating Cohybridization Standards

In order for the genomic DNA standard to be judged superior for routine use, it can be argued that it merely needs to perform as well as a mixed RNA reference in replicate experimental comparisons. The rationale is that its other merits—universality, sample-to-sample invariance, sequence comprehensivity, data integration, and low cost—make it intrinsically superior if the basic normalization performance over a complex microarray is comparable. We previously tested this experimentally and found that genomic DNA, labeled according to protocol #1 in this chapter, is comparable to the commercial standard Stratagene RNA mix in reproducibility, even for genes well represented in the RNA standard. It proved superior for many genes that are underrepresented in the RNA mix (Williams *et al.*, 2004). We also verified empirically the prediction that genomic DNA can be definitively superior to a "universal" RNA collection in array feature coverage. Substantial additional arguments in its favor are that it also provides significant information on RNA prevalence and makes integration across disparate studies possible.

Recovering Information on RNA Prevalence

In most ratiometric array studies, all information about RNA prevalence of one gene relative to a different gene, either within a sample or across samples, is eliminated from the higher order analysis. This follows from the fact that ratios—not absolute fluorescence intensities from which ratios are computed—are used in subsequent analyses, and the reason for this is the previously discussed feature variation. In principle, however, it should be possible to compute RNA prevalence information *if* one knew the absolute concentration of each RNA transcript within an RNA hybridization standard mix. Of course, the experimenter does not know the abundance of most RNAs in the reference sample (in fact, one rarely knows the prevalence of *any* RNA transcript in a reference standard). In practical terms, this means that two genes (array features) having precisely the same experimental RNA/reference RNA ratio (say 1.0) can differ from each other by two or three orders of magnitude in absolute signal intensities (in both channels). These intensity differences reflect similar degrees of difference in the prevalence of their respective transcripts in an experimental sample, but they still suffer from the effects of feature variability. As discussed in this chapter, a genomic DNA standard differs importantly from RNA standards on this point because the concentration of each sequence *is* known and it is generally uniform. This provides us with a natural way to recover prevalence information that also normalizes out feature variation. Thus, a high ratio of RNA to the genomic DNA reference reflects a high abundance of that transcript in the experimental RNA sample. A low ratio should similarly reflect low RNA abundance. Independent verification of abundance relationships can be provided by quantitative RT-PCR. We show a strong correlation between array-based abundance estimates and qRT-PCR estimates. One can therefore use genomic DNA to capture major trends in absolute RNA prevalence on global arrays that are lost from RNA normalized array designs.

Shelf Life and Quality Control for Genomic DNA Standards

A significant remaining issue for a genomic DNA standard relative to other current reference standards was experimental workload and shelf life. This is addressed later with a new indirect labeling protocol for genomic DNA (protocol #2). We also give the direct labeling protocol characterized in Williams *et al.* (2004). It had its origin in the original comparative genome hybridization protocol (Pollack *et al.*, 1999), which in our hands fell well short of providing enough signal in the context of a 70-mer oligonucleotide format microarray. The resulting modified version successfully and reproducibly boosted the specific signal intensity (the signal-to-noise ratio) and delivered comprehensive coverage in the oligonucleotide array context. However, the direct labeling protocol must be performed on the same day as cDNA labeling, followed by its immediate use. This limits the number of arrays that can be processed on the same day by a single user and also limits the amount and sophistication of quality control one can do before committing to use it on many arrays and RNA samples.

These complications were addressed by developing an indirect genomic DNA-labeling protocol (protocol #2). Genomic DNA is labeled with aminoallyl-dUTP and stored as a dried product at $-80°$ until ready for use. Once the standard is labeled with aminoallyl-dUTP and frozen, it can be coupled and cleaned up quickly, with minimal effort, while performing labeling reactions with experimental RNA samples. This means that a stock of genomic DNA standard can be prepared well in advance for a large number of experiments, and a small aliquot can be tested for label incorporation and yield or even taken through an entire microarray hybridization test prior to committing precious RNA samples and many arrays. In our own work, this is now the protocol of choice. Modifications to either of the protocols will be posted (http://woldlab.caltech.edu). This chapter also includes updated data analysis that is relevant to ratiometric array use in general and to evaluating how well a given reference standard performs.

Method

Increased Signal Strength Requires an Increase in Both Reaction Yield and Label Incorporation

Initial experiments to directly adapt the original CGH genomic DNA labeling protocol (Pollack *et al.*, 1999) for use on 70-mer microarrays produced low feature intensities on too large a fraction of a complex expression array (~16,000 features) to be useful as a hybridization standard. Low denominator intensity measurements naturally increase the variance of ratiometric measurements and, if low enough, cease to serve as a ratiometric standard and effectively turn a feature into a single channel detector without the benefit of normalization. A survey of reaction components and hybridization conditions was undertaken with the objective of improving specific signal strength on the entire array. When both label incorporation *and* overall reaction yield were increased, the protocol became suitable for routine use on large-scale 70-mer mouse expression arrays.

Genomic DNA Preparation

Label incorporation and yield of target are strongly influenced by the source of genomic DNA used as the template. In our hands, genomic DNA from several commercial suppliers gave inefficient labeling compared with companion reactions using DNA prepared directly from mouse tissue. This method allows specification of the strain, genotype, and sex of the genomic DNA standard, thereby allowing the evaluation of qualitative differences in Y-linked genes and quantitative differences in X-linked genes.

We chose the EpiCentre MasterPure DNA purification kit for in-laboratory production of mouse genomic DNA and supplemented it with three additional RNases available from EpiCentre. Mouse kidney is prepared for DNA extraction by freezing immediately after dissection in liquid nitrogen, grinding in a mortar and pestle under liquid nitrogen, and aliquoting in cryo vials stored at $-80°$. Using this protocol, careful attention should be paid to the appropriate volume of lysis buffer relative to the amount of tissue being extracted, using the manufacturer's directions for guidance. A small aliquot of the crude preparation from the kit is set aside for agarose gel electrophoresis to visualize the intact genomic DNA and any residual RNA remaining after the RNase A treatment in the kit (Fig. 1). We then use EpiCentre's Riboshredder as a potent, broad-spectrum RNase to target any residual RNA, which, if later labeled, would introduce preparation-to-preparation variation. Next, we add RNase I to degrade

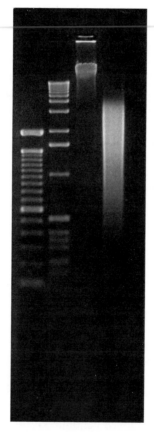

FIG. 1. Agarose gel electrophoresis of sonicated genomic DNA fragments used as a template in the labeling reaction. Lane 1, InVitrogen 100-bp ladder; lane 2, InVitrogen 1-kb ladder; and lane 3, crude preparation of genomic DNA prior to triple RNase digestion and column cleanup; material in the well is likely residual protein. There is a faint smear of material in the range of 100 to 200 bp (not visible in this exposure), which disappears after RNase treatment. Lane 4, final template, after sonication and column cleanup.

any remaining single-stranded RNAs and then RNase H to remove any remaining RNA in heteroduplex form with the genomic DNA.

Size fractionation of genomic DNA is accomplished by random shearing with a sonicator to an average size of 1–2 kb with a maximum of approximately 4 kb (Fig. 1). Our labeling and yield measurements are consistently higher for this size fragment when compared to shorter sheared products or DNA digested with *Dpn*II. Likewise, the signal intensity on array features after hybridization is consistently greater for longer sheared fragments than for shorter DNA, just as one would predict. Published results suggest that shorter fragments may increase specific signal strength by increasing target mobility (Dai *et al.*, 2002), but over the size ranges we surveyed, longer DNA gave a stronger signal.

The following protocol produces ~90 μg of DNA from 70 mg of mouse kidney tissue, which is enough for 36, 2.5-μg aliquots or 12 microarray experiments. Keep in mind that if only male mouse kidney is chosen, X-linked genes will be represented in the standard at half the abundance of autosomal genes.

Equipment and Reagents

- Liquid nitrogen
- Mortar and pestle for grinding
- Oven or heating block set at 65° that can accommodate a 14-ml centrifuge tube
- Eppendorf 5417C microcentrifuge with variable rpm/rcf adjustment at room temperature (any good variable speed laboratory microcentrifuge should work; make sure that it is capable of speeds greater than 10,000g)
- Microcentrifuge with variable rpm/rcf adjustment at 4°
- Laboratory balance
- UV spectrophotometer
- Microson XL 2007 sonicator with small tip to fit inside microcentrifuge tubes. (Note: any good sonicator with a microtip will work. You will have to experiment with the power and duty settings to achieve fragmentation of the DNA to achieve a smear topping out at about 3–4 kb, with an average size of about 2 kb, as seen on a 1% agarose gel. See Fig. 1 for an example.)
- Zymo 25-μg capacity DNA clean and concentrator columns
- Distilled H_2O adjusted to pH 8.0 with NaOH
- Equipment and reagents for standard agarose gel electrophoresis
- Microcentrifuge tubes, gloves, aerosol barrier pipette tips, dry ice, wet ice, ethanol, isopropanol

- 1 *M* NaCl solution
- 100 m*M* MgCl$_2$ solution
- 1 *M* Tris–HCl, pH 7.5, solution
- EpiCentre MasterPure complete DNA purification kit
- EpiCentre Riboshredder
- EpiCentre RNase I
- EpiCentre RNase H
- ISC BioExpress GeneMate 2-ml microcentrifuge tubes.

Freezing and Pulverization

1. Dissect mouse kidneys quickly and flash freeze in liquid nitrogen.
2. It is important to get a reasonably accurate weight of the tissue aliquots. Before preparing to grind the kidneys under liquid nitrogen, prepare about 30 tubes to hold tissue aliquots. Weigh the tubes prior to filling and mark the empty tube weight on each tube. Place the marked tubes in ground dry ice to cool.
3. Grind the kidneys in liquid nitrogen in a mortar and pestle that has been prechilled using liquid nitrogen.
4. Aliquot ground tissue at approximately 20 mg per tube. Keep on dry ice. (Caution: Do not close tube immediately, as pressure can build up due to gas release from the liquid nitrogen and cause the tube to pop.)
5. Reweigh the tubes cold to get an accurate estimate of tissue weight, but do not bother trying to readjust individual tubes by redistributing ground tissue. Keep the dry ice near when weighing to avoid freeze/thaw. Mark the filled tube weight on each tube.
6. Store aliquots separately at −80°.

Tissue Lysis

It is important to maintain the correct proportions of T&C lysis buffer and proteinase K (MasterPure kit) to the estimated weight of your tissue sample. If this is not done, you will have difficulty achieving complete lysis, which will affect your yield of genomic DNA.

1. When tissue samples are brought from the freezer, add appropriate amounts of T&C lysis buffer to each tube quickly (300 μl of T&C lysis buffer per 5 mg tissue, plus 1 μl of proteinase K per 5 mg tissue). We process 14–15, 5-mg samples at a time; therefore select aliquot tubes from the freezer that will total ~70 to 75 mg of tissue mass.

2. Incubate the samples separately for 5 min at 65° and then combine in a single 14-ml centrifuge tube.

3. Continue the incubation for another 10 min at 65°, vortexing at 5 min and again at 10 min. If lysis is not complete at 15 min, continue incubation for an additional 5 min with an additional vortexing.

4. Incubate at 37° for 5 min.

5. Add 1 μl of RNase A (MasterPure kit) per 5 mg of tissue.

6. Incubate at 37° for 30 min, vortexing once at 15 min and, more importantly, at the end of the 30-min incubation (prior to splitting). While waiting, prepare 14 separate fresh tubes and place in a 4° ice bath.

Protein Precipitation

7. Split sample into the 14 chilled tubes (~300 μl per tube). Chill on wet ice for 4 min.

8. Add 150 μl MPC protein precipitation reagent (MasterPure kit) to each tube.

9. Vortex 10 s.

10. Spin at 10,000g for 10 min at room temperature.

Genomic DNA Precipitation and Resuspension

11. Remove the supernatant to 14 *fresh* tubes.

12. Add 500 μl of isopropanol (room temperature) to each tube.

13. Invert tubes 40 times, *do not vortex.*

14. Spin at 12,000g for 10 min at 4°.

15. Aspirate and discard the supernatant.

16. Rinse pellets twice with about 300 μl 70% ethanol (vortex briefly and spin down each time).

17. Remove trace remaining ethanol with a fine tip and let the samples air dry on the benchtop for about 30 min (may take longer).

18. Resuspend pellet in each tube in 35 μl of TE (MasterPure kit).

19. Incubate at 37° for 20 min.

20. Resuspend each sample by titration with a pipette tip and then combine all 14 samples into a single tube and count 2 μl on the spectrophotometer. A typical estimated yield at this point is 20–25 μg per 5-mg sample. The 260/280 ratio will be about 1.4–1.6.

If you "park" the preparation here, approximately 4.5 h has elapsed since beginning the extraction. The sample volume is now 490 μl. The preparation can be stored at −20° overnight or at 4° if you want to avoid freeze/thaw.

RNase Digestion

   1. Add 8 μl of RiboShredder to the sample.

   2. Incubate at 37° for 30 min.

   3. Add (4.3 μl per sample) × (14 samples) = 60.2 μl ($\sim$60 μl) of 1 M NaCl.

   4. Add (2 μl per sample) × (14 samples) = 28 μl of 1:10 dilution of RNase I in dilution buffer (for 1:10 dilution, use 3 μl stock to 27 μl dilution buffer).

   5. Incubate at 37° for 30 min.

   6. Add (5 μl per sample) × (14 samples) = 70 μl 100 mM MgCl$_2$ (10 mM final).

   7. Add (2.5 μl per sample) × (14 samples) = 35 μl 1 M Tris–HCl, pH 7.5 (40 mM final).

   8. Add 8 μl of RNase H.

   9. Incubate at 37° for 30 min.

  10. Reserve 5 μl for electrophoresis here.

  11. Final volume is now $\sim$700 μl.

Sonication

  12. Split the preparation into three aliquots of 233 μl ($\sim$230 μl) each into 2-ml microcentrifuge tubes (ISC BioExpress).

  13. Place each 2-ml tube in −20° ethanol while sonicating at setting 18 on a Microson XL 2007 sonicator for 50 s.

  14. Recombine the sonicated aliquots ($\sim$700 μl total) in a 4-ml Falcon tube.

Column Cleanup

  15. The entire preparation will be cleaned up over four Zymo columns (25 μg capacity for each column).

  16. Add 2 volumes (1400 μl) of Zymo DNA-binding buffer.

  17. Vortex lightly and spin down briefly to bring liquid down from sides of tube.

  18. Split the preparation into four samples of 525 μl each.

  19. Pass the entire volume (525 μl) of each sample over each column twice in order to bind as much DNA as possible to the column. However, because the column load volume is only 400 μl, this will require four spins.

  20. Add the first volume of each sample to the Zymo column.

  21. Set the centrifuge speed to maximum and watch the rcf counter. When it hits 10,000g, begin timing for 10 s and then stop the centrifuge.

22. Repeat with same volume (the "flow through") and then discard.

23. Bind the second volume (the remainder of each sample) with two spins as described earlier.

24. Wash with 200 μl of wash buffer. Time the spin as described in step 21.

25. Repeat the wash, but this time spin for 30 s after 10,000g is reached and then stop the centrifuge.

To elute the DNA

26. Add 35 μl ddH$_2$O that has been adjusted to pH 8.0 (this is an important step).

27. Incubate for 5 min at room temperature.

28. Spin for 30 s total elapsed time with the centrifuge set at maximal speed. Do not wait until the centrifuge hits maximum speed and then start timing.

29. Repeat elution as described in step 26.

30. Count 3 μl in the UV spectrophotometer. A typical yield is 1.3 μg genomic DNA per milligram tissue input. The 260/280 should be 1.8–1.9.

31. A 2-μg aliquot run on a 1% agarose gel should show a smear topping out at about 3–4 kb, with an average size of about 2 kb (see Fig. 1).

32. Store in 2.5-μg aliquots at $-80°$.

Labeling Method I. Direct Incorporation of Cy-Labeled dCTP

The following is a detailed version of a previously published protocol for direct labeling of a genomic DNA template (Williams *et al.*, 2004). This protocol provides a strong signal and very low local background fluorescence. It is derived from the original protocol for comparative genome hybridization (Pollack *et al.*, 1999), which can be found in detail on http:// cmgm.stanford.edu/pbrown/protocols/4_genomic.html. One disadvantage of the direct labeling protocol is that the label needs to be applied to the arrays on the same day it is produced; this lengthens the workday to about 14 h and limits the number of arrays that can be processed by a single person in a single experiment. We have not tried to freeze samples for overnight storage after labeling and prior to hybridization. For a large number of samples or to introduce a break in the processing protocol, the indirect labeling method (see later) is recommended. An additional disadvantage to the direct labeling method is the cost. We estimate that the indirect labeling method cuts the cost of labeling in half.

Our experiments are currently performed on a 32 sector array, under a 22 × 40-mm glass lifter slip in a hybridization volume of 30 μl. To effectively label a single array with this standard, three aliquots (2 μg each) of sonicated mouse genomic DNA should be labeled as described later. These can then be combined for cohybridization with a single labeled cDNA sample.

The concentrated preparation of Klenow enzyme (40 U/μl) that is provided in the BioPrime labeling kit or the RadPrime labeling kit (In Vitrogen) is of primary importance for achieving a maximal reaction yield and incorporation of label. Other less concentrated preparations of Klenow tested did not produce the yield and incorporation required for strong specific signal on hybridized arrays. The labeling reaction runs for a total of 5 h, with a midincubation respike of Klenow and pyrophosphatase at 2.5 h. The reaction is incubated in a 37° hybridization oven, which produces a better yield and incorporation than incubation in a thermocycler with a heated lid. Use of Cy3 dCTP instead of Cy3 dUTP also improves reaction yield and incorporation. After testing dNTPs from five different manufacturers, we obtained consistently better results with Roche dNTPs in the lithium salt formulation. Call the Roche tech service and have them deliver the most recent lot with the longest projected shelf life.

Equipment and Reagents

Denaturation and Labeling

- Thermal cycler
- Incubation oven capable of holding steady 37°
- Invitrogen BioPrime DNA-labeling system
- Amersham Pharmacia Cy3 dCTP
- Roche Applied Science dNTP set (Note: These dNTPs work well, but call Roche to ask for the most recent lot with the most distant expiration date. Even at −80°, performance declines as time passes)

Prior to labeling reactions, the following reagents are stored in single-use aliquots at −80°.

- 20-μl aliquots of 2.5× random primer solution from BioPrime DNA labeling system
- 3-μl aliquots of 1 mM Cy3 dCTP
- 2 μg sonicated genomic DNA (see mouse genomic DNA preparation protocol)
- 10× dNTP solution for labeling: 50 μl of 100 mM dATP, 25 μl of 100 mM dCTP, 50 μl of 100 mM dGTP, 50 μl of 100 mM dTTP, and 2325 μl of ddH$_2$O. Final volume = 2500 μl (final concentrations in

the reaction are 200 μM dATP, dGTP, and dTTP, 100 μM unlabeled dCTP, and 60 μM Cy3 dCTP. Ratio of unlabeled dCTP:labeled dCTP is 5:3)

Store in single-use 20-μl aliquots at $-80°$.

Cleanup and Probe Preparation

- Eppendorf 5417C microcentrifuge with variable rpm/rcf adjustment at room temperature (any good variable speed laboratory microcentrifuge should work; make sure that it is capable of speeds greater than 10,000g)
- Qiagen Qiaquick PCR cleanup columns
- Speed-Vac
- Heating block at 85°
- Hybridization solution [50% formamide (Sigma F-9037), 5× SSC, 0.1% SDS, 1 μg/μl yeast tRNA]
- Lifter slips (Erie Scientific)

Miscellaneous

- Thin-walled PCR tubes, standard microcentrifuge tubes, and wet ice

Labeling

1. In a thin-walled PCR tube, combine 2 μg aliquot of genomic DNA (reduce the volume to 10 μl or less in a Speed-Vac) and 20 μl of random octamers in buffer (BioPrime kit; Invitrogen).

2. Heat to 97° for 3 min and 50 s in a ThermoCycler.

3. Plunge into an ice water bath (adding salt to the ice brings the temperature to $-5°$) for 3 min.

4. Spin down in centrifuge and place on wet ice.

5. Add on ice 3 μl of Cy3 dCTP 1 mM (Amersham), 5 μl of 10× dNTPs (10× concentration is 2 mM for dATP, dGTP, and dTTP; 1 mM dCTP), and X μl of chilled ddH$_2$O to bring the reaction to a volume of 49 μl. (Note: Cye dye is photosensitive; attempt to protect from light when possible, especially when heating.)

6. Mix the reagents and move to the benchtop for 2 min at room temperature.

7. Add 1 μl Klenow enzyme (BioPrime kit, high activity concentration). Mix gently with a pipette.

8. Let sit for 5 min at room temperature.

9. Place in a 37° oven for 2.5 h (Ambion, 2004).

10. Respike with 1 μl Klenow after 2.5 h. Mix gently with a pipette.

11. Place in a 37° oven for 2.5 h.

12. Stop the reaction with 5 μl stop solution (BioPrime kit).

Cleanup

13. Cleanup according to the protocol in the Qiagen QiaQuick PCR cleanup kit.

14. Pass the labeled sample over the column twice to bind as much labeled DNA as possible before washing. Spin at 10,000g for 1 min each time.

15. Wash the column at least twice with 750 μl of PE buffer. Spin at 10,000g for 1 min each.

16. Be sure to dry the column thoroughly by spinning at top speed for 2 min after the last wash.

17. Elute with 50 μl of elution buffer, incubating for 5 min on the benchtop. Spin at top speed for 1 min.

18. Repeat with another 50 μl of elution buffer.

19. Reserve 5 μl of labeled sample for counting using PicoGreen. This can be stored at −80°.

Quantify yield and incorporation efficiency using the fluorometry protocol (see later).

Prepare the Labeled Target for the Array

20. Combine three genomic DNA labeling reactions with a single cDNA labeling reaction.

21. Dry the combined labeled samples down in a Speed-Vac until just dry. Make an effort to avoid overdrying.

22. Resuspend the labeled samples in an appropriate volume of hybridization solution. We use 30 μl. Use the hybridization solution to wash down the sides of the tube to bring all of the dried product down into solution. Mix thoroughly with a pipette to disrupt all crystalline pieces.

23. Heat in the dark to 85° for 2 min.

24. Immediately spin down for 1 min in the dark to cool the probe.

25. Place the probe at hybridization temperature (46°) until ready to apply to your array.

26. Setup lifter slip on top of preblocked array and put this assembly on top of a heated block set at 46°. Do only one at a time.

27. Spin down labeled samples at top speed for 30 s immediately prior to loading the probe onto the array.

28. Load probe onto the array using a P200 pipette with slow, steady hand pressure. Try to avoid bubbles.

Fluorometric Quantification of Reaction Yield and Label Incorporation

We use a fluorometric assay for yield and incorporation that is fast, inexpensive, and consumes only a small portion of labeled product. The assay is performed in disposable plastic cuvettes in a Bio-Rad fluorometer, with custom-made filters (Chroma) for the Cy3 and PicoGreen (Molecular Probes) reagents. This method offers several advantages. A small sample of the labeling reaction (5%) is read and then discarded, avoiding the risk of cross contamination with the larger sample sizes necessary for reading on a spectrophotometer. If a strict standardization of the amount of labeled target applied to the array is required, the assay can be performed after reaction cleanup while the remaining target volume is being reduced in the Speed-Vac. Alternatively, if post hoc evaluation is sufficient, the small samples can be frozen for convenience.

Equipment and Reagents

- Bio-Rad VersaFluor Fluorometer
- Bio-Rad excitation filter for PicoGreen
- Bio-Rad emission filter for PicoGreen
- Chroma (www.chroma.com) excitation filter for Cy3 (535 nm)
- Chroma (www.chroma.com) emission filter for Cy3 (565 nm)

Quantifying Yield and Incorporation Efficiency

1. Prepare the Cy3 dUTP fluorometric standards in Bio-Rad cuvettes.

Cy3 Standards (2 ml Final Volume Each)

- 0 ng of Cy3 dUTP, 400 ng of λ DNA in 2 ml TE
- 50 ng of Cy3 dUTP, 400 ng of λ DNA in 2 ml TE
- 5 ng of Cy3 dUTP, 400 ng of λ DNA, in 2 ml TE
- 500 pg of Cy3 dUTP, 400 ng of λ DNA, in 2 ml TE
- 50 pg of Cy3 dUTP, 400 ng of λ DNA, in 2 ml TE

2. Add each 5-μl sample of labeled genomic DNA (the unknown) to 2 ml of TE in a Bio-Rad cuvette. Mix with a pipette.

3. Count the Cy3 standards and unknowns according to the directions for the Bio-Rad VersaFluor fluorometer. Be sure to insert the appropriate

Cy3 excitation and emission filters. Use the low gain setting. Zero the instrument with Cy3 standard A, and set the range of 19,999 rfu using Cy3 standard B. Continue reading the rest of the standards (which should decrease linearly from 19,999) and then read the (unknown) samples. [Note: Standard E (50 pg Cy3) is close to the limits of the machine at low gain and may be erratic. Also, the Cy3 dye will bleach if read multiple times or left in the machine for too long.]

Note: Do not add the PicoGreen fluorescent DNA-binding reagent until after the Cy3 measurements have been taken. We have found some spectral overlap between the two fluors.

4. Prepare the PicoGreen/DNA fluorometric standards in Bio-Rad cuvettes.

DNA Standards (2 ml Final Volume Each)

- 0 ng λ DNA, 1 ml TE, 1 ml 1:200 PicoGreen reagent in TE
- 400 ng λ DNA in 1 ml TE, 1 ml 1:200 PicoGreen reagent in TE
- 40 ng λ DNA in 1 ml TE, 1 ml 1:200 PicoGreen reagent in TE
- 4 ng λ DNA in 1 ml TE, 1 ml 1:200 PicoGreen reagent in TE
- 400 pg λ DNA in 1 ml TE, 1 ml 1:200 PicoGreen reagent in TE

Note: Add PicoGreen to standards and unknowns simultaneously.

5. Add the 1 ml of 1:200 diluted Picogreen reagent in TE to each of the standards and mix well.

6. Add 5 μl PicoGreen dsDNA quantitation reagent (undiluted) to each of the unknowns and mix well.

7. Incubate for 10 min in the dark at room temperature.

8. Count the standards and unknowns as before using PicoGreen excitation and emission filters, the low gain setting, DNA standard A to zero the fluorometer, and DNA standard B to set the range to 19,999.

A good yield will have rfu for Cy3 between 3000 and 5000, and rfu for PicoGreen between 8000 and 10,000.

Labeling Method II. Indirect Labeling via Incorporation of
    Amino-allyl dUTP

The indirect labeling method provides convenience, experimental consistency, and risk control, as large quantities of template can be labeled and stored at $-80°$ prior to coupling to the fluor. A small aliquot can be tested for labeling efficiency independent of when the cohybridized cDNAs are prepared. As with the direct labeling method, we have found that labeled product from three genomic DNA reactions must be combined to produce sufficient signal on a single array. When the cDNA samples are prepared, the labeled genomic DNA is quickly coupled to fluor and cleaned up. To ensure uniformity of application across all arrays, we pool our labeled genomic DNA samples and split them evenly for combination with the appropriate cDNA samples. The reactions are still performed in individual 50-μl volumes; we have not attempted to scale up the reaction volume.

Heating the genomic DNA template to 100° can cause an increase in nonspecific background signal on DNA filter arrays (Amon and Ivanov, 2003). We have replaced the standard boiling denaturation step with alkali denaturation of sonicated, double-stranded genomic DNA using NaOH, followed by neutralization, cleanup, and a final milder denaturation at 70° prior to annealing to random octamers. We now have our random octamers synthesized (Operon) with an activated amine group on the 5′ end, which can also be coupled chemically to the Cy-dye ester in the same way that incorporated amino-allyl groups are coupled. We recommend having synthesized octamers HPLC purified if this option is chosen.

First run experiments using recommended reagent concentrations (Kim et al., 2002) produced reaction yields far lower than with the direct labeling protocol. We performed a series of labeling reactions at different total nucleotide concentrations with different ratios of labeled to unlabeled nucleotides (Fig. 2). A high ratio (4:1) of labeled to unlabeled nucleotide incorporated the most fluor, but reduced the yield almost fourfold compared to the other ratios and eightfold compared to the yield normally obtained in the direct labeling reaction at 200 μM dNTPs. Ratios of labeled to unlabeled nucleotides of 3:5 or 2:3 improved the yield to approximately half of what we were achieving with the direct labeling protocol. By increasing the total dNTP concentration to 900 μM, at the 2:3 ratio, we were able to produce a yield of labeled product that is comparable to a direct labeling reaction, with an almost 30% increase in the mass of label incorporated.

We were concerned that the elevated concentration of dNTPs in the reaction might not be removed efficiently with a single cleanup column and could confound our estimates of cDNA yield. After testing the columns for dNTP retention at the elevated concentration, we improved our cleanup method by including three column cleanups prior to fluorometric quantification. The initial incorporation reaction is cleaned up over a Sepharose G50 column to remove excess dNTPs. The product is then dried down and stored at −80° until coupling to the fluor. After

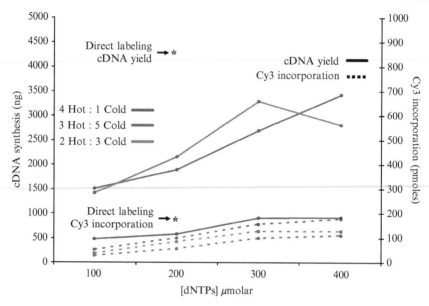

FIG. 2. Summary of indirect labeling experiments. Solid lines are plots of cDNA yield, calibrated to the *y* axis on the left-hand side; dashed lines are plots of mass of Cy3 incorporated in the reaction product calibrated to the *y* axis on the right-hand side. The *x* axis shows four different concentrations of total dNTPs that were tried. Further experiments (data not shown) extended the range of dNTPs to 1 m*M*. Different proportions of amino-allyl-labeled dUTP (hot) to unlabeled dTTP (cold) are indicated in the color legend. Asterisks indicate typical values for yield and incorporation obtained in direct labeling experiments. (See color insert.)

chemical coupling, the reaction is further cleaned of residual proteins and most of the excess fluor on a Qiagen PCR cleanup column. Any additional excess fluor is removed in a third P-30 BioGel chromatography column (Bio-Rad).

Equipment and Reagents

- Incubation oven capable of holding steady 37°
- Eppendorf 5417C microcentrifuge with variable rpm/rcf adjustment at room temperature (any good variable speed laboratory microcentrifuge should work; make sure that it is capable of speeds greater than 10,000*g*).
- Speed-Vac

- Heating block at 70 and 85°
- Zymo DNA clean and concentrator columns
- Invitrogen BioPrime DNA labeling system
- Sigma aminoallyl dUTP
- Roche Applied Science dNTP set. Note: these dNTPs work well, but call Roche and ask for the most recent lot with the most distant expiration date. Performance declines as time passes, even when stored at −80°.
- Roche Applied Science pyrophosphatase
- 100 mM sodium acetate, pH 5.2
- 100 mM Na_2CO_3, pH 9.0
- 4 M hydroxylamine solution
- USA Scientific G-50 Sephadex cleanup columns
- Qiagen Qiaquick PCR cleanup columns
- Bio-Rad P30 BioGel cleanup columns
- Bio-Rad cuvettes
- Thin-walled PCR tubes, standard microcentrifuge tubes, and wet ice
- Hybridization solution (50% formamide, 5× SSC, 0.1% SDS, 1 μg/μl yeast tRNA)
- Phosphate elution buffer (http://pga.tigr.org/sop/M004_1a.pdf): dilute 1 M KPO_4 solution to 4 mM with double distilled water
- To make 1 M KPO_4 solution, pH 8.5–8.7, combine 9.5 ml 1 M K_2HPO_4 and 0.5 ml 1 M KH_2PO_4

Before beginning labeling protocol, be sure that the following reagents are available as single-use aliquots stored at −80°.

- 2.5-μg aliquots of sonicated mouse genomic DNA
- 20-μl aliquots of 2.5× random primer solution from BioPrime DNA labeling system or 2.5× solution of random octamers with activated 5' amino group at 750 μg/ml in 125 mM Tris–HCl, pH 6.8, 12.5 mM $MgCl_2$, and 25 mM β-mercaptoethanol
- 4.5-μl aliquots of monofunctional Cy-dye ester in dimethyl sulfoxide (DMSO)

Protocol for Resuspension of Dry Cy Dye in DMSO (http://pga.tigr.org/sop/M004_1a.pdf)

1. Add 73 μl DMSO (DMSO is hygroscopic, keep in desiccator) to one tube (1 mg) monofunctional reactive dye.
2. Mix thoroughly and spin down.
3. Aliquot at 4.5 μl per tube.
4. Store in desiccator at −80°.

10× Aminoallyl-dUTP/dNTPS Labeling Mix

1. Resuspend 1 mg of aminoallyl-dUTP as a 20 mM stock solution (http://cmgm.stanford.edu/pbrown/protocols/amino-allyl.htm).
2. Combine 85 μl of ddH$_2$O and 0.8 μl of 0.1 N NaOH.
3. Add this to 1 mg of aminoallyl-dUTP.
4. Draw up 5 μl to check pH on a pH strip. pH should be approximately 7.0.
5. Prepare a 9 mM 10X stock solution of dNTPs for labeling: 40 μl of 100 mM dATP,40 μl of 100 mM dCTP, 40 μl of 100 mM dGTP, 80 μl of 20 mM AAdUTP, 24 μl of 100 mM dTTP, and 220 μl of ddH$_2$O. The final volume is 444 μl.

Final concentrations in the reaction are 900 μM dATP, dCTP, and dGTP, 360 μM AAdUTP, and 540 μM unlabeled dTTP. Ratio of (labeled AAdUTP:unlabeled dTTP) is 2:3.

6. Store in single-use aliquots at −80°.

Note: Optimization experiments for yield and incorporation indicated that Klenow enzyme works best at a 900 μM concentration with a 2:3 ratio of labeled to unlabeled precursor.

Denature Sonicated Genomic DNA Template with NaOH (Amon and Ivanov, 2003)

1. Combine 10 μl (2.5-μg aliquots) sonicated genomic DNA, 30 μl ddH$_2$O, and 10 μl 1 N NaOH. The final volume is 50 μl.
2. Incubate at room temperature for 20 min.
3. Incubate in wet ice/NaCl bath for 3 min.
4. Spin down briefly to collect evaporation. Keep on wet ice.
5. Neutralize with 10 μl 1 N HCl. This is an important step.

Cleanup over Zymo Columns (either 5 or 25 μg Capacity)

6. Add a volume of Zymo DNA-binding buffer that is equivalent to twice the sample volume to the sample. Mix lightly and spin down.
7. Bind the sample to the Zymo column. Pass the entire volume of the sample over the column twice in order to bind as much DNA as possible to the column. If the sample volume is greater than 400 μl, load the column serially, running the first portion of the sample over the column twice and

then running the second volume or remainder portion of the sample over the column twice.

8. Add the sample to the column (or the first portion of the sample, if volume is greater than 400 μl).

9. Set the centrifuge speed to maximum, and watch the rcf counter. When it hits 10,000g, begin timing for 10 s. Then stop the centrifuge.

10. Repeat with same volume (the "flow through") and then discard.

11. Bind the second volume with two spins as described earlier (if volume is greater than 400 μl).

12. Wash with 200 μl of wash buffer. Time the spin as described earlier.

13. Repeat the wash, but this time spin for 30 s after 10,000g is reached in order to dry the column and then stop the centrifuge.

14. Elute the DNA from the column.

15. Add an appropriate volume of ddH$_2$O that has been adjusted to pH 8.0 (this is an important step) to the column to give an eluate concentration approximately 200 ng/μl.

16. Incubate for 5 min at room temperature.

17. Spin for 30 s total elapsed time with the centrifuge set at maximal speed. Do not wait until the centrifuge hits maximal speed and then start timing.

18. Repeat elution as described in step 11.

19. Place on wet ice.

20. Recount on UV spectrophotometer OD of 1 for single-stranded if DNA is 40 μg/ml.

21. Realiquot approximately 2 μg per reaction.

Annealing to Random Primers

22. Combine 2 μg denatured sonicated mouse genomic DNA (approximately 10 μl) and 20 μl of 2.5× random primers.

23. Heat to 70° for 10 min.

24. Plunge into wet ice/NaCl bath for 3 min.

25. Spin down and place on wet ice.

Labeling

26. Add to primed genomic DNA on ice: 2 μl of 25 mM MgCl$_2$, 5 μl of 10× AAdUTP/dNTPs labeling mixture, 1 μl of pyrophosphatase diluted 1:10 in ddH$_2$O, and X μl of chilled ddH$_2$O to bring the reaction to a volume of 49 μl.

27. Mix and incubate on the benchtop for 2 min at room temperature.

28. Add 1 μl Klenow enzyme (BioPrime kit, high activity concentration). Mix gently with pipette.

29. Incubate 5 min at room temperature.

30. Place in a 37° oven for 2.5 h (Ambion, 2004, TechNotes).

31. Respike with 1 μl Klenow and 1 μl of freshly diluted pyrophosphatase after 2.5 h. Mix gently with pipette.

32. Place in a 37° oven for another 2.5 h.

33. Stop the reaction by adding 5 μl of stop solution (BioPrime kit).

Cleanup

Gel filtration chromatography columns are used to remove excess unincorporated nucleotides from the aminoallyl-dUTP labeling reaction. We are working with a high concentration of free nucleotides. This step is necessary to prevent column failure in the cleanup steps on the Qiagen columns.

34. Reduce the volume of the labeled sample to 25 μl in the Speed-Vac. The G-50 cleanup columns are more efficient at removing unincorporated nucleotides when the sample is in a smaller volume.

35. While waiting, prepare Sephadex G-50 columns.

36. Bring Sephadex G-50 columns to room temperature.

37. Repack the Sephadex by spinning the column at 3500 rpm (1000g) for 3 min; discard the flow-through buffer.

38. Add 500 μl of phosphate elution buffer.

39. Spin again for 3 min; discard the flow-through buffer.

40. Repeat the addition of 500 μl of phosphate elution buffer and spin for 3 min. Discard the flow-through buffer.

41. Check the volume of the labeled probe to confirm that it is 25 μl and then add the 25 μl of labeled probe to the repacked column.

42. Spin for 3 min at 3500 rpm (1000g).

43. Transfer the sample to a new microcentrifuge tube and dry down the sample until just dry in the Speed-Vac.

44. Wrap the top of the tube in Parafilm and store at −80°.

Coupling to Monofunctional Reactive Dye (http://pga.tigr.org/sop/ M004_1a.pdf)

1. Resuspend sample in 4.5 μl of 0.1 M Na$_2$CO$_3$, pH 9.0. (High pH is important for coupling.) Wash the sides of tube with this volume to resuspend all dried material.

2. Incubate at 37° for 20 min to resuspend dried DNA.

3. Spin down briefly to collect evaporation.

4. Add 4.5 μl of Cy-dye ester (resuspended in DMSO) to sample. Resuspend and mix thoroughly. Use the entire volume to wash the sides of the tube to resuspend all crystallized sample on the walls of the tube. (Note: Cye dye is photosensitive; attempt to protect from light where possible, especially when heating.)

5. Incubate for 1.5 h at room temperature in the dark.

6. Quench the reaction with 4.5 μl of 4 M hydroxylamine.

7. Incubate 15 min at room temperature in the dark.

8. Add 45 μl of 100 mM NaOAc.

Qiagen PCR Column to Remove Enzyme and Uncoupled Dye from Sample

9. Add 5× volume (300 μl) PB to the sample.

10. Vortex and spin down.

11. Bind to the Qiagen column by passing the sample volume over the column twice.

12. Add 750 μl PE wash buffer to column.

13. Spin at 10,000g for 1 min and discard flow-through wash.

14. Repeat wash steps an additional three times (steps 12 and 13).

15. After the fourth wash, discard the wash buffer and respin at top speed for 2 min to dry the column.

16. Add 50 μl of elution buffer to column. Let sit 5 min at room temperature.

17. Spin for 1 min at top speed.

18. Elute again by adding 50 μl of elution buffer to column. Let sit 5 min at room temperature.

19. Spin for 1 min at top speed.

Final Cleanup over P30 Biogel Chromatography Column

20. Reduce volume to 25 μl in Speed-Vac. The gel filtration columns are more efficient at retaining small molecules when the sample volume is reduced.

21. While waiting, prepare Bio-Rad P30 columns.

   a. Bring Bio-Rad P30 columns to room temperature.

   b. Do not use Bio-Rad collection tubes; replace with 2-ml microfuge tubes.

   c. Spin at 3400 rpm for 2 min to repack column; discard flow-through buffer.

d. Add 500 μl EB buffer (Qiagen elution buffer) to matrix.

e. Spin at 3400 rpm for 2 min to repack column; discard the flow-through buffer.

f. Add another 500 μl EB buffer and spin again at 3400 rpm for 2 min to repack column; discard flow-through buffer.

g. Add 25 μl sample to repacked P30 column.

h. Spin at 3400 rpm for 4 min.

Prepare the Labeled Target for the Array

22. Combine three labeled genomic DNA samples with a single alternately labeled cDNA or cRNA experimental sample for cohybridization.

23. Dry the combined samples down in a Speed-Vac until just dry. Make an effort to avoid overdrying.

24. Resuspend the sample in an appropriate volume of hybridization buffer, depending on the hybridization method and microarray size. We use 30 μl. Use the buffer to wash down the sides of the tube to bring all of the dried product down into solution. Mix thoroughly with a pipette to dissolve all crystalline pieces.

25. Heat in the dark to 85° for 2 min.

26. Immediately spin down for 1 min in the dark to cool the probe.

27. Place the probe at hybridization temperature until ready to apply to the array.

Quantifying Yield and Incorporation Efficiency

Carry out the fluorometric quantification procedure as described earlier. A good yield will have a rfu between 4000 and 5000 for Cy3 and a rfu between 9000 and 10,000 for PicoGreen.

Hybridization and Washing Methods

The specific signal on the array features was increased strongly by increasing the hybridization time from 16 to 72 h. Thermodynamic and kinetic modeling experiments predict that the specificity of signal in microarray experiments will be increased by a longer hybridization time and a greater concentration of labeled target, both of which act to drive the hybridization reaction in the direction of steady-state equilibrium (Bhanot *et al.*, 2003). Experimental results have confirmed that hybridization of perfect match oligonucleotides approaches equilibrium much more slowly than the accumulation of nonspecific mismatched oligonucleotides in microarray experiments (Dai *et al.*, 2002; Sorokin *et al.*, 2005) and that conditions that drive the hybridization reaction toward equilibrium increase the precision of ratiometric measurements (Dai *et al.*, 2002; Dorris *et al.*, 2003; Sartor *et al.*, 2004).

The signal-to-noise ratio is also improved by the use of Surmodics 3D-Link slides (now sold as CodeLink slides by GE Healthcare). They contain on their surface a three-dimensional gel with activated surface chemistry that is capable of linking covalently to oligonucleotides with activated amino group substitutions. 3D-Link slides were designed to closely mimic solution phase hybridization chemistry in a microarray format. This behavior was tested on arrays consisting of perfect match probes and scanning sets of 2- and 3-bp mismatch probes, hybridized against labeled perfect match targets. Fluorescence accumulation at the different probes was compared to the melting temperature behavior of the same sets of probes in a continuously monitored solution (Dorris *et al.*, 2003). The change in fluorescence intensity attributable to specific mismatches closely followed the change in melting temperature of those mismatches in solution.

We were able to further decrease spurious background signal on 3D-Link slides by including formamide (50%) in the hybridization solution, by raising the hybridization temperature to 46°, and by raising the wash temperatures to 67° (Williams *et al.*, 2004). Hybridizing the arrays under "Lifter Slips" helps promote a more even distribution of the hybridization signal on the slide.

Scanning on Axon Instruments Scanner

All data for directly labeled experiments in this chapter were produced with the Axon Instruments 4000A scanner, scanning at 100% laser power with both PMT voltages set to 600 and averaging signal over two scanned lines. Data were normalized with the global scaling algorithm available in the GenePix software, and the median of pixel-by-pixel ratios was chosen for further analysis (Brody *et al.*, 2002). Our core facility has changed our array scanner to the Axon 4000B. We have noticed that at the 100% laser power setting, the two instruments give very different results with the same PMT settings. The 4000B gives a stronger signal, but tends to increase the background, as seen in Fig. 3B and D. This may be compensated by readjustment of the PMT voltages.

Analysis of Results

Performance of the genomic DNA hybridization standard can be evaluated by many different analysis techniques, which are treated only briefly here. A vitally important part of our experimental design was the inclusion of a collection of negative control features expected to have no sequence

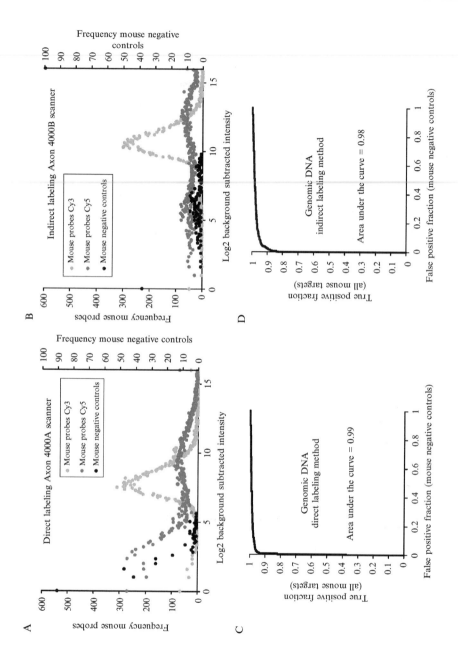

representation in the mouse genome (Operon). Improvements in feature intensity for true positive features must be judged in relation to the background intensity found on true negative features (Fig. 3). Improvements in the labeling and hybridization protocols described in this chapter produce a distribution of background-subtracted feature intensities well within the range of intensities covered by a typical experimental cDNA sample (Fig. 3A and B). The sharper peak of genomic DNA intensities relative to cDNA intensities is expected from a hybridization standard that represents most array features with an equimolar concentration. Receiver operating characteristic curve plots (Fig. 3C and D) with areas very close to 1.0 indicate a very high degree of separation between negative controls and true positives labeled with the genomic DNA standard.

The genomic DNA cohybridization standard was conceived to provide totally comprehensive hybridization coverage of any mouse array. Its performance in this sense can be judged by the percentage of features that are covered by signal intensities judged to be significantly different from negative control feature intensity. The plot shown in Fig. 4 shows that greater than 97% of features on a 16,000 probe array are covered with signal intensity that is greater than 3 SD above the median negative controls feature intensity. Comparison with the Stratagene Universal Mouse RNA (UMRR) standard shows that less than 90% of array features are covered by it. Furthermore, the percentage feature coverage at any given threshold is highly reproducible with the genomic DNA standard, as judged by the spread in coverage seen across five independent replicate arrays (Fig. 4).

FIG. 3. Intensity distributions for genomic DNA signals compared to distributions for negative controls and cDNA-derived signals. (A) Signals from a direct labeling experiment obtained from the Axon 4000A scanner. Genomic DNA labeled with Cy3 (green) produces a distribution of feature intensities well within the distribution of Cy5 signals (red) obtained from C2C12 cDNA and well separated from the distribution of intensities from "on-slide" negative controls (black). (B) Same comparison as in A, but using genomic DNA labeled via the indirect method and scanning with the Axon 4000B scanner. Note a slight increase in overlap between genomic DNA and negative control distributions. (C) Receiver operating characteristic (ROC) curve demonstrating the separation between distributions of the positive mouse probes and the mouse negative control features for the direct labeling method. On the vertical axis is the fraction of positive mouse probes at or below a given level of Cy3 intensity; on the horizontal axis is the corresponding fraction of negative control features at or below that same level of Cy3 intensity. The area under the curve indicates separation of the two distributions, with a score of 1.0 indicating perfect separation of the distributions (i.e., no overlap). (D) ROC plot for the indirect labeling method. The slight decrease in area under the curve indicates a slight increase in nonspecific signal when the indirect method is used in conjunction with the new generation Axon 4000B scanner. (See color insert.)

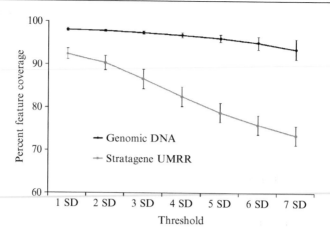

FIG. 4. Feature coverage percentage comparison. Comparison of array feature coverage for genomic DNA and Stratagene UMRR cohybridization standards. The mean percentage mouse feature coverage as a function of increasing threshold value is shown for five replicate experiments. Background thresholds are defined as the median fluorescence intensity for the group of negative control features, plus a multiple of the standard deviation value for the negative controls group (i.e., median negative controls ± 1 SD, etc.). The feature coverage percentage is defined as the number of features exceeding a given background threshold value divided by the total number of mouse features on the array multiplied by 100. Mean and standard deviation for the replicate groups at each threshold are plotted. Reprinted with permission from Williams *et al.* (2004).

To estimate the interarray signal variance in arrays cohybridized with a genomic DNA standard, we analyzed a set of five replicate hybridizations made with five aliquots of the same RNA sample from C2C12 skeletal myoblasts at 24 h in differentiation medium. The RNA samples were random primed and reverse transcribed using a direct labeling protocol (http://cmgm.stanford.edu/pbrown/protocols/4_human_RNA.html). Signal variance for all positive features on the array was estimated using the coefficient of variation (CV), expressed as a percentage, for the five ratiometric measurements of intensity for each feature. In microarray experiments, low-intensity features exhibit greater variation in replicate measurements because the lower number of photons counted results in a greater standard error for the measure of intensity (Sharov *et al.*, 2004). When CV is plotted against intensity (Fig. 5B), a moving average line plotted through all data clearly indicates this relationship (Shippy *et al.*, 2004). If the analysis is focused on features with C2C12 signals consistently greater than 250 (blue dots), the error behaves consistently at less than 20% for the majority of features. This estimate of the error also compares favorably with that from the UMRR (Fig. 5A).

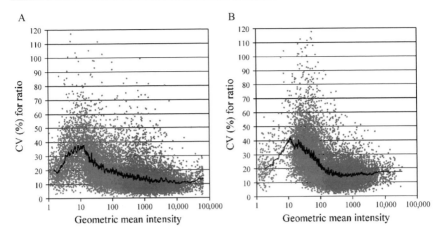

FIG. 5. Comparison of signal variance as a function of intensity. (A) Plot of signal intensity vs variation for all five Stratagene replicates. The x axis indicates the geometric mean of five normalized intensities; the y axis indicates the coefficient of variation in a standardized signal expressed as a percentage. Values for the entire data set are plotted in red. Overlaid on the full data set plot are the values for features with Cy5 intensities greater than 250 in all five Stratagene replicates (shown in blue). The black line indicates the moving average for 100 features. (B) Plot of signal intensity vs variation for all five genomic DNA replicates. In this case, values shown in blue are for the same features displaying a numerator signal greater than 250 in the Stratagene replicates. (See color insert.)

Intensity dependent and other sources of systematic error can be adjusted using a locally weighted scatter plot smoothing (Lowess) method (Cleveland, 1979; Dudoit *et al.*, 2002; Yang *et al.*, 2002a,b). For the five replicate hybridizations done with each standard, we made 10 interslide "within standard" comparisons of intensity for all array features. Two representative "self vs self" comparisons are shown in Fig. 6. Negative or zero intensity values resulting from the subtraction of local background were all set to 1.0. In this case, variance between the replicates is expressed as a ratio of normalized ratios (R). When the value of $\log_2 R$ is plotted against the $\log_2$ geometric mean of the intensities (an R-I or M-A plot), the shape of the error distribution indicates a systematic positive bias in measurement variation for features with low intensity (Fig. 6A). The Lowess adjustment using a two-step routine with 0.5 smoothing (Minitab Software) removes this bias, causing the distribution of points to lie symmetrically around the expected value of zero (Fig. 6C). When the Lowess adjustment is applied to features with numerator signals greater than 250 (Fig. 6E and G), the variance in the genomic DNA standardized signals compares favorably with the Universal Mouse Reference RNA standard (Fig. 6B, D, F, and H).

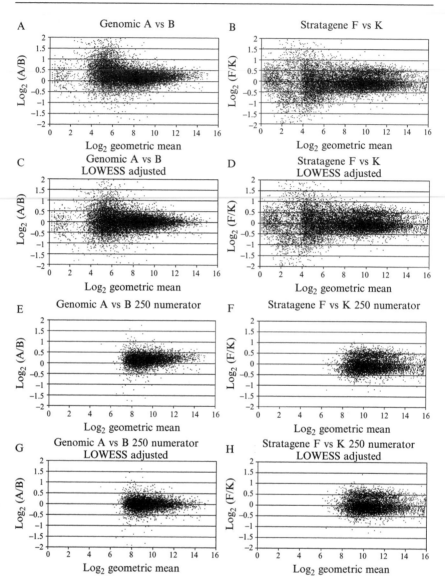

FIG. 6. Comparison of variance in replicate experiments after Lowess adjustment and replicate filtering. (A) Self-self comparison of full array data from two representative hybridizations using C2C12 cDNA in the Cy5 channel and genomic DNA in the Cy3 channel. Note the greater spread in the comparison ratio in the low-intensity range. (B) Same as in A, except using the Stratagene mixed cDNAs in the Cy3 channel. (C) Lowess-adjusted values from plot in A. (D) Lowess-adjusted values from the plot in B. (E) Self-self comparison plot for features that had consistently greater than 250 Cy5 intensity in five replicate hybridizations

A plot of the cumulative distribution of the error for all 10 pairwise comparisons for both the genomic DNA standard and the UMRR is shown in Fig. 7. The Lowess adjustment causes a pronounced shift of the error curve for the genomic DNA standard toward the expected value of zero.

The genomic DNA reference produces a much narrower distribution of signal intensity across all gene features than the UMRR standard (Fig. 8A). This follows directly from the fact that the vast majority of genes (exons) are represented in genomic DNA as a single copy sequence, while the same sequences are represented in cellular RNA populations over a range of at least three orders of magnitude. The practical upshot is nontrivial and useful: In the role of normalizing standard, genomic DNA-denominated

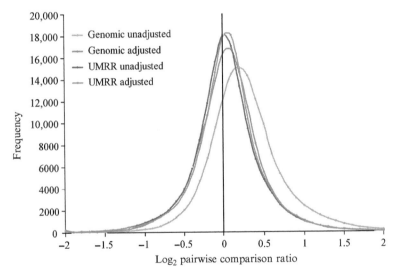

FIG. 7. Lowess adjustment reduces error in self-against-self comparisons that use genomic DNA as the denominator. Comparison of self-against-self error rates between the two standards. Five replicate slides of C2C12 cDNA (Cy5) hybridized against either genomic DNA or the Stratagene mixed cDNAs (Cy3) standard were used in studies of signal variance as a function of intensity. Log_2 ratios vs log_2 geometric mean intensity (RI plots not shown) for 10 self-against-self comparisons within each standard were made, and Lowess adjustment was applied to the log_2 ratio values. Cumulative distributions of the ratios before and after Lowess adjustment are plotted. Note that the spread in the ratios is greater than previous studies (Yang, 2002), as these self-against-self comparisons are compiled from interslide comparisons. (See color insert.)

against the Stratagene standard. In this case, the arrays were hybridized against genomic DNA. (F) Same features shown as in E, except that the arrays were hybridized against the Stratagene standard. (G) Lowess-adjusted values from plot E. (H) Lowess-adjusted values from plot F.

A

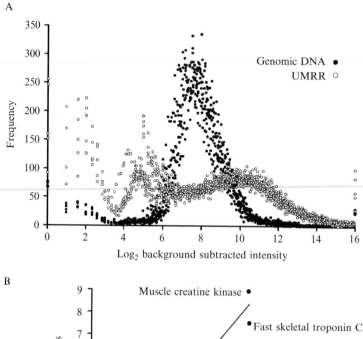

B

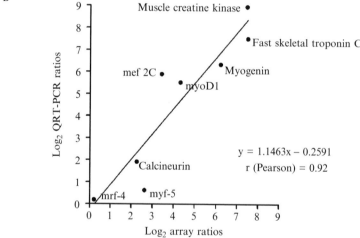

FIG. 8. (A) Comparison of the distribution of background subtracted intensities produced by the genomic DNA standard ($n = 5$ arrays) and the Stratagene Universal Mouse Reference RNA (UMRR) ($n = 5$ arrays) standard. The genomic DNA intensities are distributed uniformly within a narrow range, whereas the UMRR intensities vary widely. (B) Comparison of ratios obtained from genomic DNA-standardized microarrays with ratios obtained in quantitative RT-PCR experiments using genomic DNA as a baseline. Values for $\log_2$ array ratios are the median of five measurements; values for $\log_2$ qRT-PCR ratios are median of four measurements. mRNA from C2C12 cells at 24 h was used to produce cDNA for five replicate hybridizations on microarrays with genomic DNA as the cohybridization standard or was reverse transcribed and used as a template in four replicate qRT-PCR measurements. The amount of genomic DNA used as a template for the qRT-PCR baseline was chosen to match the proportions normally used in a hybridization experiment.

ratios directly reflect the abundance of each RNA in the experimental sample. Thus, a high ratio of RNA to the genomic DNA reference reflects a high abundance of that transcript in the experimental RNA sample. A low ratio should similarly reflect low RNA abundance. This is possible because we know quite precisely—given the entire genome sequence—the "abundance" of each and every gene sequence (exon, intergenic sequence feature, etc.) in genomic DNA, and we know that the abundance is usually a single copy per haploid genome. Where it is not single copy, one can correct appropriately at the bioinformatics level by dividing feature intensities by a suitable constant that reflects the number of complementary sequences known to be present in the genome that might hybridize under the stringency conditions employed. In contrast, RNA standards, whether prepared from a mix of tissue sources or experimental points or made from a single cell type, have unknown and widely divergent concentrations of different RNA species, which means that there is no simple way to calculate prevalence.

The uniform length of our 70-mer feature probes (which is not a feature of cDNA-based arrays), combined with random-primed labeling for the cDNA experimental sample [which is not a feature of the oligo(dT) priming protocols used in the Affymetrix system], and the use of median of pixel-by-pixel ratios measurements (Brody *et al.*, 2002) should contribute to the overall reliability of our microarray-based estimates of transcript abundance in C2C12 mRNA samples. To critically evaluate the accuracy of our estimates, we used quantitative real-time RT-PCR to independently measure transcript prevalence. We mimicked the array experiments by using experimental cDNA and genomic DNA as inputs to the qRT-PCR reactions at concentrations that reflect their relative masses in a cohybridization experiment. We focused on a subset of muscle-specific genes representing rare (mrf-4, myf5, and calcineurin), moderate (myoD1, mef2C, and myogenin), and abundant (muscle creatine kinase, fast skeletal troponin C) transcript prevalence groups. Transcript prevalence groups were defined as a percentage of the summed Cy5 intensities present on each chip. The rare transcript range was assigned to features displaying 0.001–0.005% of the summed Cy5 signal intensity, moderate transcripts were in the 0.018–0.068% range, and abundant transcripts were in the range of 0.2% or greater. The median of five replicate microarray ratio values over genomic DNA was compared to the median of four replicate qRT-PCR ratio values that used genomic DNA as a baseline comparison (Fig. 8B). The overall Pearson correlation of 0.92 indicates that use of a genomic DNA cohybridization standard in microarray measurements accurately preserves relative transcript prevalence information over a wide range of transcript abundance.

As expression microarrays have come into wide use, it has become appreciated that the data sets have extraordinary power in reanalysis and meta-analysis that continues long after a study has been initially mined and published. In the best scenarios, very large collections of array data from many studies can be assembled and interrogated many times over, often using different algorithms and different data combinations, to probe different biological questions (Bergmann *et al.*, 2004; Kim *et al.*, 2001). This potential for repeated use of data—in which summed data collections can give much greater and different leverage than their individual parts—sets microarray data apart from classical molecular biology data. Coupled with all the usual incentives to extract the most from any large and costly experiment, this motivates ongoing experimental and analytical improvements aimed at data robustness and ease of data use across disparate studies (Choe *et al.*, 2005; Consortium, 2005; Irizarry *et al.*, 2005; Larkin *et al.*, 2005).

For ratiometric array platforms, development and selection of an appropriate and general cohybridization standard is one such design issue with impact on data quality, interpretability, and usefulness in meta-analyses. A proposed data processing pathway for the linkage of large collections of microarray experiments using the same cohybridization standard is shown in Fig. 9. The linkage is formed by scaling signal intensities

Cohybridize and scan arrays
(different labs, different time frames, different experiments,
same genomic DNA denominator)

Collection of all data sets for study

Quantile normalization of denominator measurements

Bioinformatic correction of denominator values

Quantile normalization of numerator measurements

Recompute ratiometric estimates of transcript abundance

Analysis of changes from normalized ratio data

FIG. 9. Proposed data-processing pathway using genomic DNA-standardized data.

for the common reference standard across all arrays in the study, ensuring that the experimental samples are fairly compared to a common reference. This may be done by quantile normalization on the denominator channel (Bolstad *et al.*, 2003), which standardizes the ranks of denominator signals across arrays. The accuracy of ratiometric abundance measurements can be improved further with bioinformatic correction for cross hybridization of closely related gene family members and pseudo-genes in the denominator channel. If desired, the numerator channel can also be quantile normalized to control systematic error between experimental samples. From this point forward, the many different available methods for difference analysis can be used.

Conclusions

Comprehensive Reference Standard for Analysis of Multiple Microarray Experiments

Microarray profiling is now the dominant method for assaying global gene expression patterns, and similar arrays are coming into wide use for large-scale *in vivo* protein:DNA interaction studies via chromatin immunoprecipitation. From the current vantage point we anticipate that other technologies will supercede ratiometric microarrays in some, perhaps even all, applications (i.e., various hyperefficient direct DNA sequencing methods that can generate a million different DNA sequence reads per sample for direct profiling). These candidate replacement technologies are sure to have their own technical strengths and weaknesses and their cost structure issues are unknown. For these reasons, plus the simple fact that ratiometric arrays are now a major part of standard technology, it seems certain that ratiometric microarray data will continue to be generated for some time to come. Development and adoption of some kind of universal and comprehensive ratiometric reference standard will improve data integration across studies within individual labs and across data sets from entirely unrelated studies. Currently, genomic DNA is the most accessible and viable option for this purpose, as it is readily available and meets all major criteria for a universal reference standard (Williams *et al.*, 2004).

A challenge addressed here was to make a genomic DNA standard more convenient and robust for routine use by developing an indirect labeling protocol that tolerates extended frozen storage. This enables production of sizable lots and extensive quality control testing. Our hybridization performance objective was reproducible success on oligonucleotide arrays (40- to 70-mers), as these arrays likely push sensitivity requirements maximally, compared with longer PCR products or bigger DNA inserts. This objective was met by both direct and indirect labeling methods. However, the directly labeled standard is hybridized on the day of production, and quality control measurements are therefore limited to label incorporation and yield. The indirectly labeled standard can be prepared in advance, and test aliquots can be quality checked for incorporation, yield, *and* hybridization performance prior to cohybridization with experimental samples.

Transcript Prevalence Information

A useful but largely unexploited property of ratiometric measurements made with a genomic DNA standard is that they carry good information on the abundance of each transcript in the target RNA sample. This is in contrast to RNA reference standards of various kinds that obscure this information. This chapter showed that transcript prevalence information from genomic DNA-normalized ratios is well correlated with quantitative RT-PCR measurements ($r = 0.92$). This compares favorably to the validation levels seen in similar studies of Affymetrix arrays and long oligonucleotide arrays (Dallas *et al.*, 2005; Irizarry *et al.*, 2005; Shippy *et al.*, 2004). Experimental design variables that affect uniformity across genes such as the RNA-labeling protocol [random priming vs oligo(dT) priming; fragmentation of cDNA or cRNA products to short uniform length, etc.] and the length, uniformity, and design of array features (same-length oligonucleotides vs variable length PCR products) may affect how accurately microarray ratios will reflect transcript prevalence in the RNA sample. However, even when design choices are not optimized for uniformity, the overall trends in natural RNA abundance classes will still be directly represented and intuitively clear.

Previous efforts to estimate absolute transcript abundance in mammalian samples have employed either a set of "spiked-in" calibration standards of known concentration (Carter *et al.*, 2005) or the equimolar representation of array feature sequences provided by a genomic DNA standard (Shyamsundar *et al.*, 2005). A combination of these techniques would provide both internal linearity scaling within each experiment and standardization of all ratiometric measurements to a cohybridization standard of known specific activity. This would improve the accuracy of RNA prevalence measurements in mammalian samples and has already been tested in yeast (Dudley *et al.*, 2002).

Alternate Comprehensive Standards

A different class of possible standards is less universal and, at this time, not readily available, although they could be a significant improvement over RNA-based standards. These standards are labeled oligonucleotides of known specific activity and sequence content. An oligonucleotide-based reference standard complementary to a universal sequence tag that was present on each feature of a PCR-based array has been reported (Dudley *et al.*, 2002). A clear virtue of this kind of design is that one can set the reference signal level arbitrarily and conveniently by adjusting the specific activity of the standard and the amount used. There is, however, no community consensus on a universal sequence tag for all array elements of all arrays, and for long oligonucleotide arrays including such tags would require a major redesign that would also increase the oligonucleotide length for every feature. A more ambitious and powerful version of this idea would be to make an equimolar collection of oligonucleotides complementary to all sequence features on an array. A similar idea using a mixture of the PCR products that were printed on a cDNA array has been tested successfully on a modest scale (Sterrenburg *et al.*, 2002). It is tempting to think that such equimolar standards, perhaps even ones that are prelabeled for immediate use, might emerge as commercial products corresponding to specific arrays. The greatest limitation would then be that the oligonucleotide mix would have to be designed to specifically match a particular array.

Other Applications for the Genomic DNA Standard

This work was undertaken with an eye toward alternative microarray applications for which a genomic DNA standard is plausible or even required. Primary among these are microarray readouts from global chromatin immunoprecipitation assays, so called ChIP/chip experiments (Negre *et al.*, 2006). In ChIP/chip studies of global *in vivo* protein:DNA interactions, DNA sequences associated with a specific protein are enriched by immunoprecipitation of genomic DNA that has been cross-linked *in vivo*, sonicated, and ultimately labeled and ratioed against an unenriched or mock-enriched genomic DNA sample. The assay readout is on a microarray of intergenic and (sometimes) genic DNA tiling paths to map the presence of specific transcription factors, general transcription machinery, and various chromatin modifications. In many ChIP/chip protocols, a DNA amplification step is applied by the ligation-mediated polymerase chain reaction (Harbison *et al.*, 2004; Lee *et al.*, 2002; Ren *et al.*, 2000); however, by using the labeling protocols presented here, one can eliminate the amplification procedure, at least in yeast (L. A. Dunipace and B. J. Wold, unpublished results). We believe that this will be a viable strategy in the larger genomes as well. At minimum, this would simplify the ChIP/chip experiment significantly and might also reduce variability introduced by the amplification step.

A microarray alternative to SELEX-type methods (Tuerk and Gold, 1990) has also emerged that seeks to characterize all possible DNA-binding sites for a given transcription factor. This new DNA IP/chip technique (Liu *et al.*, 2005), uses cloned protein to immunoprecipitate fragmented genomic DNA and also requires a cohybridization standard capable of universal array coverage.

Acknowledgments

We are indebted to our Bioinformatics staff (Diane Trout, Brandon King, and Joe Roden) for their patient assistance with this project. This work was funded by grants to BJW from the Department of Energy (DOE), the National Aeronautic and Space Administration (NASA), and the National Institutes of Health (NIH). BAW was supported by an NIH NRSA fellowship.

References

Ambion (2004). Tips for successful RNA amplification. *TechNotes* **11**.

Amon, P., and Ivanov, I. (2003). Genomic DNA labeling for hybridization with DNA arrays. *BioTechniques* **34**, 700–704.

Bergmann, S., Ihmels, J., and Barkai, N. (2004). Similarities and differences in genome-wide expression data of six organisms. *PLoS Biol.* **2**, 0085–0093.

Bergstrom, D. A., Penn, B. H., Strand, A., Perry, R. L. S., Rudnicki, M. A., and Tapscott, S. J. (2002). Promoter-specific regulation of myoD binding and signal transduction cooperate to pattern gene expression. *Mol. Cell* **9**, 587–600.

Bhanot, G., Louzon, Y., Zhu, J., and De Lisi, C. (2003). The importance of thermodynamic equilibrium for high throughput gene expression arrays. *Biophys. J.* **84**, 124–135.

Bolstad, B. M., Irizarry, R. A., Astrand, M., and Speed, T. P. (2003). A comparison of normalization methods for high density oligonucleotide array data based on variance and bias. *Bioinformatics* **19**, 185–193.

Brody, J. P., Williams, B. A., Wold, B. J., and Quake, S. R. (2002). Significance and statistical errors in the analysis of DNA microarray data. *Proc. Natl. Acad. Sci. USA* **99**, 12975–12978.

Carter, M. G., Sharov, A. A., Van Buren, V., Dudekula, D. B., Carmack, C. E., Nelson, C., and Ko, M. S. H. (2005). Transcript copy number estimation using a mouse whole-genome oligonucleotide microarray. *Genome Biol.* **6**, R61.

Choe, S. E., Boutros, M., Michelson, A. M., Church, G. M., and Halfon, M. S. (2005). Preferred analysis methods for Affymetrix GeneChips revealed by a wholly defined control dataset. *Genome Biol.* **6**, R16.

Churchill, G. A. (2002). Fundamentals of experimental design for cDNA microarrays. *Nat. Genet.* **32**, 490–495.

Cleveland, W. S. (1979). Robust locally weighted regression and smoothing scatterplots. *J. Am. Stat. Assoc.* **74**, 829–836.

Consortium (2005). Standardizing global gene expression analysis between laboratories and across platforms. *Nat. Methods* **2**, 351–356.

Dai, H., Meyer, M., Stepaniants, S., Ziman, M., and Stoughton, R. (2002). Use of hybridization kinetics for differentiating specific from non-specific binding to oligonucleotide microarrays. *Nucleic Acids Res.* **30,** e86.

Dallas, P. B., Gottardo, N. G., Firth, M. J., Beesley, A. H., Hoffmann, K., Terry, P. A., Freitas, J. R., Boag, J. M., Cummings, A. J., and Kees, U. R. (2005). Gene expression levels assessed by oligonucleotide microarray analysis and quantitative real-time RT-PCR: How well do they correlate? *BMC Genom.* **6.**

Dorris, D. R., Nguyen, A., Gieser, L., Lockner, R., Lublinsky, A., Patterson, M., Touma, E., Sendera, T. J., Elghanian, R., and Mazumder, A. (2003). Oligodeoxyribonucleotide probe accessibility on a three-dimensional DNA microarray surface and the effect of hybridization time on the accuracy of expression ratios. *BMC Biotechnol.* **3,** 6.

Dudley, A. M., Aach, J., Steffen, M. A., and Church, G. M. (2002). Measuring absolute expression with microarrays with a calibrated reference sample and an extended signal intensity range. *Proc. Natl. Acad. Sci. USA* **99,** 7554–7559.

Dudoit, S., Yang, Y. H., Callow, M. J., and Speed, T. P. (2002). Statistical methods for identifying differentially expressed genes in replicated cDNA microarray experiments. *Stat. Sin.* **12,** 111–139.

Eisen, M. B., and Brown, P. O. (1999). DNA arrays for analysis of gene expression. *Methods Enzymol.* **303,** 179–205.

Hacia, J. G., Brody, L. C., Chee, M. S., Fodor, S. P. A., and Collins, F. S. (1996). Detection of heterozygous mutations in BRCA1 using high density oligonucleotide arrays and two-colour fluorescence analysis. *Nat. Genet.* **14,** 441–447.

Hacia, J. G., Edgemon, K., Sun, B., Stern, D., Fodor, S. P. A., and Collins, F. S. (1998). Two color hybridization analysis using high density oligonucleotide arrays and energy transfer dyes. *Nucleic Acids Res.* **26,** 3865–3866.

Harbison, C. T., Gordon, D. B., Lee, T. I., Rinaldi, N. J., Macisaac, K. D., Danford, T. W., Hannett, N. M., Tagne, J.-B., Reynolds, D. B., Yoo, J., Jennings, E. G., Zeitlinger, J., Pokholok, D. K., Kellis, M., Rolfe, P. A., Takusagawa, K. T., Lander, E. S., Gifford, D. K., Fraenkel, E., and Young, R. A. (2004). Transcriptional regulatory code of a eukaryotic genome. *Nature* **431,** 99–104.

Irizarry, R. A., Warren, D., Spencer, F., Kim, I. F., Biswal, S., Frank, B. C., Gabrielson, E., Garcia, J. G. N., Geoghegan, J., Germino, G., Griffin, C., Hilmer, S. C., Hoffman, E., Jedlicka, A. E., Kawasaki, E., Martinez-Murillo, F., Morsberger, L., Lee, H., Petersen, D., Quackenbush, J., Scott, A., Wilson, M., Yang, Y., Ye, S. Q., and Yu, W. (2005). Multiple-laboratory comparison of microarray platforms. *Nat. Methods* **2,** 345–349.

Kim, H., Zhao, B., Snesrud, E. C., Haas, B. J., Town, C. D., and Quackenbush, J. (2002). Use of RNA and genomic DNA references for inferred comparisons in DNA microarray analyses. *BioTechniques* **33,** 924–930.

Kim, S. K., Lund, J., Kiraly, M., Duke, K., Jiang, M., Stuart, J. M., Eizinger, A., Wylie, B. N., and Davidson, G. S. (2001). A gene expression map for *Caenorhabditis elegans*. *Science* **293,** 2087–2092.

Larkin, J. E., Frank, B. C., Gavras, H., Sultana, R., and Quackenbush, J. (2005). Independence and reproducibility across microarray platforms. *Nat. Methods* **2,** 337–343.

Lee, T. I., Rinaldi, N. J., Robert, F., Odom, D. T., Bar-Joseph, Z., Gerber, G. K., Hannett, N. M., Harbison, C. T., Thompson, C. M., Simon, I., Zeitlinger, J., Jennings, E. G., Murray, H. L., Gordon, D. B., Ren, B., Wyrick, J. J., Tagne, J.-B., Volkert, T. L., Fraenkel, E., Gifford, D. K., and Young, R. A. (2002). Transcriptional regulatory networks in *Saccharomyces cerevisiae*. *Science* **298,** 799–804.

Liu, X., Noll, D. M., Lieb, J. D., and Clarke, N. D. (2005). DIP-chip: Rapid and accurate determination of DNA binding specificity. *Genome Res.* **15,** 421–427.

Negre, N., Lavrov, S., Hennetin, J., Bellis, M., and Cavalli, G. (2006). Mapping the distribution of chromatin proteins by ChIP on chip. *Methods Enzymol.* **410** (this volume), 316–341.

Perou, C. M., Sorlie, T., Eisen, M. B., van de Rijn, M., Jeffrey, S. S., Rees, C. A., Pollack, J. R., Ross, D. T., Johnsen, H., Akslen, L. A., Fluge, O., Pergamenschikov, A., Williams, C. F., Zhu, S. X., Lonning, P. E., Borresen-Dale, A.-L., Brown, P. O., and Botstein, D. (2000). Molecular portraits of human breast tumours. *Nature* **406**, 747–752.

Peterson, A. W., Heaton, R. J., and Georgiadis, R. M. (2001). The effect of surface probe density on DNA hybridization. *Nucleic Acids Res.* **29**, 5163–5168.

Pollack, J. R., Perou, C. M., Alizadeh, A. A., Eisen, M. B., Pergamenschikov, A., Williams, C. F., Jeffrey, S. S., Botstein, D., and Brown, P. O. (1999). Genome-wide analysis of DNA copy-number changes using cDNA microarrays. *Nat. Genet.* **23**, 41–46.

Ren, B., Robert, F., Wyrick, J. J., Aparicio, O., Jennings, E. G., Simon, I., Zeitlinger, J., Schreiber, J., Hannett, N. M., Kanin, E., Volkert, T. L., Wilson, C. J., Bell, S. B., and Young, R. A. (2000). Genome-wide location and function of DNA binding proteins. *Science* **290**, 2306–2309.

Ross, D. T., Scherf, U., Eisen, M. B., Perou, C. M., Rees, C. A., Spellman, P., Iyer, V., Jeffrey, S. S., van de Rijn, M., Waltham, M., Pergamenschikov, A., Lee, J. C. F., Lashkari, D., Shalon, D., Myers, T. G., Weinstein, J. N., Botstein, D., and Brown, P. O. (2000). Systematic variation in gene expression patterns in human cancer cell lines. *Nat. Genet.* **24**, 227–235.

Sartor, M., Schwanekamp, J., Halbleib, D., Mohamed, I., Karyala, S., Medvedovic, M., and Tomlinson, C. R. (2004). Microarray results improve significantly as hybridization approaches equilibrium. *BioTechniques* **36**, 790–796.

Shalon, D., Smith, S. J., and Brown, P. O. (1996). A DNA microarray system for analyzing complex DNA samples using two-color fluorescent probe hybridization. *Genome Res.* **6**, 639–645.

Sharov, V., Kwong, K. Y., Frank, B., Chen, E., Hasseman, J., Gaspard, R., Yu, Y., Yang, I., and Quackenbush, J. (2004). The limits of log-ratios. *BMC Biotechnol.* **4**.

Shippy, R., Sendera, T. J., Lockner, R., Palaniappan, C., Kaysser-Kranich, T., Watts, G., and Alsobrook, J. (2004). Performance evaluation of commercial short-oligonucleotide microarrays and the impact of noise in making cross-platform correlations. *BMC Genom.* **5**.

Shyamsundar, R., Kim, Y. H., Higgins, J. P., Montgomery, K., Jorden, M., Sethuraman, A., van de Rijn, M., Botstein, D., Brown, P. O., and Pollack, J. R. (2005). A DNA microarray survey of gene expression in normal human tissues. *Genome Biol.* **6**, R22.

Sorokin, N. V., Chechetkin, V. R., Livshits, M. A., Pan'kov, S. V., Donnikov, M. Y., Gryadunov, D. A., Lapa, S. A., and Zasedatelev, A. S. (2005). Discrimination between perfect and mismatched duplexes with oligonucleotide gel microchips: Role of thermodynamic and kinetic effects during hybridization. *J. Biomol. Struct. Dynam.* **22**, 725–734.

Spellman, P. T., Sherlock, G., Zhang, M. Q., Iyer, V. R., Anders, K., Eisen, M. B., Brown, P. O., Botstein, D., and Futcher, B. (1998). Comprehensive identification of cell cycle-regulated genes of the yeast *Saccharomyces cerevisiae* by microarray hybridization. *Mol. Biol. Cell* **9**, 3237–3297.

Sterrenburg, E., Turk, R., Boer, J. M., van Ommen, G. B., and den Dunnen, J. T. (2002). A common reference for cDNA microarray hybridizations. *Nucleic Acids Res.* **30**, e116.

Tuerk, C., and Gold, L. (1990). Systematic evolution of ligands by exponential enrichment: RNA ligands to bacteriophage T4 DNA polymerase. *Science* **249**, 505–510.

Williams, B. A., Gwirtz, R. M., and Wold, B. J. (2004). Genomic DNA as a cohybridization standard for mammalian microarray measurements. *Nucleic Acids Res.* **32**, e81.

Yang, I. V., Chen, E., Hasseman, J., Liang, W., Frank, B., Wang, S., Sharov, V., Saeed, A. I., White, J., Li, J., Lee, N. H., Yeatman, T. J., and Quackenbush, J. (2002a). Within the fold: Assessing differential expression measures and reproducibility in microarray assays. *Genome Biol.* **3,** 0062.1–0062.12.

Yang, Y. H., Dudoit, S., Luu, P., Lin, D. M., Peng, V., Ngai, J., and Speed, T. P. (2002b). Normalization for cDNA microarray data: A robust composite method addressing single and multiple slide systematic variation. *Nucleic Acids Res.* **30,** e15.

Zhu, X., Hart, R., Chang, M. S., Kim, J.-W., Lee, S. Y., Cao, Y. A., Mock, D., Ke, E., Saunders, B., Alexander, A., Grossoehme, J., Lin, K.-M., Yan, Z., Hsueh, R., Lee, J., Scheuermann, R. H., Fruman, D. A., Seaman, W., Subramanian, S., Sternweis, P., Simon, M. I., and Choi, S. (2004). Analysis of the major patterns of B cell gene expression changes in response to short-term stimulation with 33 single ligands. *J. Immunol.* **173,** 7141–7149.

[13] Analysis of Sequence Specificities of DNA-Binding Proteins with Protein Binding Microarrays

By MARTHA L. BULYK

Abstract

DNA-binding proteins are important for various cellular processes, such as transcriptional regulation, recombination, replication, repair, and DNA modification. Of particular interest are transcription factors (TFs), since through interactions with their DNA binding sites, they modulate gene expression in a manner required for normal cellular growth and differentiation, and also for response to environmental stimuli. To date, the DNA-binding specificities of most DNA-binding proteins remain unknown, as earlier technologies aimed at characterizing DNA–protein interactions have been laborious and not highly scalable. New DNA microarray-based technology, termed protein binding microarrays (PBMs), has been developed that allows rapid, high-throughput characterization of *in vitro* DNA binding site sequence specificities of TFs or of any DNA binding protein. DNA binding site data from PBMs can be used to predict what genes are regulated by a given TF, what the functions are of a given TF and its predicted target genes, and how that TF may fit into the transcriptional regulatory networks of the cell.

Introduction

DNA-binding proteins play key roles in various cellular processes, including transcriptional regulation, recombination, genome rearrangements, replication, repair, and DNA modification. The interactions

METHODS IN ENZYMOLOGY, VOL. 410
0076-6879/06 $35.00
DOI: 10.1016/S0076-6879(06)10013-0

between transcription factors (TFs) and their DNA binding sites are of particular interest because they regulate the gene expression required for progression through the cell cycle, in differentiation, and in response to environmental stimuli, and thus contribute to the expression patterns observed through transcript profiling [reviewed in Lockhart and Winzeler (2000)]. However, only a small number of sequence-specific TFs have been characterized well enough such that all the sequences that they can and, just as importantly, cannot bind are known. The sparseness of these binding site sequence data is highly problematic because these data are used frequently to search for functional genomic occurrences of these TF-binding sites, with many false-positive and false-negative *cis*-regulatory elements being predicted. Earlier technologies aimed at characterizing DNA–protein interactions have been laborious and not highly scalable, and microarray readout of chromatin immunoprecipitations ("ChIP-chip" or genome-wide location analysis) requires that the given DNA binding protein be bound to its target sites when the cells are fixed [reviewed in Bulyk (2003); also see Negre *et al.* (2006)]. Thus, there is a need for technology that allows high-throughput determination of TF-binding sites.

Advances in genomics and proteomics are now permitting high-throughput functional studies of proteins. For example, overexpression and purification of genes from a variety of genomes in a high-throughput manner (reviewed in Braun and LaBaer, 2003) are now becoming more common and are permitting various large-scale biochemical studies (reviewed in Zhu *et al.*, 2003). In addition, most researchers have access to DNA microarraying facilities, if not at their own institution, then through another institution that provides microarraying services for a fee. Likewise, DNA microarray scanners are readily available in most departments or institutions.

An *in vitro* DNA microarray technology, termed protein binding microarrays (PBMs) for characterization of the sequence specificities of DNA–protein interactions, has been developed. This technology allows *in vitro* binding specificities of individual DNA-binding proteins to be determined in a single day by assaying the sequence-specific binding of a given DNA binding protein directly to double-stranded (ds) DNA microarrays spotted with a large number of potential DNA binding sites (Mukherjee *et al.*, 2004). An earlier version of the PBM technology, in which DNA binding domains were expressed on the surface of phage, which were then bound to microarrays spotted with a set of synthetic dsDNA oligonucleotides, has been described previously (Bulyk *et al.*, 2001). Switching from phage display constructs (Bulyk *et al.*, 2001) to epitope-tagged fusion constructs (Mukherjee *et al.*, 2004) avoids any problems associated with a potentially polyvalent phage. In addition, the use of genome-scale intergenic microarrays (Mukherjee *et al.*, 2004) provides representation of most, if not all, genomic DNA binding sites on the microarrays; moreover, because the binding sites are present in their native local sequence context, one could potentially also examine the binding of protein complexes to the DNAs.

Specifically, in PBM experiments, a DNA binding protein of interest is expressed with an epitope tag, purified, and then bound directly to triplicate dsDNA microarrays. The protein-bound microarrays are then washed to remove any nonspecifically bound protein and are labeled with a fluorophore-conjugated antibody specific for the epitope tag. In order to normalize PBM data by relative DNA concentration, separate triplicate microarrays from the same print run are stained with the dye SYBR Green I, which is specific for dsDNA (Figs. 1 and 2). This normalization is important to perform, as the spotted DNAs, which in the case of whole-genome yeast intergenic microarrays were PCR products, can vary greatly in the amount of DNA present per spot. Sequences corresponding to the significantly bound spots (Fig. 3A) are analyzed with a motif-finding tool in

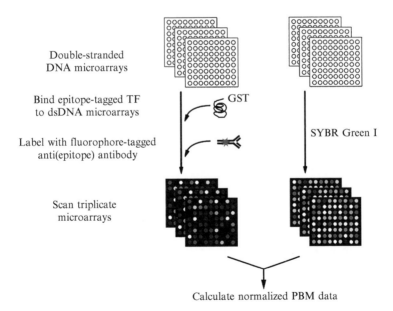

FIG. 1. Schema of protein binding microarray experiments. Reproduced from Mukherjee *et al.* (2004) with permission from Nature Publishing Group.

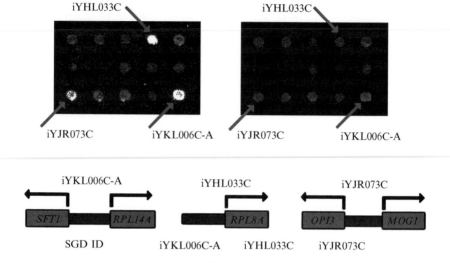

FIG. 2. Magnification of identical portions of yeast intergenic microarrays used in a PBM experiment (left) or stained with SYBR Green I (right). Fluorescence intensities are shown in false color, with white indicating saturated signal intensity, red indicating high signal intensity, yellow and green indicating moderate signal intensity, and blue indicating low signal intensity. The three labeled spots correspond to the intergenic regions depicted below, along with the *P* values derived from triplicate PBM and SYBR Green I microarray data. Reproduced from Mukherjee *et al.* (2004) with permission from Nature Publishing Group. (See color insert.)

order to identify the DNA binding site motif for the given DNA-binding protein (Fig. 3B) (Mukherjee *et al.*, 2004).

This chapter describes how to use DNA microarrays in PBM experiments and how to analyze the resulting microarray data in order to identify the significantly bound spots and the candidate DNA binding site motif. Not discussed in this chapter are methods for protein purification, as there are numerous resources available that provide standard protocols and troubleshooting tips (Coligan *et al.*, 2005). Similarly, methods for printing DNA microarrays are discussed only briefly, as they are described in detail elsewhere (Schena, 1999) (also see Hager, 2006). Various subsequent analyses that can be performed, such as determining the cross-species conservation of PBM-derived predicted genomic binding sites, comparison of PBM data versus *in vivo* (ChIP-chip) binding data, and comparison of PBM data with gene expression data, are mentioned only briefly in this

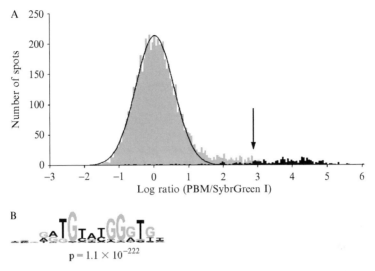

FIG. 3. Identification of the DNA binding site motif from significantly bound spots. (A) Distribution of ratios of PBM data, normalized by SYBR Green I data, for the yeast TF Rap1 bound to yeast intergenic microarrays. The arrow indicates those spots passing a *P*-value cutoff of 0.001 after correction for multiple hypothesis testing. Indicated in dark gray are spots with an exact match to a sequence belonging to the PBM-derived binding site motif. (B) Sequence logo (Schneider and Stephens, 1990) of the PBM-derived motif for the yeast transcription factor Rap1. The height of each letter is proportional to its frequency among motif matches in the set of significantly bound spots, and letters are sorted with most commonly occurring nucleotides on top; the total height at each nucleotide position indicates the information content of the sequences at that position, ranging from a minimum of 0 to a maximum possible 2 bits of information content (Schneider and Stephens, 1990). Reproduced from Mukherjee *et al.* (2004) with permission from Nature Publishing Group.

chapter, as a detailed discussion of those analyses (Mukherjee *et al.*, 2004) is beyond the scope of this chapter.

Preparation of DNA Microarrays for Use in Protein Binding Microarray Experiments

Overview

The key choice to be made in choosing what DNAs to print onto slides for use in PBM experiments is whether one wishes to synthesize a relatively low complexity microarray for directed experimentation on one or a small family of DNA-binding proteins (Bulyk *et al.*, 2001) or to synthesize a higher complexity microarray for examination of a broader set of proteins (Mukherjee *et al.*, 2004). The latter situation is discussed here, as the resulting microarrays can be used more broadly, including for uncharacterized proteins. Nevertheless, in certain situations, one might be able to restrict oneself to lower complexity microarrays that consequently are less expensive to produce.

Selection and Preparation of Double-Stranded DNAs to Be Printed

Whole-genome yeast intergenic microarrays in PBM experiments have been used in order to identify the DNA binding site specificities of the *Saccharomyces cerevisiae* TFs Rap1, Abf1, and Mig1 (Mukherjee *et al.*, 2004). These microarrays were printed with essentially all noncoding regions of the *S. cerevisiae* yeast genome (Ren *et al.*, 2000). Those genomic regions are amplified by polymerase chain reaction (PCR), with 25 cycles of 94° for 30 s, 60° for 30 s, and 72° for 90 s. The completed PCR reactions are then precipitated with 1 *M* ammonium acetate and 2 volumes of isopropanol, washed with 70% ethanol, dried overnight, and resuspended in 3× SSC printing buffer. After precipitation, the PCR products are resuspended in approximately 15 μl 3× SSC spotting buffer at a concentration between 100 and 500 ng/μl. Alternatively, PCR products may be filtered with 96-well MultiScreen PCR filter plates (Millipore, Billerica, MA) according to the manufacturer's protocols. After application of a vacuum for 10 min, plates may be air dried in a clean chemical hood. The extra filtration provided by the MultiScreen plates increases the purity of the dsDNA. Other printing buffers or additives, such as Sarkosyl or betaine, may aid in increasing the spot uniformity, thus improving the morphology of the printed spots. Different slide types will also exhibit different spot morphologies with given printing buffers; care should be taken to ensure that the chosen printing buffer is compatible with the chosen slide type.

Microarrays spotted with coding regions are also expected to aid in identifying the sequence-specific binding properties of DNA-binding proteins, despite the fact that it is currently thought that most *in vivo* functional regulatory sites will be located in nonprotein-coding regions. Because PBM experiments are *in vitro* technology, as long as there is sufficient sequence space represented on the DNA microarrays, one can expect to be able to derive a good approximation of the DNA binding site motif from PBM data. Because in essence what matters most is the representation of a large enough sequence space on the DNA microarrays, it is actually not necessary to utilize microarrays spotted with amplicons representing genomic regions from the same genome as the DNA-binding protein of interest, but rather one can use microarrays spotted with a sequence of a different genome. Nevertheless, one could of course use a genome-specific microarray, such as a promoter microarray or a CpG island microarray, matched for the protein(s) of interest, as long as such microarrays covered a sufficient amount of sequence space. Similarly, microarrays spotted with synthetic dsDNAs can also be used in PBMs (Berger *et al.*, manuscript in preparation; Bulyk *et al.*, 2001; Mukherjee *et al.*, 2004). Likewise, the dsDNAs need not be made by PCR amplification, but rather can be made by other means, such as by primer extension. PBMs have been performed successfully using microarrays spotted with PCR products whose lengths ranged from ∼60 to ∼1500 bp (Mukherjee *et al.*, 2004) and also using microarrays spotted with synthetic dsDNAs ranging from ∼35 to ∼60 bp (Bulyk *et al.*, 2001; Mukherjee *et al.*, 2004; Berger *et al.*, manuscript in preparation).

Printing and Processing of Double-Stranded DNA Microarrays

We work with a dedicated microarray facility for the production of whole-genome yeast intergenic DNA microarrays for use in PBM experiments. There, an OmniGrid 100 microarrayer (Genomic Solutions, Ann Arbor, MI) equipped with Stealth 3 pins (Telechem International, Sunnyvale, CA) is used to spot DNA onto Corning GAPS II or UltraGAPS 25 × 75-mm amino propyl silane-coated glass slides (Fisher Scientific). Approximately 0.7 nl is deposited at each spot. Any remaining, unused blank GAPS II or UltraGAPS slides from an opened package should be stored in a vacuum desiccator (Fisher Scientific) containing Drierite desiccant (Fisher Scientific). Other slide types could potentially be used. We have found that Corning GAPS II and UltraGAPS slides result in low slide background in both PBM experiments and staining with SYBR Green I.

After printing, there are a number of options for slide processing protocols. As discussed in detail later, spot morphology is very important for achieving high-quality PBM data. We have had success in achieving uniform DNA concentration within spots, as well as round spot morphology, with the following protocol. First, in order to rehydrate the spotted dsDNAs, lay slides face up and flat and blow a moisture stream across their surface for approximately 5 s using a hand-held steam humidifier (Conair, Stamford, CT). Alternatively, one can incubate the microarrays in a humid chamber at 37° for 48 h. Next, bake the microarrays in a standard laboratory oven (VWR International, West Chester, PA) at 80° for 2 h. Baking the microarrays at 80° for 2 h in a clean oven before UV cross-linking may improve intraspot uniformity. However, we caution readers that baking may also result in a decreased shelf life of the microarrays. In order to cross-link the dsDNAs to the slide surface, apply 300 mJ to each slide using a Stratalinker 1800 UV cross-linker (Stratagene, La Jolla, CA). Alternatively, amino-modified DNAs could be end-attached to amine-reactive slides, such as PDC slides (Bulyk et al., 2001). The printed, processed slides should be stored in black or orange light-protective plastic slide boxes (Fisher Scientific) placed in a vacuum desiccator equipped with Drierite desiccant.

Staining Double-Stranded DNA Microarrays

Staining the dsDNA microarrays serves two purposes: (1) it allows one to check the quality of a given print run and (2) it allows one to normalize protein-binding data by the relative amount of DNA present at each spot. We routinely stain at least one microarray from each print run and batch of processed slides in order to assess their quality; for larger printer runs, we stain one slide from early in the print run, one slide from the middle of the print run, and one slide from late in the print run.

SYBR Green I (Molecular Probes, Eugene, OR) is used to stain the dsDNA microarrays, as SYBR Green I is much more specific for double-stranded versus single-stranded DNA than ethidium bromide. To prepare the SYBR Green I staining solution, completely thaw the SYBR Green I stock solution at room temperature in the dark to prevent photobleaching and then vortex before using. Next, prepare a 1:5000 dilution of SYBR Green I in 2× SSC, 0.1% Triton X-100 (Fisher Scientific), made fresh before use, using sterile water, a stock solution of 10% Triton X-100 (Sigma), filter sterilized using a 0.22-μm filter unit (Fisher Scientific), and stored at room temperature, and a stock solution of 20× SSC, pH 7.0 (Sambrook et al., 1989), sterilized by autoclaving, and stored at room temperature. Unless otherwise noted, buffers are prepared using distilled, deionized water (ddH$_2$O). If additional slides will be stained within a week, then the prepared SYBR Green I staining solution can be wrapped in aluminum foil, stored at 4°, and reused. Mix SYBR Green I staining solution before using to stain microarrays. Because SYBR Green I is light sensitive, all possible care should be taken to avoid photobleaching in the course of the SYBR Green I staining procedure. We recommend turning off all overhead and benchtop lighting when handling these reagents, and also when handling the microarrays once they have been stained with SYBR Green I. Stain the microarrays with the SYBR Green I staining solution for 12 min by shaking in a Coplin staining jar (Fisher Scientific) for 3 × 1-in. slides at room temperature at ~100–125 rpm on a platform shaker. Shaking at speeds faster than ~125 rpm may cause the Coplin jar to tip over while shaking. Microarrays should be handled with forceps (Fisher Scientific) and gloved hands. Wash the microarrays in 2× SSC, 0.1% Triton

X-100 (Fisher Scientific) wash buffer for 5 min, and then in 2× SSC for 2 mins. All wash steps in Coplin jars are performed by shaking at room temperature at ~125 rpm on an adjustable speed platform shaker (Fisher Scientific). Immediately spin slides dry in a tabletop centrifuge by centrifuging for 5 min at 40 g [500 rpm using an IEC Centra CL3R equipped with a 224 microplate rotor (Thermo Electron, Milford, MA)].

Before scanning the stained microarrays, wipe the backs (i.e., the non-DNA sides) of the slides with a slightly dampened Kimwipe (Fisher Scientific) in order to remove any streaks or spots due to dried buffer. Very quickly blow any lint off of the spun-dry microarrays using canned air (Fisher Scientific). Scan the microarrays at a range of different laser power intensities or photomultiplier tube (PMT) gain settings per microarray using an appropriate laser and filter set [for SYBR Green I, an argon ion laser (488-nm excitation) and 522-nm emission filter]. We typically scan at ~3–6 different settings so as to capture signal intensities for even very low signal intensity spots, while ensuring that we capture subsaturation signal intensities for all spots on the microarray (Bulyk et al., 2001; Mukherjee et al., 2004). Integrating data from these multiple scans is described later in this chapter. Using a ScanArray 5000 microarray scanner (Perkin Elmer, Boston, MA), we have found the PMT gain to be optimal at 70–80%. We typically fix the PMT gain setting and vary the laser power in increments of 10–15% of total laser power so that there are no spots with saturated signal intensities in the lowest intensity scan.

Protein Binding Microarray Experiments

Overview

In PBM experiments, the epitope-tagged protein is bound directly to DNA microarrays. It is important to keep in mind the nature of the protein; if a TF is being examined, then one needs to be sure to use dsDNA microarrays. The microarrays can be spotted with either PCR products or double-stranded DNA oligonucleotides; alternatively, single-stranded oligonucleotide arrays could be made double-stranded in on-array double-stranding reactions (Berger et al., manuscript in preparation; Bulyk et al., 1999, 2001). This chapter describes the use of microarrays spotted with dsDNA, such as the whole-genome yeast intergenic microarrays used in PBM experiments on the yeast TFs Rap1, Abf1, and Mig1 (Mukherjee et al., 2004). It is also important to keep in mind that these in vitro experiments are biochemical assays that depend on the reaction conditions; the addition of certain small molecule or protein cofactors may be essential for sequence-specific binding by a given protein.

The PBM experiments themselves are neither time-intensive nor laborious and are readily completed within a single day. Indeed, much of the total experiment time is dedicated to incubations, with only relatively short periods of time being spent on reaction setups, washes, and scanning of the microarrays. Significantly more time is devoted in advance of the PBM experiments to generating the proteins to be bound to the microarrays and to analysis of the resulting microarray data.

Protein Binding Microarray Experiment Protocol

In order to minimize background, we block the DNA microarrays with a milk solution, and also include milk in both the protein-binding and the antibody-labeling reactions. First, prepare phosphate-buffered saline (PBS) according to standard protocols (Sambrook *et al.*, 1989). Next, prepare 2% and separately 4% nonfat dried milk (Sigma) by dissolving it in PBS with gentle shaking on a platform shaker at 25 rpm, either overnight or for a few hours. Sterilize the 2% and 4% milk solutions with a syringe (Fisher Scientific) equipped with a sterile 0.22-μm filter (Millipore, Billerica, MA) for the 2% milk solution or with a 0.45-μm filter (Millipore) for the 4% milk. Filtering of the 2% and 4% milk solutions serves two purposes: (1) sterilization and (2) removal of fine particulates that may contribute to speckles in the microarray images. We have found that the 4% milk solution readily clogs a 0.22-μm filter unit so we recommend using a 0.45-μm filter unit for syringe filtering of the 4% milk solution. The 2% milk solution can be syringe filtered with a 0.22-μm filter. Alternatively, 0.45-μm filters can be used for sterilizing both the 2% and the 4% milk solutions.

Before using the DNA microarrays in any reactions, first etch small grooves on the face (i.e., DNA side) of each microarray at a distance of a few millimeters beyond the borders of the printed area using a diamond-tipped glass scribe (Fisher Scientific). Microarrays should be handled by their edges with gloved hands or with a forceps; when using forceps either here or in subsequent steps, care should be taken not to scratch the microarray surface. Not only do these etches make it easier for the experimenter in subsequent steps to see where the printed microarray is on the slide surface, but they will also help confine all solutions to the center portion of the microarray throughout the PBM experiment. Note that border marks made with certain brands of laboratory markers tend to come off the slides during the various microarray reaction and wash steps.

Prepare prewetting buffer (PBS/0.01% TX-100). For these experiments, make all wash buffers fresh before use. Prepare using a stock solution of 20% Tween 20 (Sigma) or 10% Triton X-100 (Sigma), as indicated, filter sterilized using a 0.22-μm filter unit (Fisher Scientific) and stored at room temperature. Prewet the microarrays in PBS/0.01% TX-100 for at least 5 min by shaking in a Coplin staining jar (Fisher Scientific) at room temperature at ∼125 rpm on a platform shaker (Fisher Scientific). While the microarrays are being prewet, thaw the previously purified DNA binding protein of interest on ice. Purified DNA-binding protein, epitope tagged with glutathione S-transferase (GST), should have been purified according to standard protocols, aliquoted in PBS before storing at −80° in order to avoid unnecessary freeze/thaw, and stored at −80°.

Working with one microarray at a time, quickly remove the microarray from the Coplin jar, gently shake off any excess buffer, and wipe the back (i.e., the non-DNA side) and sides of the microarray with a Kimwipe. Drying the back and sides of the microarray with a Kimwipe helps prevent solution from leaking out from the edges of the coverslip. Because of the hydrophobic surface of the Corning GAPS II and UltraGAPS glass slides, drying the front of the glass slide outside the perimeter of the DNA spots, using the etched grooves as a guide, can help confine the protein or antibody solution, once dispensed onto the microarray, to the area of the microarray that contains the DNA spots. This must be done quickly so that the area containing the DNA spots does not dry. Briefly centrifuge all reaction mixtures before applying to microarrays in order to remove bubbles. When pipetting the reaction mixtures onto the microarrays, avoid pipetting any fine bubbles that may remain at the very top surface of the reaction mixtures. If a few air bubbles do become apparent once a reaction mixture has been applied to a microarray, very carefully attempt to pipette up the bubbles, while avoiding the removal of the reaction mixture. If some air bubbles still remain, they may be brought to the edge of the glass slide, and thus outside the spotted area, by gently rocking the coverslip as it is laid down on the microarray. Apply 250 μl of 2% milk solution (preblocking buffer) to the microarrays, cover with a LifterSlip™ coverslip (Erie Scientific, Portsmouth, NH), and allow to incubate in a hydration chamber for 1 h at room temperature. In applying reaction mixtures onto the microarrays, certain techniques can aid in spreading the mixture over the surface of the microarray and in increasing the homogeneity of the reaction mixture when applied to a prewet or washed microarray. The reaction mixture can be dispensed one droplet at a time, covering the entire surface where the DNA was spotted. The microarray can also be rocked back and forth to spread the reaction mixture uniformly across the spotted area. The use of LifterSlip™ coverslips helps ensure a uniform distribution of the reaction mixture over the surface of the microarray. Microarrays are incubated in a hydration chamber to prevent excessive evaporation of the reaction mixture under the coverslip. An empty pipette tip box works nicely as a hydration chamber. Lift out the tip rack, fill the bottom of the pipette tip box with about half an inch of sterile water, and replace the tip rack. Wipe off the inside of the lid and the tip rack with ethanol using a Kimwipe before every use and between reaction steps in the PBM experiments.

As soon as the preblocking of the microarray has been set up, preincubate the DNA-binding protein of interest with nonspecific competitors for 1 h at room temperature. Specifically, dilute the thawed DNA-binding protein to a 20 nM final concentration in a 100 μl protein-binding reaction mixture consisting of PBS, 50 μM zinc acetate (ZnAc) (Sigma, St. Louis, MO), 2% (w/v) nonfat dried milk (Sigma), 51.3 ng/μl salmon testes DNA (Sigma), and 0.2 μg/μl bovine serum albumin (New England Biolabs, Beverly, MA). A stock solution of 25 mM ZnAc should be stored in aliquots at $-20°$ and then added to various buffers as required; typically ZnAc is necessary only when performing PBM experiments on zinc finger proteins. The final 2% milk concentration is achieved by using a twofold dilution of the 4% milk solution that was prepared. A 100-μl reaction volume will be adequate for a printed microarray that encompasses approximately two-thirds of a 1 $\times$ 3-in. slide surface. While the microarrays and the DNA-binding protein are preblocking separately, thaw the Alexa Fluor 488-conjugated anti-glutathione-S-transferase (anti-GST) polyclonal antibody (Molecular Probes) on ice, covered with either an ice bucket lid or aluminum foil, in order to prevent photobleaching. The antibody should be stored according to the manufacturer's recommendations. For long-term storage of the Alexa Fluor 488-conjugated anti-GST antibody, we recommend aliquoting and storing at $-20°$ per the manufacturer's recommendations. Other epitope tags and/or other antibodies conjugated with other fluorophores might also be used successfully. We have not yet performed a rigorous comparison of such alternatives.

Once the 1-h preblocking step is completed, wash the microarrays once with PBS/0.1% Tween (0.1% Tween 20 in PBS) for 5 min, followed by once with PBS/ZnAc/0.01% TX100 (0.01% Triton X-100 in PBS containing 50 μM ZnAc) for 2 min. Wipe the back and sides of the microarrays with a Kimwipe, apply the protein-binding reaction mixture to the microarrays, cover with a LifterSlip™ coverslip, and allow to incubate in a hydration chamber for 1 h at room temperature. As soon as the microarray protein-binding reaction has been set up, dilute the Alexa Fluor 488-conjugated anti-GST antibody to a concentration of 0.05 mg/ml in 2% milk in 150 μl of PBS containing 50 μM ZnAc and allow to preincubate for 1 h at room temperature in the dark. As with SYBR Green I, all possible care should be taken to avoid photobleaching in the course of staining protein-bound microarrays with the Alexa Fluor 488-conjugated anti-GST antibody.

Once the 1-h protein-binding step is completed, wash the microarrays once in PBS/ZnAc/0.5% Tween (0.5% Tween 20 in PBS containing 50 μM

ZnAc) for 3 min, followed by once in PBS/ZnAc/0.01% TX100 for 2 min. Wipe the back and sides of the microarrays with a Kimwipe, apply the preincubated antibody mixture to the microarrays, cover with a Lifter-Slip™ coverslip, and allow to incubate in a hydration chamber for 1 h at room temperature, covered with either an ice bucket or aluminum foil, in order to protect from light.

Once the 1-h antibody staining is completed, wash the microarrays three times with PBS/ZnAc/0.05% Tween (0.05% Tween 20 in PBS containing 50 μM ZnAc), with each wash going for 3 min, followed by once with PBS/ZnAc for 2 min. Immediately spin slides dry in a tabletop centrifuge, wipe with Kimwipes, and blow any lint off with canned air, as described earlier. Scan the microarrays at a range of different laser power intensities or PMT gain settings per microarray using an appropriate laser and filter set [for Alexa Fluor 488, argon ion laser (488-nm excitation) and 522-nm emission filter] as described earlier.

Analysis of Protein Binding Microarray Data

Overview

The first step in PBM data analysis is to filter microarray data in order to remove noisy spots from consideration. Only then are protein-binding data normalized by SYBR Green I data. From normalized PBM data, p values are assigned to each spot in order to identify the significantly bound spots. Sequences corresponding to the significantly bound spots are then searched for candidate DNA binding site motifs with the use of a motif-finding algorithm. Additional data analyses, such as the examination of PBM-derived target genes for overrepresented functional categories of genes, comparison with ChIP-chip data, cross-species sequence conservation of the predicted binding sites, and prediction of network interactions, are beyond the scope of this chapter, and algorithms for those analyses will likely improve beyond the initial efforts described in Mukherjee *et al.* (2004).

Microarray Data Quality Control

If the DNA concentration at a particular spot is too low to allow accurate quantification of its signal intensity or if the DNA is spread nonuniformly throughout the pixels of a particular spot, accurate measurements can be more difficult. Obviously problematic microarrays can be identified visually (Fig. 4) (Berger and Bulyk, 2006). Severe problems with spot morphology, including nonuniform DNA distribution within the spots, can be attributed frequently to the choice of printing buffer and/or

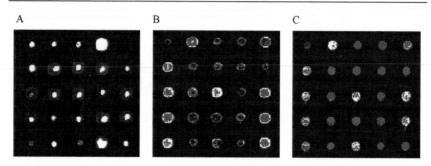

FIG. 4. Examples of DNA microarray spot quality. Identical portions of yeast intergenic microarrays printed onto Corning GAPS II slides, processed in different ways (see later) before UV cross-linking, and then stained with SYBR Green I. Images have been false colored as in Fig. 2. (A and B) Examples of microarrays with poor spot quality. In both of these cases, the DNA is distributed nonuniformly with either (A) high concentrations near the centers of spots or (B) high concentrations along spot perimeters. Both of these microarrays resulted from two separate print runs, from which microarrays were UV cross-linked without first rehydrating and baking. (C) An example of a microarray of acceptable quality. This microarray was rehydrated and then baked before being UV cross-linked. Reproduced from Berger and Bulyk (2006) with permission from The Humana Press, Inc. (See color insert.)

postprinting processing before UV-cross-linking. More subtle differences in spot quality will be identified through analysis of quantified signal intensity data. It is important to remove error-prone spots from consideration, as microarray data are subsequently used to estimate the degree of sequence-specific binding of a given DNA-binding protein to each spot. Because some spots may be noisy even after removing spots with highly variable pixel signal intensities, various additional filtering criteria can be applied to remove them from consideration.

Quantification of Microarray Signal Intensities and Quality Control

The microarray TIF images are quantified with GenePix Pro microarray analysis software (Axon Instruments, Inc.). Because the PBM experiments are one-color experiments, one can specify in GenePix to analyze the TIF image as a single-color image. Because the images are single-color images that require normalization using data from separate microarrays stained with SYBR Green I and because even after data filtering spot morphology can still be somewhat variable at times, we recommend keeping the feature size fixed in the GenePix alignment procedure. After image quantification, use Excel, GenePix, or other software to calculate the background-subtracted median intensities; for background, we recommend using the median local background in order to take into account inhomogeneous background over the microarray.

Because the microarrays were scanned at multiple laser or PMT gain settings in order to capture subsaturation signal intensity data for as many spots as possible on the microarray, relative signal intensity data over the full series of scans need to be calculated (Bulyk et al., 2001). To accomplish this task in a semiautomated fashion, we recommend using masliner (MicroArray Spot LINEar Regression) software, which combines the linear ranges of multiple scans from different scanner sensitivity settings onto an extended linear scale (Dudley et al., 2002). The dynamic range of the final PBM and SYBR Green I-stained microarrays frequently have post-masliner fluorescence intensities that span five to six orders of magnitude (Mukherjee et al., 2004).

After masliner analysis, a few normalization procedures are required. First, for microarray data for a given DNA-binding protein or SYBR Green I for each of the triplicate microarrays, remove data corresponding to any flagged spots (i.e., spots that had dust flecks, etc.). Next, normalize data from each of three triplicate microarrays according to total signal intensity so that the average spot intensity is the same for all three replicate microarrays. Then, within each individual microarray, separate data into sectors, according to their local region on the slide. For example, for whole-genome yeast intergenic arrays (Mukherjee et al., 2004), sector the spots into the 32 subgrids of the printed microarray. Normalize data again so that the mean spot intensity is the same over all the sectors. This serves to normalize for any region-specific inhomogeneities in the background or the binding and labeling reactions.

After the signal intensity normalization procedures, a number of quality control filtering criteria are applied. We have found that spots with highly variable pixel signal intensities lead to noisy PBM data; therefore, we remove from consideration any spots whose standard deviation (SD) divided by median value is greater than 2. Next, average the background-subtracted, normalized signal intensities for all spots with reliable data in at least two of the three replicate microarrays and calculate the SD/mean value. In order to eliminate spots that do not have highly reproducible data over the triplicate spots [here, triplicate microarrays, as essentially all intergenic regions were printed only once per microarray (Mukherjee et al., 2004; Ren et al., 2000)], remove from consideration any spots for which the SD/mean value is greater than 1. SYBR Green I microarray data undergo the same quality control filtering criteria, plus an additional step in which we remove from consideration any spots with fewer than 50% pixels with signal intensities greater than 2 SDs beyond the median background signal intensity; this additional criterion is applied to SYBR Green I data, as these spots presumably do not have enough DNA present to allow accurate quantification of signal intensities (Mukherjee et al., 2004), and thus this avoids the problem of dividing by noisy small values in a subsequent normalization (see next section). Although not necessary, we have found empirically that the following three additional filtering criteria help eliminate "false positive" calls (i.e., spots with no identifiable binding sites being erroneously identified as bound): (1) DNA length greater than 1500 bp; (2) low SYBR Green I raw signal intensity; and (3) low DNA density (SYBR Green I/length). These three additional filters together remove 2.7% of spots from consideration in our PBM experiments using yeast whole-genome intergenic microarrays (Mukherjee et al., 2004). Actual values for the second and third additional filtering criteria are not provided here because these values will vary somewhat among individual microarray scanners. We recommend that the user use all three of these additional criteria as suggested guidelines to employ and adjust as may be appropriate (Berger and Bulyk, 2006).

Normalization of Protein-Binding Signal Intensities by DNA Signal Intensities and Identification of Significantly Bound Spots

Once PBM and SYBR Green I microarray data have been quantified, normalized, and filtered to remove noisy data points, these separate data can then be combined in order to identify the significantly bound spots (Mukherjee et al., 2004). "Significantly bound" means those spots that are bound to a degree beyond that of nonspecific binding by the given protein.

To identify the significantly bound spots, first calculate the $\log_2$ ratio of the mean PBM signal intensity divided by the mean SYBR Green I signal intensity and create a scatter plot of those log ratios versus the spots' SYBR Green I signal intensities. Although we expect that the log ratio should be independent of DNA concentration, we have found that higher DNA concentrations, as determined by higher SYBR Green I signal intensities, appear to bind proportionately less protein. In order to restore the independence of log ratio and SYBR Green I intensity, fit the scatter plot with a locally weighted least-squares regression using the Lowess function (smoothing parameter = 0.5) (Cleveland and Devlin, 1988) of the R statistics package (www.r-project.org). Then, subtract the value of the regression at each spot from its log ratio, yielding a modified log ratio that is independent of DNA concentration. Next, plot the distribution of all log ratios as a histogram (bin size = 0.05); this distribution should resemble a Gaussian distribution, corresponding to spots bound only nonspecifically, along with a heavy upper tail, which corresponds to the spots bound specifically by the given protein.

We use the large Gaussian-like distribution to calculate P values for each spot; these P values represent the probability that the spot is contained within the distribution of nonspecifically bound spots. Thus, spots with very small P values in the heavy upper tail of the real distribution are likely to be bound sequence specifically by the given DNA-binding protein. We do not use a standard z score to calculate these P values because the heavy upper tail for some proteins can contain quite a number of spots that appear to overlap the right shoulder of the Gaussian-like distribution, thus causing the Gaussian-like distribution to be nonsymmetrical. Therefore, we instead calculate a pseudo-z score to obviate this complication. First, determine the mode of the Gaussian-like distribution by searching for the window of nine bins with the highest number of spots and taking the middle bin. Next, reflect all values less than the mode and fit these values to a Gaussian function using the Mathematica software package (Wolfram Research, Inc., Champaign, IL). This provides the mean and SD of the distribution of nonspecifically bound spots. Then, adjust the log ratios so that the peak of the distribution of nonspecifically bound spots is centered on zero. The P value for each spot is then calculated based on z, the number of SDs that the log ratio of the spot departs from the mean of the Gaussian distribution using the normal error integral (Taylor, 1997). This function is related to the probability of observing a data point greater than z SDs above the mean of a normal distribution. Thus, we are not calculating a true z score, as here we do not calculate the P values relative to all data, but rather just to the reflected left half of the distribution. These P values can be calculated easily in Microsoft Excel using the standard normal cumulative distribution function: normsdist(-z).

Finally, data must be corrected for multiple hypothesis testing (e.g., on an array with 6000 spots, at a significance level of $\alpha = 0.001$, 6 spots would be expected to be false positives just by chance alone). Therefore, we adjust all individual P values to a modified significance level using the modified Bonferroni method (Bulyk et al., 2002; Sokal and Rohlf, 1995). For significance testing of PBM data, we recommend using an initial $\alpha = 0.001$, which corresponds to α' equal to approximately 1.5×10^{-7} for the highest-ranking test case when evaluating ~6400 unique spots, which is the case for typical yeast intergenic microarray PBM data (Mukherjee et al., 2004). Spots meeting or exceeding α' are considered 'bound' at this statistically significant threshold (Fig. 3A). Users may wish to consider spots at less stringent significance thresholds accordingly.

Identification of DNA Binding Site Motif from Protein Binding Microarray Data

Once the spots that are bound at a threshold significance level have been identified, then the DNA sequences corresponding to those spots can be searched with one or more motif-finding algorithms in order to identify the likely DNA binding site motif of the given protein (Mukherjee et al., 2004). Here, "DNA binding site motif" refers to the DNA sequence specificity of the given protein; note that not all occurrences of DNA binding sites matching a given binding site motif are necessarily going to be bound (i.e., occupied) in vivo at any given time in all cells (see also later). Typically we choose to search the sequences from all the spots that have a Bonferroni-corrected *P* value less than or equal to 0.001 in order to minimize the consideration of potentially false-positive spots that would contribute noise to the motif-finding searches. For this set of input sequences, we frequently use BioProspector (Liu et al., 2001) to perform separate motif searches at each width between 6 and 18 nucleotides in order to identify the highest scoring motifs at each width. Other motif finders, such as AlignACE (Hughes et al., 2000; Roth et al., 1998), MEME (Bailey and Elkan, 1995), and MDscan (Liu et al., 2002), can also be used to identify the DNA binding site motif. We tend to choose BioProspector over other available motif-finding programs because we have found it to be the most inclusive in accepting the largest number of input sequences in the construction of yeast TF binding site motifs (Mukherjee et al., 2004). A graphical sequence logo (Schneider and Stephens, 1990) for each motif is often convenient to have available and can be generated readily (Crooks et al., 2004).

Once a motif has been identified by the given motif finder, we assess the likelihood of it being the DNA binding site motif of the given protein by considering its group specificity score (i.e., we calculate how specific the motif is to the set of bound spots, as compared to all the spots on the microarray). To perform this calculation, first identify all matches to the motif within all sequences spotted on the microarray using the program ScanACE (Hughes et al., 2000) and then calculate the group specificity score (Hughes et al., 2000) of the discovered motif(s) using MotifStats (Hughes et al., 2000); alternatively, one can perform these same calculations using the software package MultiFinder (Huber and Bulyk, 2006). If multiple motifs were discovered for a given data set, then we choose the single motif with the lowest group specificity score (i.e., most specific to the input set of spots) as the most likely TF binding site motif. In order to assess the statistical significance of the motifs resulting from analysis of the PBM experiments, repeat these calculations for computational negative control sequence sets. Specifically, perform identical motif searches on 10 separate sets of randomly selected spots from the same microarrays used for the PBM experiments, with each of the 10 random sets containing the same number of sequences as the original input set for the given PBM data set. We consider PBM-derived motifs with group specificity scores that are more significant than the group specificity scores of the corresponding computational negative control sets to correspond to the DNA binding site motif for the given DNA-binding protein (Fig. 3B). Examples of the ranges of group specificity scores for computational negative controls and for actual PBM data for yeast TFs can be found in Mukherjee *et al.* (2004).

Conclusions

PBM binding site data on the TFs Rap1, Abf1, and Mig1, determined using whole-genome *S. cerevisiae* intergenic microarrays, corresponded well (Mukherjee *et al.*, 2004) with binding site specificities determined from ChIP-chip (Lee *et al.*, 2002; Lieb *et al.*, 2001). Furthermore, comparative sequence analysis of PBM-derived binding sites indicated that many of the sites identified as bound in PBMs, including some not identified as bound in ChIP-chip data, are highly conserved in other *sensu stricto* yeast genomes and thus are likely to be functional *in vivo* binding sites that may be utilized in a condition-specific manner (Mukherjee *et al.*, 2004). Further comparisons of these data types may reveal local sequence context features that govern binding site usage *in vivo*. Improved data analysis algorithms may help to more precisely identify both the significantly bound spots and their biological relevance.

Looking beyond yeast, at this time the DNA-binding specificities of a large majority of metazoan TFs have not yet been characterized. PBM experiments may help assign functions for these proteins by determining their DNA-binding specificities and thus predicting their target regulated genes. For example, in analyzing the Gene Ontology (GO) annotation of the PBM-predicted target genes for Rap1, Abf1, and Mig1, we observed highly significant enrichment of functional categories that are consistent with the known regulatory roles of these TFs (Mukherjee *et al.*, 2004). There are hundreds of *Drosophila* TFs and thousands of mammalian TFs that could be examined in this manner. Likewise, a number of uncharacterized genes were predicted to be target genes of these yeast TFs; one may also be able to infer the biological process that these predicted target genes are involved in from the functions of the given TF. Integration of gene expression data with PBM data and GO annotations may allow the refinement of these regulatory and functional predictions, including predictions regarding in which environmental conditions a given TF is exerting an important regulatory role.

Importantly, like any other microarray experiment, multiple PBM experiments can be performed in parallel, allowing many proteins to be examined on a genome scale at once. Alternatively, instead of intergenic microarrays, one could instead create microarrays spotted with a large number of synthetic DNA sequence variants in order to examine protein–DNA interactions at a higher sequence resolution. The highly parallel nature of microarray experiments, both in terms of the spotted DNAs and the number of microarray experiments that can be performed simultaneously, provides significant cost, time, and labor savings over the use of traditional methods for examining protein–DNA interactions. The resulting data will likely contribute to the elucidation of transcriptional regulatory networks in a variety of genomes. In addition, these data may allow us to glean insights on the biophysical properties that determine protein–DNA recognition specificity. Finally, analysis of PBM data for orthologous TFs of various phylogenetic distances may provide insights into the evolution of binding sites and thus the species-specific regulatory roles of TFs.

Acknowledgments

I thank Michael F. Berger and Tom Volkert for technical assistance. This work was supported in part by National Institutes of Health grants from the National Human Genome Research Institute to M.L.B. (R01 HG002966 and R01 HG003420).

References

Bailey, T., and Elkan, C. (1995). The value of prior knowledge in discovering motifs with MEME. *Proc. Int. Conf. Intell. Syst. Mol. Biol.* **3**, 21–29.
Berger, M. F., and Bulyk, M. L. (2006). Protein binding microarrays (PBMs) for the rapid, high-throughput characterization of the sequence specificities of DNA binding proteins. *In* "Gene Mapping, Discovery, and Expression" (M. Bina, ed.). The Humana Press, Totowa, NJ.
Berger, M. F., Philippakis, A. A., Qureshi, A., He, F. S., Estep, P. W., III., and Bulyk, M. L. Determination of transcription factors' DNA binding specificities with compact, universal DNA microarrays. (Manuscript in preparation.)
Braun, P., and LaBaer, J. (2003). High throughput protein production for functional proteomics. *Trends Biotechnol.* **21**, 383–388.
Bulyk, M. (2003). Computational prediction of transcription-factor binding site locations. *Genome Biol.* **5**, 201.
Bulyk, M., Johnson, P., and Church, G. (2002). Nucleotides of transcription factor binding sites exert interdependent effects on the binding affinities of transcription factors. *Nucleic Acids Res.* **30**, 1255–1261.
Bulyk, M. L., Huang, X., Choo, Y., and Church, G. M. (2001). Exploring the DNA-binding specificities of zinc fingers with DNA microarrays. *Proc. Natl. Acad. Sci. USA* **98**, 7158–7163.
Bulyk, M. L., Gentalen, E., Lockhart, D. J., and Church, G. M. (1999). Quantifying DNA-protein interations by double-stranded DNA arrays. *Nature Biotechnology* **17**, 573–577.
Cleveland, W., and Devlin, S. (1988). Locally weighted regression: An approach to regression analysis by local fitting. *J. Am. Statist. Assoc.* **83**, 596–610.

Coligan, J., Bunn, B., Speicher, D., Wingfield, P., and Ploegh, H. (2005). "Current Protocols in Protein Science." Wiley, Edison, NJ.

Crooks, G. E., Hon, G., Chandonia, J. M., and Brenner, S. E. (2004). WebLogo: A sequence logo generator. *Genome Res.* **14,** 1188–1190.

Dudley, A., Aach, J., Steffen, M., and Church, G. (2002). Measuring absolute expression with microarrays with a calibrated reference sample and an extended signal intensity range. *Proc. Natl. Acad. Sci. USA* **99,** 7554–7559.

Hager, J. (2006). Making and using spotted DNA microarrays in an academic core laboratory. *Methods Enzymol.* **410** (this volume), 135–168.

Huber, B., and Bulyk, M. (2006). Meta-analysis discovery of tissue-specific DNA sequence motifs from mammalian gene expression data. *BMC Bioinformatics* (in press).

Hughes, J. D., Estep, P. W., Tavazoie, S., and Church, G. M. (2000). Computational identification of cis-regulatory elements associated with groups of functionally related genes in *Saccharomyces cerevisiae*. *J. Mol. Biol.* **296,** 1205–1214.

Lee, T., Rinaldi, N., Robert, R., Odom, D., Bar-Joseph, Z., Gerber, G., Hannett, N., Harbison, C., Thompson, C., Simon, I., Zeitlinger, J., Jennings, E., Murray, H., Gordon, D., Ren, B., Wyrick, J., Tagne, J., Volkert, T., Fraenkel, E., Gifford, D., and Young, R. (2002). Transcriptional regulatory networks in *Saccharomyces cerevisiae*. *Science* **298,** 799–804.

Lieb, J. D., Liu, X., Botstein, D., and Brown, P. O. (2001). Promoter-specific binding of Rap1 revealed by genome-wide maps of protein-DNA association. *Nat. Genet.* **28,** 327–334.

Liu, X., Brutlag, D., and Liu, J. (2001). BioProspector: Discovering conserved DNA motifs in upstream regulatory regions of co-expressed genes. *Pac. Symp. Biocomput.* 127–138.

Liu, X., Brutlag, D., and Liu, J. (2002). An algorithm for finding protein-DNA binding sites with applications to chromatin-immunoprecipitation microarray experiments. *Nat. Biotechnol.* **20,** 835–839.

Lockhart, D. J., and Winzeler, E. A. (2000). Genomics, gene expression and DNA arrays. *Nature* **405,** 827–836.

Mukherjee, S., Berger, M. F., Jona, G., Wang, X. S., Muzzey, D., Snyder, M., Young, R. A., and Bulyk, M. L. (2004). Rapid analysis of the DNA-binding specificities of transcription factors with DNA microarrays. *Nat. Genet.* **36,** 1331–1339.

Negre, N., Lavrov, S., Hennetin, J., Bellis, M., and Cavalli, G. (2006). Mapping the distribution of chromatin proteins by ChIP on chip. *Methods Enzymol.* **410** (this volume), 316–341.

Ren, B., Robert, F., Wyrick, J. J., Aparicio, O., Jennings, E. G., Simon, I., Zeitlinger, J., Schreiber, J., Hannett, N., Kanin, E., Volkert, T. L., Wilson, C. J., Bell, S. P., and Young, R. A. (2000). Genome-wide location and function of DNA binding proteins. *Science* **290,** 2306–2309.

Roth, F. P., Hughes, J. D., Estep, P. W., and Church, G. M. (1998). Finding DNA regulatory motifs within unaligned noncoding sequences clustered by whole-genome mRNA quantitation. *Nat. Biotechnol.* **16,** 939–945.

Sambrook, J., Fritsch, E., and Maniatis, T. (1989). "Molecular Cloning: A Laboratory Manual." Cold Spring Harbor Laboratory Press, Cold Spring Harbor, NY.

Schena, M. (1999). "DNA Microarrays: A Practical Approach." Oxford Univ. Press, New York.

Schneider, T. D., and Stephens, R. M. (1990). Sequence logos: A new way to display consensus sequences. *Nucleic Acids Res.* **18,** 6097–6100.

Sokal, R., and Rohlf, R. (1995). "Biometry: The Principles and Practice of Statistics in Biological Research." Freeman, New York.

Taylor, J. (1997). "An Introduction to Error Analysis." University Science Books, Sausalito, CA.

Zhu, H., Bilgin, M., and Snyder, M. (2003). Proteomics. *Annu. Rev. Biochem.* **72,** 783–812.

[14] Microarray Analysis of RNA Processing and Modification

By Timothy R. Hughes, Shawna L. Hiley, Arneet L. Saltzman, Tomas Babak, and Benjamin J. Blencowe

Abstract

Most RNAs are processed from precursors by mechanisms that include covalent modifications, as well as the removal of flanking and intervening sequences. Traditional methods to detect RNA processing, such as Northern blotting, reverse-transcribed polymerase chain reaction and primer extension assays, are difficult to apply on a large scale. This chapter outlines several methods for analysis of the processing and modification of RNA using microarrays. These encompass protocols for the application of homemade microarrays and custom-designed commercial inkjet microarrays and are tailored for the large-scale analysis of processing of mRNA, including alternative splicing, as well as for the analysis of processing and modification of noncoding RNA. This chapter also describes practical aspects of microarray design, sample preparation, hybridization, and data analysis.

Introduction

Analysis of RNA processing and modification using microarrays requires consideration of issues beyond those encountered in more typical microarray applications, such as quantifying mRNA abundance. These include distinguishing different RNA species with sequences that overlap with each other, using probes that often detect more than one RNA species; discriminating whether detected RNA species are cleaved or spliced and thus hybridizing only partially to the array probes; and dealing with different types of RNA modifications. These issues impact array design, experimental design, and interpretation of resulting data.

This chapter describes methods used by our groups in projects aimed at global analysis of processing and modification of noncoding RNAs (ncRNAs) [e.g., rRNA, tRNA, snoRNA (small nucleolar RNA), and miRNA (micro-RNA)] and constitutive and alternative splicing of mRNA (Babak *et al.*, 2004; Hiley *et al.*, 2005a,b; Pan *et al.*, 2004; Peng *et al.*, 2003; Xing *et al.*, 2004). We developed these methods in order to screen mutants for a variety of RNA-processing functions, to analyze the abundance of

METHODS IN ENZYMOLOGY, VOL. 410
0076-6879/06 $35.00
DOI: 10.1016/S0076-6879(06)10014-2

many different ncRNAs and splice isoforms in different cell and tissue types or in response to different conditions, and to draw correlations between different processing events. Details of specific applications are discussed in the original publications. Readers may also wish to examine the methods of other groups that have performed similar or related work (Clark *et al.*, 2002; Inada and Guthrie, 2004; Johnson *et al.*, 2003). Genomic "tiling" arrays, in which partially overlapping oligonucleotides are designed to span entire genomic regions, may also be useful for monitoring RNA-processing events in certain cases (Shoemaker *et al.*, 2001).

Microarrays for Measuring Noncoding RNA Processing
  and Modification

Noncoding RNAs are typically processed by nucleolytic cleavage to remove and degrade flanking regions in order to generate the mature, functional RNA. Many ncRNAs are also modified covalently. Figure 1 shows the rRNA processing pathway as an illustration. Individual processing intermediates are typically distinguished by Northern blotting and primer extension assays (Boorstein and Craig, 1989), while the presence or absence of modifications is usually detected by primer extension and/or HPLC. However, perturbation of most major processing defects, and many

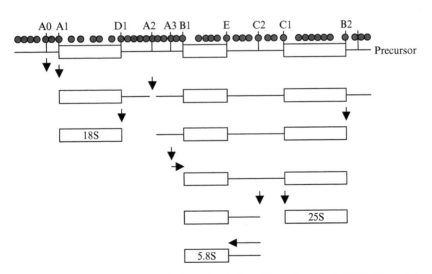

FIG. 1. Schematic of yeast rRNA processing (adapted from Peng *et al.*, 2003). Gray circles indicate oligonucleotide microarray probes. The rRNA is transcribed as a single precursor, which is processed to three mature RNA species.

covalent modifications, can also be discerned using a microarray with oligonucleotide probes detecting each of the final products, precursor fragments, and junctions (gray circles in Fig. 1). The general idea is that differences in processing and RNA abundance between two samples will result in changes in ratios in two-color microarray experiments. We have used both spotted arrays (Peng *et al.*, 2003) and inkjet arrays (Babak *et al.*, 2004; Hiley *et al.*, 2005a,b; Xing *et al.*, 2004); other array types could presumably be used, although we have not tested them. The setup costs for spotted arrays are relatively high and the costs scale with the number of probes used, but hundreds of arrays can then be manufactured at little additional cost. Inkjet arrays, which are manufactured by Agilent Technologies (Wolber *et al.*, 2006) based on user-specified designs, have a larger number of spots (currently 42,034) compared to spotted arrays and a minimal setup cost, but carry a constant per-array cost. The design principles for both are similar and could presumably be extended to other types of microarrays.

Array Design and Construction: Spotted Arrays for Monitoring ncRNA Processing and Modification

For spotted arrays, we design probes manually, essentially following Northern probe designs from the literature and using typical polymerase chain reaction (PCR) or sequencing primer design methods. Oligonucleotides spanning cleavage sites and processed RNA junctions are designed such that the two halves on either side of the cleavage site have a roughly equal T_m (melting temperature) so as to maximize the difference between binding of uncleaved and cleaved ncRNA (or unspliced and spliced RNAs). The array probes are the single-stranded reverse complement of the target RNA sequences, as covalently labeled native RNA will be hybridized directly to the array (see later). We concatenate the regions complementary to the target RNA sequences in the probes up to a total length of $\sim$60 bases to increase the likelihood of the oligonucleotides cross-linking to the poly-L-lysine coating on the slides, which in preliminary experiments appeared to be limiting. For spotting, oligonucleotides are diluted to a final concentration of 1 mg/ml in a solution of 50% dimethyl sulfoxide (DMSO), 0.1% sodium dodecyl sulfate (SDS). The DMSO/SDS solution increases spot uniformity relative to SSC buffer (Hegde *et al.*, 2000). However, it also results in larger spots, even at low temperature and humidity, so spot spacing should be adjusted accordingly ($\sim$300 μm with typical microarrayer pins). We spot directly onto homemade poly-L-lysine-coated regular microscope slides; however, other types of slides are available commercially (e.g., TeleChem SuperAmine) and can work well.

Array Design and Construction: Inkjet Microarrays

Due to the large number of spots available on inkjet-manufactured microarrays, for the purpose of monitoring ncRNA processing, we design these arrays automatically using a tiling strategy, that is, assembling a FASTA file of target sequences and using a script to select T_m-balanced probes at defined intervals, with the average T_m value set according to the specific experimental application (a script for generating array sequences is posted at http://hugheslab.med.utoronto.ca/Babak/tile_E/). For example, to detect miRNAs, which are 21–23 bases long, we use an average T_m = 54°; for detection of ncRNA processing, T_m = 53°; and for detection of covalent RNA modifications (see later) T_m ~41°. We selected these T_m values on the principle that the hybridization behavior on the array reflects that in solution, although we have not rigorously tested whether further tailoring T_m values significantly increases the performance of our arrays for the different applications. However, in general and where permissible, longer array oligonucleotides appear to be more sensitive (Hiley *et al.*, 2005a; Hughes *et al.*, 2001). Inkjet-manufactured microarrays supplied by Agilent contain oligonucleotides that are synthesized with the 3′ end covalently attached to the array; the ~30 bases at the 5′ end of a 60-mer are most important for hybridization, presumably because they are most flexible and accessible in solution (Hughes *et al.*, 2001). As mentioned earlier, for applications involving the monitoring of ncRNA processing and modification, we concatenate the target-complementary sequences in the array oligonucleotides (up to 60 bases and with no spacer), which is currently the maximum length offered for the Agilent platform. However, it is also possible to specify variable-length oligonucleotides on the arrays, and we have used this strategy in our arrays for monitoring alternative splicing (see later). Microarray oligonucleotide designs are submitted using a web-based server to Agilent Technologies, where arrays are manufactured and then delivered by courier.

Sample Preparation and Hybridization

For experiments involving yeast, total RNA samples are prepared using classical hot phenol-chloroform extraction; for mammalian samples, we use Trizol. Both methods retain low molecular weight RNAs such as tRNAs, snoRNAs, and miRNAs, which may be lost from column-based RNA isolation procedures. The size cutoff is determined primarily by the ethanol precipitation step; our current protocol (see later) recovers snoRNAs, tRNAs, and miRNAs, which are reduced greatly in many commercially obtained RNA samples. Samples are then DNase treated and labeled by direct, covalent, nonenzymatic attachment of dyes (currently purchased from Invitrogen/Molecular Probes). The attachment involves monofunctional reaction of cisplatin derivatives with guanine moieties (Wiegant et al., 1999). Our current protocol is as follows.

DNase I Treatment

1. Set up the following reaction for each sample: up to 40 μl RNA (typically 10 μg) in water, 5 μl 10× DNase reaction buffer, 5 μl DNaseI (RNase-free, 1 U/μl, Fermentas), and to 50 μl RNase-free H$_2$O.
2. Incubate at 37° for 30 min.
3. Add 150 μl of RNase-free water for a final volume of 200 μl.
4. Extract the samples with an equal volume of 25:24:1 phenol/chloroform/isoamyl alcohol (vortex, spin in microfuge for 5 min), transferring the top layer to a fresh 1.5-ml tube.
5. Precipitate the RNA with 1/10 volume 3 M NaOAc, pH 5.2, and 2.5 volumes of EtOH (mix thoroughly and freeze >30 min at –80°).
6. Thaw and spin samples at maximum speed (14K) in a microfuge at 4° (25 min).
7. Wash pellets with 70% EtOH (respin 5 min, remove EtOH); air dry.
8. Dissolve samples in 20 μl of 1× labeling buffer (TE: 5 mM Tris, 1 mM EDTA pH7.8).
9. Determine the concentration of the samples by measuring the OD$_{260}$.

Label Each of the Samples with Fluorescent Dyes

1. Set up the labeling reaction as follows (use 2–10 μg total RNA, matching the amounts for the two samples for the same array, e.g., wild type and mutant): must be <23 μl RNA in 1× buffer (TE) and to 23 μl 1× TE.
2. Heat samples to 90° for 5 min to denature RNA. Place on ice to cool.
3. Spin tubes briefly. Add 1–2 μl of the appropriate reactive dye (which is less than the current manufacturer's instructions); vortex to mix. Spin down briefly to collect sample at the bottom of the tubes. We have used Ulysis Alexa Fluor 546 (U-21652) and 647 (U-21660). Kreatech has begun selling a similar kit with Cy3 and Cy5, but we have not tested it.
4. Heat to 90° for 10 minutes. Snap cool on ice. Spin briefly to collect sample.
5. Precipitate with 1/10 volume 3 M NaOAc and 2.5 volumes EtOH (place >30 min at –80°).
6. Thaw and spin samples at 4° for 30 min at maximum speed (14K) in a microfuge. The pellets should have a clearly visible hue at this

point, which is taken as an indication that RNA is present and labeled.

7. Wash with 70% EtOH (optional if you are worried about losing the sample).
8. Air dry pellets.

If hybridizing to spotted arrays, the samples are resuspended in a combined total of ~30–50 μl hyb buffer (5× SSC, 25% formamide, 0.1% SDS). If hybridizing to Agilent arrays, the samples are resuspended and combined in a total of 500–520 μl hyb buffer [filter sterilized, containing 50 mM MES, pH 6.5 (N-morpholinoethane sulfonate [Sigma; M2933]; pH before adding to buffer), 1 M NaCl, 0.5% sodium L-sarcosine, and 33% formamide] and heat-denatured salmon sperm DNA (40 μg per 0.5 ml) added as a blocking agent. Preheating the buffer to 65° helps with dissolution; you may also need to incubate the samples at 65° for several minutes to fully solubilize the RNA.

Slides are hybridized and washed using standard protocols described in detail elsewhere (Hegde et al., 2000; Hughes et al., 2001). Either type of slide can be scanned using a commercial array scanner (we have used both Axon and Agilent scanners), and data are extracted to obtain foreground and background intensities for each spot, usually represented in an MS Excel file or the equivalent. We do not reuse either type of array in these experiments; stripping labeled RNA from the arrays seems to be more difficult than stripping DNA, presumably because of the higher stability of the RNA/DNA heteroduplex.

Data Processing and Interpretation

We initially evaluated several schemes for processing and analyzing data from RNA processing arrays (unpublished result). Regarding normalization, the major issue is that commonly used intensity-dependent smoothing methods such as Loess (Yang et al., 2002) depend on there being no inherent relationship between ratios and intensities. This assumption is not valid for ncRNA: there is a strong relationship between noncoding RNA species and abundance; for example, rRNA spots have almost always the highest intensity, whereas snoRNA spots are typically much lower in intensity. Thus, Loess smoothing would tend to minimize phenomena that impact global levels of a specific RNA species. Nonetheless, in our initial unpublished analyses, Loess smoothing resulted in higher reproducibility than when omitted and did not appear to diminish sensitivity to detect different types of RNA processing defects, including loss of rRNA and tRNA in response to Pol I and Pol III mutations. We have continued to use Loess smoothing and/or variance stabilizing normalization (Huber et al.,

2002) with tiling arrays. Both of these smoothing functions are part of the current Bioconductor suite (Gentleman *et al.*, 2004).

We analyze normalized data using standard microarray analysis approaches. Clustering analysis of mutants typically groups mutants that are involved in similar processing activities and probes that detect RNAs that are processed similarly (e.g., Hiley *et al.*, 2005a; Peng *et al.*, 2003). In addition, unlike most types of microarray data, there is an inherent ordering to the probes; one-dimensional clustering diagrams (i.e., clustering only the samples but leaving probes in the order along each RNA) can often be illuminating.

Interpretation of these array data is complicated by the fact that many of the probes can detect multiple processing intermediates and that the microarray spots do not inherently reveal the length(s) of the RNAs detected. Our attempts at automatically deconvoluting the composition of RNA species that may be present in a population of RNAs (J. Grigull and T. Hughes, unpublished result) have not been as fruitful as manually considering different types of processing defects that a given pattern is likely to represent. For example, in the rRNA processing pathway, most perturbations result in accumulation of only one or a few precursors, often accompanied by depletion of the corresponding mature RNA. This results in an increase in the ratio of these precursor fragments between mutant and wild type, as the precursors are virtually absent in the wild-type cells. Probes detecting the corresponding mature species, however, will typically show a decreased ratio, as it is usually depleted in the mutant (Fig. 2). Similar principles should

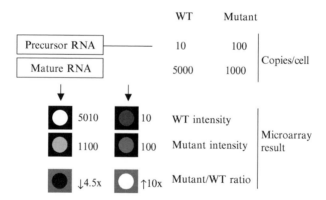

Fig. 2. Schematic of data resulting from a noncoding RNA processing array. (Top) An arbitrary RNA and its precursor. (Bottom) Expected results from microarray data representing the comparison between precursor RNA processing between a wild type (WT) and a mutant organism.

apply to virtually all processed RNAs, especially with regard to precursor accumulation. For example, accumulation of flanking sequences of tRNAs, snoRNAs, and snRNAs, as well as tRNA introns and mRNA introns, is detected in appropriate yeast RNA processing mutants (Hiley *et al.*, 2005a; Peng *et al.*, 2003).

Microarrays for Covalent RNA Modification

The principle underlying our method for detecting RNA modifications by microarrays is that binding of the modified RNA to the array probe is hindered by the base modification. Because RNA isolated from the cell is directly labeled and hybridized to the array, it still contains the modification, and in a two-color comparison between RNA that contains the modification and RNA that does not, the difference in hybridization efficiency is manifest as an elevated ratio (after normalization) because the unmodified RNA binds better to the array. We initially observed this phenomenon when analyzing a yeast dihydrouridine synthetase mutant on a spotted array design (Peng *et al.*, 2003) and have since analyzed all four yeast dihydrouridine synthetases (Xing *et al.*, 2004) and also whether all known types of RNA modifications in yeast are detectable in wild-type vs mutant comparisons using an Agilent tiling array (Hiley *et al.*, 2005b). The procedure used to survey RNA modifications is essentially the same as that used to detect RNA processing events, with the exception that in the array design, shorter probes (e.g., ~17-mers) were used in an effort to increase the differential hybridization efficiency between modified and unmodified RNAs. The types of RNA modifications that could be detected are given in Table I; the major conclusion is that, with the exception of dihydrouridine, which dramatically perturbs base architecture, virtually all modifications detectable by microarray impact Watson–Crick base pairing, as expected. Although this approach is extremely promising for rapid large-scale screening for RNA modifications, we caution that this is an immature technique; undoubtedly, the false-positive and false-negative rates are nontrivial and vary among modifications and RNA species. We suggest that any specific event of importance should be confirmed by a conventional assay.

Microarrays for Monitoring Alternative Splicing

Several different types of microarray-based strategies have been reported for monitoring alternative splicing (AS) levels in RNA isolated from cells or tissues. These encompass two different technologies for the detection of splice variants: (1) a fiber optic-based system, which monitors

TABLE I

RNA Modifications That Influence Direct Binding of RNA to Oligonucleotide Microarrays Tested in Hiley ET AL. (2004)[a]

Modification	ORF name	Gene name	Target	Detected by microarray?
Methylation	m^2_2G	YDR120C	TRM1	tRNA 26	Yes
	m^5U	YKR056W	TRM2	tRNA 54	No
	$2'O\ CH_3$	YDL112W	TRM3	tRNA 18	No
	m^5C	YBL024W	TRM4	tRNA 34, 40,38,49	No
	m^1G	YHR070W	TRM5	tRNA 37	Yes
	$2'O\ CH_3$	YBR061C	TRM7	tRNA 32,34	No
	m^7G	YDL201W	TRM8	tRNA 46	No
	mcm5U/ mcm5s2U	YML014W	TRM9	tRNA 34	No
	m^1G	YOL093W	TRM10	tRNA 9	Yes
	m^1A	YNL062C	GCD10	tRNA 58	Yes
	m^1A	YJL125C	GCD14	tRNA 58	Yes
	m^6_2A	YPL266W	DIM1	18S rRNA 3'-terminal loop	Yes
	$2'O\ CH_3$	YCL054W	SPB1	25S rRNA 2918	No
	m^5C	YNL061W	NOP2	Unknown	No
	$2'O\ CH_3$	YDL014W	NOP1	Unknown	No
Dihydrouridylation	D	YML080W	DUS1	tRNA 16/17	Yes
	D	YNR01W	DUS2	tRNA 20	Yes
	D	YLR401C	DUS3	tRNA 47	Yes
	D	YLR405W	DUS4	tRNA 20: A/20:B	Yes
Pseudouridylation	Ψ	YPL212C	PUS1	tRNA 27	No
	Ψ	YNL292W	PUS4	tRNA 55	No
	Ψ	YOR243C	PUS7	Unknown	No
	Ψ	YLR175W	CBF5	rRNA	No
Adenosine deamination		YGL243W	TAD1	tRNA 37	No
i^6A formation		YOR274W	MOD5	tRNA 37	No

[a] ORF name and gene name correspond to mutants that were compared to wild type; differential hybridization to the target RNA was assessed.

the hybridization of PCR-amplified DNA fragments from splice junction regions to oligonucleotides anchored on beads (Yeakley et al., 2002), and (2) more conventional types of microarray systems, which involve the hybridization of fluorescently labeled cDNA to oligonucleotides that have been either robot spotted (Clark et al., 2002) or synthesized in situ on glass slides (Blanchette *et al.*, 2005; Fehlbaum *et al.*, 2005; Johnson *et al.*, 2003; Le *et al.*, 2004; Pan *et al.*, 2004; Wang *et al.*, 2003). In the latter case, several groups have reported the application of microarrays that combine multiple oligonucleotide probes to both exon "body" and splice junction sequences for the detection of individual AS events. Typically, these arrays contain at least one probe per exon sequence (e.g., for the alternative exon and flanking constitutive exons) and one probe for each splice junction sequence in the mRNA variants representing inclusion and skipping of an alternative exon (see Fig. 3). Using commercial high-density microarrays such as those manufactured by Agilent Technologies or Affymetrix, it is possible to represent oligonucleotide probe sets for several thousand AS events on one array. A critical component of these microarray systems is to have an effective algorithm for accurately inferring the relative changes in splicing levels between pairs of samples or the actual percentage inclusion/exclusion levels of alternative exons in spliced transcripts.

This section describes methods for the design and application of AS arrays that have been developed (Pan *et al.*, 2004). Our system employs Agilent arrays based on our own oligonucleotide probe designs. The system employs a computer algorithm, Generative Model for AS Array Platform (GenASAP), developed by our collaborators Ofer Shai, Quaid Morris, and Brendan Frey (Department of Electrical and Computer Engineering, University of Toronto). Details of this algorithm are described elsewhere (Shai *et al.*, 2006), and we expect a user-friendly version

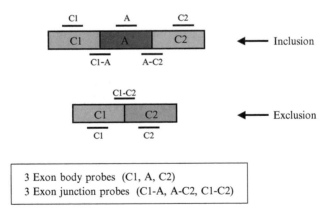

FIG. 3. Microarray oligonucleotide probes for profiling AS. A set of six probes is used to monitor each cassette exon AS event, or alternative 5'/3' splice site selection event (not shown). Data generated with these probe sets, and processed using the GenASAP algorithm, allow percentage alternative exon inclusion/exclusion levels to be inferred with considerable accuracy [(Pan *et al.*, 2004); refer to text].

to be available to the research community shortly. Interested readers are encouraged to contact Brendan Frey (frey@psi.toronto.edu) or B.J.B. (b.blencowe@utoronto.ca) for further information.

Alternative Splicing Microarray Design

Our AS microarrays typically utilize sets of six probes per AS event (Fig. 3). The AS events correspond to either "cassette"-type alternative exons (i.e., exons that are flanked by intron sequences in primary transcripts and that are either skipped or included in the final spliced mRNA) or by the alternative usage of *cis*-competing 5' or 3' splice sites. Each probe set is designed to detect a single AS event that has been identified using expressed sequence tag (EST) and cDNA sequence data. As such, our arrays are designed to monitor alternative exon inclusion/exclusion levels of sequence-validated AS events and not to identify new alternative exons, which would require a more comprehensive set of oligonucleotide probe sequences to exon and splice junction sequences. A detailed description of the steps used to identify AS events in sequence data is given elsewhere (Pan *et al.*, 2005), and our strategy is quite similar to automated methods used by others.

Oligonucleotide Probe Design

The following steps are used to obtain oligonucleotide probe sequences specific for individual exon and splice junction sequences. Exon sequences are masked using RepeatMasker and then screened for all possible sequences (overlapping by 10 bases) that have a T_m of 66–67° [the T_m is calculated using the nearest-neighbor method, essentially as described (Breslauer *et al.*, 1986) and using the values described in Sugimoto *et al.* (1996)] and with an average GC content. Splice junction probe sequences are selected to have the same overall T_m as the exon body probes and a T_m that is balanced on either side of the junction. Simple repeat sequences such as CCC and GGG are avoided. To minimize cross hybridization, probe sequences were searched against the entire RefSeq database of full-length cDNA sequences using BLAST. The ΔG difference between the top BLAST hit (the probe-candidate target sequence) and the next best hit (i.e., a similar sequence that may cross-hybridize to the probe candidate) was determined. Probe candidates were ranked according to a scoring scheme that takes into account each of the aforementioned design characteristics and the ΔG difference. AS events represented by the overall top-ranking probe sets were incorporated on the microarray. Oligonucleotide probes selected as described earlier will have an average length of

38–39 nucleotides. Finally, no more than a total of 12 T's are added to the 3′ end of each probe sequence (up to the maximum allowable length of oligonucleotide probes on Agilent microarrays of 60 nucleotides) to provide a spacer between the glass slide and the hybridizing portion of the oligonucleotide, which has been shown to facilitate efficient hybridization (Hughes et al., 2001).

In addition to unique probe sets for monitoring individual AS events, we also include ~50 duplicate probe sets using the same or slightly different probe sequences. This affords the ability to internally monitor reproducibility between probe sets on the same microarray. Several hundred negative-control probes that do not match known expressed sequences (e.g., intergenic sequences) are also included on the microarray and serve to establish background and nonspecific hybridization levels during data processing (see later). Other controls can include randomly generated sequences as well as randomly selected probes from the array with the sequence shuffled.

Labeled cDNA Preparation, Hybridization, and Processing of AS Microarrays

The following protocol has been used for the preparation of labeled cDNA for hybridization to inkjet-printed AS microarrays, manufactured by Agilent Technologies. An overview of the protocol is shown in Fig. 4. Polyadenylated [poly(A)$^+$] RNA isolated from cells or tissues is used for the synthesis of fluorescent dye (fluor)-labeled first-strand cDNA. An indirect, postsynthetic cDNA-labeling method is used and requires the incorporation of amino-allyl-modified dNTPs during cDNA synthesis. This method affords more efficient and uniform label incorporation than direct incorporation of relatively bulky Cy-dye-derivatized nucleotides. cDNA prepared from poly(A)$^+$ RNA from each cell/tissue source to be analyzed is labeled separately with each fluor dye and is hybridized on different microarrays. This permits compensation for signals arising from possible biases arising from label incorporation, hybridization, and microarray variation.

Preparation of Poly(A)$^+$ RNA

Total RNA is extracted from cell or tissue sources using Trizol (Invitrogen) according to the manufacturer's instructions. Poly(A)$^+$ RNA is isolated from total RNA by two sequential chromatographic purifications on oligo(dT) cellulose (New England Biolabs). Alternatively, a number of commercial kits are available for the purification of poly(A)$^+$ RNA. If sufficient poly(A)$^+$ RNA is isolated, a denaturing agarose gel can be used to assess the purity and integrity of poly(A)$^+$ RNA prior to cDNA

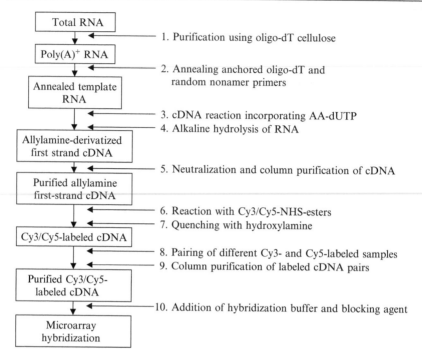

FIG. 4. Outline of the postsynthetic cDNA-labeling protocol. Poly(A)$^+$ RNA is reverse transcribed into allylamine-derivatized, first-strand cDNA. After degradation of the RNA template, amine groups on the cDNA are reacted with Cy3/Cy5-NHS esters. Excess NHS ester is neutralized with hydroxylamine. Cy3/Cy5-labeled cDNA is purified for hybridization with the microarray. Refer to the text for protocol details. poly(A)$^+$, polyadenylated; cDNA, complementary DNA; AA-dUTP, aminoallyl-dUTP; NHS, N-hydroxy succinimide.

synthesis. Pure and intact poly(A)$^+$ RNA (at least 1 μg for ethidium bromide staining) will appear as a smear from approximately 0.5 to 6 kb and will not contain prominent rRNA bands. Smaller amounts can be analyzed for purity and integrity using an Agilent bioanalyzer.

Preparation of Allylamine-Derivatized cDNA

For each reverse transcription reaction, 1–2 μg of poly(A)$^+$ RNA is primed using both an anchored oligo(dT) (T$_{18}$ VN; 0.25 μg) and a random nonamer primer (1 μg). The poly(A)$^+$ RNA–primer mixture is heated at 65° for 5 min and is then transferred to ice for 4 min. A premixed solution of the following components is added to the RNA–primer mixture to a final reaction volume of 20 μl (final concentrations are indicated): first-strand buffer (1×; Invitrogen), dithiothreitol (10 mM), dNTPs (0.5 mM), AA-dUTP [5-(3-aminoallyl)-dUTP; 0.5 mM; Ambion], and Superscript II

(200 units; Invitrogen). The reaction is incubated at room temperature for 10 min and then at 42° for 1 h.

Following first-strand cDNA synthesis, RNA is hydrolyzed by the addition of 10 μl of a 1:1 solution of 1 N NaOH and 0.5 M EDTA, pH 8.0, followed by incubation at 65° for 20 min. Prior to purification, 10 μl of 1 M Tris–Cl, pH 7.6, is added and the reaction volume is adjusted to 100 μl with sterile, deionized water. The cDNA reaction is purified using the QIAquick PCR purification kit (Qiagen) with the following modifications to the manufacturer's instructions. cDNA is applied to the column three sequential times, with centrifugation at 1500g for 1 min. Washing is performed twice using 80% ethanol, with centrifugation at 3400g for 1 min. Purified cDNA is eluted twice (40 μl each time) using sterile, deionized water preheated to 65° and allowing the column to stand for 1 min before centrifugation at 14,000g. The purified cDNA is then dried in a Speed-Vac and can be stored at –80° or labeled immediately.

CyDye Labeling of Allylamine-Derivatized cDNA

Purified allylamine-derivatized cDNA is reacted with N-hydroxy succinimide esters of Cy3 or Cy5 fluorescent dyes (Cy3 and Cy5 monofunctional dye packs, Amersham Biosciences). These reactive dyes were originally developed for protein labeling, and each vial contains enough CyDye to label 10 cDNA samples. In the protocol presented here, the contents of each Cy3 or Cy5 vial are used immediately, and any remainder is discarded, as NHS esters are readily hydrolyzed by water, and dyes that are opened and stored in aliquots have decreased activity. Single-use CyDye aliquots are available commercially; however, we have not optimized the current protocol for the use of those dyes.

Labeling buffer is prepared by dissolving the contents of one carbonate–bicarbonate capsule (Sigma) in 25 ml sterile, deionized water to yield a 200 mM carbonate–bicarbonate buffer, which is adjusted to pH 8.5–9 by the addition of HCl (approximately 120 μl; buffer stored <1 week). Dried, purified allylamine-derivatized cDNA is resuspended in 4 μl of sterile, deionized water. Each CyDye (Cy3 and Cy5) is resuspended well in anhydrous DMSO (15 μl). Labeling buffer (30 μl) is added to each resuspended CyDye, which is then mixed and aliquoted immediately (4 μl) to each cDNA sample. The labeling reactions are incubated in the dark for 1 h, with mixing at 0 and 30 min, and are then quenched by the addition of 4 M hydroxylamine (4 μl), followed by incubation in the dark for 15 min.

Labeled samples to be hybridized on the same microarray are arranged in Cy3/Cy5 pairs and combined, and the volume is adjusted to 100 μl with sterile, deionized water. Unincorporated dye is separated from labeled cDNA using the QIAquick PCR purification kit (Qiagen) as described earlier, except columns are washed three times with PE buffer (Qiagen), and purified, labeled cDNA is eluted twice (30 μl each time) with EB buffer (Qiagen). The eluate should have a purple tint. The cDNA recovery and extent of CyDye incorporation can be calculated by UV spectroscopy using the molar extinction coefficients of the dyes (Cy3: 150,000 $M^{-1}cm^{-1}$ at 550 nm; Cy5: 250,000 $M^{-1}cm^{-1}$ at 650 nm).

Microarray Hybridization of Labeled cDNA

Purified CyDye-labeled cDNA is added directly to the hybridization buffer (see earlier discussion) to a final volume of 0.5 ml. MES stock buffer is prepared in sterile, deionized water and adjusted to pH 6.5 using NaOH prior to preparation of the hybridization buffer.

After the addition of labeled control oligonucleotide probes (Agilent Technologies), which hybridize to Agilent microarray positive control spots, the sample is heated at 65° for 5 min. Samples are injected immediately into the hybridization chamber at 42° or, if necessary, kept protected from light at 42° and injected as soon as possible. After incubation for 18–22 h on a rotating platform in a hybridization oven at 42°, slides are washed (rocking ~30 s in 6× SSPE, 0.005% sarcosine, then rocking ~30 s in 0.06× SSPE) and scanned using a 4000A microarray scanner (Axon Instruments).

Conclusion

The methods described here are useful for large-scale, rapid analysis of mechanisms and regulation of ncRNA processing, RNA modification, and mRNA splicing and may also be useful for validating genome annotations, for example, by confirming predicted splice junctions. These processes and their determinants are major areas in postgenome research. The general approaches will be useful for a wide range of important questions in biomedical research and diagnostics and should be applicable to other types of microarray technologies and to the analysis of RNA processing in all organisms.

Acknowledgments

We thank Qun Pan for contribution to the development of the alternative splicing microarrays and for help with writing the chapter. We are grateful to CIHR and NCIC for funding.

References

Babak, T., Zhang, W., Morris, Q., Blencowe, B. J., and Hughes, T. R. (2004). Probing micro RNAs with microarrays: Tissue specificity and functional inference. *RNA* **10,** 1813–1819.

Blanchette, M., Green, R. E., Brenner, S. E., and Rio, D. C. (2005). Global analysis of positive and negative pre-mRNA splicing regulators in Drosophila. *Genes Dev.* **19**, 1306–1314.

Boorstein, W. R., and Craig, E. A. (1989). Primer extension analysis of RNA. *In* "RNA Processing Part A, General Methods" (J. Dahlberg and J. Abelson, eds.), Vol. 180, pp. 262–288. Elsevier Science & Technology Books.

Breslauer, K. J., Frank, R., Blocker, H., and Marky, L. A. (1986). Predicting DNA duplex stability from the base sequence. *Proc. Natl. Acad. Sci. USA* **83**, 3746–3750.

Clark, T. A., Sugnet, C. W., and Ares, M., Jr. (2002). Genomewide analysis of mRNA processing in yeast using splicing-specific microarrays. *Science* **296**, 907–910.

Fehlbaum, P., Guihal, C., Bracco, L., and Cochet, O. (2005). A microarray configuration to quantify expression levels and relative abundance of splice variants. *Nucleic Acids Res.* **33**, e47.

Gentleman, R. C., Carey, V. J., Bates, D. M., Bolstad, B., Dettling, M., Dudoit, S., Ellis, B., Gautier, L., Ge, Y., Gentry, J., Hornik, K., Hothorn, T., Huber, W., Iacus, S., Irizarry, R., Leisch, F., Li, C., Maechler, M., Rossini, A. J., Sawitzki, G., Smith, C., Smyth, G., Tierney, L., Yang, J. Y., and Zhang, J. (2004). Bioconductor: Open software development for computational biology and bioinformatics. *Genome Biol.* **5**, R80.

Hegde, P., Qi, R., Abernathy, K., Gay, C., Dharap, S., Gaspard, R., Hughes, J. E., Snesrud, E., Lee, N., and Quackenbush, J. (2000). A concise guide to cDNA microarray analysis. *Biotechniques* **29**, 548–550, 552–554, 556 passim.

Hiley, S. L., Babak, T., and Hughes, T. R. (2005a). Global analysis of yeast RNA processing identifies new targets of RNase III and uncovers a link between tRNA 5' end processing and tRNA splicing. *Nucleic Acids Res.* **33**, 3048–3056.

Hiley, S. L., Jackman, J., Babak, T., Trochesset, M., Morris, Q. D., Phizicky, E., and Hughes, T. R. (2005b). Detection and discovery of RNA modifications using microarrays. *Nucleic Acids Res.* **33**, e2.

Huber, W., von Heydebreck, A., Sultmann, H., Poustka, A., and Vingron, M. (2002). Variance stabilization applied to microarray data calibration and to the quantification of differential expression. *Bioinformatics* **18**(Suppl .1), S96–S104.

Hughes, T. R., Mao, M., Jones, A. R., Burchard, J., Marton, M. J., Shannon, K. W., Lefkowitz, S. M., Ziman, M., Schelter, J. M., Meyer, M. R., Kobayashi, S., Davis, C., Dai, H., He, Y. D., Stephaniants, S. B., Cavet, G., Walker, W. L., West, A., Coffey, E., Shoemaker, D. D., Stoughton, R., Blanchard, A. P., Friend, S. H., and Linsley, P. S. (2001). Expression profiling using microarrays fabricated by an ink-jet oligonucleotide synthesizer. *Nat. Biotechnol.* **19**, 342–347.

Inada, M., and Guthrie, C. (2004). Identification of Lhp1p-associated RNAs by microarray analysis in *Saccharomyces cerevisiae* reveals association with coding and noncoding RNAs. *Proc. Natl. Acad. Sci. USA* **101**, 434–439.

Johnson, J. M., Castle, J., Garrett-Engele, P., Kan, Z., Loerch, P. M., Armour, C. D., Santos, R., Schadt, E. E., Stoughton, R., and Shoemaker, D. D. (2003). Genome-wide survey of human alternative pre-mRNA splicing with exon junction microarrays. *Science* **302**, 2141–2144.

Le, K., Mitsouras, K., Roy, M., Wang, Q., Xu, Q., Nelson, S. F., and Lee, C. (2004). Detecting tissue-specific regulation of alternative splicing as a qualitative change in microarray data. *Nucleic Acids Res.* **32**, e180.

Pan, Q., Bakowski, M. A., Morris, Q., Zhang, W., Frey, B. J., Hughes, T. R., and Blencowe, B. J. (2005). Alternative splicing of conserved exons is frequently species-specific in human and mouse. *Trends Genet.* **21**, 73–77.

Pan, Q., Shai, O., Misquitta, C., Zhang, W., Saltzman, A. L., Mohammad, N., Babak, T., Siu, H., Hughes, T. R., Morris, Q. D., Frey, B. J., and Blencowe, B. J. (2004). Revealing global regulatory features of mammalian alternative splicing using a quantitative microarray platform. *Mol. Cell* **16**, 929–941.

Peng, W. T., Robinson, M. D., Mnaimneh, S., Krogan, N. J., Cagney, G., Morris, Q., Davierwala, A. P., Grigull, J., Yang, X., Zhang, W., Mitsakakis, N., Ryan, O. W., Datta, N., Jojic, V., Pal, C., Canadien, V., Richards, D., Beattie, B., Wu, L. F., Altschuler, S. J., Roweis, S., Frey, B. J., Emili, A., Greenblatt, J. F., and Hughes, T. R. (2003). A panoramic view of yeast noncoding RNA processing. *Cell* **113**, 919–933.

Shai, O., Morris, Q. D., Blencowe, B. J., and Frey, B. J. (2006). Inferring global levels of alternative splicing isoforms using a generative model of microarray date. *Bioinformatics* **22**, 606–613.

Shoemaker, D. D., Schadt, E. E., Armour, C. D., He, Y. D., Garrett-Engele, P., McDonagh, P. D., Loerch, P. M., Leonardson, A., Lum, P. Y., Cavet, G., Wu, L. F., Altschuler, S. J., Edwards, S., King, J., Tsang, J. S., Schimmack, G., Schelter, J. M., Koch, J., Ziman, M., Marton, M. J., Li, B., Cundiff, P., Ward, T., Castle, J., Krolewski, M., Meyer, M. R., Mao, M., Burchard, J., Kidd, M. J., Dai, H., Phillips, J. W., Linsley, P. S., Stoughton, R., Scherer, S., and Boguski, M. S. (2001). Experimental annotation of the human genome using microarray technology. *Nature* **409**, 922–927.

Sugimoto, N., Nakano, S., Yoneyama, M., and Honda, K. (1996). Improved thermodynamic parameters and helix initiation factor to predict stability of DNA duplexes. *Nucleic Acids Res.* **24**, 4501–4505.

Wang, H., Hubbell, E., Hu, J. S., Mei, G., Cline, M., Lu, G., Clark, T., Siani-Rose, M. A., Ares, M., Kulp, D. C., and Haussler, D. (2003). Gene structure-based splice variant deconvolution using a microarray platform. *Bioinformatics* **19**(Suppl. 1), i315–i322.

Wiegant, J. C., van Gijlswijk, R. P., Heetebrij, R. J., Bezrookove, V., Raap, A. K., and Tanke, H. J. (1999). ULS: A versatile method of labeling nucleic acids for FISH based on a monofunctional reaction of cisplatin derivatives with guanine moieties. *Cytogenet. Cell Genet.* **87**, 47–52.

Wolber, P. K., Collins, P. J., Lucas, A. B., De Witte, A., and Shannon, K. W. (2006). The Agilent *in situ*-synthesized microarray platform. *Methods Enzymol.* **410** (this volume), 28–57.

Xing, F., Hiley, S. L., Hughes, T. R., and Phizicky, E. M. (2004). The specificities of four yeast dihydrouridine synthases for cytoplasmic tRNAs. *J. Biol. Chem.* **279**, 17850–17860.

Yang, Y. H., Dudoit, S., Luu, P., Lin, D. M., Peng, V., Ngai, J., and Speed, T. P. (2002). Normalization for cDNA microarray data: A robust composite method addressing single and multiple slide systematic variation. *Nucleic Acids Res.* **30**, e15.

Yeakley, J. M., Fan, J. B., Doucet, D., Luo, L., Wickham, E., Ye, Z., Chee, M. S., and Fu, X. D. (2002). Profiling alternative splicing on fiber-optic arrays. *Nat. Biotechnol.* **20**, 353–358.

[15] Mapping the Distribution of Chromatin Proteins by ChIP on Chip

By Nicolas Nègre, Sergey Lavrov, Jérôme Hennetin, Michel Bellis, and Giacomo Cavalli

Abstract

The ChIP on chip method combines chromatin immunoprecipitation (ChIP) with hybridization on DNA microarrays (chip). The ChIP technique allows one to obtain a DNA sample enriched in sequences

METHODS IN ENZYMOLOGY, VOL. 410
0076-6879/06 $35.00
DOI: 10.1016/S0076-6879(06)10015-4

bound by transcription factors or chromatin-associated proteins. Usually, ChIP is used to test whether specific candidate sequences are bound by a transcription factor, but microarrays are a powerful tool that allows testing large pools of sequences at once. This chapter presents the pipeline of a ChIP on chip method that can be applied to map the binding sites of chromatin-associated proteins along *Drosophila* chromosomes at different developmental stages. This chapter provides protocols for ChIP, for quality control tests of ChIP samples, for microarray design, for hybridization of the ChIP samples onto microarrays, and for initial analysis of the data. In addition, this chapter discusses the most important steps in each of the protocols as well as the importance of bioinformatic analysis in order to extract valuable biological information from the data sets.

Introduction

The genomes of many model organisms have been completely sequenced since the beginning of the year 2000, including the human genome (Lander *et al.*, 2001) and the genome of the fly *Drosophila melanogaster* (Adams *et al.*, 2000). The postgenomic era coincides with an extensive use of microarray techniques that were initially described in 1995 for the study of gene expression in *Arabidopsis thaliana* (Schena *et al.*, 1995). There are three major types of microarray platforms available for most of the organisms studied, based on the depositing of cDNA amplicons, of oligonucleotides, and on polymerase chain reaction (PCR) amplicons (Gupta and Oliver 2003). In *Drosophila*, most of these microarrays have been used to study whole genome expression profiles by hybridizing mRNA isolated from different developmental stages (Arbeitman *et al.*, 2002; Stolc *et al.*, 2004) or mutant versus wild-type flies.

Complementary information that is needed to understand how genes are regulated at a genomic scale comes from the identification of transcription factor-binding sites. Several regulatory pathways involving transcription factors are well described genetically and, in the best cases, the *trans*-acting regulatory factors and their binding sites on the regulatory sequences of individual target genes have been identified. What is not known are the global features of gene networks involving these factors. The development of microarray technologies allows one to extend the analysis from individual genes to the whole genome and to reach an integrated understanding of gene regulatory phenomena. To this aim, one useful approach, originally developed in yeast, is the so-called ChIP on chip technique (Ren *et al.*, 2000), which combines ChIP (Orlando *et al.*, 1997) with hybridization of the ChIP samples onto microarrays.

This chapter describes a pipeline for application of the ChIP on chip method to the study of the distribution of chromatin-associated proteins along *Drosophila* chromosomes. Because this pipeline has no steps that are specific to the *Drosophila* system, except for the genomic sequence used for the design of tiling-path microarrays, it should be adaptable to ChIP samples coming from virtually any kind of biological material and organism. Our laboratory is particularly interested in understanding the role of the Polycomb (PcG) and trithorax-group (trxG) proteins in the regulation of gene expression during development (Ringrose and Paro, 2004). These proteins are well-known regulators of the patterned expression of homeotic genes. PcG proteins maintain the repressed state of homeotic genes in the segments where they were initially repressed by early acting transcription factors, whereas trxG proteins maintain active states in the appropriate regions of the body. PcG and trxG proteins act as multimeric complexes that associate to chromatin through regulatory sequences named PREs and TREs for Polycomb and trithorax response elements (Simon *et al.*, 1993), but most of these factors do not bind directly to DNA. Instead, they are recruited at target genes by other proteins and by specific histone modification marks. This is the case of the proteins forming the Polycomb Repressive Complex 1 (PRC1) (Francis *et al.*, 2001), which are recruited to PREs through the recognition of trimethylation of lysine 27 of histone H3 via the chromodomain that is present in the Polycomb (PC) protein (Min *et al.*, 2003). Although PcG and trxG proteins bind more than 100 loci on polytene chromosomes, based on immunostaining assays (Franke *et al.*, 1992), only 11 PREs have been described molecularly to date, and most of them are located in the homeotic gene complexes Bithorax and Antennapedia. Based on their sequence, no extensive homology has been identified between these PREs, and moreover their size varies, as well as their distance to target genes. This is a typical example of the type of problems encountered during the study of gene regulatory proteins. Few target genes are known, and thus their mutual relations as well as their organization along chromosomes are a mystery. Moreover, because the *in vivo* target sequences for the factors of interest cannot be predicted reliably, bioinformatic studies cannot be used in order to identify the chromosomal distribution of these factors. Therefore, ChIP on chip is the most powerful approach toward reaching this goal. For instance, this approach has been used successfully for mapping PcG protein binding to a large number of genomic sites in mouse tumor cell lines (Kirmizis *et al.*, 2004).

Chromatin Immunoprecipitation

In most ChIP applications, chemical reagents such as formaldehyde are used to cross-link proteins to DNA in a covalent manner. A specific antibody is then used to immunoprecipitate the protein of interest. The cross-linking step allows one to coimmunoprecipitate the DNA fragments bound to this protein. The final product of this assay is a pool of genomic DNA sequences, usually ranging from 200 bp to 1 kb. Ideally, only sequences specifically cross-linked to the protein of interest should be recovered after immunoprecipitation (IP), but in practice any genomic DNA sequence binds with low affinity to the beads used to recover the immunoprecipitated material. The background DNA sticking to the beads in ChIP experiments is controlled for by performing a control sample (named "mock") where no antibody is added during the IP. The specific protein targets are enriched above this background in a good ChIP sample, while the mock IP shows the same background without specific enrichments. For this reason, detection of the enriched fragments is done by comparing the IP sample with the mock sample.

Multiple agents can be used for cross-linking, such as UV (Zhang *et al.*, 2004) or methylene blue (Liu *et al.*, 2000). The most commonly used is formaldehyde (HCHO), a chemical that induces protein–protein and protein–DNA cross-links, and is particularly convenient as it allows studying not only DNA-binding proteins, but also proteins that do not bind DNA directly but associate to chromatin via other proteins.

A critical issue concerning the use of formaldehyde is the accessibility of the biological material to this agent. *Drosophila* ChIP protocols have been originally developed for cultured cells or dechorionated *Drosophila* embryos that are permeable to the formaldehyde solution. Because the other developmental stages of *Drosophila* are characterized by the presence of an impermeable cuticle, we developed a ChIP protocol that can use any tissue by directly crushing the material in the presence of formaldehyde.

Solutions and Materials

- Formaldehyde 37%
- Glycine solution 2.5 *M*
- 10% *N*-lauroylsarcosine
- Buffer A1: 60 m*M* KCl, 15 m*M* NaCl, 4 m*M* MgCl$_2$, 15 m*M* HEPES (pH 7.6), 0.5% Triton X-100, 0.5 m*M* dithiothreitol (DTT), protease inhibitors cocktail (complete, EDTA free, Roche, use following manufacturer's instructions)
- Lysis buffer: 140 m*M* NaCl, 15 m*M* HEPES (pH 7.6), 1 m*M* EDTA, 0.5 m*M* EGTA, 1% Triton X-100, 0.5 m*M* DTT, 0.1% sodium deoxycholate, 0.05% SDS, protease inhibitor cocktail

• Centricon columns (YM-100, cutoff 100 kDa from Amicon): Should be blocked by bovine serum albumin (BSA) (1 mg/ml) in phosphate-buffered saline (PBS) and then washed with PBS only before use
• Branson sonifier 250, equipped with a microtip of 5 mm diameter
• Sodium azide
• 5 mg/ml RNase A (DNase free)
• Protein A-Sepharose (PAS) suspension: 100 mg of CL-4B (Amersham, 17–0780–01) PAS should be resuspended in 1 ml of lysis buffer + 0.1 mg/ml BSA for 50% suspension. Wash in lysis buffer two to three times and equilibrate in lysis buffer for 1 h at 4° on a rotating wheel. Store up to 1 week at 4°.
• TE: 10 mM Tris–HCl (pH 8.0), 1 mM EDTA
• Elution buffer 1: 10 mM EDTA, 1% SDS, 50 mM Tris–HCl (pH 8)
• Elution buffer 2: 0.67% SDS in TE
• Proteinase K solution (250 μl): 0.5 μl of 20 mg/ml glycogen solution, 5 μl of 20 mg/ml proteinase K stock, 244.5 μl TE
• 4 M lithium chloride
• Polynucleotide kinase (Promega) 10,000 units/ml
• Klenow fragment polymerase (Promega) 5000 units/ml
• T4 DNA ligase (Promega) 400,000 units/ml with its supplied buffer
• 10 μM linker DNA: two oligonucleotides must be annealed: (i) a 24-mer of sequence 5′-AGA AGC TTG AAT TCG AGC AGT CAG (phosphorylated at 5′ end); (ii) a 20-mer of sequence 5′-CTG CTC GAA TTC AAG CTT CT. Store oligonucleotides in small aliquots at –20°. To produce the linker, mix 20 μl of 100 μM 24-mer phosphorylated primer and 20 μl of 100 μM 20-mer primer in 160 μl of TE. Incubate in a PCR machine using the following program: 5 min at 70° (remove secondary structures), 5 min at 55° (annealing). Let cool down slowly (0.01°/s) to 25°, incubate 2 h at this temperature and then cool down (0.01°/s) to 4°. Aliquot the 10 μM linker and store at –20°.
• 2 mM dNTP mix: prepare a mix with 20 μl of each dNTP (Promega, PCR grade, 100 mM) in 920 μl H_2O. Make 10× 100-μl aliquots and store at –20°.
• Taq polymerase (Promega) 5000 units/ml
• QIAquick PCR purification kit (Qiagen)

Formaldehyde Cross-Linking of Chromatin from Any Drosophila *Tissue*

1. Homogenize the material (about 150 to 200 mg of dried biological material is sufficient for four to five independent immunoprecipitations) in 5 ml of buffer A1 + formaldehyde at 1.8% final concentration (290 μl of 37% solution) at room temperature using first a Potter homogenizer and then a Dounce with type A pestle (three strokes). Wait 15 min (total time

starting from beginning of homogenization). Add glycine solution to 225 mM (540 μl of 2.5 M solution for 6 ml of cross-linked mixture), mix, and incubate 5 min. Put on ice.

2. Transfer the homogenate into a 15-ml tube. Centrifuge 5 min at 4000g at 4°. Discard the supernatant. Add 3 ml of buffer A1, resuspend the pellet, and spin down the same way. Repeat the washing step three times.

3. Wash once in 3 ml of lysis buffer without SDS. Spin down 5 min at 4000g.

4. Resuspend the cross-linked material in 0.5 ml of lysis buffer + SDS to 0.1% and N-lauroylsarcosine to 0.5%. Incubate 10 min at 4° in a rotating wheel.

5. Sonicate the chromatin to make it soluble. Parameters for the Branson sonifier are as follow: power 2, duty cycle 100%, four times 30 s with 2-min intervals. Sonication must be made in a conical-shaped tube with the tip of the sonifier going just at the limit between the cylindrical part and the conical bottom. Place the tube on melting ice during sonication to avoid excess heating of the chromatin.

6. Rotate 10 min on a rotating wheel at 4°. Transfer into Eppendorf tubes and centrifuge 5 min at room temperature at maximum speed. Transfer supernatant to a new tube. Add another 0.5 ml of lysis buffer to the pellet and rotate for 10 min. Repeat centrifugation and combine the supernatants. Centrifuge the combined supernatants 2× 10 min at maximum speed. The supernatant from this stage is the chromatin extract. Cross-linked chromatin can be stored at –80° for several months at this stage. Add sodium azide to 0.02% for storage.

7. Put the chromatin extract in a Centricon column, centrifuge 3× 40 min (or more) at 1000 g while adding lysis buffer. At least 3 volumes of lysis buffer should pass through the column. Bring the final volume of chromatin extract to 1 ml with lysis buffer.

Chromatin Immunoprecipitation Procedure

1. To an amount of chromatin corresponding to 150 mg of biological material, suspended in a final volume of 1 ml (in lysis buffer), add 100 μl of PAS suspension for preincubation. Incubate several hours or overnight at 4° and then remove PAS. Cross-linked chromatin at this stage can be stored several days at 4° or frozen at –70°.

2. Check DNA recovery as follows. From the 1-ml solution described earlier, take a 100-μl aliquot, add proteinase K up to 100 μg/ml and SDS to 1%, incubate 6 h at 60°, then 20 min at 70°, add RNase A to 50 μg/ml, and incubate for an additional 2 h at 37°. Extract the DNA with phenol-chloroform and precipitate with ethanol. After resuspension, run the sample on an agarose gel to check amount and size of DNA.

3. Separate the chromatin sample into $4\times$ 250-μl aliquots (one aliquot is sufficient for one IP). Immunoprecipitate the chromatin by adding the antibody (Ab) of interest. The amount of Ab should be determined empirically. In the case of affipure Abs, it might be in the order of 2–5 μg. If the concentration is not known, one should use the same concentration as used in regular IP experiments using soluble protein extracts. Do not forget to perform a control IP without Ab, called "mock." For microarray hybridization, one mock should be done for each different antibody, as the hybridization requires a large amount of material.

4. Incubate 4 h at 4° on a rotating wheel, add 50 μl of PAS suspension, and incubate 4 h or overnight. Spin down PAS and proceed to washes.

5. Wash PAS $4\times$ with lysis buffer, followed by $2\times$ with TE (without protease inhibitors). Each wash is for 5 min at 4°, using 1 ml of solution.

6. For elution of precipitated material, spin down PAS. Add 100 μl of elution buffer 1, mix, and incubate 10 min at 65°. Spin down PAS and transfer supernatant to a new tube. Add 150 μl of elution buffer 2 to PAS, mix, centrifuge at full speed, and transfer the eluate to a tube together with the eluate from the first centrifugation. The combined material is the "chromatin precipitate" (approximately 250 μl).

7. Incubate the chromatin precipitate 6 h (or overnight) at 65° to reverse cross-links. Add 250 μl of proteinase K solution and incubate at 50° for 2 to 3 h.

8. Add 55 μl of 4 M LiCl and 500 μl of phenol-chloroform. Mix and centrifuge at full speed at room temperature. Transfer the aqueous phase to a new tube and precipitate with 1 ml of cold 100% ethanol. Wash with 750 μl of 70% ethanol. Spin down and dry the precipitate.

9. Dissolve in 25 μl of water. This is the "native" ChIP sample. At this stage, one can quantify the amount of DNA on a slot blot hybridized with genomic DNA, including a standard curve with known amounts of genomic DNA in the blot. One should normally obtain around several nanograms in total from these samples (see Fig. 1A). Microarray hybridizations require 1 μg of DNA per slide; therefore an amplification step by blunt-end linker-mediated PCR (LM-PCR) is necessary.

10. Add 3 μl of T4 ligase buffer and 1 μl (10 U) of polynucleotide kinase (PNK). Incubate at 37° for 30 min.

11. Inactivate PNK at 68° for 20 min. Cool down to 37°. Repair staggered DNA ends by adding 1 μl of 2 mM dNTP mix and 1 μl of Klenow fragment (5 U) and incubating for 30 min at 37°. Inactivate at 75° for 10 min and cool to 4°.

12. Take 9 μl of the reaction mixture (store the remaining amount at $-20°$ for further experiments) and add 1 μl of 10 mM ATP, 1 μl of 1 μM linker, and 4 U of T4 DNA ligase. Incubate overnight at 4°.

13. Use the ligation mixture directly for PCR amplification [add to tube with ligation reaction 44 μl of bidistilled H_2O (PCR grade), 8 μl of Taq

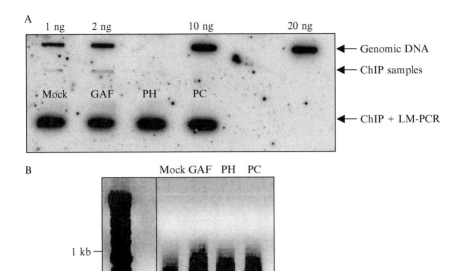

FIG. 1. Quality of ChIP samples. (A) Quantification of ChIP samples by slot blot experiment. The upper lane corresponds to a standard curve of Drosophila genomic DNA. One microliter of the ChIP samples before and after PCR amplification was deposited below. The amount of DNA present in ChIP samples before PCR amplification is 0.7 ng/μl for mock, 1.1 ng/μl for GAF, and 0.3 ng/μl for both PH and PC (the total volume of the samples being 25 μl). After PCR amplification, the amount of DNA is homogeneous in all ChIP samples and is approximately 50 ng/μl (total volume 100 μl). (B) An example of the distribution of DNA fragments obtained after PCR amplification of ChIP samples on a 1% agarose gel. The size of the fragments ranges from 200 bp to 1 kb and is centered on 500 bp.

polymerase buffer, 10 μl of 2 mM dNTP mix, 5 μl of 25 mM solution of MgCl$_2$, 0.5 μl *Taq* polymerase, 1.5 μl of 20-mer primer]. PCR amplification is as follows: 2 min at 94° (1 cycle), 34 cycles of 1 min at 94°/1 min at 55°/3 min at 72°, and 1 min at 94°/1 min at 55°/10 min at 72° (1 cycle).

14. Purify PCR products on Qiagen QIAquick columns (following manufacturer's instructions). The final elution of the DNA is in 100 μl of Qiagen "EB buffer" (10 mM Tris–HCl, pH 8.5).

Quality Control of ChIP Samples

After obtaining the ChIP sample and before using it for microarray hybridizations, several quality controls should be done.

1. Determine the concentration of the eluted DNA by a spectropho-tometer (OD_{260}).

2. Check the size and yield of DNA further by agarose gel electrophoresis (see also Fig. 1B).

3. Finally, measure the DNA amount carefully by slot blot as follows: load a standard concentration curve of *Drosophila* genomic DNA of 1, 2, 10, and 20 ng onto a slot blot apparatus using a GeneScreen Plus nylon membrane (NEN). In addition, load approximately 1 to 4 ng of the DNA samples onto the slot blot. Estimate these amounts from the OD measure and agarose gel. Note that, in general, samples out of the QIAquick columns are approximately 50 ng/μl, so they should be diluted appropri-ately. Hybridize these samples using 50 ng of radiolabeled *Drosophila* genomic DNA and measure the signals by a phosphorimager in order to evaluate DNA yields precisely (Fig. 1A).

Evaluation of ChIP Specificity by Southern Blot

Before applying the ChIP sample on microarrays, a good evaluation of ChIP quality is needed. In our case, we utilize well-characterized PREs that represent positive controls. These DNA fragments are digested by appropriate restriction enzymes and are loaded onto agarose gels that are hybridized with the DNA from the ChIP. Briefly, the procedure is as follows.

1. Run 1 μg of the different restriction digested plasmids (containing the fragments of interest) on 1% agarose gels in duplicate.

2. Transfer the gels to two nylon membranes (NEN membranes, GeneScreen +) by Southern blot overnight using 10× SSC. Then, cross-link the membranes with UV or by incubation for 1 h at 80° in a hybridization oven.

3. Label 50 ng of both mock and ChIP samples separately by incorporating ^{32}P-labeled dCTP. We typically perform random priming reactions with a PrimeaGene kit (Promega).

4. Hybridize the two membranes with the mock or the ChIP sample overnight at 65°.

5. Wash the membranes 4× 10 min in 0.1% SDS, 2× SSC at 65°.

6. Finally, expose the hybridized membranes 1 to 5 days in a Storm Phosphorimager cassette (Molecular Dynamics). Scan the cassette with the Storm Phosphorimager and quantify the signals with the ImageQuant software. For each fragment of the digested plasmid, the "fold change" is calculated as a ratio between the ChIP and the Mock samples. One example of such a test is shown in Fig. 2.

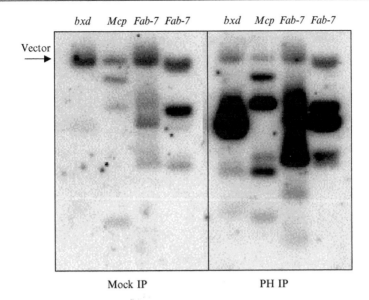

FIG. 2. Specificity of ChIP samples by Southern blot. Known PREs from the *Bithorax-Complex* (BX-C) region (*Fab-7, Mcp,* and *bxd*) were used as positive controls to test the quality of ChIP. Digested plasmids were migrated on a 1% agarose gel and then transferred onto a nylon membrane by Southern blot. One membrane is hybridized with the mock sample (left). Several fragments are enriched in the PH IP sample (right), as expected from previous analysis. One fragment also hybridizes strongly in the mock sample, illustrating that some of the fragments are overrepresented upon blunt-end ligation-mediated PCR.

Chips for ChIPs

When mapping binding sites of gene regulatory factors, cDNA arrays or arrays containing oligonucleotides spanning the coding regions of genes (typically used for transcriptome studies) are not convenient, as the binding sites of these factors are in regulatory regions that are often located far away from the coding part of the gene. Ideally, one should put the whole euchromatic genome on microarrays. This requires building tiling path arrays, where the features printed on the glass slides can be oligonucleotides or PCR fragments continuously covering large genomic regions. We assembled such a tiling path to cover 7 Mb of *Drosophila* X-chromosome euchromatin by PCR products.

Microarray Design and Production of PCR Amplicons

PCR amplicons were designed based on release 2 of the *Drosophila* genome sequence. We chose a 1.9-kb average amplicon size in order to cover a large genomic area and to keep the cost at a reasonable level. For the design, we segmented the genome sequence in fragments within which the oligonucleotide design was made. For segmentation, we applied the following set of rules. Adjacent sequences overlap on a window of 100 bp, and the oligonucleotide design for the reverse oligonucleotide of the first amplicon and the forward oligonucleotide of the second fragment is made in this overlapping window. Adjacent fragments have been designed in such a way to have different sizes of 1.7, 1.9, and 2.1 kb, respectively. This allows checking the specificity of the PCR products by visual inspection of their migration in agarose gels (Fig. 3). Finally, repeat sequences were not masked, and retrospective data mining revealed that they did not perturb the obtention of profiles. However, it must be noted

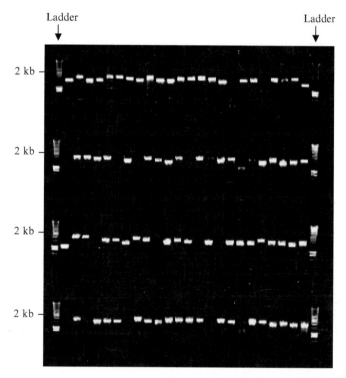

Fig. 3. Quality of PCR products deposited on tiling path arrays. An example of a 1% agarose gel loaded with 96 PCR products obtained from *Drosophila* genomic DNA is shown. In most of the cases, the size and yield of the amplicons are correct. Amplification of the fragments giving a low amount of DNA, a wrong size, or a band that is not unique is repeated under modified PCR conditions in order to recover them. If the rework fails, these amplicons are excluded from the microarrays.

here that the proteins studied do not generally bind to repeat sequences. It might be wise to separate repeats from unique sequences in the case of proteins potentially binding to DNA repeats. The quality of each PCR product was checked systematically on agarose gels for yield, size, and specificity (see Fig. 3). The percentage of success was about 84%. A total of 4153 fragments were thus obtained. In addition, the specificity of the amplicons was verified by sequencing several randomly chosen PCR fragments.

This PCR-based approach allows tiling across large genomic regions without missing any sequences and, as such, allows one to describe the distribution profile of the protein of interest. However, it involves a large amount of work, and the size of the amplicons determines the resolution of data. Although we designed amplicons of roughly 2 kb, the maximal resolution of the ChIP technique is 200 to 500 bp. Therefore, reducing the amplicon size would improve data quality, but the trade-off is a greater amount of work. Furthermore, this approach requires one to have access to a facility that can array the amplicons and print custom slides. Finally, one disadvantage regarding this PCR-based design is the heterogeneity between the fragments in terms of size, melting temperature, and CG content. This results in some degree of variability in the efficiency of hybridization of different amplicons and requires the use of good normalization and calibration tools (see later). Thus, this approach is very well suited for routine systematic analysis of large regions of interest, but extending it to the whole genome is a challenging endeavor that is better suited for a community effort than for a single laboratory.

An alternative to this approach is to use oligonucleotide tiling path arrays. The current options are short oligonucleotides (from Affymetrix) in very high-density microarrays or longer oligonucleotides that can be spotted at relatively high densities (>100,000 per slide) by maskless photolithography (Kirmizis et al., 2004; Stolc et al., 2004). The first option allows for whole-genome tiling, but the use of short oligonucleotides might be a source of specificity problems in some genomic regions. The second option warrants better specificity, but requires the design of multiple microarrays for whole-genome tiling, multiplying the amount of ChIP material required for one experiment. In addition, tiling path oligonucleotide arrays are expensive. In summary, both alternatives have advantages and drawbacks, and one should evaluate carefully which approach is best suited for the project.

Printing and Processing Microarrays Prior to Hybridization

This step strongly depends on the microarray equipment available. In our case, printing is done with two different setups, both giving satisfactory results. PCR products are deposited into 384-well plates and dried. Then, the PCR products are resuspended into 23 μl of 3× SSC or in 50% dimethyl sulfoxide using a robot. The plates are allowed to resuspend during 1 week in order to obtain a homogeneous solution before printing. The printing of microarrays is done by using (i) an Omnigrid arrayer (GeneMachines) equipped with 16 steel pins that fit the 384-well plates used to print poly-L-lysine-coated glass slides or (ii) a Lucidea array spotter (Amersham) with a 24-pen print head used to print amino-silane-coated microarray glass slides. Because most experiments are performed with slides printed with the first setup, the following protocols correspond to this type of slides. After the printing step, slides are stored inside a slide box and put into a dry chamber filled with drierite. Before using the slides, they should be processed as follows.

Solutions and Materials

- Diamond pen
- Humid chambers suitable for incubating regular glass slides (Sigma)
- Metal slide racks (VWR) and glass chambers where the metal tray can fit (VWR)
- Succinic anhydride (Sigma)
- 1,2-Dichloroethane (DCE; Acros Organics)
- 1-Methylimidazole (Fluka)
- 95% ethanol
- Drierite (Sigma)
- Slide boxes and dry chamber

Procedure

1. Mark the boundaries of each array on the back (using a diamond pen) to mark the area for deposition of the coverslip during hybridization (see later). Also, label the date and ID of each array at the side of the slide.

2. Fill the bottom of a humid chamber with the maximum volume of 4× SSC in such a way that slides do not touch the liquid once deposited in the chamber.

3. Put the humid chamber under a binocular and place a lamp on it.

4. Place arrays face down over 4× SSC and cover the chamber with its lid. This procedure will rehydrate the arrays.

5. Allow the slides to rehydrate for 5 to 15 min. With the help of the binocular, check that each spot has grown to its maximum size and is homogeneous in shape. Check also that the spots do not touch.

6. Once rehydrated, snap dry each array (face up) on an 80° inverted heat block for 3 s. Stop immediately when you see a heat wave crossing the slide.

7. Place the arrays in a metal slide rack.

8. Prepare the blocking solution. Dissolve 1.9 g of succinic anhydride in 380 ml of DCE. The solution should appear turbid.

9. Add 4.75 μl of methylimidazole to the solution, which will become immediately clear.

Transfer the clear solution to a glass chamber immediately and plunge the slide rack containing the processed arrays in this solution.

10. Shake for 1 h at room temperature on an orbital shaker. During this time, prepare 2 liters of boiling distilled water in a 4-liter beaker.

11. Discard the blocking solution appropriately (chemical waste), and fill the chamber with DCE. Wash slides for 1 min. Again, discard the DCE in the chemical waste.

12. Gently place the metallic slide rack into the boiling water and shake slowly for 2 min.

13. Remove the rack and place it into a glass chamber containing 95% ethanol. Repeat five times for 30 s each and then place the rack into a chamber containing distilled water.

14. Bring the chamber to the centrifuge and spin the rack for 5 min at 1000 rpm.

15. The slides are ready to use and can be stored in a slide box placed in a dry chamber.

Labeling and Hybridization

One microgram of the mock IP and 1 μg of the IP samples should be labeled for hybridization. Labeling consists of random priming by insertion of fluorophore-coupled nucleotides, such as Cy3 and Cy5 dCTPs. It is essential to perform labeling and double hybridization in the two color channels at the same time, as labeling and hybridization are a major source of variability (see later). Labeling and hybridization are performed as follows.

Solutions and Materials

- Bioprime DNA labeling kit: for random-priming labeling of ChIP samples do not use the dNTP mix provided in the kit
- 10× dNTP mix: prepare using PCR grade dNTPs: 1.2 mM each dATP, dGTP, and dTTP, 0.8 mM dCTP, diluted in TE
- Cy5-dCTP and Cy3-dCTP (Amersham, 1 mM stocks)
- Microcon YM-30 filter (Amicon/Millipore)
- Yeast tRNA (Invitrogen; make a 5-mg/ml stock)
- Poly(dA-dT) (Sigma; make a 1-mg/ml stock)
- TE: 10 mM Tris–HCl (pH 8.0), 1 mM EDTA

- 20× SSC
- SDS 20%
- 24 × 30-mm coverslips
- Hybridization chambers (Proteomic Solutions)
- Glass chambers and their adapted glass slide racks for the SSC washes
- Wash IA: 1.14× SSC; 0.0285% SDS
- Wash IB: 1.14× SSC
- Wash II: 0.228× SSC
- Wash III: 0.057× SSC
- Centrifuge adapted for slide racks (e.g., Jouan CR412 tabletop centrifuge with swinging rotor and adapters for 96-well plates)
- Genepix scanner 4000B (Axon Instruments)

Procedure

1. Add 1 μg DNA of the sample to be labeled into a single PCR tube (no strips).

2. Add bidistilled H_2O to bring the total volume to 21 μl. Add 20 μl of 2.5× random primer/reaction buffer mix (from the Bioprime kit). Boil 5 min (in a PCR machine) and then place on ice.

3. On ice, add 5 μl of the10× dNTP mix.

4. Add 2 μl Cy5-dCTP in the experimental IP and 2 μl Cy3-dCTP into the mock IP. As microarrays require multiple replicates, interchange the fluorophore for further replicates of the same ChIP sample. This "dye swap" step allows one to eliminate variations of signals due to fluorophores.

5. Add 1 μl of the Klenow fragment from the Bioprime kit and incubate at 37° for 1 to 2 h. Then stop the reaction by adding 5 μl 0.5 M EDTA, pH 8.0.

6. Purify the DNA probe using a Microcon column as follows. Add 450 μl TE, pH 7.4, to the stopped labeling reaction. Lay onto Microcon filter. Spin 10 min at 8000g. Invert and spin 1 min at 8000g to recover the purified probe into a new tube ($\sim$20–40 μl volume). Combine the experimental and the mock-purified probes (Cy3 and Cy5 labeled) in a new Eppendorf tube. Then add 20 μg yeast tRNA and 20 μg poly(dA-dT) [this blocks hybridization to poly(A) tails of cDNA array elements] in 400 μl TE, pH 7.4.

7. Concentrate with a Microcon column as described earlier (8000g, $\sim$18 min, then check volume every 1 min until volume is 26.9 μl or less). Then adjust the volume of the probe mixture to exactly 26.9 μl with bidistilled H_2O.

At this step, it is possible to bring the labeled probe immediately to the DNA array facility (or one can keep it frozen for later use). For immediate hybridization, proceed as follows.

8. For a total volume of 33 μl, covering a 24 × 30-mm coverslips, add 4.95 μl of 20× SSC (final concentration of about 3×) to the 26.9 μl of the probe and mix by finger tapping. Then add 1.21 μl of 5% SDS. Boil the sample for 2 min and spin down in a small tabletop centrifuge to recover condensed droplets. Allow the sample to cool down to room temperature before applying to arrays.

9. Pipette two droplets (15 μl) of 3× SSC in the bottom of the hybridization chamber. Pipette 33 μl of the hybridization mixture onto a coverslip and then place the microarray glass slide (with the DNA side facing down) on the drop of hybridization mixture. Flip the slide quickly. Try to avoid air bubbles. If any bubbles form, remove them by tapping gently with forceps. Stick two thin Parafilm strips at the left and right sides of the coverslip and put the slide in the hybridization chamber (DNA face up). Add some small droplets of 3× SSC at the external side of each Parafilm strip in order to humidify the chamber. Take particular care to avoid these droplets from mixing with the hybridization mixture below the coverslip (the Parafilm strips should effectively isolate the coverslip). Close the hybridization chamber and incubate in a water bath at 42° for 15 min to adjust the temperature. Then place the chambers into a water bath at 64° for 16 to 18 h for hybridization.

10. After hybridization, wash the arrays several times in SSC. Dry the hybridization chamber with a towel. Unscrew the chamber and remove all traces of water.

11. Place the arrays, singly, in a rack inside the wash IA bath. Without any movement, wait for the coverslip to slip down by itself. If necessary, use forceps to carefully remove the coverslip. Avoid scrapping the slide. Agitate vigorously for 2 min.

12. Remove arrays from the rack and rinse in wash IB bath without a rack to remove traces of SDS. Then place the arrays in a rack in the wash II bath. Agitate vigorously for 2 min and transfer the rack into wash III bath. Again, agitate vigorously for 2 min. Then transfer the rack in an appropriate centrifuge already balanced and centrifuge for 2 min at 1000g in order to dry the slides (proceed rapidly to avoid uneven drying of the slides).

13. Store the slides in a dark chamber to avoid decay of the signal. Preferably scan the arrays immediately using a Genepix scanner.

Data Acquisition

1. Before scanning, do a prescan and examine the histogram of the intensity distribution of signals for each channel. The two channels should show overlapping curves. If this is not the case, change the voltage settings

of the scanner for each channel in order to obtain comparable signals. Once the settings are good, scan the array in the selected area of the slide at maximum resolution and save the picture.

2. Open the .gal file generated during the spotting of the slides that gives the ID of each feature on the slide with its coordinates. It generates a .gps file that corresponds to the virtual grid of the slides. Save each .gps file generated for each microarray.

3. Adjust the grid of the .gps file to the picture of the array. Each feature of the grid should correlate to the physical feature on the slide. The Genepix software calculates the signal intensity of each feature inside the virtual feature of the grid. It also calculates the median values for each feature and several normalization factors. We use the ratio between the average median intensities in the two channels as the normalization factor.

4. Once the grid is well set up, press the "analyze" button. It will generate a text tab separated file that can be saved as a .gpr file. This file gives all the necessary information of single microarray experiments, for example, date and name of the array, and, for each feature, its ID, coordinates, intensity in each channel, and ratio of medians. The normalization factor is given in a separate line in this file.

5. Open the .gpr file with Microsoft Excel software.

6. In order to obtain the normalized ratio for each feature ID, also known as "fold change," create a new column, paste in it the value of the normalization factor, and then multiply it by the ratio of medians.

7. In the Excel file, reorder the features by their position in the genome. In our case, the ID number of the PCR fragments corresponds to the absolute coordinate of the sequence in the *Drosophila* genome. Use the graphic assistant to generate the distribution profile of the protein along the chromosomes. One example of such a distribution is seen in Fig. 4, which presents distribution of the chromatin proteins PC, Polyhomeotic (PH), and GAGA factor (GAF), over a genomic region containing the gene *polyhomeotic*, a well-known target of all these factors (Bloyer *et al.*, 2003). A further example is shown in Fig. 5 for the PH protein in the whole 7 Mb tiling path of the X chromosome.

One important point should be noted here. One might expect that all signals of individual spots on the chips hybridized with the mock IP sample are equal, as they should represent background hybridization to spotted fragments of roughly the same size (2 kb in our case). In fact, the background sequences are not totally represented randomly. The PCR efficiency, the spotting procedure, and the sequence complexity of each DNA fragment affect the efficiency of hybridization of each spot. However, the level of nonspecific attachment of each fragment to the IP beads

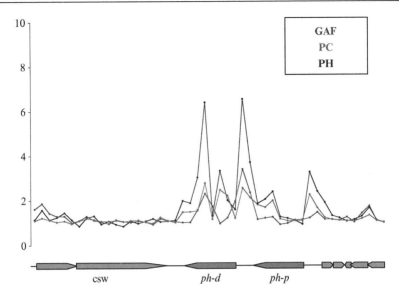

FIG. 4. Profile of chromatin proteins in the *ph-d ph-p* region. This graph represents the distribution of three chromatin components (GAF in red, PH in blue, and PC in light blue) on the *ph-d ph-p* region. The *ph* gene is duplicated in this region and encodes for the PH protein itself. Strong binding sites for the three proteins occur upstream and downstream of the *ph-d* gene. A binding site for PC and PH but not for GAF is observed upstream of the *ph-p* gene. (See color insert.)

and minor skews in the ligation-mediated PCR amplification procedure may affect the amount of fragment present in the ChIP sample. All these sources of variation lead to signal intensities varying 5- to 10-fold for different fragments, even in control IP samples without antibody. An example of this variation is seen in Fig. 5, where the signal levels for the mock IP sample do not form a flat curve, but have peaks and valleys throughout the tiling path. Therefore, one should not be surprised to find signal variation in the mock IP channel. Fortunately, however, the ChIP technique is reproducible, and the dynamic range of detection in microarray experiments is large. Therefore, this variation in signal intensity does not prevent specific enrichments from being detected reproducibly, as can be seen in Fig. 5 (bottom).

Statistical Analysis of Data

At this point, basic data, as well as a graphic display of the protein distribution profile, are available. However, this is not the end of the analysis. It is crucial to determine how significant the profiles that have

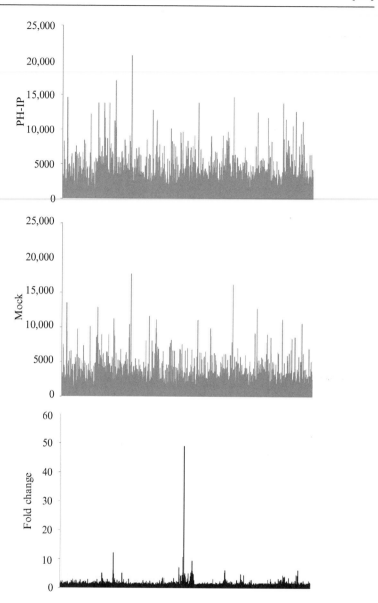

Fɪɢ. 5. Distribution of the hybridization signal on the microarrays. These graphs present the distribution of signals on the X chromosome. The normalized signal intensity for both ChIP (top) and mock (center) samples is shown in gray. The two graphs show a very well-correlated profile, consistent with most of the PH/mock ratios being close to 1. Only a minor fraction of the signals is enriched in the PH-ChIP sample, giving rise to the fold change graph (bottom).

been obtained are and what chromosomal features correspond to the peaks and valleys seen in the profiles. A large body of knowledge exists concerning statistical analysis of microarray data in the field of transcriptional studies. Even if no particular method, algorithm, or software prevails upon others for detecting significant variations, trivial methods have been found to be generally inappropriate and should be dismissed. The "fold change" is one of these simple methods. In this method, ratios greater than an arbitrary threshold (often set at twofold) are considered significant. Unfortunately, the fold change technique is inefficient in detecting small variations in the upper part of the signal range (generating false negatives); however, it selects a number of nonsignificant variations in the lower part of the signal range (generating false positives). This method is still used because it is simple to apply and because results are expressed in an intuitive way. Several alternative approaches have been proposed to correct for its obvious defects, but they are generally not applied to ChIP on chip analysis.

In order to improve robustness, specificity, and sensitivity of the analysis, the rank difference analysis of microarray (RDAM) method was adapted to ChIP on chip analysis (Nègre *et al.*, 2006). RDAM (Martin *et al.*, 2004) replaces raw signals by their rank, expressed on a 0–100 scale, which is a powerful normalizing procedure. Also, RDAM does not reduce replicated signals to their means, but instead considers variations, expressed as rank differences, between individual experimental points. Finally, RDAM estimates the total number of truly varying signals, assigns a p value to each signal variation, characterizes the selection of a signal using the false discovery rate in order to estimate the expected amount of false positive signals that may be present in the selected sample, and estimates the percentage of truly varying signals included in the selection (sensitivity). A detailed description of the RDAM method is found in Martin *et al.*, 2004. Its application to ChIP on chip data has shown that this method is superior to the fold change, that is, setting an arbitrary threshold, for example, 2 in the fold change method would incur both in false positive and in false negative estimations that are not statistically significant when analyzed by RDAM. For this reason, we recommend analyzing ChIP on chip data using methods that are able to estimate the statistical significance of the enrichment of each feature, such as RDAM.

The statistical comparison of microarray samples also shows an interesting feature of ChIP on chip samples, namely that sample labeling and hybridization vary in independent experiments. For technical reasons, one set of the slides has been hybridized in autumn 2003 and the end of the replicates during winter 2004 in a different laboratory. We applied a variant of the method (Hennetin and Bellis, 2006) to display the relationships between all experimental points by tracing a dendrogram. Instead of doing comparisons between an experimental point and the "median" point and selecting points by the p value, we made internal comparisons inside each condition (IP channel vs mock channel) and retained the 200 most varying probes in each comparison. Then we calculated the distance between any two comparisons as follows: d = 1 − #com/200, where #com is the number of common probes between the two 200 top lists. As observed on the corresponding dendrogram (Fig. 6), the pool of 2004 experiments is generally apart from the 2003 set. This type of diagram can help select a subset of points used and to conduct further statistical analysis and to discard

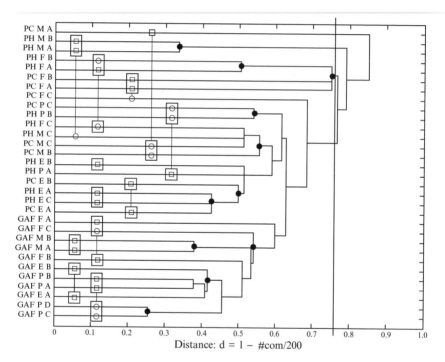

FIG. 6. Dendrogram displaying the correlation between different microarray replicates. Each experimental point displayed on the ordinate axis refers to a comparison between the IP channel and the mock channel. Names are abbreviated by designating the protein used in ChIPs (PH, PC, or GAF), the developmental stage analyzed (E, embryos; P, pupae; F, adult females; M, adult males), and the replicate experiment (A to D). Open circles and squares indicate, respectively, 2004 and 2003 samples. All experimental points belonging to the same condition are linked and aligned vertically. Boxes indicates those points selected for further statistical analysis by applying a threshold distance of $d = 0.76$, indicated by the vertical line. Black circles indicate the distance between the two best replicates in each condition.

outlier experiments. The retained samples can be subsequently analyzed by the RDAM method to estimate the significance of enrichments for each sample in each microarray feature. As an example we applied a threshold distance of 0.76, which is the minimal value allowing one to select at least two replicates for each condition. Other possibilities would have been to select in each condition the two replicates separated by the minimal distance, as indicated by closed circles in Fig. 6, or to take the whole set of replicates for subsequent analysis.

Graphic Comparison of ChIP on Chip Data with Genome Annotations

After applying a statistical analysis on microarray data, a list of binding sites for the protein of interest is obtained. Once the absolute coordinates of the binding sites along the genome are known, they can be visualized and compared to the genomic annotations using a variety of graphic interfaces. One of these tools, which is of simple use, is the GBrowse interface. In the case of the *Drosophila* genome, this is available at http://flybase.bio.indiana.edu/cgi-bin/gbrowse_fb/dmel. This interface allows external users to upload and display their own genomic annotations in the field: "Upload/Remote Annotations." A set of simple rules, described in the "Help" section of the web site, can be applied to create a text file from microarrays data. This file must be formatted as in the following example:

> *reference = chrX*
> *PHBindingSites BindingSite1 1979305-1981305*
> *PHBindingSites BindingSite2 2001696-2003696*
> *etc.*

The first line indicates the genomic entity (BACs, scaffolds, chromosomes) to which the absolute coordinates of the annotations apply. Here, chrX indicates the X chromosome. Then, each of the following lines refers to an annotation corresponding to the binding sites detected by ChIP on chip. For each annotation, the generic description of the data set (PHBindingSites), the name of a particular feature (BindingSite1), and the absolute start and end coordinates of the feature (1979305–1981305) should be given. Each field must be space separated. When the text file is created and uploaded into GBrowse, a category line named "PHBindingSites" appears at the bottom of the graphic interface and displays squares for each described features at their appropriate position on the X chromosome.

Perspectives

To date, major limitations for ChIP on chip studies are the biological material, the efficiency of ChIP, and the microarray substrates. Although the ChIP method described here can be used with any *Drosophila* tissue, we are still unable to perform ChIP in specific cell types, and a major future challenge will be to isolate sufficient amounts of pure individual cell types for ChIP. One could imagine tagging specific cell types with GFP, dissociate cells from tissues, and then sort them by FACS. An alternative approach might rely on *in vivo* protein biotinylation (de Boer *et al.*, 2003). In this approach, tissue-specific expression of the BirA bacterial biotin ligase (by appropriate transgenes) coupled to expression of a fusion protein between the protein of interest and a target peptide tag can lead to *in vivo* biotinylation of the target protein. In conjunction with formaldehyde cross-linking, this might allow one to recover the biotinylated protein of interest in the cell type to be studied by using the strong and highly specific biotin–streptavidin interaction.

Another technical point that needs to be improved is the amount of biological material needed for ChIP. For the moment, no less than tens of milligram amounts of cells are needed for one IP, which is equivalent to more than 10^7 cells. Perhaps this might be scaled down by improving the efficiency of immunoprecipitation and/or by developing a microarray fluidic technology allowing much smaller volumes (i.e., amounts) of labeled DNA to be hybridized.

A further aspect that is likely to improve with time concerns the microarrays themselves. It is becoming possible to analyze large genomic regions or even whole genomes, but at the moment this requires large amounts of work, and in most cases one genome can only be contained in several chips. Future improvements in the array density should allow one to obtain whole genome data from a single array, to improve reproducibility to a point where data from different labs can be compared directly, and to reduce the cost of the experiments. In summary, disposing of a large number of directly comparable data on the whole-genome transcription profiling as well as protein location for multiple, evolutionarily related organisms will be of invaluable importance in order to understand how gene networks established by transcription factors modulate gene expression throughout development and evolution.

Acknowledgments

We are very grateful to K.P. White (Yale University) for giving N.N. access to his microarray platform and to L.V. Sun for training him in microarray production and processing. We thank the Montpellier L-R Genopole for arraying, spotting, and processing custom microarrays. N.N is supported by the Ministère de l'Enseignement Supérieur, the European Molecular Biology Organisation (EMBO), and the Association pour la Recherche sur le Cancer. J.H. was supported by a postdoctoral fellowship from the CNRS. G.C. was supported by grants from the CNRS (Programme "Puces à ADN"), the Association pour la Recherche sur le Cancer, the Human Frontier Science Program Organization, the European Union FP6 (Network of Excellence The Epigenome and STREP 3D Genome), the Indo-French Centre for Promotion of Advanced Research, and the Ministère de l'Enseignement Supérieur, ACI BCMS.

References

Adams, M. D., Celniker, S. E., Holt, R. A., Evans, C. A., Gocayne, J. D., Amanatides, P. G., Scherer, S. E., Li, P. W., Hoskins, R. A., Galle, R. F., George, R. A., Lewis, S. E., Richards, S., Ashburner, M., Henderson, S. N., Sutton, G. G., Wortman, J. R., Yandell, M. D., Zhang, Q., Chen, L. X., Brandon, R. C., Rogers, Y. H., Blazej, R. G., Champe, M., Pfeiffer, B. D., Wan, K. H., Doyle, C., Baxter, E. G., Helt, G., Nelson, C. R., Gabor, G. L., Abril, J. F., Agbayani, A., An, H. J., Andrews-Pfannkoch, C., Baldwin, D., Ballew, R. M., Basu, A., Baxendale, J., Bayraktaroglu, L., Beasley, E. M., Beeson, K. Y., Benos, P. V., Berman, B. P., Bhandari, D., Bolshakov, S., Borkova, D., Botchan, M. R., Bouck, J., Brokstein, P., Brottier, P., Burtis, K. C., Busam, D. A., Butler, H., Cadieu, E., Center, A., Chandra, I., Cherry, J. M., Cawley, S., Dahlke, C., Davenport, L. B., Davies, P., de Pablos, B., Delcher, A., Deng, Z., Mays, A. D., Dew, I., Dietz, S. M., Dodson, K., Doup, L. E., Downes, M., Dugan-Rocha, S., Dunkov, B. C., Dunn, P., Durbin, K. J., Evangelista, C. C., Ferraz, C., Ferriera, S., Fleischmann, W., Fosler, C., Gabrielian, A. E., Garg, N. S., Gelbart, W. M., Glasser, K., Glodek, A., Gong, F., Gorrell, J. H., Gu, Z., Guan, P., Harris, M., Harris, N. L., Harvey, D., Heiman, T. J., Hernandez, J. R., Houck, J., Hostin, D., Houston, K. A., Howland, T. J., Wei, M. H., Ibegwam, C., Jalali, M., Kalush, F., Karpen, G. H., Ke, Z., Kennison, J. A., Ketchum, K. A., Kimmel, B. E., Kodira, C. D., Kraft, C., Kravitz, S., Kulp, D., Lai, Z., Lasko, P., Lei, Y., Levitsky, A. A., Li, J., Li, Z., Liang, Y., Lin, X., Liu, X., Mattei, B., McIntosh, T. C., McLeod, M. P., McPherson, D., Merkulov, G., Milshina, N. V., Mobarry, C., Morris, J., Moshrefi, A., Mount, S. M., Moy, M., Murphy, B., Murphy, L., Muzny, D. M., Nelson, D. L., Nelson, D. R., Nelson, K. A., Nixon, K., Nusskern, D. R., Pacleb, J. M., Palazzolo, M., Pittman, G. S., Pan, S., Pollard, J., Puri, V., Reese, M. G., Reinert, K., Remington, K., Saunders, R. D., Scheeler, F., Shen, H., Shue, B. C., Siden-Kiamos, I., Simpson, M., Skupski, M. P., Smith, T., Spier, E., Spradling, A. C., Stapleton, M., Strong, R., Sun, E., Svirskas, R., Tector, C., Turner, R., Venter, E., Wang, A. H., Wang, X., Wang, Z. Y., Wassarman, D. A., Weinstock, G. M., Weissenbach, J., Williams, S. M., Woodage, T., Worley, K. C., Wu, D., Yang, S., Yao, Q. A., Ye, J., Yeh, R. F., Zaveri, J. S., Zhan, M., Zhang, G., Zhao, Q., Zheng, L., Zheng, X. H., Zhong, F. N., Zhong, W., Zhou, X., Zhu, S., Zhu, X., Smith, H. O., Gibbs, R. A., Myers, E. W., Rubin, G. M., and Venter, J. C. (2000). The genome sequence of *Drosophila melanogaster*. *Science* **287**(5461), 2185–2195.

Arbeitman, M. N., Furlong, E. E., Imam, F., Johnson, E., Null, B. H., Baker, B. S., Krasnow, M. A., Scott, M. P., Davis, R. W., and White, K. P. (2002). Gene expression during the life cycle of *Drosophila melanogaster*. *Science* **297**(5590), 2270–2275.

Bloyer, S., Cavalli, G., Brock, H. W., and Dura, J. M. (2003). Identification and characterization of polyhomeotic PREs and TREs. *Dev. Biol.* **261**(2), 426–442.

de Boer, E., Rodriguez, P., Bonte, E., Krijgsveld, J., Katsantoni, E., Heck, A., Grosveld, F., and Strouboulis, J. (2003). Efficient biotinylation and single-step purification of tagged transcription factors in mammalian cells and transgenic mice. *Proc. Natl. Acad. Sci. USA* **100**(13), 7480–7485.

Francis, N. J., Saurin, A. J., Shao, Z., and Kingston, R. E. (2001). Reconstitution of a functional core polycomb repressive complex. *Mol. Cell* **8**(3), 545–556.

Franke, A., DeCamillis, M., Zink, D., Cheng, N., Brock, H. W., and Paro, R. (1992). Polycomb and polyhomeotic are constituents of a multimeric protein complex in chromatin of *Drosophila melanogaster*. *EMBO J.* **11**(8), 2941–2950.

Gupta, V., and Oliver, B. (2003). Drosophila microarray platforms. *Brief Funct. Genom. Proteom.* **2**(2), 97–105.

Hennetin, J., and Bellis, M. (2006). Clustering methods for analyzing large data sets: Gonad development, a study case. *Methods Enzymol.* **411**, 387–407.

Kirmizis, A., Bartley, S. M., Kuzmichev, A., Margueron, R., Reinberg, D., Green, R., and Farnham, P. J. (2004). Silencing of human polycomb target genes is associated with methylation of histone H3 Lys 27. *Genes Dev.* **18**(13), 1592–1605.

Lander, E. S., Linton, L. M., Birren, B., Nusbaum, C., Zody, M. C., Baldwin, J., Devon, K., Dewar, K., Doyle, M., FitzHugh, W., Funke, R., Gage, D., Harris, K., Heaford, A., Howland, J., Kann, L., Lehoczky, J., LeVine, R., McEwan, P., McKernan, K., Meldrim, J., Mesirov, J. P., Miranda, C., Morris, W., Naylor, J., Raymond, C., Rosetti, M., Santos, R., Sheridan, A., Sougnez, C., Stange-Thomann, N., Stojanovic, N., Subramanian, A., Wyman, D., Rogers, J., Sulston, J., Ainscough, R., Beck, S., Bentley, D., Burton, J., Clee, C., Carter, N., Coulson, A., Deadman, R., Deloukas, P., Dunham, A., Dunham, I., Durbin, R., French, L., Grafham, D., Gregory, S., Hubbard, T., Humphray, S., Hunt, A., Jones, M., Lloyd, C., McMurray, A., Matthews, L., Mercer, S., Milne, S., Mullikin, J. C., Mungall, A., Plumb, R., Ross, M., Shownkeen, R., Sims, S., Waterston, R. H., Wilson, R. K., Hillier, L. W., McPherson, J. D., Marra, M. A., Mardis, E. R., Fulton, L. A., Chinwalla, A. T., Pepin, K. H., Gish, W. R., Chissoe, S. L., Wendl, M. C., Delehaunty, K. D., Miner, T. L., Delehaunty, A., Kramer, J. B., Cook, L. L., Fulton, R. S., Johnson, D. L., Minx, P. J., Clifton, S. W., Hawkins, T., Branscomb, E., Predki, P., Richardson, P., Wenning, S., Slezak, T., Doggett, N., Cheng, J. F., Olsen, A., Lucas, S., Elkin, C., Uberbacher, E., Frazier, M., Gibbs, R. A., Muzny, D. M., Scherer, S. E., Bouck, J. B., Sodergren, E. J., Worley, K. C., Rives, C. M., Gorrell, J. H., Metzker, M. L., Naylor, S. L., Kucherlapati, R. S., Nelson, D. L., Weinstock, G. M., Sakaki, Y., Fujiyama, A., Hattori, M., Yada, T., Toyoda, A., Itoh, T., Kawagoe, C., Watanabe, H., Totoki, Y., Taylor, T., Weissenbach, J., Heilig, R., Saurin, W., Artiguenave, F., Brottier, P., Bruls, T., Pelletier, E., Robert, C., Wincker, P., Smith, D. R., Doucette-Stamm, L., Rubenfield, M., Weinstock, K., Lee, H. M., Dubois, J., Rosenthal, A., Platzer, M., Nyakatura, G., Taudien, S., Rump, A., Yang, H., Yu, J., Wang, J., Huang, G., Gu, J., Hood, L., Rowen, L., Madan, A., Qin, S., Davis, R. W., Federspiel, N. A., Abola, A. P., Proctor, M. J., Myers, R. M., Schmutz, J., Dickson, M., Grimwood, J., Cox, D. R., Olson, M. V., Kaul, R., Raymond, C., Shimizu, N., Kawasaki, K., Minoshima, S., Evans, G. A., Athanasiou, M., Schultz, R., Roe, B. A., Chen, F., Pan, H., Ramser, J., Lehrach, H., Reinhardt, R., McCombie, W. R., de la Bastide, M., Dedhia, N., Blocker, H., Hornischer, K., Nordsiek, G., Agarwala, R., Aravind, L., Bailey, J. A., Bateman, A., Batzoglou, S., Birney, E., Bork, P., Brown, D. G., Burge, C. B., Cerutti, L., Chen, H. C., Church, D., Clamp, M., Copley, R. R., Doerks, T., Eddy, S. R., Eichler, E. E., Furey, T. S., Galagan, J., Gilbert, J. G., Harmon, C., Hayashizaki, Y., Haussler, D., Hermjakob, H., Hokamp, K., Jang, W., Johnson, L. S., Jones, T. A., Kasif, S., Kaspryzk, A., Kennedy, S., Kent, W. J., Kitts, P., Koonin, E. V., Korf, I., Kulp, D., Lancet, D., Lowe, T. M., McLysaght, A., Mikkelsen, T., Moran, J. V., Mulder, N., Pollara, V. J., Ponting, C. P., Schuler, G., Schultz, J., Slater, G., Smit, A. F., Stupka, E., Szustakowski, J., Thierry-Mieg, D., Thierry-Mieg, J.,

Wagner, F., Wallis, J., Wheeler, R., Williams, A., Wolf, Y. I., Wolfe, K. H., Yang, S. P., Yeh, R. F., Collins, F., Guyer, M. S., Peterson, J., Felsenfeld, A., Wetterstrand, K. A., Patrinos, A., Morgan, M. J., de Jong, P., Catanese, J. J., Osoegawa, K., Shizuya, H., Choi, S., and Chen, Y. J. (2001). Initial sequencing and analysis of the human genome. *Nature* **409**(6822), 860–921.

Liu, Z. R., Sargueil, B., and Smith, C. W. (2000). Methylene blue-mediated cross-linking of proteins to double-stranded RNA. *Methods Enzymol.* **318**, 22–33.

Martin, D. E., Demougin, P., Hall, M. N., and Bellis, M. (2004). Rank difference analysis of microarrays (RDAM), a novel approach to statistical analysis of microarray expression profiling data. *BMC Bioinformat.* **5**(1), 148.

Min, J., Zhang, Y., and Xu, R. M. (2003). Structural basis for specific binding of Polycomb chromodomain to histone H3 methylated at Lys 27. *Genes Dev* **17**(15), 1823–1828.

Negre, N., Hennetin, J., Sun, L. V., Lavrov, S., Bellis, M., White, K. P., and Cavalli, G. (2006). Chromosomal distribution of PcG proteins during *Drosophila* development. *PLOS Biol.* **4**, e170.

Orlando, V., Strutt, H., and Paro, R. (1997). Analysis of chromatin structure by *in vivo* formaldehyde cross-linking. *Methods* **11**(2), 205–214.

Ren, B., Robert, F., Wyrick, J. J., Aparicio, O., Jennings, E. G., Simon, I., Zeitlinger, J., Schreiber, J., Hannett, N., Kanin, E., Volkert, T. L., Wilson, C. J., Bell, S. P., and Young, R. A. (2000). Genome-wide location and function of DNA binding proteins. *Science* **290** (5500), 2306–2309.

Ringrose, L., and Paro, R. (2004). Epigenetic regulation of cellular memory by the Polycomb and Trithorax group proteins. *Annu. Rev. Genet.* **38**, 413–443.

Schena, M., Shalon, D., Davis, R. W., and Brown, P. O. (1995). Quantitative monitoring of gene expression patterns with a complementary DNA microarray. *Science* **270**(5235), 467–470.

Simon, J., Chiang, A., Bender, W., Shimell, M. J., and O'Connor, M. (1993). Elements of the Drosophila bithorax complex that mediate repression by Polycomb group products. *Dev. Biol.* **158**(1), 131–144.

Stolc, V., Gauhar, Z., Mason, C., Halasz, G., van Batenburg, M. F., Rifkin, S. A., Hua, S., Herreman, T., Tongprasit, W., Barbano, P. E., Bussemaker, H. J., and White, K. P. (2004). A gene expression map for the euchromatic genome of *Drosophila melanogaster*. *Science* **306**(5696), 655–660.

Zhang, L., Zhang, K., Prandl, R., and Schoffl, F. (2004). Detecting DNA-binding of proteins in vivo by UV-crosslinking and immunoprecipitation. *Biochem. Biophys. Res. Commun.* **322** (3), 705–711.

[16] DamID: Mapping of *In Vivo* Protein–Genome Interactions Using Tethered DNA Adenine Methyltransferase

By FRAUKE GREIL, CELINE MOORMAN, and BAS VAN STEENSEL

Abstract

A large variety of proteins bind to specific parts of the genome to regulate gene expression, DNA replication, and chromatin structure. DamID is a powerful method used to map the genomic interaction sites of these proteins *in vivo*. It is based on fusing a protein of interest to *Escherichia coli* DNA adenine methyltransferase (dam). Expression of this fusion protein *in vivo* leads to preferential methylation of adenines in DNA surrounding the native binding sites of the dam fusion partner. Because adenine methylation does not occur endogenously in most eukaryotes, it provides a unique tag to mark protein interaction sites. The adenine-methylated DNA fragments are isolated by selective polymerase chain reaction amplification and can be identified by microarray hybridization. We and others have successfully applied DamID to the genome-wide identification of interaction sites of several transcription factors and other chromatin-associated proteins. This chapter discusses DamID technology in detail, and a step-by-step experimental protocol is provided for use in *Drosophila* cell lines.

Introduction

Hundreds of nuclear proteins control transcription, replication, and chromatin structure. To understand the functions and molecular mechanisms of these proteins, it is essential to know in detail where they bind in the genome.

Although studies of protein–DNA interactions *in vitro* can be informative, they often cannot substitute for studies *in vivo* (Biggin, 2001). DamID is a method that allows the identification of the *in vivo* genomic binding sites of a wide range of proteins. So far, DamID has been used successfully to generate genome-wide binding maps for a number of sequence-specific factors, such as GAGA factor (van Steensel *et al.*, 2001, 2003), the Max family of transcriptional regulators (Orian *et al.*, 2003), and the transcriptional repressor Hairy (Bianchi-Frias *et al.*, 2004), as well as coregulators and chromatin proteins such as HP1, dSir2, dCtBP, and

METHODS IN ENZYMOLOGY, VOL. 410
0076-6879/06 $35.00
DOI: 10.1016/S0076-6879(06)10016-6

Groucho (Bianchi-Frias *et al.*, 2004; Greil *et al.*, 2003; van Steensel *et al.*, 2001). Most of the published DamID maps were generated using cultured *Drosophila* cells. With some minor modifications of the protocol, DamID may also be used to obtain protein-binding maps of whole flies and larvae (de Wit *et al.*, 2005) or even dissected tissues, and it should, in principle, be transferable to other biological systems, such as mammalian cell lines (Song *et al.*, 2004).

Principle of DamID

The basic principle of DamID is outlined in Fig. 1. For simplicity, we use the term "chromatin protein" to denote any protein that interacts with specific parts of the genome (either directly or via other proteins). As a first step, a fusion protein is constructed consisting of *E. coli* DNA adenine methyltransferase (dam) and a chromatin protein of interest. Dam is a single polypeptide of 32 kDa that methylates the 6 position of adenines in double-stranded DNA within the sequence GATC (Barras and Marinus, 1989). Importantly, dam retains its methyltransferase activity when fused to other proteins at either its C or its N terminus (van Steensel and Henikoff, 2000). Adenine methylation does not occur endogenously in *Drosophila* and most other eukaryotes, hence dam provides a unique tagging system to mark the genomic-binding sites of chromatin proteins. Additionally, adenine methylation has only minor effects on DNA topology (Barras and Marinus, 1989), and toxicity studies in *Drosophila* have shown that it does not affect viability, fertility, or development (Boivin and Dura, 1998; Wines

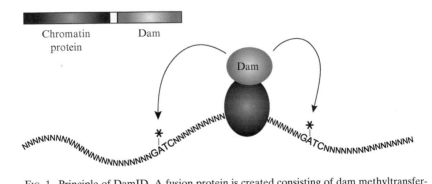

FIG. 1. Principle of DamID. A fusion protein is created consisting of dam methyltransferase and a protein of interest. When expressed in *Drosophila*, dam is targeted to the native binding sites of its fusion partner, resulting in local methylation of adenines in GATC sequences close to the binding sites. The genomic methylation pattern, which can be mapped using DNA microarrays, provides information on the location of the protein binding sites.

et al., 1996), suggesting that it does not interfere with normal genome function.

Upon *in vivo* expression of the dam fusion protein, dam is targeted to the native genomic binding sites of its fusion partner. The resulting high local concentration of dam leads to preferential methylation of adenines in GATC sequences that are close to the chromatin protein binding sites. The spreading of dam methylation of up to 5 kb from a binding site (van Steensel and Henikoff, 2000) and the high frequency of GATC sites in *Drosophila* (on average once every 200–300 bp) facilitate the probing of almost every region in the genome.

After a sufficient time span of expression of the dam fusion protein, targeted dam methylation can be detected in individual genomic loci using Southern blotting or quantitative polymerase chain reaction (PCR) assays (van Steensel and Henikoff, 2000). A far more powerful application of DamID involves the use of DNA microarrays to detect adenine methylation simultaneously in thousands of loci. Here, the adenine methylated regions are first cleaved by the restriction endonuclease *Dpn*I, which selectively recognizes G^mATC sites. Previously, we have used sucrose gradient centrifugation to isolate the methylated fragments based on their small size compared to unmethylated and therefore uncleaved genomic regions (van Steensel *et al.*, 2001). This original protocol has been improved upon by replacing the sucrose gradient step with a methyl fragment-specific PCR reaction, which is less time-consuming and results in an increased sensitivity of DamID (Greil *et al.*, 2003). The isolated methylated fragments are subsequently labeled with a fluorescent dye and hybridized to microarrays containing DNA fragments or oligonucleotides representing genomic regions of interest. From the detected methylation pattern, the binding sites of the chromatin protein can be inferred.

Correcting for Untargeted Binding of Dam

An important aspect of methylation by dam fusion proteins is that targeting of dam to the binding sites of the chromatin protein is not perfect. Inevitably, some of the fusion protein is freely diffusing in the nucleus, and the high intrinsic affinity of dam for GATC sequences throughout the genome causes considerable levels of background methylation. Additionally, methylation by dam is modulated by the local structure of chromatin: DNA in condensed chromatin, such as heterochromatin, is generally less accessible to dam than DNA in transcriptionally active, decondensed euchromatin (Boivin and Dura, 1998; Gottschling, 1992; Singh and Klar, 1992; Wines *et al.*, 1996).

This problem is illustrated in Fig. 2. If chromatin accessibility was uniform throughout the genome, methylation would be strongest near the binding site of the chromatin protein and gradually decay over a distance of a few kilobases. Further away from the binding site a relatively low, constant level of methylation would occur due to the nontargeted activity of dam (Fig. 2A).

Figures 2B and 2C illustrate a more realistic scenario. Due to local variations in chromatin structure, some loci are more accessible than others (Fig. 2B). The degree of chromatin accessibility significantly affects the levels of background methylation, and possibly also the levels of targeted methylation (Fig. 2C). In fact, the background methylation in a nontarget locus with an open chromatin structure may reach similar or even higher levels than the targeted methylation of a binding site of the chromatin protein of interest. Thus, without a correction of the observed methylation levels for chromatin accessibility, "open" chromatin regions will be mistaken for targets of the protein of interest.

The correction for background methylation is done by performing a control experiment in which the methylation levels of the same probed loci are measured after the expression of *unfused* dam (Fig. 2D). Data obtained with unfused dam are then used to correct data obtained with the dam fusion protein. In practice, the best way to perform this correction is to calculate the ratio (methylation by dam fusion:methylation by unfused dam). With microarray platforms that support the use of two-color hybridizations, the dam fusion and dam-only samples can be labeled with different dyes and cohybridized, and the dye ratio measured for each probed sequence can be taken as the methylation ratio. After normalization, a ratio >1 indicates binding of the protein of interest, whereas a ratio of ~1 indicates lack of binding (Fig. 2E). Examples of experimental data illustrating this correction procedure can be found elsewhere (Orian *et al.*, 2003; van Steensel and Henikoff, 2000; van Steensel *et al.*, 2001, 2003).

Expression Levels of Dam and Dam Fusion Proteins

Dam is highly active when expressed in *Drosophila* or mammalian cells. To avoid saturating levels of methylation in the genome, which would corrupt the quantitative approach outlined earlier, expression levels of dam fusion proteins and unfused dam must be kept very low. This can be achieved by using an extremely weak promoter to drive expression of the dam proteins. We have found that the hsp70 promoter in *Drosophila*, in the absence of heat shock, gives an appropriately low expression level for DamID (van Steensel and Henikoff, 2000). At these levels, we have been unable to detect dam fusion proteins by immunofluorescence microscopy

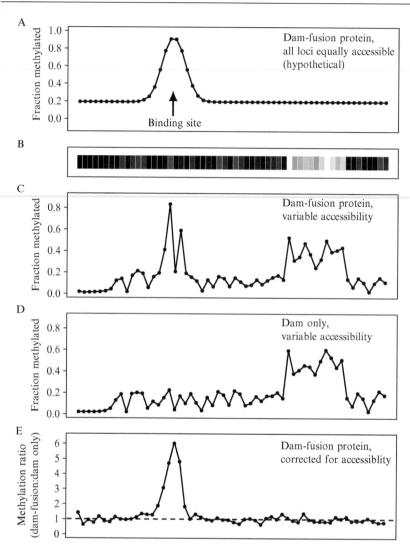

Fig. 2. Simulation of methylation patterns that may be expected in a DamID experiment. Methylation levels are plotted along an imaginary genomic region of ~60 kbp containing one binding site for the chromatin protein of interest. (A) In the hypothetical situation where DNA accessibility is equal along the entire genome, expression of a dam fusion protein will lead to strong methylation in the vicinity of the binding site (arrow). In addition, due to the intrinsic affinity of dam for DNA, a constant level of nontargeted methylation will occur in more distal regions. (B) In a more realistic scenario, DNA accessibility varies along the chromosomal region (dark, inaccessible; light, accessible). (C) The accessibility differences

or Western blotting (van Steensel and Henikoff, 2000), indicating that only trace amounts are present in the cell. An additional advantage of these low expression levels is that the dam fusion protein is less likely to interfere with the function of the endogenous chromatin protein or its target loci.

The importance of low expression levels should not be underestimated. For DamID in *Drosophila* cells we advise not to use other expression vectors than those listed in this chapter. Other promoters, such as the natural promoter of the gene encoding the chromatin protein of interest, typically drive expression levels that are too high.

Despite the low nuclear concentration, the dam fusion protein appears to be readily able to compete for binding sites with the typically much more abundant endogenous chromatin protein. This may be explained by the extremely dynamic behavior of most chromatin proteins: photobleaching experiments have shown that virtually all chromatin proteins exchange within minutes between genome-bound and free-diffusing states (Phair *et al.*, 2004). Thus, exposure of cells to a dam fusion protein for several hours ($\sim$24 h in a typical DamID experiment) will allow the fusion protein molecules to visit and methylate most of their target loci.

Comparison of DamID and ChIP

Currently, there are two methods available for the detection of *in vivo* protein–chromatin interactions in higher eukaryotes: chromatin immunoprecipitation (ChIP) (Bernstein *et al.*, 2004; Buck and Lieb, 2004; Ren and Dynlacht, 2004) and DamID. In ChIP, protein–DNA interactions are fixed by cross-linking, and the sites of protein–DNA association are identified by purification of the cross-linked DNA fragments with an antibody against the protein of interest. DamID and ChIP each have distinct advantages and disadvantages.

First, ChIP requires an antibody with high specificity and avidity for the protein under investigation. DamID does not rely on an antibody, but it does require that the tethered dam does not interfere with the function of the chromatin protein, which has to be tested for each dam fusion protein. Second, DamID is not suitable for the detection of posttranslational modifications, such as histone modifications, in contrast to ChIP. Third, DamID can easily be used to study the effects of mutations on the targeting specificity of a chromatin protein, while this is often more difficult with

strongly affect the methylation pattern (compare to A). (D) Accessibility can be estimated directly by measuring the methylation pattern after expression of dam only. (E) Calculation of the methylation ratio [dam fusion:dam only] reveals the protein binding site. Ratios are normalized by setting the median ratio of all probed loci to 1 (dotted line).

ChIP. Fourth, the success of ChIP may depend on the "cross-linkability" of a protein (in most cases by formaldehyde), which is unpredictable. In addition, due to the cross-linking approach, ChIP might not be efficient in detecting transient protein–DNA interactions or interactions of proteins that do not bind to DNA directly but via other proteins. DamID has been used successfully to map binding sites of chromatin proteins that do not bind directly to the DNA (Bianchi-Frias *et al.*, 2004; Greil *et al.*, 2003) and may be more efficient in detecting transient protein–genome interactions because the adenine methylation mark remains present after the protein has dissociated from its target DNA. Fifth, because cross-linking occurs within minutes, ChIP is more suited to monitor rapid changes in protein binding than DamID, which requires expression of the dam fusion protein for at least several hours. Sixth, the mapping resolution of ChIP and DamID appears to be roughly comparable. For ChIP, fragment size depends on the degree of DNA shearing and is typically $\sim$1 kb. For DamID we have originally reported a resolution of 2–5 kb (van Steensel and Henikoff, 2000; van Steensel *et al.*, 2003). However, unpublished observations using high-density tiling arrays show that the resolution obtained with the DamID protocol described here can be higher ($\sim$1 kb) and may thus be close to the resolution of ChIP. Systematic experimental comparisons of DamID and ChIP have not yet been reported.

Materials

Construction of Vectors Encoding Dam Fusion Proteins

1. Vectors pNDamMyc and pCMycDam. These plasmids can be obtained from the authors.

Transfection of Kc Cells

All materials should be sterile.

1. Cell culture medium, e.g., BPYE (Shields and Sang M3 insect medium supplemented with 2.5 g/liter bacto peptone, 1 g/liter yeast extract, and 5% heat-inactivated fetal calf serum)
2. Cell-culture grade plastic 10-cm petri dishes
3. Electroporation device, e.g., GenePulser II with capacitance extender module (Bio-Rad, Hercules, CA)
4. Electroporation cuvettes, 4-mm gap width (e.g., from Eppendorf, Hamburg, Germany)
5. $T_{10}E_{0.1}$: 10 mM Tris, 0.1 mM EDTA, pH 7.5

6. Sterile stocks of dam expression vectors (pNDamMyc and a vector encoding the dam fusion protein of interest) adjusted to at least 1 μg/μl in 10 mM Tris, 0.1 mM EDTA, pH 7.5. Note: the pCMycDam plasmid cannot be used for expression of dam because it does not contain a start codon.

Purification and Digestion of Genomic DNA

1. Qiagen DNeasy tissue kit (Qiagen, Hilden, Germany)
2. $T_{10}E_{0.1}$: 10 mM Tris, pH 7.5, 0.1 mM EDTA
3. 3 M sodium acetate, pH 5.2
4. *Dpn*I restriction enzyme with NEB restriction buffer 4 (New England Biolabs, Ipswich, USA)
5. Isopropanol
6. 70% ethanol

Ligation of Adaptors

1. Oligonucleotides[1]:
   100 μM AdRt 5′ CTAATACGACTCACTATAGGGCAGCGTG
   GTCGCG GCCGAGGA 3′
   100 μM AdRb 5′ TCCTCGGCCG 3′
2. To generate double-stranded adaptor dsAdR, mix equal volumes of AdRt and AdRb, heat to 100° for 1 min, and let cool down slowly (in about 20 min) to room temperature.
3. T4 DNA ligase, 5 units/μl, and ligase buffer with ATP (Roche, Basel, Switzerland)

*Dpn*II *Digest*

1. *Dpn*II restriction enzyme and *Dpn*II buffer (New England Biolabs, Ipswich, USA)

Polymerase Chain Reaction

1. Advantage cDNA Polymerase mix (enzyme and buffer mix) (Clontech, Mountain View, CA)
2. 50 μM AdR_PCR 5′ GGTCGCGGCCGAGGATC 3′
3. dNTPs (PCR grade, 2.5 mM each)

---

[1] The 5′ ends of the adaptor oligonucleotides are not phosphorylated. This prevents self-ligation of the adaptors. The oligonucleotide AdRt may be longer than is strictly necessary; however, shorter versions have not been tested.

4. QIAquick PCR purification kit (Qiagen, Hilden, Germany)
5. PCR thermocycler
6. PCR tubes

Methods

For a flowchart of the experimental steps, see Fig. 3.

Construction of Plasmids Encoding Dam Fusion Proteins

The plasmids pNDamMyc and pCMycDam contain a small number of restriction sites that can be used for insertion of the open reading frame of a chromatin protein of interest. Sequences and maps can be downloaded from http://www.nki.nl/nkidep/vansteensel. Dam can be fused either to the N-terminus (pNDamMyc) or to the C-terminus (pCMycDam) of the chromatin protein, and it is recommended to try both possibilities. In both cases a myc epitope tag serves as a linker peptide between dam and its fusion partner. This tag can be used to detect the fusion proteins by Western blotting or immunofluorescence microscopy to confirm their correct size and localization. Note that in order to obtain a detectable amount of protein, the fusion protein should be overexpressed by heat shock induction. If possible, additional tests should be done to verify that the chromatin protein is still functional when fused to dam.

The plasmid pNDamMyc without any additional insert is used to express myc-tagged dam only. The plasmids are based on the standard P-element plasmid pP{CaSpeR-hs} and can be used for transient transfection of cultured cell lines as well as for germline transformation.[2] This chapter focuses on transfection of the embryonal Kc cell line.

Transfection of Cultured Drosophila *Kc Cells*

An essential requirement for the success of DamID in cultured cells is sufficient transfection efficiency. For some proteins, transfection efficiency can be monitored easily by immunofluorescence microscopy. These are proteins that either localize to a defined region of the nucleus or proteins that are very stable. Proteins with a more diffuse nuclear localization, such as dam itself, or proteins with a high turnover rate are more difficult to

---

[2] Some difficulties have been reported with obtaining germline transformants when using pNDamMyc and pCMycDam vectors. Recloning of the dam fusion protein open reading frame into the vector pP{UAST} solved these difficulties. The pP{UAST} vector is suited for obtaining the required low expression levels (B. van Steensel, unpublished results; Wines *et al.*, 1996).

A

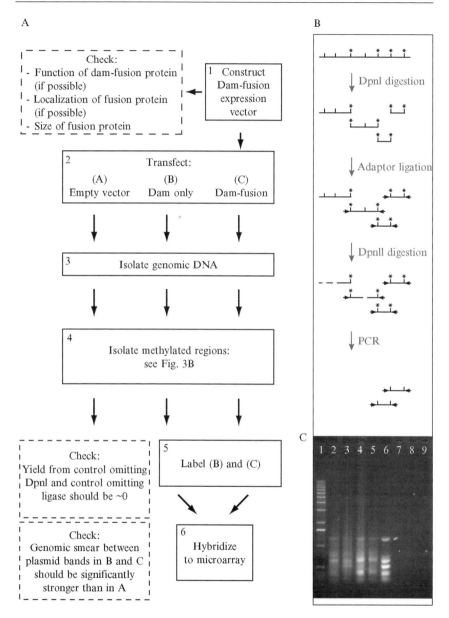

FIG. 3. Outline of the DamID method. (A) Flowchart of a typical DamID experiment. Dashed boxes indicate quality control points. (B) Diagram depicting the steps during the isolation of methylated regions. Genomic DNA is depicted as a horizontal line, and each vertical line marks a GATC site. Asterisks mark methylated GATC sites, and arrows depict

detect. If transfection rates cannot be measured, successful DNA methylation by the fusion protein can still be estimated from the products of the
methyl-PCR (see later).

In each experiment, three different transfections are carried out: (A)
empty plasmid pP{CaSpeR-hs}, (B) pNDamMyc, and (C) the plasmid
encoding the dam fusion protein. A protocol is provided here for the
electroporation of Kc cells. Because there are several different sublines
of Kc cells, it may be necessary to optimize various parameters, such
as cell growth conditions, plasmid concentration, electroporation field
strength, and pulse length.[3] We have not tested this protocol on other
cell lines. All materials should be sterile and all steps should be carried
out in a cell culture flow hood under sterile conditions. All steps
are performed at room temperature because cold negatively affects transfection efficiencies.

Two days before transfection, split cells to 5×10^5 cells/ml in culture
medium in a 10-cm dish (total volume: 10–12 ml). Prepare one dish for each
transfection.

On the day of transfection:

1. For each transfection pipette 10 ml medium into a 10-cm dish. Allow
   medium to adjust to room temperature.
2. Pipette 20 μg plasmid DNA for each transfection into a separate
   cuvette. Do not forget to take along a control with 20 μg pP{CaSpeR-hs}
   DNA (empty plasmid).
3. Count the cells of one to two plates to ensure that the density is
   $\sim 2 \times 10^6$ cells/ml. If the cell density is lower, do not proceed because
   transfection efficiency might decrease and the DNA yield may be
   too low. Collect and pool the cells in 50-ml tubes. Spin cells at 800g
   for 3 min.
4. Remove the supernatant and resuspend cells in 900 μl culture

---

ligated adapters used to specifically amplify methylated fragments in the PCR step. (C)
Typical agarose gel of methyl PCR products originating from cells transfected with plasmids
containing different dam fusions (lanes 2–4), dam only (lane 5), or empty vector (lane 6). The
remaining lanes contain control reactions of the sample amplified in lane 2 omitting $DpnI$
(lane 7) or ligase (lane 8). In lane 9, no template was added to the PCR reaction.

[3] Occasionally, we have experienced times when the transfection efficiency of Kc cells
dropped dramatically. This seemed to be related, at least in part, to certain batches of culture
medium. Switching to a different batch or to a different type of medium may solve the
problem. Other factors that may affect the transfection efficiency are cell density and purity
of the plasmid preparations. Plasmid DNA should be taken up in $T_{10}E_{0.1}$, not in water, to
avoid degradation. Some laboratories have observed an increase in transfection efficiency
when 10 mM glutamine is added to the culture medium.

medium per transfection (e.g., 2.7 ml for three transfections).
5. Add 800 μl of this cell suspension to each cuvette and pipette up and down two to three times to mix.
6. Electroporate the cells at 1000 μF and 250 V.
7. Wait 5 min.
8. Transfer 700 μl to each 10-cm dish, avoiding the viscous layer of dead cells floating at the top of the cuvette.
9. Incubate the cells at 23.5° for ~24 h.

Isolation and Digestion of Genomic DNA

Genomic DNA (gDNA) is isolated using the DNeasy kit from Qiagen as specified by the manufacturer. To obtain the methylated fragments, the genomic DNA is digested with *Dpn*I, a restriction enzyme that cleaves only $G^{m6}ATC$ sites and not unmethylated GATC sites.

1. Twenty-four hours after transfection,[4] spin down the transfected cells in 15-ml tubes (800g; 10 min). Note that these cells were *not* subjected to heat shock. It is possible to freeze the cell pellets for long-term storage at this step.[5]
2. Remove supernatant and use cells for genomic DNA isolation with the DNeasy tissue kit from Qiagen. It is essential to always process samples from the three transfections (A, B, and C) in parallel. Add RNase A as described in the DNeasy protocol. To avoid shearing, do not vortex the DNA solution at any time, but mix well by inverting the tubes.
3. Precipitate the genomic DNA (gDNA) by adding 2 volumes 100% EtOH, 0.1 volume sodium acetate (3 M) and mix. Centrifuge at 4° for 20 min at 14,000 rpm/ 21,000g in a tabletop centrifuge. Remove the supernatant and carefully wash the pellet with 70% EtOH. Air dry the pellet and then resuspend the pellet in the appropriate volume of $T_{10}E_{0.1}$ (10–20 μl) for a DNA concentration of 200–400 ng/μl.
4. Digest 2.5 μg gDNA of each transfection sample with 10 units *Dpn*I

---

[4] We have not varied the duration between electroporation and the isolation of genomic DNA. Hence, all protein-binding data obtained with this method represent the average binding over ~24 h. In theory, a shorter time span could provide greater time resolution, which could be used to study changes in protein binding throughout the cell cycle, for example. Longer time spans have not been tested.

[5] It is possible to freeze the cells 24 h after transfection and perform the gDNA isolation at a later stage, for example, after assessing the transfection efficiency by immunofluorescence microscopy. To freeze, pellet the cells by centrifugation, remove the supernatant, and snap freeze in a dry ice-ethanol bath. Store the cells for up to several weeks at –80°. For subsequent gDNA isolation, treat the cells as described in the DNeasy manual.

in NEB buffer 4.[6] The final volume should be 10 μl. Include a control with 2.5 μg gDNA of transfection sample C omitting DpnI. This will be taken along during all subsequent steps and serve as a control for the G^{m}ATC specificity of DpnI. Digest overnight at 37°. Perform digestion in either a PCR thermocycler with a heated lid or Eppendorf tubes submerged in a water bath to prevent condensation of the sample in the lid of the tube.

5. Inactivate DpnI for 20 min at 80°.

Ligation of Adaptors to Methylated DNA Fragments

After digestion of genomic DNA with DpnI, a double-stranded adaptor oligonucleotide is ligated to the cleaved DNA ends.

1. Take 5 μl of the DpnI-digested gDNA and add 2 μl 10× ligation buffer (Roche), 0.8 μl double-stranded AdR (50 μM), 1 μl T4 ligase (5 U/μl, Roche), and 11.2 μl H$_2$O. Include a DpnI-digested control sample of C omitting T4 ligase. This will be taken along during all subsequent steps and serve as a control for the adapter specific amplification of methylated fragments. Ligate for 2 h at 16° in a PCR thermocycler (overnight ligation is also possible).

2. Inactivate T4 ligase for 10 min at 65°.

DpnII Digest

Following ligation, the DNA is treated with DpnII (or with an iso-schizomer such as MboI or NdeII), which only cuts unmethylated GATC. Like DpnI, DpnII displays high specificity with regard to the adenine methylation status of GATC sequences. The sequential use of DpnI and DpnII creates a double selection for methylated DNA fragments: only methylated GATC sequences are cut by DpnI and therefore ligated to the adaptors, and only fragments in which *all* GATCs are methylated are resistant to degradation by DpnII and can therefore be amplified. We find that this double selection allows for highly selective amplification of methylated genomic regions.

1. Add 8 μl 10× DpnII buffer, 0.5 μl DpnII enzyme (10 U/μl), and 51.5 μl H$_2$O to the 20 μl ligation reaction. Mix and digest for 1 h at 37°.

---

[6] In principle, this protocol may be scaled down further or may be tested for use with dam and dam fusion transformed flies, embryos, etc. Note that the reliability and reproducibility should be tested vigorously when adjustments are made to this protocol.

Polymerase Chain Reaction

For PCR amplification of the methylated fragments a primer is used that is complementary to the adaptor sequence. Because of the involvement of PCR, this part of the protocol is potentially sensitive to artifacts. Although some sequences are undoubtedly amplified less efficiently than others in the PCR reaction, we have optimized the entire protocol to ensure that sequence biases affect the experimental and the reference samples equally.

After PCR amplification of the methylated fragments, the PCR products should be analyzed on an agarose gel. The most prominent products will be bands from amplified methylated plasmid DNA, but a smear of genomic methylated fragments should be clearly visible. It is not possible to avoid amplification of the plasmid by isolating the plasmids from dam⁻ bacteria: we found that methylation of the plasmid still occurs by leaky expression of the dam (-fusion) from the plasmid. If the DNA smear from cells infected with the fusion protein and the dam-only (samples B and C) is clearly stronger than in the empty plasmid control (sample A), the experiment can be continued.

1. Perform PCR reactions for all samples in parallel and include a negative control without template. Use 20 μl *Dpn*II-digested gDNA and add 8 μl 10× cDNA PCR reaction buffer (Clontech), 1.25 μl primer Adr-PCR (50 μM), 4 μl dNTPs (2.5 mM each), 1.6 μl PCR Advantage enzyme mix (50×, Clontech), and 45.15 μl H$_2$O to a final volume of 80 μl. Run the following PCR program:
   1. 68° 10 min
   2. 94° 1 min
   3. 65° 5 min
   4. 68° 15 min
   5. 94° 1 min
   6. 65° 1 min
   7. 68° 10 min
   8. Repeat steps e–g three times
   9. 94° 1 min
   10. 65° 1 min
   11. 68° 2 min
   12. Repeat steps i–k 14 to 17 times[7]
   13. 4°
2. Digest 5 μl PCR product with 5 units of *Dpn*II for 2 h and analyze

[7] For Kc cells grown in BPYE medium, the optimum amount of cycles is 17. For other cells/medium this number may need to be optimized. Ideally, the PCR should yield 4–7 μg DNA.

on a 1% agarose gel. The dam-only and dam fusion reactions should show a smear of genomic fragments that is clearly stronger than the background smear from sample A (empty vector), but all other samples should be negative (see Fig. 3C).[8,9]

3. Purify samples with the QIAquick PCR purification kit and quantify with a spectrophotometer. The final yield is typically between 4 and 7 μg.

Sample Labeling and Microarray Hybridizations

The sample labeling and hybridization protocols depend on the type of microarray that is used. Ideally, microarrays covering genic and intergenic regions are used, such as the genomic tiling arrays now offered by several companies (e.g., Affymetrix, Santa Clara, CA; Agilent, Palo Alto, CA; NimbleGen, Madison, WI). However, because of the spreading of targeted dam methylation, cDNA arrays have also been employed successfully in detecting protein binding to regulatory elements in the vicinity of coding regions (Orian et al., 2003; van Steensel and Henikoff, 2000; van Steensel et al., 2001, 2003). When using cDNA arrays, it is not possible to distinguish between binding to regulatory elements in the vicinity of coding regions and binding in the coding region itself so care should be taken when interpreting results. Oligonucleotide arrays used for expression profiling are often designed to probe only the 3' ends of genes, hence such arrays are probably less suitable for the detection of protein binding to promoter regions.

For probing of cDNA and genomic tiling arrays, the amplified methylated fragments are labeled using the protocol described by Pollack et al. (1999).[10] Labeling reactions are performed in parallel for methylated DNA from transfections B and C, each with a different Cy dye.[11,12]

---

[8] In the control reactions (no DpnI, no ligase, and no PCR template), the PCR should yield no product.

[9] During the ligation step, often two or more DNA fragments are concatenated randomly between adaptor molecules. This causes the banded pattern of coamplified plasmid fragments in the PCR products to appear rather fuzzy. These bands become sharper when the PCR product is postdigested with DpnII.

[10] Other labeling protocols may be used, but we find this protocol to be extremely robust.

[11] It is important to perform "dye swaps," that is, to perform two separate array hybridizations, one with the orientation of the dyes for the experimental and reference sample switched to rule out dye-bias effects in the microarray hybridizations.

It should be mentioned that in some cases the large amounts of methylated fragments originating from the coamplified transfected plasmid DNA might cause background problems during hybridization. This background may occur if sequences present on the microarray have partial homology to sequences present in the plasmid and if the hybridization conditions are not sufficiently stringent. This problem may be identified by sequence alignment of candidate "target" loci to the plasmid sequence. To suppress this plasmid cross-hybridization it is recommended to include 25 μg *Dpn*I-digested and subsequently purified plasmid DNA (i.e., the plasmid used in transfection C) in the hybridization mix. If this is not sufficient, genomic DNA may be separated from the plasmid DNA on an agarose gel prior to *Dpn*I digestion. Alternatively, the generation of stable cell lines expressing dam and the dam fusion should be considered.

Data Processing and Analysis

The parameter that provides the best estimate of protein binding is the ratio (methylation by dam fusion:methylation by dam only). For the microarray-based detection as described here, this methylation ratio is approximated by the fluorescence ratio (transfection C:transfection B) of each spot on the array.

Typically, the measured fluorescence ratios are background corrected and normalized. Ideally, data would be normalized relative to one or more control loci that are known not to bind the protein under investigation. In most cases such loci are not known with certainty, and therefore data are often normalized to the median ratio of all spots on the array. Here, the underlying assumption is that the vast majority of loci are not bound by the chromatin protein. More advanced methods of normalization have been described (Quackenbush, 2002).

Because the cells are typically exposed to the dam fusion protein for ∼24 h, the fluorescence ratios reflect the average binding of the chromatin protein over the entire cell cycle (Kc cells have a doubling time of ∼24 h at room temperature). Moreover, the fluorescence ratio of a probed locus depends not only on the number of chromatin protein molecules bound,

---

[12] To test the reproducibility of the protocol (particularly of the PCR reaction), it is recommended to carry out the following control experiment: perform the entire protocol from the ligation step to the PCR reaction in duplicate, starting from a single sample of *Dpn*I-digested DNA from either transfection B or transfection C. Label one of the duplicate PCR products with Cy3 and the other with Cy5 and perform a microarray hybridization. Such a "self-versus-self" experiment provides a good estimate of the *technical* reproducibility of the protocol. Note that *biological* reproducibility should also be addressed by repeating the entire experiment several times on different days.

but also on the distance of the binding sites from the probed region. Hence, caution should be taken when interpreting the results. For practical purposes, it is often desirable to identify loci with statistically significantly increased ratios. Several algorithms have been described for the statistical analysis of microarray data (Nadon and Shoemaker, 2002; Slonim, 2002) and most of these can be applied to DamID data sets. Bioinformatics tools are also available for the identification of sequence motifs that correlate with the binding pattern of a chromatin protein, for example, the REDUCE algorithm (Orian et al., 2003; van Steensel et al., 2003).

Acknowledgments

We thank members of our laboratory, as well as Amir Orian, Inhua Muijrers-Chen, and Helen Pickersgill for helpful comments, and Daan Peric-Hupkes for help with editing. Celine Moorman is supported by the Human Frontier Science Program.

References

Barras, F., and Marinus, M. G. (1989). The great GATC: DNA methylation in E. coli. Trends Genet. 5, 139–143.

Bernstein, B. E., Humphrey, E. L., Liu, C. L., and Schreiber, S. L. (2004). The use of chromatin immunoprecipitation assays in genome-wide analyses of histone modifications. Methods Enzymol. 376, 349–360.

Bianchi-Frias, D., Orian, A., Delrow, J. J., Vazquez, J., Rosales-Nieves, A. E., and Parkhurst, S. M. (2004). Hairy transcriptional repression targets and cofactor recruitment in Drosophila. PLoS Biol. 2, E178.

Biggin, M. D. (2001). To bind or not to bind. Nat. Genet. 28, 303–304.

Boivin, A., and Dura, J. M. (1998). In vivo chromatin accessibility correlates with gene silencing in Drosophila. Genetics 150, 1539–1549.

Buck, M. J., and Lieb, J. D. (2004). ChIP-chip: Considerations for the design, analysis, and application of genome-wide chromatin immunoprecipitation experiments. Genomics 83, 349–360.

de Wit, E., Greil, F., and van Steensel, B. (2005). Genome-wide HP1 binding in Drosophila: Developmental plasticity and genomic targeting signals. Genome Res. 15, 1265–1273.

Gottschling, D. E. (1992). Telomere-proximal DNA in Saccharomyces cerevisiae is refractory to methyltransferase activity in vivo. Proc. Natl. Acad. Sci. USA 89, 4062–4065.

Greil, F., van der Kraan, I., Delrow, J., Smothers, J. F., de Wit, E., Bussemaker, H. J., van Driel, R., Henikoff, S., and van Steensel, B. (2003). Distinct HP1 and Su(var)3–9 complexes bind to sets of developmentally coexpressed genes depending on chromosomal location. Genes Dev. 17, 2825–2838.

Nadon, R., and Shoemaker, J. (2002). Statistical issues with microarrays: Processing and analysis. Trends Genet. 18, 265–271.

Orian, A., Van Steensel, B., Delrow, J., Bussemaker, H. J., Li, L., Sawado, T., Williams, E., Loo, L. W., Cowley, S. M., Yost, C., Pierce, S., Edgar, B. A., Parkhurst, S. M., and

Eisenman, R. N. (2003). Genomic binding by the *Drosophila* Myc, Max, Mad/Mnt transcription factor network. *Genes Dev.* **17,** 1101–1114.

Phair, R. D., Scaffidi, P., Elbi, C., Vecerova, J., Dey, A., Ozato, K., Brown, D. T., Hager, G., Bustin, M., and Misteli, T. (2004). Global nature of dynamic protein-chromatin interactions *in vivo*: Three-dimensional genome scanning and dynamic interaction networks of chromatin proteins. *Mol. Cell. Biol.* **24,** 6393–6402.

Pollack, J. R., Perou, C. M., Alizadeh, A. A., Eisen, M. B., Pergamenschikov, A., Williams, C. F., Jeffrey, S. S., Botstein, D., and Brown, P. O. (1999). Genome-wide analysis of DNA copy-number changes using cDNA microarrays. *Nat. Genet.* **23,** 41–46.

Quackenbush, J. (2002). Microarray data normalization and transformation. *Nat. Genet.* **32** (Suppl.), 496–501.

Ren, B., and Dynlacht, B. D. (2004). Use of chromatin immunoprecipitation assays in genome-wide location analysis of mammalian transcription factors. *Methods Enzymol.* **376,** 304–315.

Singh, J., and Klar, A. J. (1992). Active genes in budding yeast display enhanced *in vivo* accessibility to foreign DNA methylases: A novel *in vivo* probe for chromatin structure of yeast. *Genes Dev.* **6,** 186–196.

Slonim, D. K. (2002). From patterns to pathways: Gene expression data analysis comes of age. *Nat. Genet.* **32**(Suppl.), 502–508.

Song, S., Cooperman, J., Letting, D. L., Blobel, G. A., and Choi, J. K. (2004). Identification of cyclin D3 as a direct target of E2A using DamID. *Mol. Cell. Biol.* **24,** 8790–8802.

van Steensel, B., Delrow, J., and Bussemaker, H. J. (2003). Genome wide analysis of *Drosophila* GAGA factor target genes reveals context-dependent DNA binding. *Proc. Natl. Acad. Sci. USA* **100,** 2580–2585.

van Steensel, B., Delrow, J., and Henikoff, S. (2001). Chromatin profiling using targeted DNA adenine methyltransferase. *Nat. Genet.* **27,** 304–308.

van Steensel, B., and Henikoff, S. (2000). Identification of *in vivo* DNA targets of chromatin proteins using tethered dam methyltransferase. *Nat. Biotechnol.* **18,** 424–428.

Wines, D. R., Talbert, P. B., Clark, D. V., and Henikoff, S. (1996). Introduction of a DNA methyltransferase into *Drosophila* to probe chromatin structure *in vivo*. *Chromosoma* **104,** 332–340.

[17]   Whole-Genome Genotyping

By Kevin L. Gunderson, Frank J. Steemers, Hongi Ren,
Pauline Ng, Lixin Zhou, Chan Tsan, Weihua Chang, Dave Bullis,
Joe Musmacker, Christine King, Lori L. Lebruska, David Barker,
Arnold Oliphant, Kenneth M. Kuhn, and Richard Shen

Abstract

We have developed an array-based whole-genome genotyping (WGG) assay (Infinium) using our BeadChip platform that effectively enables unlimited multiplexing and unconstrained single nucleotide polymorphism (SNP) selection. A single tube whole-genome amplification reaction is used

METHODS IN ENZYMOLOGY, VOL. 410
Copyright 2006, Elsevier Inc. All rights reserved.
0076-6879/06 $35.00
DOI: 10.1016/S0076-6879(06)10017-8

to amplify the genome, and loci of interest are captured by specific hybridization of amplified gDNA to 50-mer probe arrays. After target capture, SNPs are genotyped on the array by a primer extension reaction in the presence of hapten-labeled nucleotides. The resultant signal is amplified during staining and the array is read out on a high-resolution confocal scanner. We have employed our high-density BeadChips supporting up to 288,000 bead types to create an array that can query over 100,000 SNPs using the Infinium assay. In addition, we have developed an automated BeadChip processing platform using Tecan's GenePaint slide processing system. Hybridization, washing, array-based primer extension, and staining are performed directly in Tecan's capillary gap Te-Flow chambers. This automation process increases assay robustness and throughput greatly while enabling laboratory information management system control of sample tracking.

Introduction

Most major diseases have a complex genetic component in which a number of "predisposition" genes contribute to the disease etiology. Finding these genes is an important goal for improving human health. Linkage disequilibrium (LD)-based association studies have been proposed as the most viable approach for finding these predisposition genes (Hirschhorn and Daly, 2005). In this approach, genetic markers are genotyped across cases and control populations, and regions in the genome in which the markers show statistically significant (χ^2 test) differences in allele frequency are flagged as regions harboring a potential candidate disease locus (Carlson et al., 2004a). The development of technology to genotype a large number of markers rapidly across large sample sets will accelerate the study of these genetic factors greatly in human health and disease.

Single nucleotide polymorphisms (SNPs) are the markers of choice for high-density array-based genotyping assays, with over 10 million SNPs (>1%) present in the human genome (Consortium, 2003). In most of the genome, these SNPs occur in blocks of strong LD such that only a small subset of haplotype "tagging" SNPs (htSNPs) suffices to characterize these haplotype blocks (Johnson et al., 2001). Alternatively, rather than characterizing the LD of the genome by haplotype blocks, the LD can be characterized by selected "tagSNPs" that represent LD bins (Carlson et al., 2004b). The major goal of the International HapMap project is to define these haplotype blocks and associated htSNPs or tagSNPs for the three major ethnic groups (www.hapmap.org). Data from Hinds et al. (2005) indicate that about 300,000 tagSNPs will provide approximately 70% LD coverage ($r^2 > 0.8$) of the genome in Caucasians and Asians. They demonstrate that 300,000 tagSNPs is roughly equivalent in power to about 1,000,000 randomly chosen SNPs, exemplifying the power of the tag SNP approach. An alternative SNP selection approach is to focus on putative functional SNPs, as well as SNPs in or near coding and conserved regions. In particular, it is estimated that about 98K (~3.3 SNPs/gene) tagSNPs would be needed to span the Caucasian gene-centric HapMap and 144K SNPs (~4.8 SNPs/gene) for the corresponding African HapMap (Stephens et al., 2001).

Despite the improvements offered by intelligent SNP selection, improvements in genotyping technology are required to practically enable large-scale genotyping studies. In contrast to gene expression studies, the use of array technology for large-scale genotyping has been muted by the inability of sample preparation approaches to multiplex at the scale of high-density arrays. Several approaches to overcoming this multiplexing bottleneck have been described in the literature, but high-level multiplexing with unconstrained SNP selection remains to be demonstrated (Fan et al., 2003; Hardenbol et al., 2003; Kennedy et al., 2003; Matsuzaki et al., 2004). Ideally, a large-scale genotyping assay would mirror a gene expression experiment in which a single tube sample preparation (total cDNA and in vitro transcription amplification/labeling) is hybridized to a single array, allowing readout of the entire transcriptome. Using this paradigm, we designed the Infinium whole-genome genotyping (WGG) assay, which genotypes array-hybridized genomic targets directly (Gunderson et al., 2005).

WGG Assay Design

The Infinium WGG assay is a simple, robust, single tube, scalable, genome-wide SNP genotyping assay that consists of four modular steps: (1) whole-genome amplification (WGA), (2) hybridization capture to an array of 50-mer oligonucleotide probes, (3) array-based enzymatic SNP scoring, and (4) an antibody-based sandwich detection assay (Fig. 1). The WGA reaction is used to amplify, in a relatively unbiased manner, the input gDNA (a few hundred nanograms) by over 1000–2000×, leading to almost a milligram of final amplified DNA product. This relatively high concentration of target (low pM, 150 μl volume) drives the hybridization capture of target loci to the probes on the array. The probe consists of an 80-mer oligonucleotide sequence 5′ end immobilized of which the first 30 bases (5′ sequence) are used for decoding and the remaining 50 bases constitutes the query sequence for the locus (gene) of interest. After capture of the target loci, an array-based polymerase extension step is used

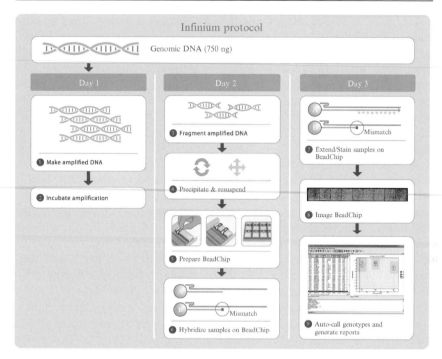

Fig. 1. Whole-genome genotyping on DNA arrays. The DNA sample is whole-genome amplified in an overnight isothermal reaction (steps 1 and 2). The amplification is relatively unbiased with no appreciable allelic partiality. The DNA is amplified 1000–2000× starting with 250–750 ng. The final product concentration is relatively insensitive to the starting amount. After amplification, the product is fragmented enzymatically in a controlled process (step 3). After isopropanol precipitation and resuspension (step 4), the samples are applied to a BeadChip mated to a capillary gap Te-Flow chamber (step 5). The samples are hybridized overnight to a BeadChip array containing 50-mer capture probes to SNP loci. After hybridization capture, the SNPs are scored using either an allele-specific primer extension (ASPE) step or a single base extension (SBE) reaction. The ASPE assay uses two bead types per SNP and a single-color readout, whereas the SBE assay uses a single bead type per SNP and a two-color readout (step 7). Biotinylated/DNP nucleotides are incorporated during the primer extension step and these labels are stained and amplified in a immunohistochemistry-based multilayer sandwich assay (ref.). After staining, the BeadChips are read on a high-resolution confocal scanner (0.84-μm resolution). Genotype calls are made automatically by GenCall software, generating reports of the results. (See color insert.)

to score the SNP directly on the array surface. The partitioning of the assay into an independent capture step and separate SNP scoring step enhances discrimination greatly. Finally, after completion of the primer extension step, an immunohistochemistry-based staining and signal amplification step is used to boost the overall signal to noise of the assay even further (Mokry, 1996; Pinkel *et al.*, 1986).

For SNP allele discrimination, both allele-specific primer extension (ASPE) and single base extension (SBE) approaches have proven effective for array-based SNP scoring assays (Pastinen *et al.*, 1997, 2000; Shumaker *et al.*, 1996). Our first commercial product, the Sentrix Human-1 Genotyping BeadChip employs a single-color ASPE reaction using two probe pairs (designated bead types A and B corresponding to A and B alleles) per locus (Fig. 2A). The two probes are identical except at their terminal 3' base, which is designed to perfectly complement one allele or the other. The perfect match probe allows efficient primer extension, whereas a single mismatch reduces extension efficiency greatly, generating the desired allelic discrimination.

In an alternative primer extension approach, the ASPE scoring step can be replaced with single base extension (SBE) depicted in Fig. 2B (Steemers

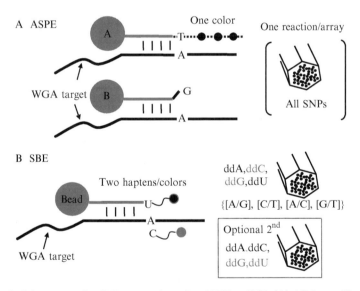

Fig. 2. Primer extension SNP genotyping using ASPE or SBE. (A) Allele-specific primer extension (ASPE) uses two bead types and one color per SNP assay. The incorporated label is read in a single color sandwich staining assay. (B) Single base extension (SBE) employs a single bead type and two haptens (colors) per SNP assay. The probes are 50 bases in length and hybridize immediately adjacent to the SNP query site, followed by extension by a single base. Allelic discrimination is achieved by using two color differentially labeled terminators. The use of only two colors in SBE restricts the number of SNP classes (four out of six) that can be assayed in any given reaction as shown. All 100% of SNPs can be assayed using two separate arrays.

et al., 2006). SBE has several advantages over ASPE. First of all, only a single probe is required per SNP, rather than two, effectively doubling the information content of the BeadChip. The second advantage of SBE is its inherently robust biochemistry. SBE, unlike ASPE, is an end point assay in which labeled dideoxynucleotides are competitively incorporated into the probe. We observed that up to fivefold changes in nucleotide or polymerase concentrations had minimal impact on the assay. One disadvantage of SBE is that multiple colors, rather than multiple bead types, are required to read out the SNP alleles. The use of multiple colors requires more sophisticated signal amplification schemes, as well as color channel balancing. In the simplest SBE reaction, two colors are used to read out the genotype. Use of a two-color read-out scheme (on a single array) restricts the number of SNP classes to four out of six (A/G, C/T, A/C, and G/T). The judicious assignment of SNPs allows over 83% of dbSNP to be queried. Construction of a tagSNP product is not limited; there will be an abundance of tagSNPs available with the completion of phase II of the International HapMap project. Moreover, if so desired, the remaining 17% of the SNPs can be assayed in a second SBE reaction, preferably on a multisample BeadChip. Given the improved assay density available with SBE, we are implementing this assay for our 250 and 500K tagSNP BeadChip products.

BeadChip Design

The high density of the Human-1 BeadChip was achieved using a MEMS-patterned slide substrate (82.5 mm length vs standard 75 mm) supporting over 288,000 different bead types (Fig. 3). The BeadChip design consists of 12 sections (stripes) into which beads are assembled from a pool containing ~24,000 different bead types. Each stripe receives a different bead pool, thus a different set of 24,000 bead types for a total of 288,000 bead types. Each stripe also holds up to 890,000 beads, generating an average redundancy of greater than 30 beads per bead type. A decoding process that maps the identity of each bead is described by Gunderson *et al.* (2004). Using this BeadChip layout, up to 144,000 assays are supported in the ASPE assay and up to 288,000 assays in the SBE assay. A further reduction in bead size and spacing can increase the number of assays greatly. For instance, by decreasing our current bead diameter from 3 to 1.5 μm (and corresponding spacing), the bead density increases by a factor of four to over 1 million bead types/array.

The choice of slide substrate has also contributed to the success of the Infinium assay, with the BeadChip format, providing an ideal substrate for array-based enzymatic assays. Beads provide a better substrate than slides for bulk surface modifications and immobilization of assay oligonucleotides.

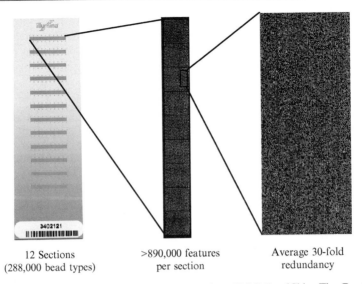

| 12 Sections | >890,000 features | Average 30-fold |
| (288,000 bead types) | per section | redundancy |

FIG. 3. Substrate design for a high-density Infinium WGG BeadChip. The BeadChip consists of 12 sections (stripes) each containing over 890,000 beads. Within each section, up to 24,000 bead types, with average 30-fold redundancy, can be loaded. The total number of bead types supported by this platform is ~288,000.

These bulk processes, in which beads of a particular type are created in "one" immobilization event, improve array-to-array feature consistency greatly. This is particularly important, as ASPE-based genotyping employs a ratiometric comparison between two bead types with this ratio relatively invariant from one array to another. The built-in redundancy of bead arrays, in which each feature (bead type) is composed of measurements from 20 to 30 individual beads from the same immobilization event, further improves the robustness of the ratiometric measurement.

Commercial Products

Our Sentrix Human-1 Genotyping BeadChip consists of over 100,000 SNP assays deployed on the described high-density BeadChip platform. It is a simple single tube, single array assay. Overall genotyping quality, as assessed by measuring genotyping quality parameters across 138 different DNA samples (not a single failure in DNA amplification), is remarkably high. The call rate was >99.4% for almost 14 million genotype calls, the reproducibility was >99.99%, and the concordance with HapMap data was >99.65% (manuscript in preparation). A typical genotype plot of

θ values for all 100,000 loci for a given sample is shown in Fig. 4A. Impressively, three clusters were easily observed across the collection of 100,000 loci. A representative genotyping plot for a single locus across all 138 samples is shown in Fig. 4B and C. For easier clustering and visualization, the rectangular coordinates (A bead type vs B bead type signal) are transformed into polar coordinates (log-normalized R intensity, and a θ ratiometric measurement). In the Genoplot, the three SNP clusters are clearly distinguishable, allowing accurate genotype calls.

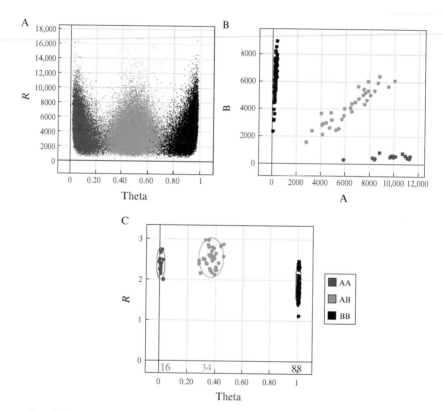

FIG. 4. Genotype clustering on high-density Infinium WGG BeadChips. (A) Genoplot of genotyping data (θ values) for over 100,000 SNPs from a single BeadChip. Note the three distinct clusters corresponding to the three genotype categories (A/A, A/B, and B/B). (B and C) A representative GenoPlot of both raw and normalized polar coordinate genotyping data for a single locus across 138 different DNA samples. The encircled data points indicate the locations of the archetypal clusters generated by Illumina GenTrain software. R is normalized intensity and θ is $(2/\pi)^*\arctan(B/A)$. These GenoPlots were generated by proprietary GenCall software.

One of the strengths of this assay is the relatively unconstrained ability to select "high-value" SNPs. Using this feature, the content of the Human-1 BeadChip was developed around a combination of putative functional SNP markers and exon-centric SNPs. In fact, over 85% of the exons in the genome (~90% of RefSeq exons) are within 10 kb of a SNP marker, and over 23,000 SNPs reside directly in transcripts, including many putative functional SNPs (nonsynonymous, splice sites, synonymous, etc.). As an example, the EGFR gene (188 kb) has 16 SNPs in and around its exons, providing good coverage for a LD-based detection of variants. In summary, it is likely that a large fraction of disease-related variants, including non-synonymous, splice site and regulatory variants, will lie in conserved and exon-rich regions.

The flexibility of the BeadChip substrate allows a variety of different product offerings, including both single sample and multisample platforms (Fig. 5A). Our initial WGG product employs over 100,000 SNPs across a single sample; however, multisample products are planned such that 10,000–20,000 loci can be analyzed from across 6–12 samples. We already employ multisample BeadChips in our whole-genome gene expression platform in which multiple samples are analyzed in parallel (Kuhn *et al.*,

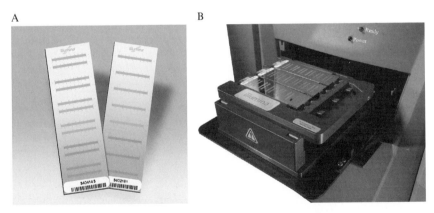

A   B

FIG. 5. Multisample BeadChips in a BeadArray reader. (A) Multisample BeadChips for six or eight samples across one or two stripes, respectively. We have designed two genome-scale array formats for gene expression analysis using these BeadChip substrates. The Human-6 BeadChip contains six arrays (two stripes per array) on a single BeadChip; each array contains >46,000 probes, and the HumanRef-8 BeadChip contains eight arrays (one stripe per array); each array contains >24,000 probes derived from the RefSeq database. (B) The Illumina BeadArray reader accommodates three BeadChips in a single reading at 0.84-μm resolution. The adapter tray facilitates reading the Sentrix array matrix and different types of BeadChip products.

2004). Although specific whole-genome gene expression protocols are not listed in this chapter, more information can be obtained by contacting technical support at www.illumina.com. Both single sample and multisample BeadChips are scanned on the same Bead Array reader. A special adapter tray accommodates three BeadChips (up to 24 arrays) and can be scanned in a single loading (Fig. 5B).

Methods

The Infinium WGG assay is straightforward to implement in most molecular biology laboratories. The four modular assay elements are robust and easily performed. The first two modular steps can be performed manually or robotically with similar ease, whereas the last two modules are performed more easily using a pipetting robot (XStain process), but could be performed manually if desired. We have formulated reagents into aliquoted single-use tubes and automated the assay using Tecan robotics (Genesis and Evo) and their GenePaint slide processing system (www.tecan.com). GenePaint employs a capillary gap Te-Flow Through chamber (Fig. 6) to enable easy washing and reagent exchanges on the BeadChip. The capillary gap is created using a 70-μm spacer, utilizing capillary action to retain reagent within the gap. The Te-Flow Through chambers are placed in a temperature-controlled Te-Flow chamber rack, allowing precise temperature control of all extension and staining steps. Finally, a Tecan robot performs all reagent transfer steps, including pipetting of wash

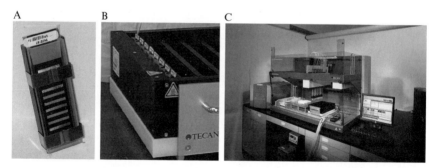

FIG. 6. Automated BeadChip processing. (A) A Tecan capillary gap Te-Flow Flow Through chamber is used for easy reagent exchanges. (B) A temperature-controlled Te-Flow chamber rack and Tecan robot are used for the washing, primer extension, and staining (detection) steps of the Infinium assay. (C) A complete automated robotics solution for processing the BeadChips.

solutions, blocking mixes, extension reagents, and staining reagents. In summary, the complete assay has been automated from sample preparation to posthybridization slide processing.

Sample Amplification

Contamination is much less of an issue in WGG compared to polymerase chain reaction-based genotyping assays, as the amplification level in WGG is generally on the order of 1000–2000× rather than a millionfold or greater. The current protocol requires 250–750 ng for single sample WGG analysis and ~50–200 ng (smaller amplification reaction) in the multisample mode. The use of high-quality DNA in the assay is important, as degraded DNA (<1 kb average size) amplifies less efficiently. The best method to access the quality of DNA is by gel analysis using ~100 ng of gDNA run out on a 6% PAGE gel and stained with SYBR Gold nucleic acid stain (Invitrogen). The major smear of DNA should be above the 1-kb size marker, although we have obtained good performance on samples with an average size as small as 500 bp.

Before starting the experiments, all prealiquoted reagents should be thawed at room temperature immediately before use, mixed by inversion 10 times, and spun down at 280g. If using automation, the reagents are placed on the robot bed in accordance with the "bed map" in the user manual.

In the first step of the assay, whole-genome amplification is used to amplify the genome by a factor of ~1000–2000× in a relatively unbiased manner. The Infinium assay uses our preformulated WGA amplification kit (MP1 and AMM) for compatibility with downstream fragmentation and processing. Amplification proceeds at 37° for ~20 h. Incubation for shorter periods of time reduces yield slightly, whereas overincubation (up to 44 h) appears to have minimal impact on the assay. After amplification, a fragmentation protocol is used to reduce the fragment size to 200–300 bp. This fragmentation improves both resuspension and hybridization efficiency. After fragmentation, the reaction is stopped using a mix of ammonium acetate and EDTA. The EDTA serves the dual purpose of stopping the fragmentation reaction and helping dissolve precipitates formed during amplification, which could interfere in downstream processes. One volume of isopropanol is added to precipitate the final product. This reduces the carryover of dNTPs and concentrates the DNA sample. The DNA pellet is resuspended in a formamide-containing hybridization buffer. Incubation of the sample at 48° for 1 h and subsequent vortexing aid greatly in this resuspension step.

Amplification Protocol

1. Resuspend DNA samples at 50 ng/μl using PicoGreen quantitation (Invitrogen).

2. Add 15 μl 0.1 N NaOH to a 15-μl DNA sample placed in a MIDI (0.8 ml) 96-well storage plate (ABgene, AB-0859) and incubate for 10 min at room temperature. For single sample applications, up to 24 samples can be processed in a MIDI plate. Only three columns (1, 5, and 9) are used for samples, as the remaining columns will be used to accommodate the extra volume required in the isopropanol precipitation step.

3. Add 270 μl primer/neutralization mix (MP1) to each sample well.

4. Add 300 μl amplification master mix (AMM) to each sample well.

5. Seal plate with a "nonautoclavable" cap mat (ABgene, AB-0566). This type of cap map seals more effectively.

6. Invert plate 10 times to mix contents and spin down briefly at 280g.

7. Place MIDI plate in oven at 37° for ~20 h.

8. After 37° overnight incubation, a white flocculent material (magnesium pyrophosphate by-product, $Mg_2P_2O_7 = Mg_2PPi$) should be visible in the wells, indicating that amplification was effective (Mori et al., 2001).

9. Briefly spin down plate at 50g for 1 min (a light spin avoids compacting precipitate).

10. Split the 600-μl amplification reaction into three additional wells for a total of four wells per sample, each with 150 μl (Col 1 → 2, 3, and 4; Col 5 → 6, 7, and 8; Col 9 → 10, 11, and 12). The extra volume in the well is needed for isopropanol ppt.

11. Add 50 μl fragmentation mix (FRG) to each well, seal, and vortex at 1600 rpm (signature high-speed microplate shaker, VWR International, 13,500–890, 110 V) for 1 min, spin down at 50g for 1 min, and then incubate at 37° for 1 h. This plate shaker employs Velcro straps for securing the plate.

12. Briefly spin down plate at 50g for 1 min.

13. Add 100 μl precipitation mix (PA1) to each well, seal, and vortex at 1600 rpm for 1 min, spin down at 50g for 1 min, and incubate 5 min at 37°.

14. Add 300 μl isopropanol to each well.

15. Seal plate with a fresh "nonautoclavable" cap mat (make sure cap mat and plate are dry around rims to ensure a good seal, as isopropanol may leak during inversion). Mix plate by inversion 10 times.

16. Incubate the amplification plate for 30 min at 4°.

17. Spin down plate at 3000g for 20 min at 4°.

18. Immediately after centrifugation, remove supernatant by slowly inverting plate over sink and pouring the isopropanol out. Blot inverted plate

on a paper towel. Blue pellets (due to a blue additive in the precipitation step) should be visible in the bottom of each well at this point.

19. Dry inverted plate on a tube rack for 1 h at room temperature.

20. Add 42 μl hybridization buffer (RA1) to each well and seal plate with heat seal foil sheet (AB-gene, AB-0559). Incubate for 1 h at 48°.

21. Resuspend sample by vortexing at 1800 rpm for 1 min and then centrifuge at 280g for 1 min.

22. Recollect the four sample wells into the one original well for a total volume of 160 μl per original sample well.

23. Denature samples at 95° for 20 min in heat block (SciGene, Hybex heating base, 1057-30-0, and custom aluminum block for MIDI plate, Illumina No. 21119).

24. Remove samples from the heat block and place on bench for ~5–10 min before loading samples onto BeadChips. Alternatively, the samples can be frozen down and stored at –20° for up to several months. Immediately before hybridization, thaw and denature samples again at 95° for 10 min on heat block.

BeadChip Hybridization

The average yield from the WGA reaction is ~1.5 μg/μl, and the final hybridization concentration is 5–6 μg/μl. After resuspension, the sample is denatured at 95° for 20 min to be used immediately or frozen down (–20 or –80°) and stored for at least several months before use. Immediately before hybridization, the sample should be denatured again at 95° for ~10 min and allowed to cool on the bench for 5–10 minutes before applying to the BeadChip array.

1. Assemble BeadChip into the Te-Flow Through chamber (Tecan, 760–810) as recommended by the Infinium assay system manual (Illumina).

2. Prepare Illumina hybridization chambers by pipetting 200 μl humidifying buffer (PB2) into troughs as described in the Infinium user manual. (This buffer has the same vapor pressure as the sample to prevent evaporation.)

3. Slowly dispense 150 μl of 100% formamide into an assembled Te-Flow Through chamber placed upright in a Te-Flow rack (equilibrated to room temperature) (Tecan, 760–800). Make sure no bubbles have formed within the capillary gap. If bubbles are present, disassemble, wash slide in low salt buffer (PA1), spin dry slide (280g), and repeat assembly and formamide loading.

4. Dispense 150 μl of hybridization buffer (RA1) into the reservoir of the Te-Flow Through chamber and allow to flow through. Repeat with another 150 μl of hybridization buffer (RA1).

5. Dispense 150 μl DNA sample into the BeadChip as described earlier.

6. Remove Te-Flow Through chamber from rack, briefly blot residual hybridization buffer from the bottom of the Te-Flow Through chamber, and place horizontally in Illumina hybridization enclosures according to manufacturer's recommendations.

7. Incubate BeadChips overnight at 48° for 16–24 h.

Extension/Staining (XStain)

This process involves pipetting of various reagents to the Te-Flow Through chambers placed on the GenePaint Te-Flow chamber rack, equilibrated to 44°. The processes can either be performed manually or via a Tecan pipetting robot. The robot saves in manual labor; however, it is recommended to use a multidispense pipettor if performing the assay manually.

Array-Based Primer Extension

1. Remove BeadChip Te-Flow chambers from hybridization chamber and place in Te-Flow chamber rack equilibrated to 44°.

2. Add 450 μl hybridization buffer (RA1) to Te-Flow Through reservoir to wash BeadChip and allow to drain. Repeat five times.

3. Add 450 μl blocking buffer (XB1) and incubate 10 min.

4. Add 450 μl equilibration buffer (XB2) and incubate 5 min.

5. Add 200 μl extension master mix (EMM) and incubate 15 min.

6. Add 450 μl 95% formamide/10 mM EDTA and incubate for 1 min. Repeat once.

Multilayer IHC Sandwich Staining

7. Add 450 μl wash buffer (XB3) and drain. Repeat twice.

8. Add 250 μl staining solution (LMM) and incubate 10 min.

9. Repeat wash in step 6.

10. Add 250 μl antistain solution (ASM) and incubate 10 min.

11. Repeat wash in step 6.

12. Add 250 μl staining solution (LMM) and incubate 10 min.

13. Repeat wash in step 6.

14. Add 250 μl antistain solution (ASM) and incubate 10 min.

15. Repeat wash in step 6.

16. Add 250 μl staining solution (LMM) and incubate 10 min.

17. Repeat wash in step 6.

18. Remove Te-Flow chambers from GenePaint rack and disassemble.

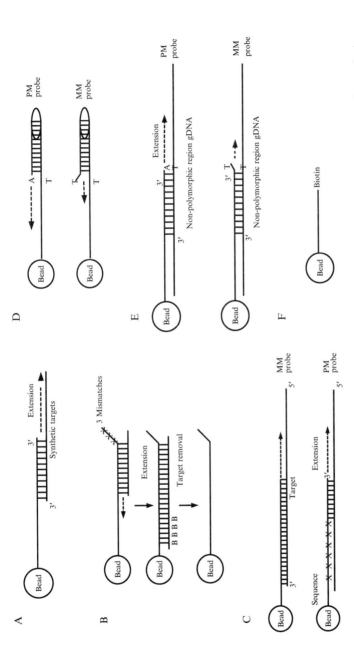

Fig. 7. Controls for Infinium WGG assay. (A) Hybridization controls test the overall performance of the entire assay using synthetic targets instead of amplified DNA. These synthetic targets complement the sequence on the array perfectly, allowing the probe to extend on the synthetic target as template. The synthetic targets are present in the hybridization buffer (RA1) at three levels: high (5 p*M*), medium (1 p*M*), and low (0.1 p*M*) concentration targets. (B) Target removal controls in the hybridization buffer (RA1) test the efficiency of the stripping step after the extension reaction. During extension, the control oligonucleotides are extended/labeled using the probe sequence as a template. The signal of the target removal controls should be low compared to the hybridization controls, indicating effective target stripping. (C) Hybridization stringency controls are used to test the specificity of hybridization. The probes are designed such that the 3′ end of the probe is available for extension, with mismatches (0 or 6) introduced into the body of the probe to affect hybrid stability. (D) Extension controls test the efficiency of extension from a

19. Place BeadChips in slide rack in 250-ml wash tray containing low salt wash buffer (PB1). Wash slides in rack by dunking about 20 times.

20. Spin slides dry in rack by centrifugation at 280*g* for 1 min.

21. Store slides in a desiccated, light-protected container until scanning (scan within 24 h of dry-down step).

Data Analysis

BeadChips are scanned at 0.84-μm resolution (one or two colors) on a BeadArray reader that accommodates up to three BeadChips. The intensity values on the BeadChip are extracted after automated registration and grid placement (Galinsky, 2003a,b). The individual intensities for each bead within a bead type are averaged ($\sim$30 beads/type) to generate composite A bead type and B bead type intensities (or A and B channel intensities for SBE). Genotyping analysis is performed using proprietary GenCall software that generates archetypal clusters from a set of training data based on a population of gDNA samples. The A and B intensity information is converted to normalized polar coordinates. Genotype calls are made using a probabilistic model in which probabilities are assigned for membership into the three clusters (AA, AB, or BB). Data points falling between clusters have a low probability of belonging to any one cluster and are scored as a no call (GenCall score threshold of 0.2–0.25). Genotyping data are output in a tabular comma delimited format for further downstream analysis.

In addition to genotyping data, a set of assay controls is included on every array to assess most steps of the assay from sample preparation through staining. These controls monitor amplification, hybridization, extension, stripping, and staining (see Fig. 7). All these controls can be displayed across all assayed samples in a GenCall Controls "dashboard." Monitoring of these controls is useful for optimization and for troubleshooting.

Conclusions

We have described a whole-genome genotyping technology that has the characteristics of scaling indefinitely (with number of features on array) and almost unconstrained SNP selection. These two characteristics render

---

hairpin probe and are therefore sample independent. The perfect match hairpin controls should result in high signal, and the mismatch probes should result in low signal. (E) Nonpolymorphic controls test the overall performance of the assay, from amplification to detection, by querying a particular base in a nonpolymorphic region of the genome. Nonpolymorphic controls allow sample-to-sample performance comparison. (F) Staining controls (bead types labeled with different biotin levels) are used to examine the efficiency of the staining step. These controls are independent of the hybridization and extension step.

the assay truly whole genome. The intrinsic scalability of BeadArray technology will enable rapid density improvements without complicated array fabrication technology. This will enable the development of arrays with which more than 1 million SNPs can be assayed. This density will have a profound impact on association studies. In addition, these high-resolution arrays have a number of applications beyond standard genetic studies, including DNA copy number/loss-of-heterozygosity studies, allele-specific expression, and ChIP on chip analysis, as well as other array-based studies.

Acknowledgments

We thank the many scientists and engineers at Illumina involved in the creation and optimization of arrays and assays. We appreciate the critical reading of this chapter and constructive suggestions provided by John Stuelpnagel, Nicky Espinosa, Melanie Smith, and Bill Craumer. We also thank Lori Lebruska for critical review and preparation of manuscript, and thank Judith Guest for providing graphics from the Infinium Assay System Manual. This work was supported in part by grants from the NIH/NCI 1 R43 CA103406–01, 1 R43 CA108391-01, and 2 R44 CA103406-02.

References

Carlson, C. S., Eberle, M. A., Kruglyak, L., and Nickerson, D. A. (2004a). Mapping complex disease loci in whole-genome association studies. *Nature* **429,** 446–452.
Carlson, C. S., Eberle, M. A., Rieder, M. J., Yi, Q., Kruglyak, L., and Nickerson, D. A. (2004b). Selecting a maximally informative set of single-nucleotide polymorphisms for association analyses using linkage disequilibrium. *Am. J. Hum. Genet.* **74,** 106–120.
Consortium, T. I. H. (2003). The international HapMap project. *Nature* **426,** 789–796.
Fan, J. B., Oliphant, A., Shen, R., Kermani, B. G., Garcia, F., Gunderson, K. L., Hansen, M., Steemers, F., Butler, S. L., Deloukas, P., Galver, L., Hunt, S., McBride, C., Bibikova, M., Rubano, T., Chen, J., Wickham, E., Doucet, D., Chang, W., Campbell, D., Zhang, B., Kruglyak, S., Bentley, D., Haas, J., Rigault, P., Zhou, L., Stuelpnagel, J., and Chee, M. S. (2003). Highly parallel SNP genotyping. *Cold Spring Harb. Symp. Quant. Biol.* **68,** 69–78.
Galinsky, V. L. (2003a). Automatic registration of microarray images. I. Rectangular grid. *Bioinformatics* **19,** 1824–1831.
Galinsky, V. L. (2003b). Automatic registration of microarray images. II. Hexagonal grid. *Bioinformatics* **19,** 1832–1836.
Gunderson, K. L., Kruglyak, S., Graige, M. S., Garcia, F., Kermani, B. G., Zhao, C., Che, D., Dickinson, T., Wickham, E., Bierle, J., Doucet, D., Milewski, M., Yang, R., Siegmund, C., Haas, J., Zhou, L., Oliphant, A., Fan, J. B., Barnard, S., and Chee, M. S. (2004). Decoding randomly ordered DNA arrays. *Genome Res.* **14,** 870–877.
Gunderson, K. L., Steemers, F. J., Lee, G., Mendoza, L. G., and Chee, M. S. (2005). A genome-wide scalable SNP genotyping assay using microarray technology. *Nat. Genet.* **37,** 549–554.
Hardenbol, P., Baner, J., Jain, M., Nilsson, M., Namsaraev, E. A., Karlin-Neumann, G. A., Fakhrai-Rad, H., Ronaghi, M., Willis, T. D., Landegren, U., and Davis, R. W. (2003). Multiplexed genotyping with sequence-tagged molecular inversion probes. *Nat. Biotechnol.* **21,** 673–678.

Hinds, D. A., Stuve, L. L., Nilsen, G. B., Halperin, E., Eskin, E., Ballinger, D. G., Frazer, K. A., and Cox, D. R. (2005). Whole-genome patterns of common DNA variation in three human populations. *Science* **307,** 1072–1079.

Hirschhorn, J. N., and Daly, M. J. (2005). Genome-wide association studies for common diseases and complex traits. *Nat. Rev. Genet.* **6,** 95–108.

Johnson, G. C., Esposito, L., Barratt, B. J., Smith, A. N., Heward, J., Di Genova, G., Ueda, H., Cordell, H. J., Eaves, I. A., Dudbridge, F., Twells, R. C., Payne, F., Hughes, W., Nutland, S., Stevens, H., Carr, P., Tuomilehto-Wolf, E., Tuomilehto, J., Gough, S. C., Clayton, D. G., and Todd, J. A. (2001). Haplotype tagging for the identification of common disease genes. *Nat. Genet.* **29,** 233–237.

Kennedy, G. C., Matsuzaki, H., Dong, S., Liu, W. M., Huang, J., Liu, G., Su, X., Cao, M., Chen, W., Zhang, J., Liu, W., Yang, G., Di, X., Ryder, T., He, Z., Surti, U., Phillips, M. S., Boyce-Jacino, M. T., Fodor, S. P., and Jones, K. W. (2003). Large-scale genotyping of complex DNA. *Nat. Biotechnol.* **21,** 1233–1237.

Kuhn, K., Baker, S. C., Chudin, E., Lieu, M. H., Oeser, S., Bennett, H., Rigault, P., Barker, D., McDaniel, T. K., and Chee, M. S. (2004). A novel, high-performance random array platform for quantitative gene expression profiling. *Genome Res.* **14,** 2347–2356.

Matsuzaki, H., Loi, H., Dong, S., Tsai, Y. Y., Fang, J., Law, J., Di, X., Liu, W. M., Yang, G., Liu, G., Huang, J., Kennedy, G. C., Ryder, T. B., Marcus, G. A., Walsh, P. S., Shriver, M. D., Puck, J. M., Jones, K. W., and Mei, R. (2004). Parallel genotyping of over 10,000 SNPs using a one-primer assay on a high-density oligonucleotide array. *Genome Res.* **14,** 414–425.

Mokry, J. (1996). Versatility of immunohistochemical reactions: Comprehensive survey of detection systems. *Acta Med. (Hradec Kralove)* **39,** 129–140.

Mori, Y., Nagamine, K., Tomita, N., and Notomi, T. (2001). Detection of loop-mediated isothermal amplification reaction by turbidity derived from magnesium pyrophosphate formation. *Biochem. Biophys. Res. Commun.* **289,** 150–154.

Pastinen, T., Kurg, A., Metspalu, A., Peltonen, L., and Syvanen, A. C. (1997). Minisequencing: A specific tool for DNA analysis and diagnostics on oligonucleotide arrays. *Genome Res.* **7,** 606–614.

Pastinen, T., Raitio, M., Lindroos, K., Tainola, P., Peltonen, L., and Syvanen, A. C. (2000). A system for specific, high-throughput genotyping by allele-specific primer extension on microarrays. *Genome Res.* **10,** 1031–1042.

Pinkel, D., Straume, T., and Gray, J. W. (1986). Cytogenetic analysis using quantitative, high-sensitivity, fluorescence hybridization. *Proc. Natl. Acad. Sci. USA* **83,** 2934–2938.

Shumaker, J. M., Metspalu, A., and Caskey, C. T. (1996). Mutation detection by solid phase primer extension. *Hum. Mutat.* **7,** 346–354.

Steemers, F. J., Chang, W., Lee, G., Shen, R., Barker, D. L., and Gunderson, K. L. (2006). Whole-genome genotyping (WGG) using single-base extension (SBE) array. *Nat. Methods* **3,** 31–33.

Stephens, J. C., Schneider, J. A., Tanguay, D. A., Choi, J., Acharya, T., Stanley, S. E., Jiang, R., Messer, C. J., Chew, A., Han, J. H., Duan, J., Carr, J. L., Lee, M. S., Koshy, B., Kumar, A. M., Zhang, G., Newell, W. R., Windemuth, A., Xu, C., Kalbfleisch, T. S., Shaner, S. L., Arnold, K., Schulz, V., Drysdale, C. M., Nandabalan, K., Judson, R. S., Ruano, G., and Vovis, G. F. (2001). Haplotype variation and linkage disequilibrium in 313 human genes. *Science* **293,** 489–493.

[18] Mapping *Drosophila* Genomic Aberration Breakpoints with Comparative Genome Hybridization on Microarrays

By JEREMY N. ERICKSON and ERIC P. SPANA

Abstract

Chromosomal aberrations are genetic "reagents" that are commonly used in *Drosophila* research. Stocks containing chromosomes carrying large deletions of DNA (deficiency stocks, designated Df) as well as stocks carrying an extra copy of a chromosomal region (duplication stocks, designated Dp) are essential for a variety of genetic analyses. The extent of what is deleted or duplicated has typically been determined cytologically by salivary gland polytene chromosome squashes, which identify the edges of the aberration (so-called breakpoints) of each Df or Dp at low resolution. The margin of error for this technique can be quite high, however, because it is dependent on the quality of the squash and the experience of the scientist interpreting the data. Comparative genome hybridization on microarrays provides a precise molecular method to identify which regions of the genome are deleted or duplicated in these stocks by examining a change in chromosomal ploidy across the whole genome. Furthermore, this technique allows genetic data obtained with these strains to be placed in a molecular genomic context.

Introduction

The use of chromosomal aberrations provides an important tool in *Drosophila* research. Especially useful are chromosomes with regions of deleted DNA, called deficiencies (or Df), and chromosomes with regions present in an extra copy, called duplications (or Dp). FlyBase (www.flybase.org) documents more than 5600 deficiency lines that have been used in *Drosophila* genetics, and the Bloomington *Drosophila* stock center (flystocks.bio.indiana.edu) maintains over 1700 Df stocks and 634 Dp stocks. The Drosophila community uses Df and Dp stocks in multiple areas of research, making these aberrations extremely valuable. Common uses of deficiency stocks include complementation analysis of new alleles, dominant interaction screens to identify new genes in a pathway or process, and homozygous phenotype examination. In each of these cases, knowing the exact molecular breakpoints of the deficiencies would allow the researcher

METHODS IN ENZYMOLOGY, VOL. 410   0076-6879/06 $35.00

to move more quickly from phenotype to gene because it would provide a precise account of the genes affected by the chromosomal aberration.

It is of such importance to have molecularly identified breakpoints for each deficiency stock that two groups have initiated research programs to generate new Df kits with molecularly defined lesions. One group, a collaboration between Exelixis, Inc. and the Bloomington *Drosophila* stock center, is using multiple methods to generate deletions between two transposable elements (P-elements and piggyBac inserts) (Parks *et al.*, 2004). A second, independent group, the DrosDel consortium, is using FRT-mediated recombination within P-elements to generate deletions (Ryder *et al.*, 2004). These stocks also have the advantage of a clean, isogenic background free from extraneous deletions. When completed, these stocks will represent an additional resource for the *Drosophila* community.

Despite the benefits of the new molecularly defined deficiencies, older deficiencies without defined molecular lesions are still commonly used. The lesions of most chromosomal aberrations have been determined by examining the salivary gland polytene chromosomes (Ashburner, 1989). In this method, the salivary glands of third instar larvae, which are heterozygous for the Df, are flattened (or squashed) under a coverslip, fixed, and stained. Because each region of the genome has a well-characterized banding pattern, one can identify which bands are lost in the Df. However, because there is no correlation of cytological band size to kilobases, a cytological band can be as small as 2 kb or as large as 25 kb. It is common for a breakpoint to have a multiband range for breakpoints, which in some cases can encompass hundreds of kilobases. This time-tested method can be quite precise when done well; however, precise preparation of squashes, as well as interpretation of results, is difficult. The precision of the method is also influenced by the number of bands in the region of genome where the deletion resides. Once the potential breakpoints are identified cytologically, complementation with mutant alleles of genes can further refine the breakpoints by placing genes either within the deficiency region or outside the region.

Since the publication and distribution of the complete genome sequence (Adams *et al.*, 2000), the polymerase chain reaction (PCR) has been used to identify which regions are lost in deficiency strains. Genomic DNA from embryos homozygous for the deficiency (identified by their lack of a GFP marked balancer chromosome) is used as the template for PCR reactions. The PCR primers are designed to make small (300–500 bp) products at regular intervals across the breakpoint predicted by cytology (Myster *et al.*, 2004; Wei *et al.*, 2003). This method is rapid, as all the PCR reactions can be done at once. Drawbacks include the required production of homozygous Df genomic DNA, which may be difficult to isolate if the lethal phase is early during embryogenesis. Additionally, the method must be based on predicted breakpoints determined cytologically to make the PCR primers.

We have found that comparative genome hybridization on microarrays (array CGH) (Pollack *et al.*, 1999) is an excellent alternative to chromosome squashes and PCR. The theory of array CGH is conceptually similar to a quantitative Southern blot (which has also been used for mapping aberrations). For a quantitative Southern blot, one labels a small DNA fragment and hybridizes it to immobilized genomic DNA from aberration and wild-type controls and looks for a change in hybridization intensity. Array CGH is similar to a Southern blot, but probe and target are inverted. Instead of labeling a small fragment and hybridizing to genomic DNA, in array CGH, genomic DNA from wild-type and aberration lines are labeled with different fluorescent dyes and hybridized simultaneously to immobilized small fragments and look for a change in hybridization intensity (see Fig. 1). Array CGH methods may generate terminology issues, as the word "probes" is frequently used interchangeably to describe the immobilized DNA on the microarray as well as the labeled DNA. This chapter uses the

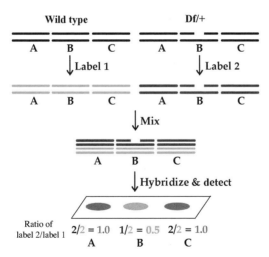

FIG. 1. Comparative genome hybridization on microarrays. The diagram depicts genomic DNA for three genomic regions: A, B, and C in wild type and a heterozygous deficiency stock. Locus B is deleted in the deficiency. The genomic DNA for each genotype is labeled with a different fluorescent dye and is mixed together. Because locus B is missing in the deficiency, it has 50% less corresponding label than A or C. After hybridization to the microarray, quantitative scanning of the fluorescent labels that hybridized to the corresponding target probe shows that locus B has a lower ratio of deficiency to wild-type fluorescence. The lower ratio places the locus in the deficiency.

phrase "hybridization probe" to describe the labeled genomic DNA and the phrase "target probe" to describe the DNA on the microarray. This chapter details the procedure we have adapted from Pollack *et al.* (1999) to identify deficiency breakpoints in *Drosophila* by array CGH using long oligonucleotide-printed microarrays. This procedure can additionally be adapted easily to duplication stocks as well.

Materials and Methods

Fly Stocks

Because deficiency stocks are maintained as balanced stocks, and the balancer chromosome may themselves contain small duplications or deletions, it is necessary to isolate DNA from flies in which the balancer has been replaced by a wild-type chromosome. We use the stock w^{1118}(Bloomington *Drosophila* stock center stock number 3605) as our source of wild-type chromosomes in both heterozygous-deficiency flies and wild-type DNA. The simple cross of 5 Df/Balancer males to 10 w^{1118}virgin females will produce enough Df/+ males for the production of genomic DNA. We prefer to use adult males as the source of the genomic DNA because females undergo overreplication of genomic regions during oogenesis (Claycomb *et al.*, 2004), which has the potential to complicate the results. If males cannot be used, we recommend carefully staging the age of the females for making genomic DNA so that the ages of the experimental and wild-type control females are the same.

Genomic DNA Purification

We use approximately 50 adult male flies for isolation of genomic DNA. We have found that the DNeasy kit from Qiagen provides a clean, reproducible method for extraction. We recommend making one sample of wild type, control DNA for each experimental sample. Following is a detailed description of Qiagen's DNeasy genomic DNA isolation protocol adjusted for adult *Drosophila* tissue.

Place anesthetized flies in a 1.5-ml microcentrifuge tube and add 180 μl of solution ATL and grind with a plastic pestle until no body parts are recognizable. Add 20 μl of the DNeasy proteinase K solution and mix by vortexing. Incubate at 55° overnight. Add 4 μl of RNase A (100 mg/ml solution) and incubate at room temperature for 2 min. Centrifuge at 20,000g for 10 min to pellet the cell debris. After transferring the supernatant to a new tube, add 200 μl of buffer AL and vortex for 15 s. Incubate at 70° for 10 min. Add 200 μl of 100% ethanol and mix by vortexing. Pipette the mixture into a DNeasy column and centrifuge for 1 min at 6000*g*. Move the column to a new collection tube, add 500 μl of wash buffer AW1, and centrifuge for 1 min at 6000*g*. Again move the column to a new collection tube, add 500 μl of wash buffer AW2, and centrifuge for 3 min at 20,000*g*. Discard flow through and collection tube. Place column in a 1.5-ml microcentrifuge tube and put 200 μl of elution buffer AE directly onto the DNeasy membrane. Incubate at room temperature for 1 min and then elute the genomic DNA by centrifugation at 6000*g* for 1 min. We have found that this procedure isolates 15–20 μg of genomic DNA.

Labeling

Each Df genotype will be examined by analysis of a total of four microarrays, and the hybridization probe for each microarray is a mixture of labeled wild type and labeled Df DNA. To make the hybridization probe, we use random primed labeling on genomic DNA that is reduced in length with a four-cutter restriction enzyme. We fluorescently label approximately 2 μg of genomic DNA for each microarray slide and because restriction digestion of genomic DNA is a relatively complete reaction, we combine the digests for all four arrays into one large reaction. After the digest, each labeling reaction is done individually: one labeling reaction per genotype per microarray. Hybridizations are performed in duplicate and are also performed with the dyes reversed to compensate for any dye incorporation effects (the dye flip). Thus, for the four microarrays, we make two hybridization probes with Cy3-labeled wild type and Cy5-labeled Df and two hybridization probes with Cy3-labeled Df and Cy5-labeled wild type. Following is a description of the restriction digestion, clean up, and random primed labeling procedure.

Digest 8 μg of genomic DNA from the experimental and wild type, respectively, with 20 units of Msp1 in a suitable reaction volume for at least 3 h. Following restriction digestion we use a Qiagen PCR cleanup kit to change volume and exchange buffers. Add 5 volumes of solution PB to the restriction digest and mix. Add 1 volume of isopropanol and mix. Repeatedly add 700 μl of the mixture to a Qiaquick column and centrifuge at 13,000*g* for 3 min until the entire mixture has been applied to column, discarding the flow through. Add 700 μl of wash buffer PE and centrifuge for 3 min at 13,000*g* and discard flow through. Centrifuge again for 3 min at 20,000*g* to dry the column and move the column into a 1.5-ml microcentrifuge tube. Elute the digested DNA by adding 50 μl of EB directly to the column matrix, waiting 3 min, and then centrifuging for 3 min at 13,000*g*.

To fluorescently label the genomic DNA to create the hybridization probe, we use components of the BioPrime labeling kit from Invitrogen.

WET-BENCH PROTOCOLS [18]

Combine 10 μl of digested genomic DNA eluted from the previous step, 20 μl of random octamers (from BioPrime labeling kit), and 11 μl of dH$_2$O in a tube, heat at >95° for 5 min, and then place on ice. Following denaturation, add 5 μl of a dNTP mix (1.2 mM each dATP, dGTP, and dTTP, 0.6 mM dCTP in 10 mM Tris, pH 8.0, 1 mM EDTA), 3 μl of the fluorescently labeled dCTP nucleotide (1 mM, from Amersham Biosciences), and 1 μl of DNA polymerase, Klenow fragment (from the BioPrime labeling kit) and incubate at 37° for 2 h. When the labeling reactions are complete, there are eight tubes: two tubes of Cy3-labeled wild type, two tubes of Cy5-labeled wild type, two tubes of Cy3-labeled experimental, and two tubes of Cy5-labeled experimental.

Hybridization and Detection

The next step of generating the hybridization probe is to combine the labeled wild type and experimental hybridization probes for each microarray. For the four microarray slides, four unique tubes of hybridization probe are prepared: two tubes containing both Cy3-labeled wild type and Cy5-labeled experimental and two tubes containing Cy5-labeled wild type and Cy3-labeled experimental.

To make the hybridization mix, add 50 μl of the Cy3 wild type, 50 μl of the Cy5 experimental (or vice versa), 400 μl of TE, and 4 μl of yeast tRNA from a 25-mg/ml stock solution. Move this solution into a Microcon YM-30 tube (Millipore) and centrifuge for 12 min at 13,000g. Continue to centrifuge until the retained volume is less than 13.5 μl. When sufficient flow through has been achieved, remove the column and place inverted in a 1.5-ml microcentrifuge tube. Centrifuge for 3 min to recover the retained solution. Bring up the retained hybridization probe in 60 μl of hybridization solution with a final concentration of 5× SSC, 0.1% SDS, and 50% formamide.

The hybridization, washes, and detection described are performed at the Duke University Microarray Core Facility. The arrays we use are the *Drosophila* Array-Ready Oligo Set from Operon. This set consists of 14,593 70-mer-long oligonucleotides representing 13,664 genes printed on Corning UltraGap slides and are available from the Duke Microarray Core Facility (mgm.duke.edu/genome/dna_micro/).

Handling of the microarrays is essentially the same as for any printed microarray. Bake the microarray slides at 80° for 80 min and then cross-link at 200 mJ in a Stratagene Stratalinker 2400. After outlining the array on the back with a diamond tip pencil, prehybridize the microarray slides in 5× SSC, 0.1% SDS, 0.5% bovine serum albumin at 42° in a Coplin jar for 1 h. After prehybridization, move the microarray slides to a slide rack and washing jar and perform two 1-min washes in deionized water with agitation. Dry slides by centrifuging for 2 min at 1500 rpm.

Denature the hybridization probe for 2 min at 95° and then centrifuge briefly for 20–30 s to cool. While denaturing samples, warm the microarray slides to 80° and then place in humidified hybridization chamber. Apply the hybridization probe to the array and cover with a 22 × 60-mm coverslip. Incubate the hybridization chamber in a 42° water bath overnight.

Following hybridization, wash the microarray slides in 1× SSC, 0.2% SDS at 42° in a Coplin jar to remove the coverslip. Move the slides into a metal slide holder and wash in 1× SSC, 0.2% SDS at 42° for 1 min with shaking followed by a 1-min wash in 0.2× SSC, 0.1% SDS at room temperature with shaking and for 2 min in 0.2× SSC at room temperature with shaking. The final wash is for 1 min in 0.05× SSC at room temperature. Dry the slides by centrifugation for 2 min at 1000 rpm. Store the slide in a dark box until scanning. Scanning is performed on a Molecular Devices Genepix 4000B scanner as directed by the manufacturer.

Data Interpretation

The method we use to analyze data and identify the breakpoints of the aberrations is simply to examine the fluorescent ratio of aberration to wild type for all of the target probes. Theoretically, a haploid region should have a ratio of 0.5 compared to a 1.0 ratio for diploid regions as seen in Fig. 1. To determine breakpoints rapidly, however, it is absolutely essential to know the order of the oligonucleotides as they appear across the genome. In the set of Operon *Drosophila* Array-Ready Oligos, not all of the oligonucleotides are contiguous with genomic DNA, as some span splice sites. A Microsoft Excel spreadsheet that details the level of identity of each target probe in the genome, the order the oligonucleotides appear across the genome, and the transcript associated with each oligonucleotide is available at www.operon.net/arrays/omad.php. It is possible that target probes that are not contiguous with the genome may exhibit abnormal hybridization properties.

This section details the procedure used to analyze data generated by the Genepix 4000B scanner. Output data from the GenePix software come in the form of a .gpr file, which is the basis for all of our analysis. To examine the ratios for each target probe, it is best to create a spreadsheet containing data on gene identity and position, as well as hybridization data from the .gpr file for each microarray. In the .gpr file, two columns are used to determine the ratio of experimental to wild type: the net fluorescent signal for Cy5 is labeled "F635Mean-B635" and the signal for Cy3 is labeled as "F532Mean-B532." The values in these columns correspond to the

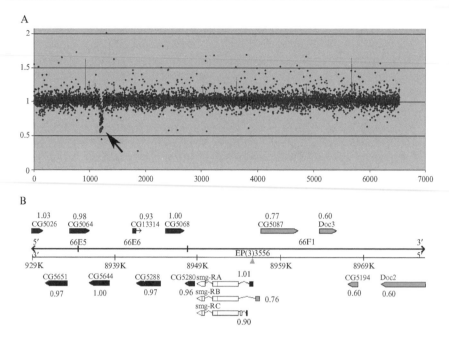

Fig. 2. Array CGH results from Df(3L)BSC35. (A) A graph of the fluorescent ratio of deficiency to wild type (Y axis) versus chromosomal position (X axis) on chromosome 3 for the strain Df(3L)BSC35. Ratios depicted are an average of four microarrays. This chromosome contains one deletion (arrow). The line is the 5 position moving average. (B) The genomic region at the distal breakpoint of Df(3L)BSC35 shows which genes are lost in the deficiency. Genes shown in gray have P values <0.05 as determined by the two-step mixed model analysis, and black indicates genes that are not significantly different from wild type. The three-digit number associated with each gene is the average ratio of four microarrays of BSC35/wild type. For this deficiency, the target probe corresponding to the unique smg-RA exon is not different from wild type. However, the target probe corresponding to the unique smg-RB exon has a P value <0.05 and is likely lost in this deficiency. The triangle represents the position of the P-element EP(3)3556 from which Df(3L)BSC35 was generated. The five genes to the right of EP(3)3556 are removed in the deficiency, whereas those to the left are unaffected by the deletion.

background fluorescence of a target probe subtracted from the mean fluorescent intensity of the target probe. We then divide the net fluorescent signal of the experimental by the net fluorescent signal of the wild type for each target probe to generate the Df/wild-type ratio. This procedure is repeated for all four microarrays and in this manner we determine the average ratio of Df/wild type for each target probe. A simple way to visualize the deficiency region is to sort the spreadsheet in ascending order by Df/wild-type ratio and fill the cells with a ratio less than 0.7 with a color (we use red) and ratios between 0.7 and 0.8 with another color (we use orange). When the spreadsheet is resorted according to target probe position in the genome, the colored region will cluster in the deficiency. It can help to look at the values for each array as well as the average ratio to see any potential dye effects as well as outlier values that may affect the average ratio. We only accept fluorescent signal values that are greater than 70% of the background plus 1 SD. Specifically, we calculate the Df/wild-type ratio for target probes that have values in the "%>B532 ± 1SD" and "%>B635 ±1 SD" columns that are greater than or equal to 70. A graph of the fluorescent ratio versus the chromosomal position can help visualize the deficiency region as seen in Fig. 2.

Examination of the ratios is a purely qualitative treatment. For a quantitative measure, we have used the more complicated two-step mixed model analysis (Gibson and Wolfinger, 2004), which compares the fluorescent signal of the experimental and wild type after accounting for fixed and random sources of variation. The output of the two-step mixed model analysis is a P value that can be used to determine if the two fluorescent signals are significantly different from one another. As seen in Fig. 2B, there is a strong correlation between P value and ratio. The breakpoints determined by array CGH can be verified by PCR or complementation tests if necessary.

Acknowledgments

We are extremely grateful to Erica M. Selva (University of Delaware) and Chandra Tucker (Duke University) for critical reading of this manuscript. Thanks also to Jamie Roebuck and Alicia Carlucci (Model System Genomics, Duke University) for support and flies and to Laura-Leigh Rowlette (Duke Microarray Facility) for microarray help.

References

Adams, M. D., Celniker, S. E., Holt, R. A., Evans, C. A., Gocayne, J. D., Amanatides, P. G., Scherer, S. E., Li, P. W., Hoskins, R. A., Galle, R. F., George, R. A., Lewis, S. E., Richards, S., Ashburner, M., Henderson, S. N., Sutton, G. G., Wortman, J. R., Yandell, M. D., Zhang, Q., Chen, L. X., Brandon, R. C., Rogers, Y. H., Blazej, R. G., Champe, M., Pfeiffer, B. D., Wan, K. H., Doyle, C., Baxter, E. G., Helt, G., Nelson, C. R., Gabor, G. L., Abril, J. F., Agbayani, A., An, H. J., Andrews-Pfannkoch, C., Baldwin, D., Ballew, R. M., Basu, A., Baxendale, J., Bayraktaroglu, L., Beasley, E. M., Beeson, K. Y., Benos, P. V., Berman, B. P., Bhandari, D., Bolshakov, S., Borkova, D., Botchan, M. R., Bouck, J., Brokstein, P., Brottier, P., Burtis, K. C., Busam, D. A., Butler, H., Cadieu, E., Center, A., Chandra, I., Cherry, J. M., Cawley, S., Dahlke, C., Davenport, L. B., Davies, P., de Pablos, B., Delcher, A., Deng, Z., Mays, A. D., Dew, I., Dietz, S. M., Dodson, K., Doup, L. E., Downes, M., Dugan-Rocha, S., Dunkov, B. C., Dunn, P., Durbin, K. J., Evangelista, C. C., Ferraz, C., Ferriera, S., Fleischmann, W., Fosler, C.,

Gabrielian, Z., Garg, N. S., Gelbart, W. M., Glasser, K., Glodek, A., Gong, F., Gorrell, J. H., Gu, Z., Guan, P., Harris, M., Harris, N. L., Harvey, D., Heiman, T. J., Hernandez, J. R., Houck, J., Hostin, D., Houston, K. A., Howland, T. J., Wei, M. H., Ibegwam, M. H., *et al.* (2000). The genome sequence of *Drosophila melanogaster*. *Science* **287,** 2185–2195.

Ashburner, M. (1989). "Drosophila: A Laboratory Handbook." Cold Spring Harbor Laboratory Press, Cold Spring Harbor, NY.

Claycomb, J. M., Benasutti, M., Bosco, G., Fenger, D. D., and Orr-Weaver, T. L. (2004). Gene amplification as a developmental strategy: Isolation of two developmental amplicons in Drosophila. *Dev. Cell* **6,** 145–155.

Gibson, G., and Wolfinger, R. D. (2004). Gene expression profiling using mixed models. *In* "Genetic Analysis of Complex Traits Using SAS" (A. M. Saxton, ed.), pp. 251–278. SAS Users Press, Cary, NC.

Myster, S. H., Wang, F., Cavallo, R., Christian, W., Bhotika, S., Anderson, C. T., and Peifer, M. (2004). Genetic and bioinformatic analysis of 41C and the 2R heterochromatin of *Drosophila melanogaster*: A window on the heterochromatin-euchromatin junction. *Genetics* **166,** 807–822.

Parks, A. L., Cook, K. R., Belvin, M., Dompe, N. A., Fawcett, R., Huppert, K., Tan, L. R., Winter, C. G., Bogart, K. P., Deal, J. E., Deal-Herr, M. E., Grant, D., Marcinko, M., Miyazaki, W. Y., Robertson, S., Shaw, K. J., Tabios, M., Vysotskaia, V., Zhao, L., Andrade, R. S., Edgar, K. A., Howie, E., Killpack, K., Milash, B., Norton, A., Thao, D., Whittaker, K., Winner, M. A., Friedman, L., Margolis, J., Singer, M. A., Kopczynski, C., Curtis, D., Kaufman, T. C., Plowman, G. D., Duyk, G., and Francis-Lang, H. L. (2004). Systematic generation of high-resolution deletion coverage of the *Drosophila melanogaster* genome. *Nat. Genet.* **36,** 288–292.

Pollack, J. R., Perou, C. M., Alizadeh, A. A., Eisen, M. B., Pergamenschikov, A., Williams, C. F., Jeffrey, S. S., Botstein, D., and Brown, P. O. (1999). Genome-wide analysis of DNA copy-number changes using cDNA microarrays. *Nat. Genet.* **23,** 41–46.

Ryder, E. J., Blows, F., Ashburner, M., *et al.* (2004). The DrosDel collection: A set of P-element insertions for generating custom chromosomal aberrations in *Drosophila melanogaster*. *Genetics* **167,** 797–813.

Wei, H. C., Shu, H., and Price, J. V. (2003). Functional genomic analysis of the 61D-61F region of the third chromosome of *Drosophila melanogaster*. *Genome* **46,** 1049–1058.

[19] Performing Quantitative Reverse-Transcribed Polymerase Chain Reaction Experiments

By GEORGES LUTFALLA and GILLES UZE

Abstract

Quantitative polymerase chain reaction (PCR) is as old as PCR, but it has had to wait for the introduction of real-time PCR instruments to become widely used. These instruments allow monitoring of the PCR reaction *on line*; they involve the use of a fluorescent probe that allows quantification of the amplified DNA. Different fluorescent formats and

METHODS IN ENZYMOLOGY, VOL. 410
Copyright 2006, Elsevier Inc. All rights reserved.

0076-6879/06 $35.00
DOI: 10.1016/S0076-6879(06)10019-1

different applications have been developed for quantitative PCR, but this chapter focuses on the use of the SYBR Green label for the quantification of specific cDNAs in reverse transcription mixes: RT-PCR. We propose optimal reaction conditions for the reactions to be performed on the different available instruments and discuss the important parameters for setting up experiments: specificity, efficiency, and reproducibility. We also introduce the reader to the problems of relative quantification.

Introduction

Polymerase chain reaction was born in the early 1980s when K. Mullis realized that it was possible to "amplify" a DNA fragment using a pair of primers (Saiki *et al.*, 1985). It immediately appeared to many investigators that measuring the amount of amplified material at each cycle would allow this method to become quantitative. The breakthrough came 10 years later from the development of specially designed instruments that combined a thermocycler with an optical system that could measure a fluorescent signal in each sample throughout the amplification procedure. This chapter focuses on the use of quantitative PCR (Q-PCR) to measure the relative amounts of different mRNAs in complex RNA samples. Because PCR cannot be applied to RNA, these are first reverse transcribed (RT) to cDNAs that are used as PCR templates. Hence the name RT-PCR.

In quantitative RT-PCR, the aim is to compare the amount of a given mRNA in two different samples. For each sample, the amount of a specific mRNA will be given relative to the amount of a reference mRNA that is supposed to be constant in the two samples. The ratio specific/reference will be used to compare both samples. There is no universal reference cDNA. In each chosen experimental system, one has to test different candidates (rRNA, GAPDH, ribosomal proteins, splicing factors) and choose the one that seems to vary the least relative to the calculated amount of RNA input in the RT. The amount of the specific cDNA in the sample is measured by comparing its amplification pattern to that of standards with given amounts of the specific template. The same measure will be performed on the reference template to calculate the ratio specific/reference. In the case of RT-PCR, it is recommended to perform a single RT and to use it to measure the different matrixes (reference and specifics) rather than performing "one tube" RT-PCR for each sample, as the latter introduces a new source of variations: the efficiency of the RT reaction. We recommend priming the RT reaction using random hexamers rather than oligo(dT). This is critical in measuring the relative abundance of cDNAs with alternative 5′ ends or alternative splicings.

Choice of the Fluorescence Format

Many fluorescent formats exist that are compatible with Q-PCR. Some would give fluorescent signals only in the presence of a specific DNA fragment, and some are just dyes specific for DNA. DNA-specific dyes are more convenient for small-scale experiments. They have to be used to apprehend all the amplification events occurring in the sample. Sequence-specific probes are a good choice for routine experiments such as diagnostic tests, but they just allow the measure of what is wanted; any amplification of an unspecific product remains undetected. We will see that amplification of a given template in a reaction interferes with the amplification of other templates. It is therefore critical that sequence-specific probes be used only in amplification reaction where one is sure that a single template is amplified. For this reason, setting up experiments with sequence-specific probes implies preliminary experiments using DNA-specific dyes. This chapter deals with the setting up of quantitative RT-PCR for confirmation of arrays data. As this by essence is not a routine activity, we will restrict our topic to the use of DNA-specific probes. Preliminary experiments have been done using ethidium bromide (Higuchi et al., 1993) but it was abandoned for two reasons: (1) because of its low discrimination for single- versus double-stranded DNA and (2) because as an intercalating agent it was interfering with polymerization. Investigators have shifted to another DNA-specific dye, SYBR Green I, that binds to the small groove of the double-stranded DNA helix (Schneeberger et al., 1995). This chapter deals only with SYBR Green I.

Choice of the Enzymatic System

Let us see the requirements in terms of enzymes to set up the experiment. The first requirement is for an enzyme that would have a high processivity so that all the molecules present in the experiment would be entirely copied at each cycle. This can be obtained easily when addressing small fragments (less than 300 bp) and using Taq DNA polymerase. The fidelity of the polymerase is not limiting for Q-PCR.

The second requirement is a "hot start" system (Chou et al., 1992; Nuovo et al., 1991). This is because quantitative PCR is much more stringent than conventional PCR, which can afford some primer dimers formation, unacceptable in quantitative PCR. Two main systems are available for "hot start." The first is an autoactivable enzyme that is chemically modified to be inactive at low temperature. Due to the pH shift at high temperature, the chemical modification is unstable at 95° and the enzyme is "autoactivated" during the denaturation steps of the PCR reaction (Birch, 1996).

The only drawback of this strategy is that the incubation times at 95° need to be lengthened. The other system consists of using a *Taq* DNA polymerase in complex with an antibody that blocks its activity (Kellogg *et al.*, 1994). The antibody is denatured completely during the initial 3-min denaturation. The present contribution is illustrated with experiments done with the antibody hot start system, but similar results are obtained with the autoactivable system (although more slowly).

Amplification and Denaturation Curves

A typical Q-PCR experiment happens in three successive stages (Fig. 1): an initial denaturation stage (95°) that is important for perfect sample denaturation and for enzyme activation (see "hot start"), an amplification stage with a repetition of denaturation/hybridization/elongation cycles (98/65/72°), and a terminal progressive denaturation stage whose function is to apprehend the nature of the amplified fragments.

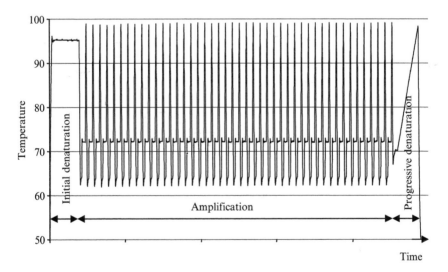

FIG. 1. Temperature cycling during a typical experiment. Temperature (°C) in the reaction device is represented as a function of time during the experiment. In the presented experiment, the initial denaturation temperature is 95° for 3 min in order to release the polymerase from its antibody (hot start). During amplification, the samples are denatured at 98°, annealed at 65°, and elongation is performed at 72°. After amplification, T_m values of the products are analyzed by raising the temperature progressively from 70 to 98° in order to monitor their denaturation.

In order to measure the amount of amplified DNA, the fluorescence is measured in each sample at the end of each elongation cycle. This is shown in Fig. 2 with a range of initial amounts of the matrix to amplify: a sample together with successive dilutions of a factor of 10. Each sample goes through four successive steps: in a first step, the signal cannot be distinguished from the background level; during the second step, for a limited number of cycles, the fluorescent signal increases exponentially; and during the third step, the signal increases linearly before reaching a plateau (fourth step). The amplification curves presented in Fig. 2 are "corrected" curves: the value measured at the first cycle (background) has been subtracted to all the measured values. This is the "arithmetic correction."

During the progressive denaturation stage (Fig. 1), double-stranded DNA samples are progressively heated and fluorescence is measured permanently. Evolution of the fluorescence as a function of the temperature is shown in Fig. 3A. In the case where a single DNA fragment has been amplified (Fig. 3, curve 1), fluorescence first decreases slowly (due to pH shift) before dropping rapidly, indicating that the DNA molecule is being denatured. For this reason, these curves are called "fusion curves." The inflexion point of this denaturation curve (the maximum derivative,

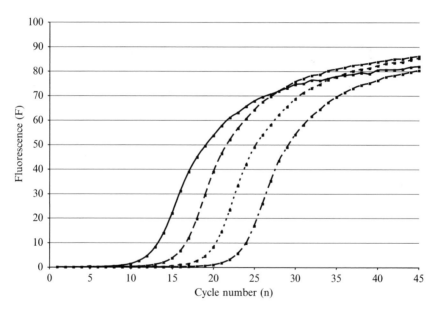

FIG. 2. Amplification stage. Fluorescence as a function of cycle number. Fluorescence (arbitrary units) is measured in four different samples (serial dilutions) at the end of the elongation step of each cycle.

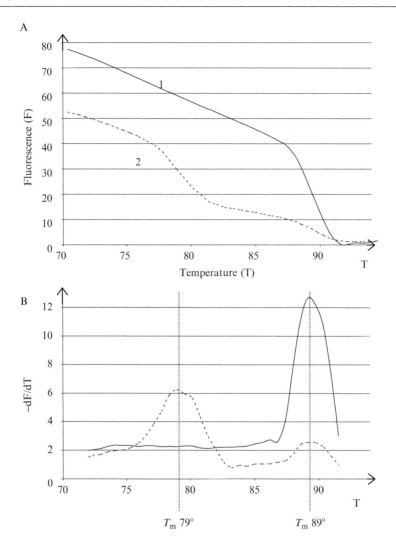

FIG. 3. Denaturation stage. Evolution of fluorescence as a function of temperature (T). (A) Fluorescence as measured on line during the temperature rise from 70 to 98°; arbitrary unit. In the case of curve 1, the sharp drop around 89° represents denaturation of the molecule: SYBR Green binds on the small groove of the dsDNA helix but not on ssDNA. The preceding slow decrease is an indirect consequence of the raising temperature: it induces a pH raise that changes the fluorescent properties of SYBR Green. In the case of curve 2, the two sharp drops reveal the presence of two types of amplified DNA molecules. (B) Peaks of the first derivatives of fluorescence as a function of temperature are used to calculate the T_m of the amplified molecules (vertical lines).

Fig. 3B) is considered the T_m of the amplified molecules. In real-time PCR, the measure of the T_m of the amplified fragments is similar to the measure of the size of the amplified molecules in "traditional PCR." If different products are amplified (Fig. 3, curves 2), one can easily observe the two steps of denaturation and the derivative reveals as many peaks as different amplified products (if they have different T_m values; two in the case shown in Fig. 3, curve 2). A new dye has now been developed that allows much more sensitive analysis of the denaturation step that can be used to discriminate different alleles (Wittwer et al., 2003; Zhou et al., 2004).

Determination of the Correct Annealing Temperature: Specificity

When setting up a Q-PCR experiment, the first step is the choice of a pair of primers for the PCR reaction. With these primers in hand, it is necessary to test them for their compatibility with a robust quantification. The first step consists in finding conditions where the primer pair would amplify a single DNA fragment and where the fragment is the wanted one. It is therefore necessary to start by testing different annealing temperatures; typically with 20-mers, in the range 60–70°. One usually performs three different runs like the one depicted in Fig. 1 at, respectively 60, 65, and 70° as annealing temperatures. Samples should include positive and negative controls. A primer pair is considered acceptable if it is possible to find an annealing temperature at which we will have amplification of a single specific product only with positive controls. In cases where more than one product is amplified, denaturation curves are as in Fig. 3 (curves 2). It is often necessary to test different primer pairs before finding one that would be correct. When these conditions have been met, it is necessary to sequence the amplified product to check its identity. The next step is then analysis of the yield of the amplification.

Checking the Amplification Yield: Efficiency

Figure 2 shows amplification curves with fluorescence as a function of cycle number. Using this representation, it clearly appears that amplification is exponential during a limited number of cycles and that factors become limiting rapidly, leading to a plateau. Obviously for quantification, it is necessary to consider the amplification reaction before any factor (other than initial concentration of matrix) becomes limiting, that is, during the exponential part. In order to better apprehend this part, let us use a logarithmic representation of the amplification curves (Fig. 4A). The ordinate is the amount of amplified material as measured by the fluorescent

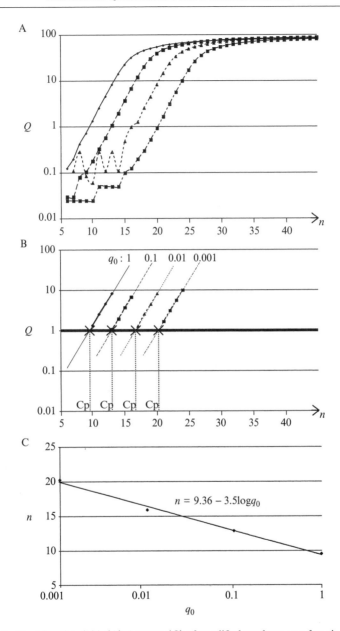

FIG. 4. Checking the yield. (A) Amount (Q) of amplified products as a function of cycle number (n): The ordinate scale is logarithmic in order to highlight the exponential part of the amplification that appears linear in this representation. (B) Calculating crossing points (Cp): regression lines are drawn for the linear parts of the amplification (in this semilogarithmic

signal (arithmetic corrections, arbitrary unit). Curves are bumpy under the fluorescence level "1." This is due to the discriminating power of the light detection system. It is therefore necessary to limit the analysis of the reaction to the linear part of this logarithmic representation. In the example shown, the curves are linear for four cycles with arbitrary fluorescent values between 1 and 10. Figure 3B shows, for each sample, the four corresponding points. A linear regression is shown for each of these four points.

Let us see how to use this representation for quantitative analysis. Denoting q_0 the initial amount of matrix in a sample, after n cycles, the amount Q of amplified product should theoretically be $Q = q_0 2^n$. In practice, no reaction has a yield of 1; it is therefore not possible to consider that the amount of amplified material would double at each cycle. We shall thus write $Q = q_0(1+\rho)^n$, where $\{B\}\rho$ is the yield of the PCR reaction. As fluorescence is proportional to the amount of amplified material, this is the equation of the curves linking fluorescence to cycle numbers. Logarithmic representation (Fig. 4B) would be

$$\log Q = \log q_0 + n\log(1 + \rho).$$

This means that the slopes of the regression lines in Fig. 4B are directly linked to the yield of the reaction: the greater the slope, the better the yield (up to 1). Quantification will be possible only if, for a given experiment, all the regression lines are parallel, meaning the same yield for all samples. As it is not possible to have exactly the same yield for each sample in a range of standards, it is more convenient to adopt another method to calculate the yield.

Let us consider the number of cycles necessary to reach an arbitrary level of fluorescence from samples with different initial amounts of matrix: the crossing points between the regression lines and a given horizontal line. These crossing point (Cp) are given as cycle numbers. Let us start again from $Q = q_0(1+\rho)^n$ to deduce the relation linking n to q_0 for a given value of Q: $\log Q = \log q_0 + n\log(1 + \rho)$.

$$n\log(1 + \rho) = \log Q - \log q_0 \quad \text{then} \quad n = \frac{\log Q}{\log(1 + \rho)} - \frac{\log q_0}{\log(1 + \rho)}$$

---

representation) and the abscissas of the intersections with the horizontal line $Q = 1$ are called Cp. (C) Correlation between Cp and initial amounts of matrix. For each standard, the Cp is plotted as a function of the initial amount of matrix (dots in the plot). A regression line is drawn using these dots. It represents the function linking the number of cycles necessary to reach the amount of fluorescence "1" to the initial amount of matrix.

We can then plot (Fig. 4C) the four Cp to obtain the representation of n as a function of q_0. The linear regression of these four points has a slope that is a direct measure of the yield of the reaction.

$$-\frac{1}{\log(1 + \rho)}$$

A theoretical yield of 1 would give a slope of -3.32. In the case reported in Fig. 4, the slope is -3.35, which corresponds to a PCR yield of 0.93, which is very good. It is important to have a good yield because the better the yield, the more reproducible the results.

Primer pairs for which it is not possible to find conditions giving a yield higher than 0.8 cannot be used for accurate quantifications. It is often possible to change the yield by changing the annealing temperature, but be careful with specificity!

Relative Quantification

Quantitative PCR is not an absolute quantification, it is a relative quantification where one compares the amplification with a given sample to that with a range of standards.

Let us consider an experiment as that in Fig. 4, with four standards with successive 10-fold dilutions and two samples of unknown initial amounts of matrix. At the end of the amplification reactions, it is necessary to check that all the amplified products are the same using the denaturation curves. Let us come back to the experiment shown in Fig. 3. Both samples have the same initial amount of the matrix, giving the amplified product with T_m 89°. Sample 2 (Fig. 3, curves 2) also has a second matrix to be amplified with the same primer pair, but giving an amplified fragment with T_m 79°. It appears that the amplification of the product with T_m 79° has interfered with the amplification of the product with T_m 89°. Due to this problem of interference, it is not possible to run quantitative PCR experiments if different amplicons are amplified in the same reaction. We will see later how it is possible to adjust the conditions to run multiplex PCR.

For quantification, the same representation as that in Fig. 4B is used but including the samples together with the standards as shown in Fig. 5. Representation of the amplification for sample A is parallel to that for the standards. This is not the case for sample B. For that sample, because the yield has not been the same as for the standards, it is not possible to use that sample for quantification. Most often, this is due to quality problems with the sample. This can usually be overcome either by diluting the sample so that impurities do not interfere or by purifying the sample.

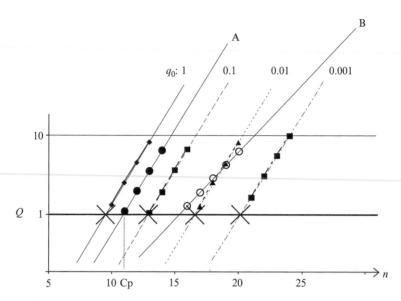

Fig. 5. Relative quantification. Regression lines for the exponential parts (linear in the semilog scale) of the amplification curves of two unknown samples (A and B) are plotted in a representation similar to that in Fig. 4B. Sample B gives a line that is not parallel to those of the standards: it cannot be analyzed. Sample A can be analyzed. The line is used to calculate the Cp that is used according to the plot in Fig. 4C to calculate the initial amount of matrix.

To determine q_0, the initial amount of matrix in sample A, the abscissa of the corresponding Cp is measured as n, the number of cycles. Using the representation shown in Fig. 4C, the corresponding q_0 is measured. The software provided with real-time PCR instruments proposes such quantifications. As any measure, it is spoilt by errors and no serious result can be displayed without at least three replicates allowing the calculation of a standard deviation and a standard error at a given confidence limit when compared to the reference cDNA using, for example, a student's t test.

Different authors have proposed other methods for the quantitive analysis of real-time PCR experiments (Livak and Schmittgen, 2001; Pfaffl, 2001). According to the software provided with real-time PCR instruments, this method is the most convenient. This is exemplified elsewhere (Coccia *et al.*, 2004; Dondi *et al.*, 2003; Lutfalla *et al.*, 2003). This method can be adapted to quantify allele ratio in cDNA samples (Weber *et al.*, 2003).

Multiplex Polymerase Chain Reaction

When using sequence specific probes, it is very appealing to use different probes with different colors to simultaneously measure different matrixes in the same sample.

As shown previously, it is difficult to be quantitative when amplifying simultaneously different matrixes in the same sample. This is because the PCR reaction rapidly reaches conditions where the limiting factor is no more the initial amount of matrix but the available nucleotides or primers in the reaction mix. This will lead to a situation where the more abundant molecules will use up the reagents before the less abundant molecules would be amplified to a detectable level.

For this reason, the widest domain of application of multiplex PCR is genotyping. The relative abundance of the loci to be amplified is within a range of two. Using pairs of primers with similar yields can lead to the simultaneous amplification of numerous loci that can be identified using sequence specific probes of different colors.

The other solution for multiplex Q-PCR is to use limiting amounts of primers. This is in the range of application of quantitative RT-PCR. If primers are in limiting amounts, the amplification of one matrix using one primer pair will not interfere with the amplification of another matrix with a different primer pair. The drawback of this solution is that limiting amounts of primer usually reduce the yield of the PCR reaction, leading to less reproducibility (larger confidence intervals); this can be critical for the quantification of low abundance matrixes. For this reason, the widely used application of multiplex quantitative RT-PCR consists of measuring both a very abundant reference cDNA (rRNA or mRNA for a ribosomal protein) and a less abundant cDNA of interest in the same sample. In that case, primers for the reference are used in limiting amounts and primers for the cDNA of interest are used in excess. Amplification of the reference will not exhaust the amplification mix. The slightly lower yield for the reference cDNA will not interfere seriously with the accuracy of the quantification because amplification will be performed in a very limited number of cycles.

Proposed Reaction Medium for SYBR Green Q-PCR

As shown in Fig. 4, to do correct quantification, it is necessary that the reaction be exponential for three to four cycles after having reached the threshold value at which the dose/response of the light detection system is linear. The ideal situation would be that amplification reactions would stop in the sample as soon as these three to four cycles of correctly detectable exponential amplification would be performed. This is because PCR is so powerful that with an increasing number of cycles, nonspecific material can be amplified, leading to complex denaturation curves that are difficult to interpret. The best way to limit the amplification once enough material has been amplified is to use limiting amounts of nucleotides. After different tests, we have reached the conclusion that a 30 μM final concentration is fine. The same holds for SYBR Green, which is used in limiting amounts so that its low affinity to single-stranded material does not interfere with PCR. As a "hot start" enzyme, we believe that the *Taq* DNA polymerase complexed to an antibody is the best solution because it does not need extended incubation times at 95 or 98°. As a biological buffer, we use 2-amino-2-methyl-1,3-propanediol because it has a $\Delta pK_a/\Delta T$ lower than the other PCR-compatible biological buffers. To make our conditions compatible with all platforms, including glass capillaries, we add bovine serum albumin (BSA). Be careful not to use acetylated BSA! We also include W-1 as a mild PCR-compatible detergent. Glycerol is added in order to allow the correct freezing of the master mix. This mix can be prepared as a $10\times$ solution that has to be frozen at –80° prior to any use. It is then stored frozen at –20° and unfrozen aliquots can be kept in the fridge for a month. It can travel at room temperature without any harm. We recommend preparing large lots to minimize lot-to-lot problems of reproducibility.

The $10\times$ reaction mix for Q-RT-PCR is as follows.

W1 (polyoxyethylene ether W1 Sigma p7516) 0.24%
BSA crist (Sigma a-4378) 500 μg/ml
dNTP (Invitrogen 10216,10217,10218, and 10219) 300 μM
KCl (Merck proanalysis 1.04936.1000) 50 mM
MgCl$_2$ (Merck proanalysis 5833.1000) 30 mM
SYBR Green (FMC Bioproducts 50513) 1/3000
Glycerol (ICN 800687) 16.24%
2-Amino-2-methyl-1,3-propanediol (Sigma A-9074)
Buffer to pH 8.3 using HCl (Merck proanalysis 1.00317.1000) 400 mM
0.4 U/μl platinium *Taq* DNA polymerase (Invitrogen 10966-034)
Water used is UltraGenetic quality from Elga (18.2 MΩ)

Note: W1 can be replaced by a combination of two detergents: Brij 56 (0.09%, Sigma P5759) and Brij 58P (0.15%, Sigma P5884).

Conclusion

Quantitative PCR is a very powerful technique that is perfectly adapted to the analysis of the relative abundance of cDNAs in reverse transcription mixes, but critical quality controls over the experiments are necessary to ensure that quantification means something: specificity of the amplification products, yield of the reactions, and replicates.

In a typical research laboratory, SYBR Green I is more adapted than sequence-specific probes for two reasons: (1) ease in implementing the experiments and (2) multiplex analysis (the best advantage of sequence-specific probes) is not very well adapted to RT-PCR quantification. As shown in this chapter, it is easy to set up multiplex RT-PCR analyses with a highly abundant reference matrix and a less abundant specific matrix. However, in most research laboratories, investigators want to measure different cDNAs against the same reference, as they do not know a priori which of their specific cDNAs will be more abundant.

References

Birch, D. E. (1996). Simplified hot start PCR. *Nature* **381**, 445–446.

Chou, Q., Russell, M., Birch, D. E., Raymond, J., and Bloch, W. (1992). Prevention of pre-PCR mis-priming and primer dimerization improves low-copy-number amplifications. *Nucleic Acids Res.* **20**, 1717–1723.

Coccia, E. M., Severa, M., Giacomini, E., Monneron, D., Remoli, M. E., Julkunen, I., Cella, M., Lande, R., and Uze, G. (2004). Viral infection and Toll-like receptor agonists induce a differential expression of type I and lambda interferons in human plasmacytoid and monocyte-derived dendritic cells. *Eur. J. Immunol.* **34**, 796–805.

Dondi, E., Rogge, L., Lutfalla, G., Uze, G., and Pellegrini, S. (2003). Down-modulation of responses to type I IFN upon T cell activation. *J. Immunol.* **170**, 749–756.

Higuchi, R., Fockler, C., Dollinger, G., and Watson, R. (1993). Kinetic PCR analysis: Real-time monitoring of DNA amplification reactions. *Biotechnology (NY)* **11**, 1026–1030.

Kellogg, D. E., Rybalkin, I., Chen, S., Mukhamedova, N., Vlasik, T., Siebert, P. D., and Chenchik, A. (1994). TaqStart antibody: "Hot start" PCR facilitated by a neutralizing monoclonal antibody directed against Taq DNA polymerase. *Biotechniques* **16**, 1134–1137.

Livak, K. J., and Schmittgen, T. D. (2001). Analysis of relative gene expression data using real-time quantitative PCR and the 2(-Delta Delta C(T)) method. *Methods* **25**, 402–408.

Lutfalla, G., Crollius, H. R., Stange-Thomann, N., Jaillon, O., Mogensen, K., and Monneron, D. (2003). Comparative genomic analysis reveals independent expansion of a lineage-specific gene family in vertebrates: The class II cytokine receptors and their ligands in mammals and fish. *BMC Genom.* **4**, 29.

Nuovo, G. J., Gallery, F., MacConnell, P., Becker, J., and Bloch, W. (1991). An improved technique for the *in situ* detection of DNA after polymerase chain reaction amplification. *Am. J. Pathol.* **139**, 1239–1244.

Pfaffl, M. W. (2001). A new mathematical model for relative quantification in real-time RT-PCR. *Nucleic Acids Res.* **29**, e45.

Saiki, R. K., Scharf, S., Faloona, F., Mullis, K. B., Horn, G. T., Erlich, H. A., and Arnheim, N. (1985). Enzymatic amplification of beta-globin genomic sequences and restriction site analysis for diagnosis of sickle cell anemia. *Science* **230**, 1350–1354.

Schneeberger, C., Speiser, P., Kury, F., and Zeillinger, R. (1995). Quantitative detection of reverse transcriptase-PCR products by means of a novel and sensitive DNA stain. *PCR Methods Appl.* **4**, 234–238.

Weber, M., Hagege, H., Lutfalla, G., Dandolo, L., Brunel, C., Cathala, G., and Forne, T. (2003). A real-time polymerase chain reaction assay for quantification of allele ratios and correction of amplification bias. *Anal. Biochem.* **320**, 252–258.

Wittwer, C. T., Reed, G. H., Gundry, C. N., Vandersteen, J. G., and Pryor, R. J. (2003). High-resolution genotyping by amplicon melting analysis using LCGreen. *Clin. Chem.* **49**, 853–860.

Zhou, L., Vandersteen, J., Wang, L., Fuller, T., Taylor, M., Palais, B., and Wittwer, C. T. (2004). High-resolution DNA melting curve analysis to establish HLA genotypic identity. *Tissue Antigens* **64**, 156–164.

[20] The Application of Tissue Microarrays in the Validation of Microarray Results

By Stephen M. Hewitt

Abstract

Microarray experiments produce a large volume of data that require validation. The validation of these experiments can be carried out in many fashions. In the reduction to clinical utility, the use of tissue microarrays has become a common tool to both validate and generalize the results of microarray experiments. A tissue microarray is a collection of tissue specimens presented on a glass slide in a grid layout. Tissue microarrays contain between tens and hundreds of samples allowing the generalization of microarray findings to a large number of samples. Tissue microarrays can be used for *in situ* hybridization and immunohistochemical analysis, confirming the results of microarray experiments at both the transcriptional and the proteomic level. This chapter reviews the theory and application of tissue microarrays in the validation of microarray experiments.

Introduction

Recent advances in molecular biology have centered on increases in throughput and quantification of biologic phenomena. No longer is experimental design focused on one gene or one protein, but rather on tens to hundreds of genes, proteins, or tissues on one analytical platform (MacBeath 2002). Some researchers have claimed that science has moved from hypothesis-driven to discovery-driven research. More accurately, science has moved from hypothesis testing to empiric approaches; however, the role of the hypothesis has not diminished, only changed.

METHODS IN ENZYMOLOGY, VOL. 410   0076-6879/06 $35.00
Copyright 2006.   DOI: 10.1016/S0076-6879(06)10020-8

The development of high-throughput analytic platforms utilizing nucleic acids for analysis of cells and tissues has resulted in the demand for higher throughput platforms for the confirmation of these findings. Analysis of tissue for the expression of genes, either at the nucleic acid level or at the protein level, remains the gold standard validation of experimental data obtained by other methods. The two most common methods of validation have been *in situ* hybridization (ISH) and immunohistochemistry (IHC) on tissue with interpretation of the histomorphology to discern the complexity of expression patterns that cannot be determined from methods that rely on the extraction of biomolecules. Even before the development of the microarray, the examination of tissue for validation and generalization of nucleic acid-based discoveries had become a bottleneck for both basic and clinical research. Development of the tissue microarray (TMA) provides a high-throughput approach for the histomorphologic examination of tissue for the expression of RNA and protein. A TMA is a slide of tens to hundreds of tissue samples that can be probed with an antibody or nucleic acid probe in contrast to an expression microarray where thousands of nucleic acid probes are presented for interrogation by a single sample. The tissue microarray is a form of condensed histopathology, where cells and tissue are presented in a miniature multiplex platform for analysis. Originally developed by Kononen *et al.* (1998), the methods of construction and applications of TMAs continue to expand rapidly. As with all high-throughput platforms, the gain in throughput is offset with a loss of experimental control or completeness of data. Although TMAs retain the histology of tissue (i.e., the presence of cells in relationship to each other resulting in tissue), the condensed nature of the specimens represents a limitation that cannot be overlooked.

The origins of TMAs can be traced to the efforts of Battifora (1986) to embed multiple tissues into a single paraffin block, which could be sectioned and stained for simultaneous analysis. These original efforts were limited in the orientation and density of specimens. Only with the development of instruments that could place cores of tissue in a recipient block did the utility of a multitissue platform come about. Original efforts in the constructions of TMAs involved the use of paraffin-embedded tissue; however, frozen tissue was soon applied to the basic methodology (Fejzo and Slamon, 2001). There have been a variety of new methodologies developed to construct TMAs (Hidalgo *et al.*, 2003; LeBaron *et al.*, 2005); however, they all result in a similar platform.

Along with the expansion of methods for constructing TMAs, there has been expansion in what is routinely presented for analysis. The vast majority of TMAs are constructed of formalin-fixed, paraffin-embedded (FFPE) surgical specimens of human disease, primarily cancer. However, normal tissue and other physiologic and pathologic processes are arrayed routinely. Animal tissue has been widely utilized for the construction of TMAs, including mouse and rat, as well as other animals used as models of human disease, such as canine models of human cancer (Khanna et al., 2002). In addition to tissue, xenograft tumors are routinely utilized in the construction of TMAs, referred to as XMAs. Cell lines can be reduced to a similar platform by a variety of means; these platforms are referred to as CMAs (Braunschweig et al., 2005; Moskaluk and Stoler, 2002). The use of TMAs in biomedical research has exploded. Originally TMAs were used primarily for confirmation of data from microarrays. Subsequently, TMAs have expanded to function as a starting point for biomarker discovery and validation and in support of molecular epidemiology (Goodman et al., 2005). Other applications of TMAs include expression profiling of cancer by IHC, often with phosphor-specific antibodies. TMAs have been utilized for spectroscopic analysis of tissue, opening new avenues of research into the molecular differences of different pathologic states (Fernandez et al., 2005).

This chapter focuses on the use of TMAs to confirm the results of microarrays. Because TMAs can be used in any fashion that a whole section of tissue can be applied they have great utility in confirming the results of any nucleic acid-based experimental platform. TMAs are used routinely to confirm the results of expression microarrays by either ISH or IHC (discussed in detail later in this chapter). In addition, FISH for DNA amplifications and deletions can be applied to TMAs based on findings from chip-based comparative genome hybridization (Huang et al., 2004). The field of TMAs is too broad to cover all detailed steps of construction and staining of TMAs in a single chapter. This chapter focuses on the steps and pitfalls encountered in using TMAs in the confirmation of microarray results, as well as points the reader to appropriate resources for the detailed methodologies. Fortunately, the construction, staining, and analysis of TMAs are built on the basic technologies of histopathology for which excellent resources are available.

Ethical Concerns in the Use of Tissue in Biomedical Research

The use of any source of tissue in biomedical research requires the appropriate regulatory and ethical oversight approval. The administrative requirements vary widely depending on the type of tissue used (human, primate, nonprimate, and rodent), as well as the source of the material (archival diagnostic material versus tissue collected for research purposes). Issues concerning the ethical use of tissue in biomedical research are beyond the scope of this chapter, and investigators should consult local authorities. In some instances, institutional review boards have required that they approve the use of TMAs obtained from commercial- or government-sponsored sources, even when these TMAs have already received broad approval from the originating organization or companies (Braunschweig et al., 2005). In addition to the ethical approvals that are always required, if the material is obtained from a second party, a material transfer agreement (MTA) should be sought. An MTA documents the intellectual property rights concerning the use of the tissue and any inventions that may be made with the TMA. These issues can be complex, depending on the nature of the material (human tissue, animal tissue, cell lines, or xenografts). Commonly an MTA is required to obtain a TMA.

Determining What Tissue to Analyze

Once the decision to use a TMA as a platform for validation has been reached, the next issue is what TMA to use. The use of samples limited to those analyzed in the microarray experiments is often desired, however, neither essential nor sufficient for adequate validation of the results. This is an important step; however, equally important is validation on a large set of samples, related and unrelated to demonstrate the universality of the observation (Nishizuka et al., 2003). Validation of microarray results on the same samples only provides a confirmation of the observation. Depending on the method of validation, in particular ISH, this validation provides little more than confirmation of the experimental platform and specificity of probes. *The goal of utilizing a TMA for validation of microarray results is to add information that was not obtained in the original experiment.* This added information can come in the nature of confirmation on a larger sample number, a more diverse sample population, at the protein rather than the mRNA level, or potentially at the functional level when activated (phosphorylated) forms of the protein are interrogated. As a result, the issue the researcher often encounters is to build or to buy a TMA for validation of results.

It is not uncommon that the original material used for the microarray experiments is unavailable for construction of a TMA. Constraints in amount of material available, intellectual property, and ethical approvals frequently stand in the way. Often only frozen material is available. If only frozen tissue is available, there are two alternatives: construction of a frozen TMA (Fejzo and Slamon, 2001) or fixation and paraffin embedding a sample of the frozen tissue for construction of a paraffin-embedded TMA (Braunschweig et al., 2005). Although it is appealing to many investigators to construct a frozen TMA, the limitations of this approach typically outweigh the potential benefits.

Frozen TMAs are not an economical use of tissue, requiring larger cores of tissue than paraffin-embedded TMAs. The capacity to target selection of tissue can be impaired, and the histology on the resultant arrays is typically of limited quality. Instrumentation for construction of frozen TMAs is improving; however, these arrays remain technically challenging to construct. Ultimately a frozen TMA block is more challenging to archive and interrogate at a later time than a TMA constructed of paraffin-embedded tissue. Benefits of frozen TMAs include the application of ISH and antibodies for ICH that do not work in FFPE. Rather than confront the challenges of frozen TMAs, we have utilized ethanol fixation and paraffin embedding of a sample of the frozen tissue for the construction of TMAs from frozen tissue. We use 70% ethanol, instead of formalin, as a fixative as a balance between frozen and FFPE tissue (Braunschweig et al., 2005; Gillespie et al., 2002). Tissue fixed in 70% ethanol is as stable as formalin-fixed tissue, with good histologic features and superior nucleic acid recovery and a lack of cross-linking, which hampers IHC with many antibodies. This is not to suggest that 70% ethanol fixation and paraffin embedding produce tissue with the same capacity of frozen tissue; however, the benefits of improved histology, greater economy of tissue, and archival nature of the resultant TMA should be considered.

Construction of a Tissue Microarray

Many investigators will assume that construction of a TMA is the best solution to their needs. Although the capacity to include the original specimens and control over the design and use of the TMA are appealing, it is essential to balance these factors with the investment in constructing a TMA. Construction of a high-quality TMA is not trivial. Even with annotated specimens in hand, the cost of staff and equipment to construct the TMA may suggest that collaboration is the best solution. It cannot be emphasized enough that a TMA can be no better than the tissue that it is built from, and often it is not as good as the original tissue. Failures in the quality of a TMA can include poor construction and sectioning of the array, poorly optimized and performed assays on the array, and poor interpretation of the results. The author has seen all of these failures in the hands of researchers who falsely believed they could "short-cut" the process and had not mastered the skills of histotechnology, immunohistochemistry, and histopathology.

The steps of construction are well summarized elsewhere (Hewitt, 2004; Kononen et al., 1998). This section highlights the decision tree and key elements of construction of a useful array. It is essential that a histopathologist assist in reviewing and selection of the tissue to be arrayed. This is a time-consuming process, but will impact all data obtained from the array. There are many pathologic entities that can only be accurately diagnosed by examination of the whole section and, when presented only as small cores on a TMA, may not be classified accurately. As a result, the design of the TMA and construction of the database begin with the review of the material for the TMA. It is important that Hematoxylin and Eosin (H&E) sections of the TMA are reviewed to determine that the tissue on the array matches the design of the array; however, this is far easier than inferring the diagnosis from the arrayed tissue.

Construction of a TMA *de novo* requires access to specialized equipment and training. A minimum of a microtome is required for sectioning of the array. Most TMAs are constructed with an instrument such as a manual tissue arrayer MTA-1 by Beecher Instruments. Additionally, low-density arrays can be constructed with the use of spinal needles only (Hidalgo *et al.*, 2003). Instrumentation for the construction of tissue microarrays is beginning to diversify; however, the majority of tissue arrayers are manual and require some technical expertise in handling paraffin-embedded tissue. High-density arrays utilizing smaller cores are best constructed by histotechnologist or others with some training in working with paraffin-embedded tissue. The same is true for sectioning of TMAs. Many researchers can perform the basic steps but lack the expertise to array or section well. The best solution for most groups interested in construction of a TMA is collaboration. There are numerous groups with extensive experience in TMA construction who will happily teach the methodology hands-on or will construct the TMA as part of collaboration or for a fee.

There are no hard and fast guidelines for the sampling of tissue for a custom array for use in the validation of microarray results. A minimum goal of twice as many samples as were used for the microarray experiments should be used for a TMA. Ratios of 4 to 1 or higher are desirable. More sophisticated designs with the inclusions of tumor samples of different stages or grades can enhance the utility of an array significantly. Inclusion of similar entities can be useful, such as determining if a new biomarker is specific or general. As an example, the original microarray samples for a project included 19 squamous cell carcinomas of the esophagus (Su *et al.*, 2003), where the TMA constructed for validation included both well and poorly differentiated squamous cell carcinomas (350) as well as a sampling of adenocarcinomas of the esophagus (on a second array). Depending on the nature of the project, rather than a single array, a family of arrays may be the best approach for studying a disease process. Inclusion of normal tissues, both of the disease process and other normal tissues as controls, is essential. If cell lines are available, they are a useful control as well.

There has been a great deal of debate on appropriate sampling strategies and sample numbers for validation with TMAs. The answer is, of course, more is better; however, rarely is more always available and the majority of TMAs are constructed based on what is available, not what is desired. More is a measure of both the number of samples and the size of the cores (volume of the tissue). Again there are no hard and fast guidelines on what appropriate sampling of any individual marker is optimal; rather it is dependent on the expression pattern of the marker. This is best demonstrated in breast cancer, where extensive data have demonstrated optimal sampling for Estrogen Receptor (ER), Progesterone Receptor (PR), and c-erb-B2 (Fergenbaum et al., 2004; Simon et al., 2001). Unfortunately, the optimal sampling for one marker does not predict the optimal sampling for any other marker. TMAs are only a sample of a sample and cannot be definitive for expression and are not used routinely in clinical care, other than as controls for staining.

Obtaining a Tissue Microarray from Other Sources

Because a TMA is only a sampling of a sampling, development of a potential biomarker is accomplished by repetitive validation on different cohorts of specimens to demonstrate its utility (Braunschweig et al., 2005). These different cohorts start with the samples used for the original microarray experiments and expand to samples collected at the same time. However, for validation with utility, the sample size must be expanded to include other samples in the same laboratory, samples from other laboratories, and so on, until a biomarker is examined in a population-based cohort on specimens that would be utilized in diagnostics (Goodman et al., 2005; Hewitt et al., 2004). Even with biomarkers of proven utility, the prevalence of a given biomarker will vary based on the cohort examined.

The value of a TMA is based on demand. Although there is a cost associated with the construction of a TMA, the value of the array is based on the tissue in the array and the associated annotation. The better the annotation, other molecular data, treatment history, and outcome (survival) of the patients, the greater utility of the TMA. This value accrues over time until the final outcome of all the patients is known. However the value of the TMA is not realized if it is not shared with other investigators. Rarely will an investigator go to the efforts of constructing a TMA and not share it with collaborators. Issues of ethical use of tissue and material transfer agreements must be addressed for these collaborative efforts, and appropriate consideration of the quality of the hypothesis and assay being applied is essential.

It is always desirable to determine if a biomarker has clinical utility as a prognostic or predictive marker; however, it is rarely appropriate to test a new marker on a sample set that is highly annotated. It is preferable to develop a biomarker through a series of stages, to confirm the validity of the assay and general expression pattern of the marker, before application to a TMA with clinical end-point annotation (Hewitt *et al.*, 2004). When seeking TMAs, a staged approach, testing on a screening array of multiple tumor types, is a good starting place, and based on these preliminary data, seeking more annotated specific arrays for confirmation of the findings from there. This approach will conserve resources (annotated tissue) and reduce the number of dead ends explored. All biomarkers require validation. Aspects of the validation include hardening of the assay. All assays have performance parameters. Although sensitivity and specificity of an assay may not be known, the reproducibility of an assay, nature of appropriate specimens, and anticipated type of data that will be obtained must be defined. One step of assay validation is to perform the assay on (i) samples from the investigator's own laboratory in his laboratory, (ii) samples from a collaborator's laboratory in the investigator's laboratory, (iii) the investigator's samples in the collaborator's laboratory, and ultimately (iv) the collaborator's samples in the collaborator's laboratory. With TMAs and the associated assays of ISH and IHC, this is easier than other assay types.

There are numerous academic groups that have produced high-quality TMAs from different diseases. Many of these groups offer these arrays through collaborations. Other groups have consolidated samples to construct TMAs, which are distributed at a cost-recovery basis. These arrays are limited in number and require an application, including preliminary data, to obtain TMAs. Numerous companies offer TMAs for sale—price typically is an indicator of quality—both in tissue and in annotation. This is not to suggest that commercial TMAs are not a useful resource, rather that bargain TMAs are bargain TMAs.

Confirmation, Validation, and Determining Experimental Approach

Determining the assay to be performed on a TMA is typically straightforward. It is recommended that when possible, IHC be the assay of choice. Although expression microarrays measure mRNA levels, FFPE tissue and clinical environment make the measurement of proteins by IHC preferable to ISH. IHC-based assays have the benefit of measuring the protein, and potentially the active form of a protein by the use of phospho-specific antibodies, which is mechanistically preferable. The primary downside of IHC assays includes the absence of correlation between mRNA levels and protein levels. Discordance between mRNA and protein levels of a particular gene are very common and can be explained by many biologic mechanisms, including translational controls and protein turnover, as well as experimental challenges such as antibody affinity (Braunschweig *et al.*, 2005). IHC has the benefit of a more clinically relevant assay, as the vast majority of measurements of gene products in clinical care are of proteins by IHC, and ISH is clinically limited to viral genes that have dramatic overexpression.

There are instances in which IHC is not a viable approach. Typically these are instances where no specific antibody that performs well in FFPE tissue is available. The failure to have an adequate antibody can occur for many reasons. In some instances, protein homology prevents the development of an antibody that detects the target gene product. This is common among transcription factors that have a great deal of sequence homology and mediate their different functions by small differences in the DNA-binding motifs. In some instances, development of an antibody that performs in a Western blot is possible; however, an antibody of IHC, specifically for FFPE, can be a limitation. This can be an issue of tertiary conformation of the protein and the binding affinity of the protein. In rare instances, the issue is that the transcript is not translated to a peptide, such as H19 (Arney, 2003).

Immunohistochemical Assays for Validation

The complexity of IHC assays is beyond the scope of this chapter. The diversity of antigen retrieval conditions, titration of primary antibody, and application of secondary antibodies, as well as detection systems, result in a multiparameter assay that must be optimized. Taylor and Cote (2005) provide and overview specific guidance in the performance of IHC. For validation of TMA results, a few considerations are worth discussion. The key issues in using IHC are the choice of the primary antibody, the detection method, and the pattern of staining. Under all circumstances, consultation with a histopathologist is the most useful recommendation. IHC is an art of pattern recognition, and a good collaborating histopathologist can save hours of labor and frustration. In the postgenomic, proteomic focus research environment it is less common today to need to develop a custom antibody for a gene of interest. There are a plethora of companies offering antibodies against nearly the entire human proteome, including splice variants and phospho-specific antibodies. Confronted with a choice of antibodies against a particular protein, it is not easy to determine which antibody to choose. If an antibody has been demonstrated to work in FFPE, then it is a natural choice, but in some instances there will be multiple antibodies that work in FFPE.

So what does matter in choosing an antibody for validation? The antibody must produce a single band on a Western blot and produce a *reproducible* immunostaining pattern that matches the anticipated pattern of expression in both cell type and cell localization. The most important steps are reviewing the primary literature about the protein of interest and the antibody that is being utilized. What is less important are issues of polyclonal vs monoclonal, choice of antigen retrieval method, and final titer of the antibody applied. There is a great deal of myth about antibodies and too little fact. The one fact is that an antibody that detects multiple bands in a Western blot will not improve in specificity in IHC. Often the choice becomes a matter of what company the user has had better experiences with. If the protein target is a common diagnostic protein, such as c-erb-B2, c-kit, or even an intermediate filament such as desmin, it is probably better to choose an antibody from a diagnostic antibody supplier rather than a research antibody source. In some instances, more than one company sells the same clone of an antibody under license from one another; however, there are other peptides for which there are multiple antibodies available, and will result in different results, based on slightly different antibody affinities. In instances like this, use of the more common clinical antibody will benefit the translation to a clinical environment. This is not to suggest that the other antibody may not perform better; however, it is essential to document the difference if the goal is to develop a clinically relevant biomarker. This is a challenge that was encountered in the development of the clinical IHC assay for c-erb-B2 and has been observed with antibodies against p53. In the absence of a clear choice of antibody, helpful guidelines include those created against the entire or larger fragments of the peptide, as opposed to those against small polypeptides. It is not uncommon that none of the antibodies have been demonstrated to work in FFPE. The "porting" of an antibody from a Western blot to IHC is not easy, and approximately only one in three antibodies will eventually produce an acceptable IHC stain. A helpful indication of this is the demonstration that the antibody will stain frozen tissue. No specific guidelines for "porting" an antibody are available, and consultation with an experienced immunochemist/pathologist is recommended for guidance on adjusting titers and developing an appropriate antigen retrieval method.

Detection methods for IHC break down into two primary groups: chromagenic, so-called "brown stains," and fluorescent detection methods. Although fluorescent detection methods offer increased dynamic range of staining and the capacity to perform multiple stains at one time, they are not as popular with FFPE tissue. Fluorescent methodologies require specialized equipment, expertise in analysis beyond that of the typical histopathologist, and can be troubled by autofluorescence. There has been progress with the use of fluorescence in IHC of TMAs (Camp et al., 2002); however, it is not widely used. Chromagenic stains, or so-called "brown stains," are the most common approach to detection of an antibody:antigen complex in FFPE. Interpretation requires a standard light microscope, with results that are relatively permanent. Contrast agents imparted by histochemical stains allow easy recognition of the staining pattern at both cytologic and histologic levels. The primary limitation of chromagenic stains is a lack of dynamic range of staining. Interpretation is typically qualitative of the nature of positive/negative or 0,1,2,3,4 scoring systems. This can be frustrating to the molecular biologist. Sophisticated instruments that provide image analysis and quantitative analysis of chromagenic stains are available. They do provide high-quality reproducible data, but are expensive and require a histopathologist who is comfortable with image analysis to be of benefit (Tan et al., 2004). Because the clinical environment is predominated by chromagenic stains that are scored as positive or negative, or in a minority of cases scored with a simplified system of 0,1,2,3, the choice to use this approach in validation assures greater relevance than an approach that cannot be simplified to the demands of the clinical environment.

As mentioned earlier, the pattern of staining is a key component of IHC. Each protein has a normal pattern of expression both at the cellular level (cytology) and at the tissue level (histology). Working out the conditions of correct staining for IHC are complex and rigorous controls are challenging if not lacking for some proteins. Too often the quality of a stain is determined by how it looks to the histopathologist. This is an uncomfortable situation for many researchers, which the author acknowledges, from both sides of the coin as a molecular biologist and as a histopathologist; however, often the staining pattern at the cytologic and histologic pattern is undefined. It is essential that the staining pattern make sense based on the protein target, localization motifs, function, and cell type. There are some proteins for which IHC is challenging. Two particularly difficult groups of proteins are transcription factors and cytokines. Transcription factors are challenging because many are present in very low copy number per cell and hard to detect by IHC. Cytokines are challenging because they tend to be small proteins that are secreted from cells and diffuse widely and easily in tissue, often appearing to lack specificity of a staining pattern.

In Situ Assays

For all of the complexity of IHC assays, they are simple compared to ISH. The basic methodology of ISH predates IHC and was used in mapping the first human genes (Harper et al., 1981); however, the rigor of the experimental conditions and detection systems has seen ISH replaced by IHC whenever possible. Key factors in ISH assays are the probe and detection system. Considerations of a probe are relatively simple: specificity of sequence, GC content, and length are the primary parameters. Detection methods were originally the incorporation of radioactive nucleotides, which remain the "gold standard" of detection of a transcript, with an unmatched sensitivity and linear detection range (Geiszt et al., 2003). Unfortunately, few laboratories are prepared to use radioactivity in ISH, and more commonly used methods include fluorescent and chromagenic probes. The advent of tirimide amplification protocols has improved sensitivity (Yang et al., 1999). Although these directions make ISH sound simple, the basic technique has always been challenging and require a meticulous technique for reproducibility and low background. Interpretation is much the same as IHC; however, cytologic criteria are less important, as the signal is typically localized to the nucleus and cytoplasm. As a result, most ISH is interpreted as positive/negative and is rarely qualified/quantified.

Collection and Interpretation of Data

For many investigators, performance of the assay is straightforward, but collection and interpretation of histomorphologic-based data are stumbling blocks. As mentioned previously, a TMA is nothing but condensed pathology, and interpretation of the results should reflect this. Interpretation of data is based on recognition of its complexity and development of a process to reduce this complexity to a biologically and/or clinically meaningful interpretation. The general parameters of data, regardless of if it is IHC or ISH, are the intensity, the localization, and the proportion of cells of interest that meet the first two criteria. Regardless of the means of collecting data, these three parameters must be considered. Collection of data is typically by review of the slide under a microscope by a trained eye. Some stains and tissues are straightforward, and general researchers should be able to interpret the staining pattern. Others are extremely complex and may require a histopathologist. The bulk probably lies in between, where the general researcher can interpret data; however, the quality will be improved with *consultation* by a pathologist. The key improvements that a consultation will improve are delineation of cell types and discussion of cell localization (issues of cytoplasmic staining overlaying the nucleus are an example). There are an increasing number of automated instruments to aid in the collection of data. Although extremely useful, they do not obviate the role of the researcher to define the staining pattern and are more an aid for cataloging, presenting, and automated scoring of the stain after an interpretation and scoring system has been devised by the researcher. For these automated systems, there are two key factors: stain quality and image quality. Optimal staining conditions are essential. If the image captured does not contain the information desired at adequate image resolution, the extraction of this information will be inaccurate.

It is not possible to define the interpretation of the staining pattern *a priori*, nor is there always a definitive interpretation for each stain. Again, much like choosing the antibody, when defining a scoring system it is beneficial to review the literature. For some antibodies, such as ER, PR, and c-erb-B2, there are well recognized scoring systems that make correlation of data with other studies as well as the review process much simpler (Fergenbaum *et al.*, 2004). In the absence of an existing scoring system, the researcher needs to consider what information they desire and then define the *simplest* system to achieve this goal. Toward this end, the researcher should consider the intensity, distribution, and localization of the stain and determine which parameters appear to be meaningful. If one of the parameters is invariant, it can be discarded easily. An example can be localization—if the stain is always nuclear or always membranous, then this parameter is not worth collecting. Intensity is typically not a parameter of value in nuclear stains, and only the distribution (percentage of cells of interest that are positive) needs to be determined. Some scoring systems are more complex, depending on both the percentage of cells staining and the intensity of the stain. Again, simplicity is best. The vast majority of clinical IHC stains is interpreted positive or negative, and not scored. For those that are scored, such as MIB-1 and c-erb-B2, it is typically a single parameter scoring method. The less complexity the scoring system has, the greater intra- and interobserver reproducibility, which is essential for a clinically relevant biomarker.

Once a slide is stained it should be reviewed for the quality of staining, to determine that the stain is applied evenly, has the desired staining quality (dynamic range), and that known positive and negative controls are stained appropriately. The slide should then be scored. This is typically performed on a grid or map of the array and not directly into a spreadsheet, as this provides a check against applying the wrong score to the wrong tissue. This information is then "dearrayed" into a spreadsheet for analysis.

Analysis of Data

Analysis of results is a goal-oriented function. In a clinically oriented environment this is often an analysis of outcome (Kaplan–Meier curve) or correlation with diagnosis or outcome (Braunschweig *et al.*, 2005). In proteomic profiling, it is often a correlation with other proteins that are expressed in the same tissue to define the pathway or other process. With confirmation of an expression array, it is less clear. Expression arrays often provide data that a gene is under- or overexpressed in relation to a different condition or tissue. Resultant data from a TMA often do not match the parameters of the expression array, and it is up to the researcher to place this information into a relevant context. This guidance is shallow to many researchers; however, it is honest and provides a great deal of flexibility depending on the question at hand. Our approach has been to examine a panel of specimens and demonstrate that the gene product is expressed in the bulk of the specimens and that this expression is in greater magnitude than found either in appropriate normal tissues or in other disease processes. We then examine other correlations to determine if the gene product has utility in diagnosis, prognosis, or prediction of response to therapy. This approach is to treat the microarray results as *discovery* and the validation on tissue as the next step in the *discovery process* toward the development and *validation* of a useful biomarker (Hewitt *et al.*, 2004).

For all the methods of data interpretation that can be applied to TMA data, there is one that should only be applied with great caution: unsupervised cluster analysis. This analytic approach can run into two separate problems with TMAs. One problem is that unsupervised cluster analysis of a limited data set is not appropriate if all the markers were obtained from a previous unsupervised cluster analysis. The second and probably more important issue is that the choice of antibodies for a TMA is not unsupervised; in fact it is oversupervised. The antibodies are chosen for their capacity to discriminate. As an example, eight antibodies could segregate the eight tumor types of the TARP1 array into appropriate groups without any effort, but this result is of no significance to any pathologist familiar with the immunohistochemical profiles of these tumors.

Other Applications of TMAs Relating to Microarrays

It would be enough for TMAs to be used for verification and validation of results from a variety of experimental platforms. However, access to specimens has become so critical in biomedical research that the general platform of TMAs has become a source of material for other assays. Laser capture microdissection (LCM) can be performed on TMAs, and although this appears redundant, for some researchers, it is the best means to obtain a large collection of material for analysis. Alternatively, others are using needles used in the construction of TMAs to extract cores of tissue that can then be extracted for DNA, RNA, or proteins. This approach is certain to expand as it provides an efficient means of extracting a subpopulation of tissue from a block for analysis. The approach may not offer the purity that is obtainable with LCM; however, it is faster, less expensive, and does less damage to the donor block than serial sectioning of the block to obtain sufficient material for LCM.

Conclusions

The TMA has become a critical tool in the confirmation and translation of expression microarray results. The use of TMAs is not different than the use of whole sections; rather they represent a condensed high-throughput platform that is accelerating the process from bench to bedside.

References

Arney, K. L. (2003). H19 and Igf2: Enhancing the confusion? *Trends Genet.* **19,** 17–23.

Battifora., H. (1986). The multitumor (sausage) tissue block: Novel method for immunohistochemical antibody testing. *Lab. Invest.* **55,** 244–248.

Braunschweig, T., Chung, J.-Y., and Hewitt, S. M. (2005). Tissue microarrays: Bridging the gap between research and the clinic. *Exp. Rev. Prot.* **2,** 325–336.

Camp, R. L., Chung, G. G., and Rimm, D. L. (2002). Automated subcellular localization and quantification of protein expression in tissue microarrays. *Nat. Med.* **8,** 1323–1327.

Fejzo, M. S., and Slamon, D. J. (2001). Frozen tumor tissue microarray technology for analysis of tumor RNA, DNA, and proteins. *Am. J. Pathol.* **159,** 1645–1650.

Fergenbaum, J. H., Garcia-Closas, M., Hewitt, S. M., Lissowska, J., Sakoda, L. C., and Sherman, M. E. (2004). Loss of antigenicity in stored sections of breast cancer tissue microarrays. *Cancer Epidemiol. Biomark. Prev.* **13,** 667–672.

Fernandez, D. C., Bhargava, R., Hewitt, S. M., and Levin, I. W. (2005). Infrared spectroscopic imaging for histopathologic recognition. *Nat. Biotechnol.* **23,** 469–474.

Geiszt, M., Lekstrom, K., Brenner, S., Hewitt, S. M., Dana, R., Malech, H. L., and Leto, T. L. (2003). NAD(P)H oxidase 1, a product of differentiated colon epithelial cells, can partially replace glycoprotein 91phox in the regulated production of superoxide by phagocytes. *J. Immunol.* **171,** 299–306.

Gillespie, J. W., Best, C. J., Bichsel, V. E., Cole, K. A., Greenhut, S. F., Hewitt, S. M., Ahram, M., Gathright, Y. B., Merino, M. J., Strausberg, R. L., Epstein, J. I., Hamilton, S. R., Gannot, G., Baibakova, G. V., Calvert, V. S., Flaig, M. J., Chuaqui, R. F., Herring, J. C., Pfeifer, J., Petricoin, E. F., Linehan, W. M., Duray, P. H., Bova, G. S., and Emmert-Buck, M. R. (2002). Evaluation of non-formalin tissue fixation for molecular profiling studies. *Am. J. Pathol.* **160,** 449–457.

Goodman, M. T., Hernandez, B. Y., Hewitt, S., Lynch, C. F., Cote, T. R., Frierson, H. F., Jr., Moskaluk, C. A., Killeen, J. L., Cozen, W., Key, C. R., Clegg, L., Reichman, M., Hankey, B. F., and Edwards, B. (2005). Tissues from population-based cancer registries: A novel approach to increasing research potential. *Hum. Pathol.* **36,** 812–820.

Harper, M. E., Ullrich, A., and Saunders, G. F. (1981). Localization of the human insulin gene to the distal end of the short arm of chromosome 11. *Proc. Natl. Acad. Sci. USA* **78,** 4458–4460.

Hewitt, S. M. (2004). Design, construction, and use of tissue microarrays. *Methods Mol. Biol.* **264,** 61–72.

Hewitt, S. M., Dear, J., and Star, R. A. (2004). Discovery of protein biomarkers for renal diseases. *J. Am. Soc. Nephrol.* **15,** 1677–1689.

Hidalgo, A., Pina, P., Guerrero, G., Lazos, M., and Salcedo, M. (2003). A simple method for the construction of small format tissue arrays. *J. Clin. Pathol.* **56,** 144–146.

Huang, H. E., Chin, S. F., Ginestier, C., Bardou, V. J., Adelaide, J., Iyer, N. G., Garcia, M. J., Pole, J. C., Callagy, G. M., Hewitt, S. M., Gullick, W. J., Jacquemier, J., Caldas, C., Chaffanet, M., Birnbaum, D., and Edwards, P. A. (2004). A recurrent chromosome breakpoint in breast cancer at the RG1/neuregulin 1/heregulin gene. *Cancer Res.* **64,** 6840–6844.

Khanna, C., Prehn, J., Hayden, D., Cassaday, R. D., Caylor, J., Jacob, S., Bose, S. M., Hong, S. H., Hewitt, S. M., and Helman, L. J. (2002). A randomized controlled trial of octreotide pamoate long-acting release and carboplatin versus carboplatin alone in dogs with naturally occurring osteosarcoma: Evaluation of insulin-like growth factor suppression and chemotherapy. *Clin. Cancer Res.* **8,** 2406–2412.

Kononen, J., Bubendorf, L., Kallioniemi, A., Barlund, M., Schraml, P., Leighton, S., Torhorst, J., Mihatsch, M. J., Sauter, G., and Kallioniemi, O. P. (1998). Tissue microarrays for high-throughput molecular profiling of tumor specimens. *Nat. Med.* **4,** 844–847.

LeBaron, M. J., Crismon, H. R., Utama, F. E., Neilson, L. M., Sultan, A. S., Johnson, K. J., Andersson, E. C., and Rui, H. (2005). Ultrahigh density microarrays of solid samples. *Nat. Methods* **2,** 511–513.

MacBeath, G. (2002). Protein microarrays and proteomics. *Nat. Genet.* **32**(Suppl.), 526–532.

Moskaluk, C. A., and Stoler, M. H. (2002). Agarose mold embedding of cultured cells for tissue microarrays. *Diagn. Mol. Pathol.* **11,** 234–238.

Nishizuka, S., Chen, S. T., Gwadry, F. G., Alexander, J., Major, S. M., Scherf, U., Reinhold, W. C., Waltham, M., Charboneau, L., Young, L., Bussey, K. J., Kim, S., Lababidi, S., Lee, J. K., Pittaluga, S., Scudiero, D. A., Sausville, E. A., Munson, P. J., Petricoin, E. F., 3rd, Liotta, L. A., Hewitt, S. M., Raffeld, M., and Weinstein, J. N. (2003). Diagnostic markers that distinguish colon and ovarian adenocarcinomas: Identification by genomic, proteomic, and tissue array profiling. *Cancer Res.* **63,** 5243–5250.

Simon, R., Nocito, A., Hubscher, T., Bucher, C., Torhorst, J., Schraml, P., Bubendorf, L., Mihatsch, M. M., Moch, H., Wilber, K., Schotzau, A., Kononen, J., and Sauter, G. (2001). Patterns of her-2/neu amplification and overexpression in primary and metastatic breast cancer. *J. Natl. Cancer Inst.* **93,** 1141–1146.

Su, H., Hu, N., Shih, J., Hu, Y., Wang, Q. H., Chuang, E. Y., Roth, M. J., Wang, C., Goldstein, A. M., Ding, T., Dawsey, S. M., Giffen, C., Emmert-Buck, M. R., and Taylor, P. R. (2003). Gene expression analysis of esophageal squamous cell carcinoma reveals consistent molecular profiles related to a family history of upper gastrointestinal cancer. *Cancer Res.* **63,** 3872–3876.

Tan, A. R., Yang, X., Hewitt, S. M., Berman, A., Lepper, E. R., Sparreboom, A., Parr, A. L., Figg, W. D., Chow, C., Steinberg, S. M., Bacharach, S. L., Whatley, M., Carrasquillo, J. A., Brahim, J. S., Ettenberg, S. A., Lipkowitz, S., and Swain, S. M. (2004). Evaluation of biologic end points and pharmacokinetics in patients with metastatic breast cancer after treatment with erlotinib, an epidermal growth factor receptor tyrosine kinase inhibitor. *J. Clin. Oncol.* **22,** 3080–3090.

Taylor, C. R., and Cote, R. J. (eds.) (2005). "Immnomicroscopy," 3rd Ed. Saunders, Philadephia.

Yang, H., Wanner, I. B., Roper, S. D., and Chaudhari, N. (1999). An optimized method for *in situ* hybridization with signal amplification that allows detection of rare mRNAs. *J. Histochem. Cytochem.* **47,** 431–445.

[21] Mapping Histone Modifications by Nucleosome Immunoprecipitation

By DAVID J. CLARK and CHANG-HUI SHEN

Abstract

Studies of histone modification patterns and their role in gene regulation have led to the proposal that there is a "histone code." We have developed a method for nucleosome immunoprecipitation that can precisely identify the specific nucleosomes that carry a posttranslational modification of interest. The process involves the isolation and micrococcal nuclease digestion of minichromosomes to generate nucleosome core particles. These are then used in immunoprecipitation reactions with an antibody directed against the histone modification of interest. Subsequently, nucleosome core particle DNA is purified and end labeled. The original locations of the nucleosomes in the immunoprecipitate can be determined at low resolution (using a modified Southern blot hybridization procedure) or at maximal resolution (using the monomer extension method). Using the latter method, the positions of specific nucleosomes that carry the posttranslational modification of interest can be identified precisely. This method is sensitive, provides maximal resolution, and is inexpensive. The approach described here may serve as a paradigm for the study of histone-modifying patterns.

Introduction

In eukaryotes, every process that utilizes DNA as a template, including transcription, replication, recombination and repair, is influenced by the packaging of DNA into chromatin. The basic structural repeat unit of chromatin is the nucleosome which contains 147 bp of DNA wrapped in about 1.75 superhelical turns around a central octamer composed of two each of the four core histones H2A, H2B, H3 and H4. Targeted covalent modification of the amino-terminal tails of the core histones in nucleosomes has emerged as an important mechanism in the regulation of transcriptional activation and repression. These core histone covalent modifications include acetylation, methylation, phosphorylation and ubiquitylation. It is believed that certain modifications occur in a temporal sequence in which the sequence and combination of modifications may yield an ordered series of events for gene expression (Cheung *et al.*, 2000; Lo *et al.*, 2000, 2001; Sun and Allis, 2002).

METHODS IN ENZYMOLOGY, VOL. 410   0076-6879/06 $35.00
DOI: 10.1016/S0076-6879(06)10021-X

Currently, the protocol commonly used to determine the distribution of histone modifications *in vivo* is chromatin immunoprecipitation (ChIP), now usually coupled with quantitative PCR in real-time (Henry *et al.*, 2003; Lo *et al.*, 2005). Since the real time system can provide easy and precise nucleic acid detection, it has become a powerful and widely applied technique for detecting the association of individual proteins with specific genomic regions *in vivo*. However, one major concern for this analysis is that the resolution of the ChIP method is limited by the size of the chromatin fragments generated by sonication, which are usually about 400–500 bp. Since two or even three nucleosomes may occupy DNA fragments of this size, the CHIP method cannot resolve the distribution of histone modifications at the individual nucleosome level. This is important because neighboring nucleosomes are sometimes modified differently, particularly at promoters.

We have developed new methods to specifically quantitate and identify individual nucleosomes that contain modified residues (Shen *et al.*, 2002). These assays are based on Southern blotting (Clark and Felsenfeld, 1992) and monomer extension analysis (Kim *et al.*, 2004; Shen and Clark, 2001; Shen *et al.*, 2001; Yenidunya *et al.*, 1994). Briefly, the protocol involves the following steps: (1) preparation of chromatin; (2) digestion to nucleosome core particles with micrococcal nuclease (MNase); (3) purification and end-labeling of nucleosome core particle DNA; (4) mapping the distribution of modified nucleosomes in the chromatin either (a) at lower resolution using the labeled nucleosomal DNA to probe Southern blots of gels with a set of plasmids carrying 200–400 bp subcloned regions of interest; or (b) at maximal resolution using the monomer extension method to identify the precise positions of the nucleosomes in the immunoprecipitate (Fig. 1).

These assays can be used for experiments aiming at characterizing specific subpopulations of nucleosomes, for example nucleosomes containing H3 acetylated at Lys 9 and Lys 14. They can also be applied to investigate the coupling of different modifications on a precise subset of nucleosomes, including phosphorylation, methylation or acetylation on other residues. In this section, we emphasize the basic application of our method using purified plasmid chromatin ("minichromosomes"). The first few steps, including the construction of minichromosomes in budding yeast, the choice of yeast strain, and the method for isolation of minichromosomes have been described previously (Alfieri and Clark, 1999; Shen *et al.*, 2002). The procedures described below are developed in our studies of the *TRP1ARS1CUP1* (TAC) minichromosome (Shen *et al.*, 2001), a 2468-bp yeast plasmid based on *TRP1ARS1* (Thoma and Simpson, 1985; Zakian and Scott, 1982).

It has been demonstrated recently that nucleosome immunoprecipitation can be combined with tiling DNA microarray technology to determine both the distribution of nucleosomes on a significant fraction of the yeast

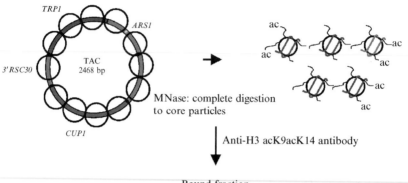

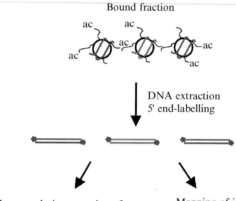

Bound fraction

DNA extraction
5' end-labelling

Low resolution mapping of   Mapping of immunoprecipitated
nuclesome positions using   nucleosomes at maximal resolu-
Southern blot hybridization   tion by monomer extension analysis

FIG. 1. Schematic diagram of histone modification mapping by nucleosome immunoprecipitation. The TAC minichromosome is based on the yeast plasmid *TRP1ARS1*; *TRP1* is a
selection marker, and *ARS1* is the origin of replication. *TRP1ARS1* also contains the
upstream region of neighboring *GAL3*. *CUP1* was inserted with the 3'-flanking region of
RSC30, the gene neighboring *CUP1* in the genome. Adapted from Shen *et al.* (2002).

genome (Yuan *et al.*, 2005) and their various post-translational modifications (Liu *et al.*, 2005). We focus here on the preparation of the nucleosomal sample for immunoprecipitation.

Preparation of Core Particles: MNase Digestion of Minichromosomes

MNase digestion of chromatin is a relatively simple procedure for
obtaining information about the locations of nucleosomes along DNA
strands. When nuclei or permeabilized cells are exposed to MNase in the

presence of calcium, the enzyme makes double-stranded cuts between nucleosomes. MNase cleaves initially at favored sites in linker DNA to produce chromatin fragments (Horz and Altenberger, 1981). These are gradually reduced in size, generating the MNase "ladder," indicating the presence of regularly spaced nucleosomes. MNase will continue to cut between nucleosomes and to trim the chromatin fragments until the entire chromatin preparation has been converted to nucleosome core particles. The fully trimmed core particle contains 147 bp and is relatively stable. However, if digestion is continued, MNase will begin to nick DNA within the nucleosome and will finally destroy it. For this experiment, it is important to adjust the MNase digestion conditions such that intact core particles are produced. Ideally, time courses of MNase digestion should be carried out, but sometimes material is limiting, as it often is with yeast plasmid chromatin, and only one or two time points are possible.

To prepare core particles from minichromosomes, the first consideration is to determine the correct amount of MNase to be used. The activity of MNase is highly dependent on the concentration of enzyme. We use a twofold concentration interval to determine the effective amount of MNase for core particle production. Another concern in MNase digestion of minichromosome is the presence of EDTA, which chelates the Ca^{2+} required to activate MNase. The solution should contain 1 mM excess Ca^{2+} over EDTA concentration to eliminate the effect of EDTA chelation.

1. Prepare MNase at 200 units/μl (Worthington) in 10 mM Tris–HCl, pH 8.0, 0.05 mM CaCl$_2$.
2. Set up a 100-μl reaction in a 1.5-ml microcentrifuge tube on ice: Dilute 45 ng of yeast minichromosomes prepared in 40 μl 40 mM Tris–acetate, pH 8.3, 2 mM Na-EDTA into 10 mM Tris–HCl (pH 8.0), 35 mM NaCl, 3 mM CaCl$_2$, 1.5 μM trichostatin A (WAKO). Mix thoroughly and then add MNase to 25 units/ml.
3. Incubate the reaction mix for 2 min at 30°.
4. Stop the reaction by adding 1 μl 0.5 M EDTA.
5. Place the tube on ice and proceed to the immunoprecipitation step as soon as possible.

We include trichostatin A, a histone deacetylase inhibitor, to prevent histone deacetylation. This is particularly important if the aim is to immunoprecipitate acetylated histone, but in any case represents a sensible precaution. If the aim is to study histone phosphorylation, a protein phosphatase inhibitor such as microcystin LR should be added to the preparation. It is recommended to proceed immediately to the immunoprecipitation step because MNase is still present (although blocked by EDTA).

Immunoprecipitation of Nucleosomes

After MNase digestion, core particles are incubated with a specific antibody of choice, and then the antibody is immobilized on protein A or protein G coupled to Sepharose beads. The choice of protein A or protein G depends on the antibody used.

The following is an example of a protocol using an antibody directed against H3 acetylated on both Lys9 and Lys14. The core particles are incubated with the antibody and immunoprecipitated nucleosomes are collected using protein A-Sepharose and washed. The core particle DNA in the immunoprecipitate and in the supernatant are purified for analysis.

1. Mix the MNase-digested core particle sample (100 μl) with an equal volume (100 μl) of IP buffer [10 mM Tris–HCl (pH 8.0), 5 mM Na-EDTA, 0.565 M NaCl, 0.2 mg bovine serum albumin (BSA)/ml, 5 μg leupeptin/ml, 0.1 mM 4-(2-aminoethyl)benzenesulfonyl fluoride (AEBSF), 15 μg pepstatin A/ml, and 1.5 μM trichostatin A] on ice. For a total input control, collect the same amount of MNase-digested core particle sample (100 μl), adjust to 500 μl with 1.25% SDS in 1.25 mM Tris–HCl (pH 8.0), 0.0125 mM Na-EDTA, and extract the DNA as in step 8.

2. Add 5 μl antibody. For a negative control, perform a no-antibody incubation.

3. Incubate overnight at 4° with rotation.

4. Add 100 μl 1:1 slurry of protein A-Sepharose CL-4B (Amersham) in 0.3 M NaCl, 1 mM Tris–HCl (pH 8.0), 0.01 mM Na-EDTA to the core particle mix and incubate for 1.5 h at 4° with rotation.

5. Collect the resin: 2000 rpm for 1 min at 4°; retain the supernatant that contains unbound core particles (purify the DNA: step 8).

6. Wash the resin three times with 0.5 ml 0.5 M NaCl, 20 mM Tris–HCl (pH 8.0), 0.05% Triton X-100.

7. Elute the DNA bound to the resin by resuspension in 500 μl 1% SDS, 1 mM Tris–HCl (pH 8.0), 0.01 mM Na-EDTA.

8. Add 5 M potassium acetate to 1 M and extract the supernatant with phenol-chloroform (1:1) and precipitate with ethanol in the presence of 20 μg glycogen (Roche) as coprecipitant.

9. Purify the core particle DNA from a 3% agarose gel (140–160 bp). Normally, 20–30 ng of core particle DNA is obtained from a typical experiment.

In step 1, IP buffer containing 565 mM NaCl is mixed with MNase-digested chromatin containing 35 mM NaCl to give a final NaCl concentration of 300 mM. The high salt concentration is used to reduce the binding of histone tails to nucleosomal DNA, making them more available to the antibody. We also include BSA in the reaction as a blocking agent to prevent nonspecific binding. To prevent histone degradation, we include the protease inhibitors leupeptin, AEBSF, and pepstatin A.

Radioactive End Labeling of Nucleosome Core Particle DNA

The purified nucleosomal DNA is 5′ end labeled using T4 polynucleotide kinase.

1. Dissolve the purified core particle DNA in 39 μl TE.
2. Set up a 5′ end-labeling reaction in a volume of 50 μl: 39 μl core particle DNA, 5 μl 10× 5′ kinase buffer, 5 μl [γ-^{3}2P]ATP(3000 Ci/mol), and 1 μl T4 polynucleotide kinase.
3. Incubate at 37° for 1 h.
4. Collect the labeled core particle DNA using a Sephadex G-50 spin column (Amersham-Pharmacia).
5. Adjust the volume of core particle DNA to 50 μl with TE and store the labeled DNA at −20°.

MNase cleaves DNA to yield a 5′-hydroxyl group and a 3′-phosphate. The 5′ end is therefore suitable for end labeling using T4 polynucleotide kinase; the 3′ phosphate is removed by the 3′-phosphatase activity of T4 kinase.

Analysis of Immunoprecipitated Nucleosomes

There are two methods available for analyzing nucleosomal DNA in the immunoprecipitate: the low-resolution mapping method is based on a modified Southern blot procedure, requires less material, and is technically less demanding. Its resolution is limited by the sizes of the insert target fragments (250–450 bp). The high-resolution method can resolve very complex chromatin structures at maximal resolution (potentially single base pair resolution but, in practice, is limited by how closely the nucleosome core particles are trimmed to 147 bp by MNase).

Low-Resolution Mapping of Nucleosome Positions Using Southern Blot Hybridization

Distribution of the immunoprecipitated modified nucleosomes relative to the input nucleosomes can be determined using Southern blot analysis in which the radiolabeled core particle DNA is used as the probe and the region of DNA to be mapped is divided into small contiguous regions blotted onto a membrane (Clark and Felsenfeld, 1992; Shen et al., 2002).

In the example discussed here, the distribution of nucleosomes containing diacetylated H3 was determined in our plasmid chromatin of interest, the TAC minichromosome. TAC was divided into seven biologically relevant regions 240–469 bp in size, which were subcloned to create a set of seven plasmids (pTAC1–7) (Fig. 2A). After gel electrophoresis, the plasmids, together with the vector control (no insert) and a plasmid carrying the entire TAC sequence as a total signal control, are transferred to a GeneScreen Plus membrane. Labeled core particle DNA derived from the input or the immunoprecipitated core particles is used as a probe. The signals for each TAC region are quantified using a phosphorimager and are compared with input nucleosome DNA to determine which regions of the TAC minichromosome are enriched in diacetylated H3 containing nucleosomes.

The first step is to produce a Southern membrane that contains all template DNAs in the area of interest. The sizes of the inserts determine the resolution of the nucleosome-mapping experiment. It is important to subclone rather than using the isolated fragments to avoid size-dependent binding of DNA to the membrane (the effect of size on membrane retention is quite strong with short DNA fragments). The plasmids are linearized prior to electrophoresis because supercoiled and linear DNAs behave differently in Southern hybridization. Five hundred nanograms of plasmid is electrophoresed in a 0.8% agarose gel and transferred to GeneScreen Plus membranes (Fig. 2A). After transfer, the DNA is fixed to the membrane by UV cross-linking. The membrane can be stored at 4° for a short time prior to hybridization.

The following is based on the method described by Church and Gilbert (1984).

1. Incubate the membrane in hybridization buffer (250 mM Na$_2$HPO$_4$, pH 7.2, 7% SDS, 1 mM EDTA, pH 8.0) at 60° for 30 min in a hybridization oven with a rotator.
2. Change the buffer and continue for 10 min to allow it to reach 60°.
3. Add the heat-denatured labeled core particle DNA probe.
4. Hybridize at 60° overnight.
5. Wash twice with wash buffer (20 mM Na$_2$HPO$_4$, pH 7.2, 1% SDS, 1 mM EDTA).
6. Air dry and quantify using a phosphorimager.

A hybridization temperature of 60° gives better results than the usual 65°, presumably because the probe is relatively short (147 bp). There

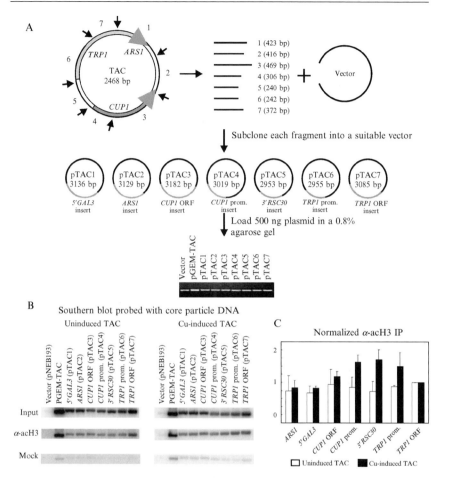

FIG. 2. Mapping modified nucleosomes in the minichromosome using the Southern blotting method (lower resolution). (A) The TAC plasmid was divided into seven biologically relevant regions, which were subcloned: (1) pTAC1: *ARS1*; (2) pTAC2: *5'GAL3*; (3) pTAC3: *CUP1* ORF; (4) pTAC4: *CUP1* promoter; (5) pTAC5: 3' *RSC30*; (6) pTAC6: *TRP1* promoter; and (7) pTAC7: *TRP1* ORF. The size of each fragment is indicated. Equal amounts (500 ng) of each plasmid were loaded in gel, which was blotted. (B) Hybridization of labeled core particle DNA to Southern blots of gels like that shown in A. Core particle DNAs derived from input core particles, core particles immunoprecipitated with the antiacetylated H3 antibody, and mock immunoprecipitates (no antibody) were used as probes. (C) Quantitative analysis of IP data from B. TAC nucleosomes from uninduced (white bars) and from copper-induced cells (black bars) were analyzed; bands were quantified and normalized to the *TRP1* ORF, which is set equal to 1. These numbers were then divided by the input values normalized identically. Adapted from Shen *et al.* (2002).

should be no signal in the vector control, indicating that there is no nonspecific hybridization under these conditions.

Nucleosomes might not be distributed equally in the plasmid chromatin (e.g., nucleosomes tend to be depleted from active promoters). This will be apparent from analysis of the input signals, which represent nucleosome density. The input signal should be normalized to insert size to compensate for differences in insert size. For comparison of experimental conditions (e.g., to compare plasmid chromatin from induced and uninduced cells), an internal control is required: a region in the minichromosome that is expected to be unaffected by induction of the gene. The value of the signal from each region is divided by the signal from the internal control to obtain the normalized signal. Comparison of the normalized signals yields the nucleosome density for each sample. In our case, we chose the coding region of *TRP1* (pTAC7) as the internal control because the activity of *TRP1* is expected to be independent of the activity of the neighboring *CUP1* gene. Our analysis showed that the nucleosome density varied from 80 to 120% of the *TRP1* signal, indicating that nucleosomes are distributed quite evenly in TAC chromatin.

To determine the distribution of the histone modification of interest, the input signals and IP signals are normalized to their respective internal control signal. The normalized immunoprecipitate signal is divided by its normalized counterpart input signal. This corrects directly for insert size and nucleosome density, yielding a ratio representing the relative degree of histone modification in each subcloned region. A ratio of 1 indicates no enrichment. In the example presented here, signals for the copper-induced *CUP1* promoter and the 3' *RSC30* region were enhanced relative to the signals from the rest of TAC, indicating that H3 acetylation was targeted to the *CUP1* promoter and the neighboring upstream (3' *RSC30*) region (Fig. 2B and C).

Mapping Immunoprecipitated Nucleosomes at Maximal Resolution by Monomer Extension

End-labeled core particle DNA can also be employed in monomer extension analysis to identify and quantify the specific nucleosomes carrying the posttranslational modification of interest. The monomer extension method was developed to determine the positions of nucleosomes in complex chromatin structures without ambiguity (Davey *et al.*, 1995, 1997; Shen and Clark, 2001; Shen *et al.*, 2001; Yenidunya *et al.*, 1994). The principle and procedure have been described in detail elsewhere (Kim *et al.*, 2004). Immunoprecipitated core particle DNA can be used to determine the positions and relative amounts of nucleosomes containing the posttranslational modification of interest.

In the example discussed here, the distribution of nucleosomes containing diacetylated H3 was determined in TAC chromatin. Briefly, immunoprecipitated, end-labeled core particle DNA is used as primer in a primer extension reaction using *single-stranded* TAC DNA as a template. Single-stranded DNA is used to ensure that only one of the two labeled strands of core DNA can anneal and act as primer. The extension products are cleaved with a restriction enzyme having a unique site in the plasmid and the products are resolved in a sequencing gel. The length of the extension product is equal to the distance of the farthest nucleosome border to the restriction site (Fig. 3A).

The entire region to be mapped should be inserted into a phagemid such as a pGEM vector (Promega) so that single-stranded DNA can be prepared by virtue of the f1 phage replication origin. Because the orientation of the f1 origin, designated (+) or (−), determines which of the two strands of the plasmid will be packaged, both orientations of the insert are useful.

Alkali Denaturation of Immunoprecipitated, Labeled Core Particle DNA

The labeled core particle DNA is denatured using NaOH, neutralized with ammonium acetate, and precipitated with ethanol.

1. Add 13 μl H$_2$O and 7 μl 2 M NaOH to 50 μl end-labeled core DNA.
2. Incubate at room temperature for 10 min.
3. Add 28 μl 5 M ammonium acetate and 42 μl H$_2$O.
4. Add 350 μl absolute ethanol and incubate the solution at −20° overnight.
5. Precipitate the DNA and dissolve it in 5 μl H$_2$O immediately before use.

Alkali denaturation of core particle DNA usually gives better results than heat denaturation for monomer extension.

Monomer Extension

Briefly, the denatured core DNA is mixed with the single-stranded DNA template and extended with the Klenow fragment, followed by restriction enzyme digestion. The products are resolved in a 6% sequencing gel.

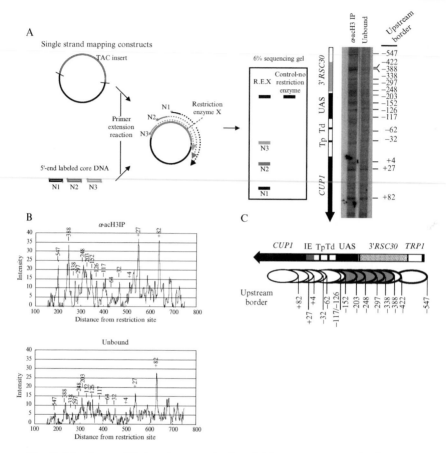

FIG. 3. Mapping modified nucleosomes in the minichromosome at maximal resolution. (A) Schematic diagram of the monomer extension protocol. N1, N2, and N3 represent different nucleosomes purified by immunoprecipitation. An example of a 6% acrylamide gel analysis of monomer extension products is shown. Positions of the nucleosomes enriched in the immunoprecipitate obtained from induced TAC plasmid chromatin using an antibody against diacetylated H3 (lys-9 and -14) were compared with those in the unbound fraction (not acetylated on H3 at lys-9 and -14). The size of each band equals the distance of the farthest border of each positioned nucleosome from the unique SspI site. Nucleosome positions are given with respect to the major upstream start site of *CUP1* (+1). Tp, proximal TATA box; Td, distal TATA box. (B) Quantitative phosphorimager analysis of the gel shown in A. The intensities of the bands in each monomer extension track were normalized to the band at +82 (in the *CUP1* ORF), and the ratio of the normalized band in the IP to the normalized band in the unbound fraction was calculated for each nucleosome. (A ratio of 1.0 indicates that this nucleosome is not enriched in the IP.) (C) Schematic showing the positions of the immunoprecipitated nucleosomes, indicated by their upstream borders relative to *CUP1*. Nucleosomes enriched in di-acetylated H3 (lys-9 and -14) are shaded. Adapted from Shen *et al.* (2002).

1. Set up the annealing mix in a PCR tube on ice. Mix end-labeled core DNA with the single-stranded DNA template in a molar ratio of approximately 1:2.5 in 25 μl 10 mM Tris–HCl, 50 mM NaCl, 10 mM MgCl$_2$, and 1 mM dithiothreitol (pH 7.9).

2. Incubate as follows in a thermal cycler: 95° for 3 min and 80° for 30 s. Cool to 55° at a rate of 45 s per degree.

3. Cool on ice.

4. After annealing, dilute to 50 μl by adding dNTPs (10 μM final concentration each of dATP, dCTP, dTTP, and dGTP), BSA (100 μg/ml final concentration), 5 units of *Escherichia coli* DNA polymerase (Klenow fragment; Amersham), and 20 units of an appropriate restriction enzyme and water. As a control, omit the restriction enzyme.

5. Incubate at 37° for 1 h and purify the DNA using phenol:chloroform and chloroform:isoamyl alcohol extractions followed by sodium acetate/ethanol precipitation.

6. Wash the product with 70% ethanol and dissolve in 10 μl sequencing stop mix (90% deionized formamide, 9 mM EDTA, pH 8, 1× TBE, 0.04% xylene cyanol, and 0.04% bromophenol blue).

7. Heat denature the samples and electrophorese in a 40-cm-long 6% (19:1) denaturing polyacrylamide gel. Use ^{32}P end-labeled *Hin*FI and *Dde*I digests of phage λ as size standards. Run the gel at 50–60 W for 6–7 h to resolve bands in the 300- to 1000-bp range.

It is important to mix denatured core DNA with the single-stranded DNA template on ice for 5 min before the addition of annealing mix; this can increase the annealing efficiency. Although we use the Klenow fragment for the extension reaction, reverse transcriptase can also be used (Kim *et al.*, 2004). Size markers can be made in a variety of ways, but *Hin*FI and *Dde*I digests of phage λ DNA are effective. It should be noted that bands below 300 bp might originate from the annealing of complementary monomer strands derived from overlapping positioned nucleosomes and subsequent filling in by the polymerase; bands of this size should be ignored in the analysis. It is essential to use a restriction enzyme with a single site in the phagemid insert for this analysis.

Analysis of the Histone Modification Map

The length of the fragment defines the distance from the far border of each nucleosome to the chosen restriction site. Use of the (+) strand as a template will identify the location of one border of each nucleosome; use of the (−) strand generates complementary data by mapping the opposite borders. The distance between the two positions should reflect the size of the nucleosomal DNA and should be close to 147 bp; both data sets should give the same nucleosome map. The average distance between the nucleosome borders determines the degree to which the two sets of positioning data are consistent.

Quantitative densitometer scans are obtained for each extension reaction after PhosphorImager analysis of the dried gels (Fig. 3B). The equation derived from plotting the marker band sizes against their mobilities should be fitted to a sixth or higher order polynomial with a correlation coefficient greater than 99.9%. The lengths of the extension products can be determined using this equation. The control lane (no restriction enzyme) reveals any bands due to pausing or dissociation of the Klenow enzyme; there are usually no bands in the control.

An example of a mapping experiment for immunoprecipitated and unbound core particles is shown in Fig. 3. Each band represents the border of a nucleosome. The appearance of multiple bands in any 147-bp region is evidence of overlapping nucleosome positions. Because nucleosomes cannot physically overlap, this is evidence for alternative positions in chromatin; the method is so sensitive that it reveals the heterogeneity of the chromatin structure. High-resolution mapping experiments usually reveal complex chromatin structures both *in vitro* and *in vivo* (Shen and Clark, 2001; Shen *et al.*, 2001). For quantitative comparison of the band signals between samples (unbound and immunoprecipitated), it is necessary to normalize to a control band corresponding to a region where there is no change in modification. In the example shown, a nucleosome in the open reading frame was chosen as a control. The immunoprecipitate and the unbound fraction are compared by calculating the ratio of each normalized band to its counterpart in the unbound fraction. Thus, a ratio of 1.0 indicates no enrichment in the IP. In our example, nucleosomes with upstream borders at −152, −203, −248, −297, −338, and −388 were all enriched in the immunoprecipitate (Fig. 3C). These acetylated nucleosomes cover the UAS elements but not the TATA boxes; the nucleosomes targeted by the histone acetylase responsible for acetylation of H3 on lysines 9 and 14 were identified precisely.

Concluding Remarks

We have described methods designed to examine the distribution of histone modifications in native yeast chromatin. We have not applied the method to chromatin from higher organisms but there is no reason in principle why it should not work. Because this method employs nucleosome core particles, its potential resolution is significantly higher than in the standard ChIP methods, in which resolution is limited by the size of the sonicated chromatin fragments (generally much larger than the nucleosome). The importance of the increased resolution obtained by using nucleosomes generated with MNase as probe has been illustrated recently by some elegant microarray experiments. These involved tiling arrays corresponding to a significant fraction of the yeast genome. In these experiments, nucleosomes were prepared using MNase and the DNA derived from the nucleosomes was used to probe the arrays (Yuan *et al.*, 2005). In effect, this is a more sophisticated version of the Southern blot method described above. The quality of the data was impressive and it was concluded that a large fraction of yeast nucleosomes are well-positioned (Yuan *et al.*, 2005). (However, it is important to note that it is nucleosome density that is being measured, not nucleosome positions.) This approach was recently extrapolated to immunoprecipitated nucleosomes carrying various histone modifications, leading to some interesting conclusions (Liu *et al.*, 2005).

It should be noted that, unlike the Southern blotting and array tilling methods which measure nucleosome density, the monomer extension method provides maximal (base pair) resolution of nucleosome positions in native chromatin. It reveals the full complexity of chromatin structure because it can resolve overlapping nucleosome positions, which are probably common in native chromatin, but cannot be detected using most other nucleosome mapping methods.

Acknowledgments

This work was supported in part by NIH (NICHD) intramural funds awarded to DJC. This work was supported in part by NIH grant 1R15GM67730-01 to CS.

References

Alfieri, J. A., and Clark, D. J. (1999). Isolation of minichromosomes from yeast cells. *Methods Enzymol.* **304**, 35–49.

Cheung, P., Tanner, K. G., Cheung, W. L., Sassone-Corsi, P., Denu, J. M., and Allis, C. D. (2000). Synergistic coupling of histone H3 phosphorylation and acetylation in response to epidermal growth factor stimulation. *Mol. Cell* **5**, 905–915.

Church, G., and Gilbert, W. (1984). Genomic sequencing. *Proc. Natl. Acad. Sci. USA* **81**, 1991–1995.

Clark, D. J., and Felsenfeld, G. (1992). A nucleosome core is transferred out of the path of a transcribing polymerase. *Cell* **71**, 11–22.

Davey, C., Pennings, S., and Allan, J. (1997). CpG methylation remodels chromatin structure *in vitro*. *J. Mol. Biol.* **267**, 276–288.

Davey, C., Pennings, S., Meerseman, G., Wess, T. J., and Allan, J. (1995). Periodicity of strong nucleosome positioning sites around the chicken adult ß-globin gene may encode regularly spaced chromatin. *Proc. Natl. Acad. Sci. USA* **92**, 11210–11214.

Henry, K. W., Wyce, A., Lo, W. S., Duggan, L. J., Emre, N. C., Kao, C. F., Pillus, L., Shilatifard, A., Osley, M. A., and Berger, S. L. (2003). Transcriptional activation via sequential histone H2B ubiquitylation and deubiquitylation, mediated by SAGA-associated Ubp8. *Genes Dev.* **17,** 2648–2663.

Horz, W., and Altenburger, W. (1981). Sequence specific cleavage of DNA by micrococcal nuclease. *Nucleic Acids Res.* **9,** 2643–2658.

Kim, Y., Shen, C.-H., and Clark, D. J. (2004). Purification and nucleosome mapping analysis of native yeast plasmid chromatin. *Methods* **33,** 59–67.

Liu, C. H., Kaplan, T., Kim, M., Buratowski, S., Schreiber, S. L., Friedman, N., and Rando, O. J. (2005). Single-nucleosome mapping of histone modifications in *S. cerevisiae. PLOs Biology* **3,** 1753–1769.

Lo, W. S., Trievel, R. C., Rojas, J. R., Duggan, L., Hsu, J. Y., Allis, C. D., Marmorstein, R., and Berger, S. L. (2000). Phosphorylation of serine 10 in histone H3 is functionally linked *in vitro* and *in vivo* to Gcn5-mediated acetylation at lysine 14. *Mol. Cell.* **5,** 917–926.

Lo, W. S., Duggan, L., Emre, N. C., Belotserkovskya, R., Lane, W. S., Shiekhattar, R., and Berger, S. L. (2001). Snf1: A histone kinase that works in concert with the histone acetyltransferase Gcn5 to regulate transcription. *Science* **293,** 1142–1146.

Lo, W. S., Gamache, E. R., Henry, K. W., Yang, D., Pillus, L., and Berger, S. L. (2005). Histone H3 phosphorylation can promote TBP recruitment through distinct promoter-specific mechanisms. *EMBO J.* **24,** 997–1008.

Shen, C.-H., and Clark, D. J. (2001). DNA sequence plays a major role in determining nucleosome positions in yeast *CUP1* chromatin. *J. Biol. Chem.* **276,** 35209–35216.

Shen, C.-H, Leblanc, B. P., Alfieri, J. A., and Clark, D. J. (2001). Remodeling of yeast *CUP1* chromatin involves activator-dependent repositioning of nucleosomes over the entire gene and flanking sequences. *Mol. Cell. Biol.* **21,** 534–547.

Shen, C.-H., Leblanc, B. P., Neal, C., Akhavan, R., and Clark, D. J. (2002). Targeted histone acetylation at the yeast *CUP1* promoter requires the transcriptional activator, the TATA boxes, and the putative histone acetylase encoded by *SPT10. Mol. Cell. Biol.* **22,** 6406–6416.

Sun, Z.-W., and Allis, C. D. (2002). Ubiquitination of histone H2B regulates H3 methylation and gene silencing in yeast. *Nature* **418,** 104–107.

Thoma, F, and Simpson, R. T. (1985). Local protein-DNA interactions may determine nucleosome positions on yeast plasmids. *Nature* **315,** 250–252.

Yuan, G. C., Liu, Y. S., Dion, M. F., Slack, M. S., Wu, L. F., Altschuler, S. J., and Rando, O. J. (2005). Genome-scale identification of nucleosome positions in *S. cerevisiae. Science* **309,** 626–630.

Yenidunya, A., Davey, C., Clark, D. J., Felsenfeld, G., and Allan., J. (1994). Nucleosome positioning on chicken and human globin gene promoters *in vitro*: Novel mapping techniques. *J. Mol. Biol.* **237,** 401–414.

Zakian, V.A, and Scott, J. F. (1982). Construction, replication, and chromatin structure of TRP1 R1 cicrcle, a multiple-copy synthetic plasmid derived from *Saccharomyces cerevisiae* chromosomal DNA. *Mol. Cell. Biol.* **2,** 221–232.

Author Index

A

Aach, J., 274, 275, 293
Aamodt, E. J., 21
Abernathy, K., 74, 100, 109, 111, 131, 146, 204, 302, 305
Abola, A. P., 3, 84, 317
Abril, J. F., 317, 378
Acharya, T., 361
Adams, M. D., 317, 378
Adelaide, J., 402
Affymetrix Technical Note, 19, 21
Agarwala, R., 3, 84, 317
Agbayani, A., 317, 378
Agilent, 31, 32, 33, 34, 35, 47, 50, 51, 52
Aguilar, F., 146
Ahram, M., 225, 404
Ainscough, R., 3, 84, 317
Aitchison, J., 86, 169, 175
Akashi, K., 20
Akhavan, R., 417, 418, 422, 423, 426
Akslen, L. A., 240
Albert, T. J., 75, 86
Albertson, D. G., 29, 30, 74, 125, 229
Aldape, K. D., 88
Alexander, A., 240
Alexander, J., 403
Alexandre, I., 140
Alfieri, J. A., 417, 424, 428
Alizadeh, A. A., 29, 30, 240, 242, 243, 249, 356, 379, 380
Allan, J., 417, 424
Allis, C. D., 416
Allison, D. B., 163
Alsobrook, J., 264, 274
Altenburger, W., 419
Altschul, S. F., 80
Altschuler, S. J., 29, 33, 300, 301, 302, 306, 307
Amanatides, P. G., 317, 378
Ambion, 251, 260
Amin, D. N., 166

Amon, P., 255, 258
An, H. J., 317, 378
Anastasiadou, E., 20
Anders, K., 240
Anderson, C. T., 378
Andersson, E. C., 401
Andrade, R. S., 378
Andrews, J., 74, 99, 119
Andrews-Pfannkoch, C., 317, 378
Angelo, M., 29
Angenent, G. C., 21
Anne, B., 169
Anniek De Witte, A., 169
Ansorge, W., 75, 77, 173, 230, 232
Aparicio, O., 99, 171, 275, 284, 293, 318
Apgar, J., 229
Apidianakis, Y., 21
Appella, D. H., 189, 191
Aragon, A. D., 206
Aravind, L., 3, 84, 317
Arbeitman, M. N., 317
Ares, M., 84, 309
Ares, M., Jr., 301, 308
Armour, C. D., 19, 29, 33, 83, 170, 301, 309
Armstrong, S. A., 4, 20
Arney, K. L., 408
Arnheim, N., 229, 387
Arnold, K., 361
Artiguenave, F., 3, 84, 317
Ashburner, M., 317, 378
Astola, J., 86
Astrand, M., 87, 273
Athanasiou, M., 3, 84, 317
Atkins, D., 68, 69
Atkins, P., 94
Atsma, D., 29
Aubin, R. A., 223
Auburn, R. P., 78, 85
Au-Young, J., 206
Avila, P. C., 29
Azevedo, M. H., 4

Subject Index

A

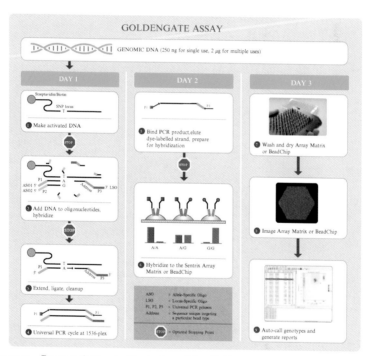

FAN *ET AL.*, CHAPTER 3, FIG. 3. Illustration of the GoldenGate genotyping assay process.

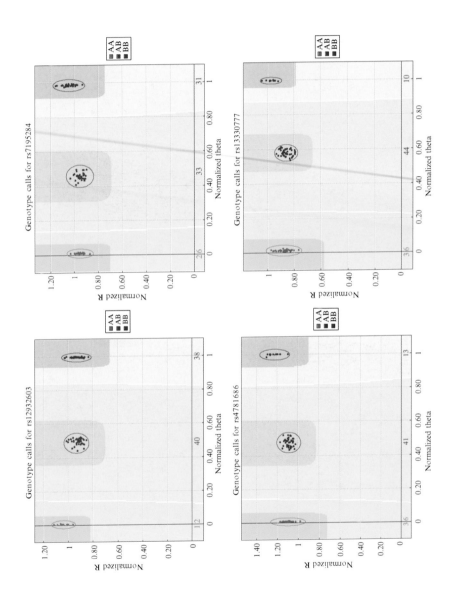

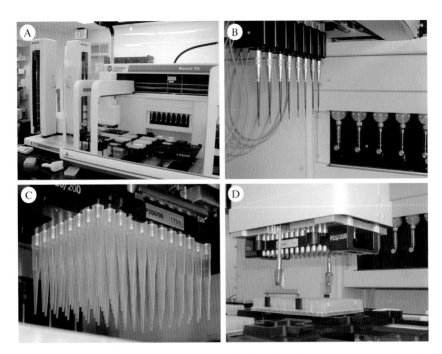

BURR *ET AL.*, CHAPTER 5, FIG. 1. Liquid handling robot: (A) image of the Beckman Coulter Biomek FX laboratory automation workstation, (B) individual probes, (C) 96-arrayed probes, and (D) gripper hand.

FAN *ET AL.*, CHAPTER 3, FIG. 4. GenCall software-produced plots of four randomly selected loci of 90 samples in polar coordinate representation. Each image is a graph of a single locus (assayed simultaneously with ~1400 other loci) with 90 "dots" representing individual DNA samples. The y axis is normalized intensity (sum of intensities of the two channels), and the x axis is the "theta" value. Theta values near 0 (left side of graph) are homozygotes for allele "A," and theta values near 1 (right side of graph) are homozygotes for allele "B." The GenCall software automatically grouped the 90 DNAs for each locus into two homozygote clusters (left and right) and the heterozygote cluster (middle).

A 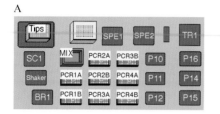 B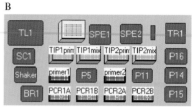

1. LOAD: TIP
2. TRANSFER: 92 μl MIX to PCR1A
3. Repeat steps 2 to add master-mix to PCR1B – PCR4A
4. UNLOAD: TIP
5. Manually transfer master-mix to final plate PCR4B

6. LOAD: Tip1prim
7. MIX: Primer1
8. TRANSFER: 8 μl PRIMER1 to PCR1A
9. TRANSFER: 8 μl PRIMER1 to PCR1B
10. UNLOAD: TIP1prim
11. Repeat 6–10 using Tip2prim, primer2, PCR2A and PCR2B
12. Check that primers are mixed with master mix
13. LOAD: Tip1mix
14. MIX: PCR1A
15. MIX: PCR1B
16. UNLOAD: Tip1mix
17. Repeat 13–16 using Tip2mix, PCR2A and PCR2B
18. Repeat 6–17 using PCR3A-PCR4B, PRIMER3–4, Tip3prim-Tip4prim, and Tip3mix-Tip4mix

BURR *ET AL.*, CHAPTER 5, FIG. 3. Automated liquid handling protocol PCR setup. Maps of the liquid handling robot deck are shown above the program steps. (A) Dispensing master mix. TIPS, P200 filter tips; MIX, reagent reservoir with master mix; PCR1A-PCR4B, empty PCR plates. (B) Adding primers. TIP1prim-TIP2mix, P200 tips; PRIMER1–2, plates of primers; PCR1A-PCR2B, PCR plates containing master mix. Notes: The mix at step 7 is 30 μl at 100% speed repeated 10 times. At steps 8 and 9, dispensing is made above the liquid level with a tip touch to avoid contaminating the tips with the master mix. Step 12 is used to ensure that the hanging droplet of primers has flowed down the wall of the well to mix with the master mix. At steps 14 and 15 the mix is 30 μl at 50% speed repeated five times.

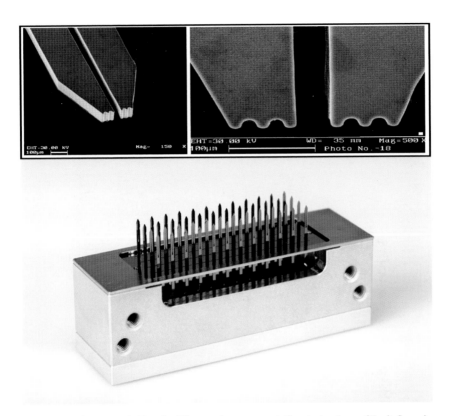

GEORGE, CHAPTER 6, FIG. 2. Silicon microarray printing technology. (Top) Scanning electron microscope images of a silicon print tip, with a tip area of 200 × 200 μm. (Bottom) A silicon microarray print head with 48 tips installed. Images courtesy of Parallel Synthesis Technologies, Inc. (Santa Clara, CA).

BURR *ET AL.*, CHAPTER 5, FIG. 5. Quality control of PCR products. (A) An electrophoresis gel image for one 96-well plate of purified PCR products. (This is half of the gel.) (B) The same image as A with an overlay by Kodak 1D software, identifying bands on each lane of one row on the gel. (C) An excerpt from tab-delimited data generated by Kodak 1D analysis. The first row identifies the lane of the gel. The molecular weight and mass of each band in each lane are recorded under the lane headings. (D) An excerpt of consolidated quality control data stored as a tab-delimited file. Included in the file is sample identification, number of bands (PCR products), molecular weight, mass, and expected size. The molecular weight and mass of only the dominant PCR product are included. (E) An example of quality control summary information. For each 96-well plate of amplicons, the number of samples failing quality control is listed categorically. They are identified as missing products (QC1), weak products (QC2), multiple products (QC3), and products of unexpected size (QC4). From these quality control statistics, the following calculations are recorded: the total number of samples passing quality control, the number of samples flagged during quality control, and the percentage of samples flagged. (F) An example of a scatter plot generated during quality control analysis. For one 96-well plate, the expected molecular weight of the amplicons (Y axis) is plotted against the observed molecular weight (X axis) of the dominant PCR product. Regression analysis is used to identify samples that deviate from their expected molecular weight.

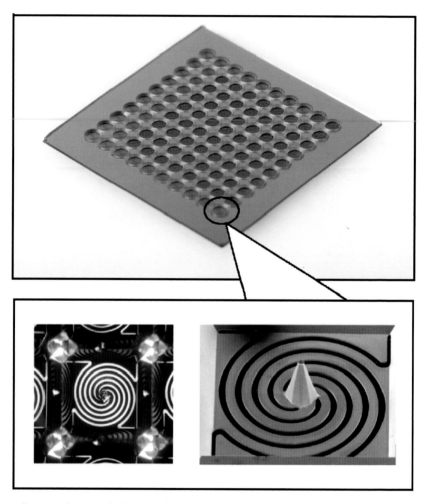

GEORGE, CHAPTER 6, FIG. 4. Printing microtiter plate from Parallel Synthesis Technologies, Incorporated. Photograph and electron micrographs of a printing microtiter plate with 100 print tips on an 1536 SBS format. Images courtesy of Parallel Synthesis Technologies, Inc. (Santa Clara, CA).

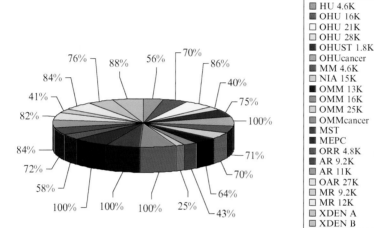

HAGER, CHAPTER 7, FIG. 1. Percentage printing success rate for all slide types with quality acceptable for distribution produced between 2000 and 2005 in the W.M. Keck Microarray Resource. A total of 17,950 microarray slides were printed and the success rate ranges from 25 to 100% depending on the slide type. The average success rate for all slides printed is 72%. Subquality sides are used for technical training in array use. HU, human cDNA; OHU, human operon 70-mer oligonucleotide; MM, mouse cDNA; NIA, Mouse National Institute of Aging cDNA; OMM, mouse operon 70-mer oligonucleotide; MST, mouse operon oligonucleotide; MEPC, Mouse Endocrine Pancreas Consortium cDNA; ORR, rat compugen oligonucleotide; AR, Arabidopsis MSU cDNA; OAR, Arabidopsis operon oligonucleotide; MR, mouse retinal cDNA; XDEN, rice operon oligonucleotide array.

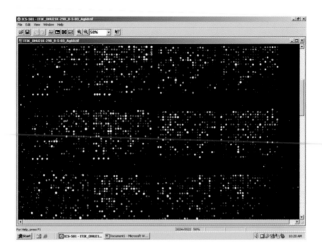

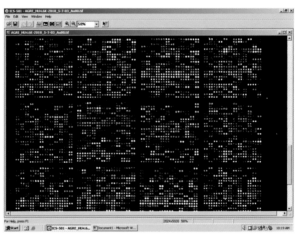

HAGER, CHAPTER 7, FIG. 3. The Resource has applied an amino-allyl approach for RLS labeling instead of directly incorporating the more expensive biotin or fluorescein d-UTP during reverse transcription. Five micrograms total HeLa and HL60 RNAs was amino-allyl labeled during cDNA creation using standard DeRisi protocols to provide less bias between samples than by direct incorporation. cDNAs were linked to monofunctional biotin or fluorescein and reacted with antibiotin- or antifluorescein-tagged gold or silver particles and hybridized against a human 21K 70-mer oligonucleotide array (top) and human 4.6K cDNA array (bottom).

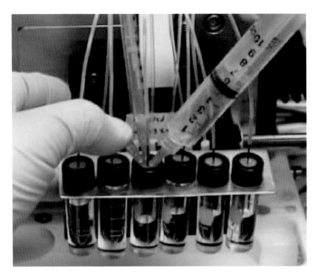

LAUSTED *ET AL.*, CHAPTER 8, FIG. 6. Loading the inkjet print head. A vent needle, a 1-ml syringe with phosphoramidite solution, and a 5- to 10-ml syringe of nitrogen gas are used to load the supply vial for bank #3.

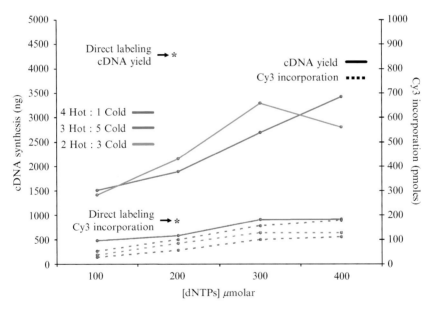

WILLIAMS *ET AL.*, CHAPTER 12, FIG. 2. Summary of indirect labeling experiments. Solid lines are plots of cDNA yield, calibrated to the *y* axis on the left-hand side; dashed lines are plots of mass of Cy3 incorporated in the reaction product calibrated to the *y* axis on the right-hand side. The *x* axis shows four different concentrations of total dNTPs that were tried. Further experiments (data not shown) extended the range of dNTPs to 1 m*M*. Different proportions of amino-allyl-labeled dUTP (hot) to unlabeled dTTP (cold) are indicated in the color legend. Asterisks indicate typical values for yield and incorporation obtained in direct labeling experiments.

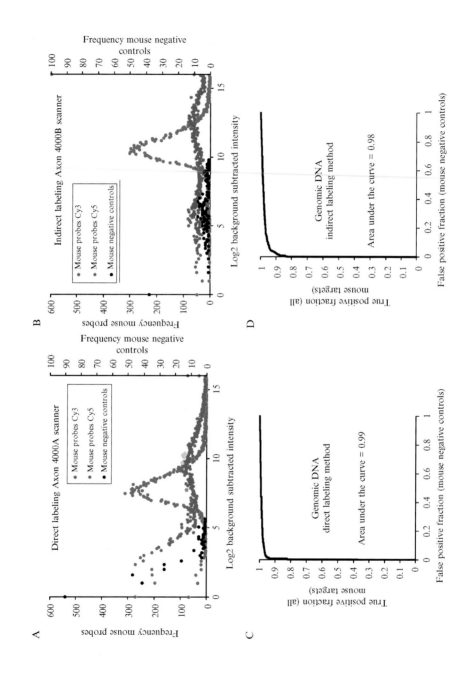

A   B

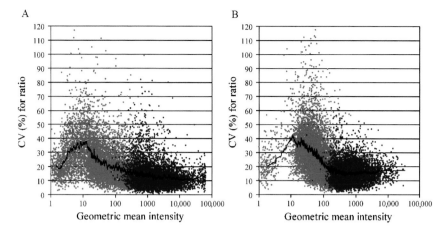

WILLIAMS *ET AL.*, CHAPTER 12, FIG. 5. Comparison of signal variance as a function of intensity. (A) Plot of signal intensity vs variation for all five Stratagene replicates. The x axis indicates the geometric mean of five normalized intensities; the y axis indicates the coefficient of variation in a standardized signal expressed as a percentage. Values for the entire data set are plotted in red. Overlaid on the full data set plot are the values for features with Cy5 intensities greater than 250 in all five Stratagene replicates (shown in blue). The black line indicates the moving average for 100 features. (B) Plot of signal intensity vs variation for all five genomic DNA replicates. In this case, values shown in blue are for the same features displaying a numerator signal greater than 250 in the Stratagene replicates.

WILLIAMS *ET AL.*, CHAPTER 12, FIG. 3. Intensity distributions for genomic DNA signals compared to distributions for negative controls and cDNA-derived signals. (A) Signals from a direct labeling experiment obtained from the Axon 4000A scanner. Genomic DNA labeled with Cy3 (green) produces a distribution of feature intensities well within the distribution of Cy5 signals (red) obtained from C2C12 cDNA and well separated from the distribution of intensities from "on-slide" negative controls (black). (B) Same comparison as in A, but using genomic DNA labeled via the indirect method and scanning with the Axon 4000B scanner. Note a slight increase in overlap between genomic DNA and negative control distributions. (C) Receiver operating characteristic (ROC) curve demonstrating the separation between distributions of the positive mouse probes and the mouse negative control features for the direct labeling method. On the vertical axis is the fraction of positive mouse probes at or below a given level of Cy3 intensity; on the horizontal axis is the corresponding fraction of negative control features at or below that same level of Cy3 intensity. The area under the curve indicates separation of the two distributions, with a score of 1.0 indicating perfect separation of the distributions (i.e., no overlap). (D) ROC plot for the indirect labeling method. The slight decrease in area under the curve indicates a slight increase in nonspecific signal when the indirect method is used in conjunction with the new generation Axon 4000B scanner.

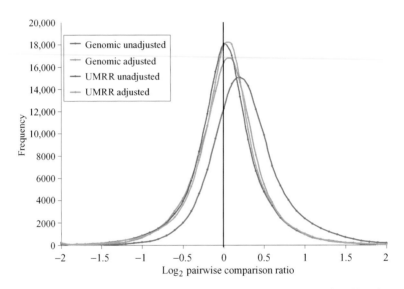

WILLIAMS *ET AL.*, CHAPTER 12, FIG. 7. Lowess adjustment reduces error in self-against-self comparisons that use genomic DNA as the denominator. Comparison of self-against-self error rates between the two standards. Five replicate slides of C2C12 cDNA (Cy5) hybridized against either genomic DNA or the Stratagene mixed cDNAs (Cy3) standard were used in studies of signal variance as a function of intensity. Log_2 ratios vs log_2 geometric mean intensity (RI plots not shown) for 10 self-against-self comparisons within each standard were made, and Lowess adjustment was applied to the log_2 ratio values. Cumulative distributions of the ratios before and after Lowess adjustment are plotted. Note that the spread in the ratios is greater than previous studies (Yang, 2002), as these self-against-self comparisons are compiled from interslide comparisons.

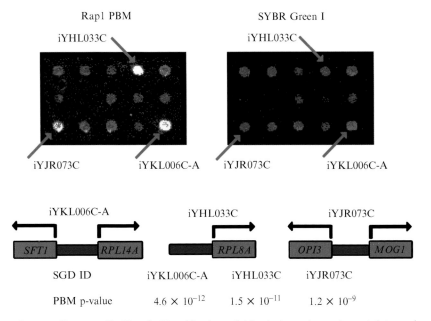

BULYK, CHAPTER 13, FIG. 2. Magnification of identical portions of yeast intergenic microarrays used in a PBM experiment (left) or stained with SYBR Green I (right). Fluorescence intensities are shown in false color, with white indicating saturated signal intensity, red indicating high signal intensity, yellow and green indicating moderate signal intensity, and blue indicating low signal intensity. The three labeled spots correspond to the intergenic regions depicted below, along with the P values derived from triplicate PBM and SYBR Green I microarray data. Reproduced from Mukherjee *et al.* (2004) with permission from Nature Publishing Group.

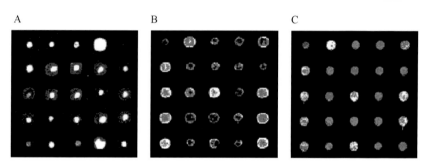

BULYK, CHAPTER 13, FIG. 4. Examples of DNA microarray spot quality. Identical portions of yeast intergenic microarrays printed onto Corning GAPS II slides, processed in different ways (see later) before UV cross-linking, and then stained with SYBR Green I. Images have been false colored as in Fig. 2. (A and B) Examples of microarrays with poor spot quality. In both of these cases, the DNA is distributed nonuniformly with either (A) high concentrations near the centers of spots or (B) high concentrations along spot perimeters. Both of these microarrays resulted from two separate print runs, from which microarrays were UV cross-linked without first rehydrating and baking. (C) An example of a microarray of acceptable quality. This microarray was rehydrated and then baked before being UV cross-linked. Reproduced from Berger and Bulyk (2006) with permission from The Humana Press, Inc.

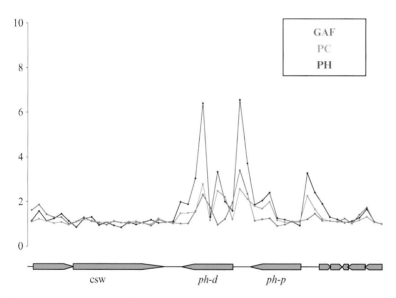

NÈGRE *ET AL.*, CHAPTER 15, FIG. 4. Profile of chromatin proteins in the *ph-d ph-p* region. This graph represents the distribution of three chromatin components (GAF in red, PH in blue, and PC in light blue) on the *ph-d ph-p* region. The *ph* gene is duplicated in this region and encodes for the PH protein itself. Strong binding sites for the three proteins occur upstream and downstream of the *ph-d* gene. A binding site for PC and PH but not for GAF is observed upstream of the *ph-p* gene.

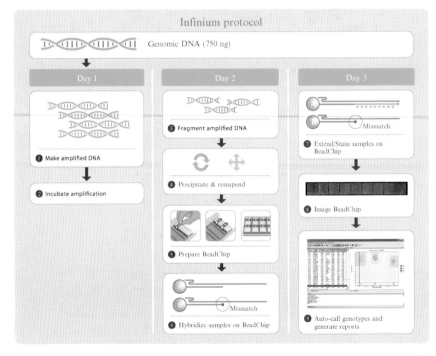

GUNDERSON *ET AL.*, CHAPTER 17, FIG. 1. Whole-genome genotyping on DNA arrays. The DNA sample is whole-genome amplified in an overnight isothermal reaction (steps 1 and 2). The amplification is relatively unbiased with no appreciable allelic partiality. The DNA is amplified 1000–2000× starting with 250–750 ng. The final product concentration is relatively insensitive to the starting amount. After amplification, the product is fragmented enzymatically in a controlled process (step 3). After isopropanol precipitation and resuspension (step 4), the samples are applied to a BeadChip mated to a capillary gap Te-Flow chamber (step 5). The samples are hybridized overnight to a BeadChip array containing 50-mer capture probes to SNP loci. After hybridization capture, the SNPs are scored using either an allele-specific primer extension (ASPE) step or a single base extension (SBE) reaction. The ASPE assay uses two bead types per SNP and a single-color readout, whereas the SBE assay uses a single bead type per SNP and a two-color readout (step 7). Biotinylated/DNP nucleotides are incorporated during the primer extension step and these labels are stained and amplified in a immunohistochemistry-based multilayer sandwich assay (ref.). After staining, the BeadChips are read on a high-resolution confocal scanner (0.84-μm resolution). Genotype calls are made automatically by GenCall software, generating reports of the results.